DISCARD

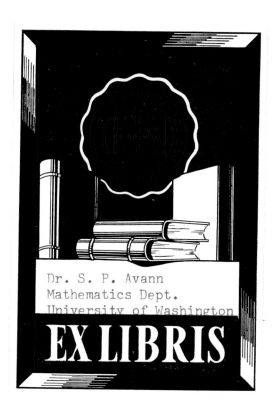

Elements of Abstract and Linear Algebra

Elements of Abstract and Linear Algebra

Hiram Paley
University of Illinois

Paul M. Weichsel
University of Illinois

Holt, Rinehart and Winston, Inc.
New York   Chicago   San Francisco   Atlanta
Dallas   Montreal   Toronto   London   Sydney

To our mothers
Zelda Paley
and
Sophie Weichsel

Copyright © 1972 by Holt, Rinehart and Winston, Inc.
All Rights Reserved
ISBN: 0-03-081311-5
Library of Congress Catalog Card Number: 76-160668
Printed in the United States of America
2 3 4 5   0 3 8   9 8 7 6 5 4 3 2 1

Preface

It is now universally accepted that a substantial course in abstract and linear algebra should be an integral part of the undergraduate mathematics curriculum. Further, the requirements of the physical, biological, and social sciences suggest that such a course is valuable in a wide variety of curricula. The present text is designed to meet the requirements of the mathematics major and those in other disciplines.

The order of topics and the level of presentation follow from two principles strongly ascribed to by the authors. First, the concepts of abstract algebra should be presented to the student in a gentle transition from concrete and relatively familiar mathematical experiences to the more abstract and newer material. Thus, for example, we have included an extensive chapter on number theory in order to provide some concrete experience upon which to base the abstract concepts of groups, rings, and fields. Second, the subject matter of elementary linear algebra is essentially geometric and so we introduce the concept of a linear transformation rather early, and base much of the work on linear algebra and matrices on linear transformations.

Throughout the text, in addition to the formal material, a great many illustrative examples and exercises are included. Each new concept is generously illustrated when it is introduced, and there is an abundance of exercises both to reenforce the notions introduced in the text and to allow the student to test his knowledge and develop his skills.

The text begins with a chapter on sets in order to establish notation and help the student fill in elementary gaps in his background. For many readers, much of Chapter 1 will be review and could be gone over very quickly. Chapter 2 provides a foundation in number theory. It proceeds through the fundamental theorem of arithmetic, congruences, and the Euler φ-function. Chapter 3 is a short presentation of results on permutations that are useful in the work on groups and matrices.

A relatively complete account of the elementary theory of groups, including some classification of groups of small orders, is presented in Chapter 4. Chapter 5 gives an introduction to rings and fields, including polynomial rings and unique factorization domains.

A development of the real and complex number systems is the topic of Chapter 6. The real number system is developed from the point of view of ordered fields and Dedekind cuts. While the material of Chapter 6 does not appear to fit into the general theme of the text, it is included as an important topic of mathematics lying somewhere between algebra and analysis which has frequently found its way into the undergraduate algebra course.

Chapter 7 begins with the work in linear algebra with the introduction of vector spaces, bases, and dimension. Linear transformations and matrices are introduced in Chapter 8, and Chapter 9 considers determinants, methods for computing determinants, and the inverses of nonsingular matrices.

The different concepts of rank are discussed in Chapter 10, and a complete account of the theory of systems of linear equations with coefficients in a field is given. Finally, Chapter 11 has an account of characteristic values and vectors, orthogonal transformations and matrices, and some diagonalization problems.

Our decision to introduce the concepts of abstract algebra before those of linear algebra was based in large part on considerations of taste, along with the desire to formulate the concepts of linear algebra in terms of fields generally. This book can be used, however, to present linear algebra before abstract algebra simply by the instructor's detailing the field axioms, or by his replacing the word "field" in Chapters 7 through 11 by "real numbers" or "complex numbers." A very few illustrative examples would have to be ignored.

Those readers who are familiar with our earlier book, *A First Course in Abstract Algebra*, will notice a general similarity between the elementary chapters of that text with Chapters 1–5 of the present text. The point of view represented there is largely adhered to here. In addition to the inclusion of linear algebra in this text, we have been able to utilize the extensive classroom experience of many colleagues to improve on the material common to both books. The result has been a substantial increase in the number of examples and exercises, as well as some rethinking in the presentation of topics. Some of our friends will be pleased to note that we now write the function symbol to the left of elements.

We have benefited greatly from the experiences of many students and colleagues. We thank all of them, both at the University of Illinois and elsewhere, for their thoughtfulness in communicating with us. We would like to express our special thanks to Mr. Gary Hensler, who read and commented on the complete manuscript. A special citation is due to Pat Coombs for the many weekends and evenings she spent typing the manuscript. We also note with gratitude the excellent cooperation afforded us by the staff of Holt, Rinehart and Winston, particularly Dorothy Garbose and Lyn Peters.

September 1971

Hiram Paley, Urbana, Illinois
Paul M. Weichsel, Jerusalem, Israel

Contents

Chapter 1 Set Theory 1

- 1.1 Concept of Set, Examples, and Notation 1
- 1.2 Union and Intersection 4
- 1.3 The Empty Set 11
- 1.4 Universal Sets and Complements 12
- 1.5 Cartesian Product 17
- 1.6 Functions 18
- 1.7 Composition of Functions 24
- 1.8 Equivalence Relations and Partitions 28

Chapter 2 Number Theory 33

- 2.1 The Arithmetic Properties of the Integers 33
- 2.2 The Order Properties of the Integers 37
- 2.3 Absolute Value 42
- 2.4 Well Ordering and Mathematical Induction 44
- 2.5 Other Forms of Induction 51
- 2.6 Elementary Concepts: Divisibility and the Division Algorithm 56
- 2.7 Greatest Common Divisor and Euclidean Algorithm 59
- 2.8 Prime Numbers 64
- 2.9 The Fundamental Theorem of Arithmetic 67
- 2.10 Congruences 74
- 2.11 Residue Classes 79
- 2.12 Residue-Class Arithmetic 82
- 2.13 Linear Congruences and Chinese Remainder Theorem 84
- 2.14 Euler φ-Function 86
- 2.15 The Theorems of Fermat and Euler 89

Chapter 3 Permutations 91

- 3.1 Permutations: Introduction 91

x   CONTENTS

    3.2   Cycles and Cyclic Decomposition   94
    3.3   Parity of Permutations   99

Chapter 4   The Theory of Groups   104

    4.1   Definition of Group and Examples   104
    4.2   Elementary Consequences of the Definition of a Group   113
    4.3   Multiplication Table for a Finite Group   114
    4.4   Isomorphism   116
    4.5   The Generalized Associative Law and the Law of Exponents   120
    4.6   Subgroups   124
    4.7   Cyclic Groups   128
    4.8   Cosets and Lagrange's Theorem   137
    4.9   Homomorphisms and Normal Subgroups   143
    4.10 Factor Groups and the First Isomorphism Theorem   152
    4.11 Permutation Groups and Cayley's Theorem   159
    4.12 Direct Products   162
    4.13 Automorphisms   167
    4.14 Classification of Groups of Small Order   176

Chapter 5   The Theory of Rings   180

    5.1   Definition of Ring and Examples   180
    5.2   Isomorphism of Rings   184
    5.3   Elementary Properties of Rings   187
    5.4   Some Special Types of Rings   190
    5.5   Homomorphisms, Kernels, and Ideals   194
    5.6   Quotient Rings   198
    5.7   Embedding Theorems and Fields of Quotients   201
    5.8   Polynomial Rings   206
    5.9   Division Algorithm for Polynomials   214
    5.10 Consequences of the Division Algorithm   216
    5.11 Prime and Maximal Ideals   220
    5.12 Euclidean Domains and Unique Factorization   225
    5.13 Direct Sums of Rings   228

Chapter 6   The Real and Complex Number Fields   231

    6.1   The Need for the Real Number System   231
    6.2   Ordered Fields   233
    6.3   Dedekind Cuts   236
    6.4   The Arithmetic of the Real Numbers: $(\mathbf{R},+)$ Is an Abelian Group   239
    6.5   The Arithmetic of Real Numbers: Multiplication and Order   243
    6.6   The Arithmetic of the Real Numbers: $\mathbf{R}$ Is a Field   246
    6.7   The Real Numbers as Ordered Field Extension of the Rational Numbers   248
    6.8   Completeness of the Real Numbers   250
    6.9   The Real Number System and Decimal Notation   253

CONTENTS   xi

   6.10   The Complex Numbers **C**; Polynomials over **R** and **C**   255
   6.11   Algebraic and Transcendental Numbers   258

Chapter 7   Vector Spaces   263

   7.1   Introduction to Vectors   263
   7.2   Definition, Examples, and Elementary Properties of Vector Spaces   268
   7.3   Linear Combinations and Spanning Sets   275
   7.4   Linear Dependence and Independence   278
   7.5   Bases   282
   7.6   Dimension   287
   7.7   Theorems Relating Dimensions   291

Chapter 8   Linear Transformations and Matrices   296

   8.1   Examples of Linear Transformations and Elementary Results   296
   8.2   Range Space, Null Space, and Quotient Space of a Linear Transformation   306
   8.3   The Arithmetic of Linear Transformations: Linear Combinations   309
   8.4   The Arithmetic of Linear Transformations: Composition   313
   8.5   Matrices: Introduction   318
   8.6   Matrices: Linear Combinations   323
   8.7   Matrices: Products   330
   8.8   Identity Matrices and Inverse Matrices; Transformation of Coordinates   339
   8.9   Change of Basis   348

Chapter 9   Determinants   357

   9.1   Determinants   357
   9.2   Computing Determinants   362
   9.3   Cofactors   372
   9.4   Determinant of a Product of Two Matrices   377
   9.5   Inverses of Matrices; Cramer's Rule   381
   9.6   Determinant of a Linear Transformation   388

Chapter 10   Rank, Equivalence, and Systems of Linear Equations   391

   10.1   Column Rank of a Matrix A; Rank of a Linear Transformation T 391
   10.2   Elementary Row and Column Operations; Elementary Matrices   395
   10.3   Column Reduced Echelon Matrices and Column Rank   403
   10.4   Row Rank and Determinantal Rank; Rank   410
   10.5   Normal Forms of Matrices   420
   10.6   Systems of Homogeneous Linear Equations   430
   10.7   Systems of Nonhomogeneous Equations   438

Chapter 11 Characteristic Vectors, Orthogonality, and Diagonalization 444

- 11.1 Characteristic Roots and Characteristic Vectors   444
- 11.2 Inner Products   454
- 11.3 Length, Distance, and Angle; the Cauchy-Schwarz Inequality   462
- 11.4 Orthogonality; the Gram-Schmidt Process   466
- 11.5 Orthogonal Transformations and Orthogonal Matrices   474
- 11.6 Diagonalization of Hermitian and Symmetric Matrices   482

References 492

Index 493

Elements of
Abstract and
Linear Algebra

1 Set Theory

1.1 Concept of Set, Examples, and Notation

The words "set," "collection," "family," "bunch," and "herd" are all familiar to us. These usually denote a number of objects about which we wish to think collectively. We speak, for example, of a herd of dairy cows, a set of bone china dinner plates, and the members of some man's family. Although, as these examples show, the English language is rich in descriptive words for this basic concept, we shall generally use the word "set." Hence in the illustration above we could have spoken of the set of dairy cows rather than the herd of dairy cows.

In any logical investigation certain terms must remain undefined. An attempt to define every term would eventually lead to circular definitions. Thus in this book the term *set* and the concept of *membership* in a set will be taken as primitive undefined concepts.[1] The description of a set that we shall give below will be adequate for our main objective—an investigation of abstract algebraic structures.

For our purposes it will be sufficient to think of a set as being determined by the objects of which it is composed. Moreover, given any object and any set, then either the object is in the set, or the object is not in the set. Alternatively, if a set is described as the collection of objects that satisfy a certain property, then for any object exactly one of these two statements must be true: (a) the object has this property or (b) the object does not have this property.

To illustrate, if P is the property that an object is the cube of a positive integer, then it is clear that any object must either satisfy this property or fail to satisfy it. There is no middle ground. Thus all those objects that satisfy P form a set, namely, the set of third powers of positive integers.

---

[1] For a more formal treatment of set theory see, for example, P. R. Halmos, *Naive Set Theory*. New York: Van Nostrand-Reinhold Company, 1960.

Before we consider some examples of sets, it is convenient to agree on terminology and notation. Capital letters A, B, C, ... will usually be used to denote sets; lower-case letters a, b, c, ... will generally stand for the objects of sets. If a is an object in the set A, we shall also say that a is an *element* of A or that a is a *member* of A or that a *belongs to* A. This will be denoted $a \in A$. If a is not an element of A, we write $a \notin A$. Frequently we shall use the shorthand

$$A = \{x \mid x \text{ has property } P\},$$

which is to be read "A is the set of all objects x such that x has property P." At times, when it is possible to list the elements of A, or at least to list sufficiently many to make clear just what are the elements of A, we shall write $A = \{x, y, z, \ldots\}$.

Example 1.1.1 The set of all integers:

$$Z = \{\ldots, -3, -2, -1, 0, 1, 2, 3, 4, \ldots\}.$$

Given the four real numbers, 0, -2, $\frac{1}{2}$, and π, clearly $0 \in Z$, $-2 \in Z$, $\frac{1}{2} \notin Z$, $\pi \notin Z$.

Example 1.1.2 The set of all positive integers: $P = \{1, 2, 3, 4, 5, \ldots, n, \ldots\}$. Alternatively, we may write $P = \{x \mid x \text{ is an integer and } x > 0\}$.[2] Notice here that the defining property of P is that an element x is both an integer and greater than zero. In this way it is possible to combine several properties into a single property.

Example 1.1.3 The set A of all triangles in the plane that are not isosceles. If $\triangle_1$ stands for a 45-, 45-, 90-degree triangle, $\triangle_1 \notin A$, whereas triangle $\triangle_2$ with sides 3, 4, and 5 does belong to A.

Example 1.1.4 $B = \{x \mid x \text{ is an even integer}\}$. Alternatively,

$$B = \{\ldots, -4, -2, 0, 2, 4, 6, 8, \ldots\}.$$

Clearly, $2 \in B$, $3 \notin B$.

Example 1.1.5 $C = \{(x, y) \mid x \text{ and } y \text{ are real numbers satisfying the equation } x^2 + y^2 = 1\}$. In analytic geometry the set C is called the locus of all points satisfying $x^2 + y^2 = 1$. C is, of course, the circle with center at the origin of the plane and radius 1.

Example 1.1.6 $D = \{x \mid x \text{ is an incorporated city in the United States having a population in the 1970 census of more than 300,000}\}$. Chicago

---

[2] Throughout this text we shall make frequent use of symbolism and concepts that the reader has previously encountered in calculus and analytic geometry. We shall rely on such background material, however, only in the illustrative examples.

1.1 CONCEPT OF SET, EXAMPLES, AND NOTATION

and New York City are in the set D, whereas the city of Urbana, Illinois, although incorporated, is not in the set D.

Example 1.1.7 $F = \{2,4,6,8,\ldots,100\} = \{x \mid x$ is an even integer and $2 \leq x \leq 100\}$.

Since it is possible to think of a set as being a single entity, we may form a set whose elements are themselves sets. For example, although Z and P of Examples 1.1.1 and 1.1.2 are sets, we may still form a set $G = \{Z,P\}$. We note that G consists of precisely two elements, namely the sets Z and P. And although $1 \in Z$ and $1 \in P$, $1 \notin G$.

A set of sets will usually be called a "family of sets" or a "collection of sets." Clearly, "family of collections of sets" is preferred to "set of sets of sets."

EXERCISE

1. Use the notations described in this section to represent each of the following:
   (a) The set A whose elements are the first four positive integers.
   (b) The set B whose elements are the first three positive odd integers.
   (c) The set C of all real numbers whose square is 2.
   (d) The set D of all nonnegative real numbers whose cube is less than 27.
   (e) The set E of all odd integers.
   (f) The set F of all real numbers whose square is greater than 5.
   (g) The set G of all real numbers that are not integers.
   (h) The set H of all points interior to the square in the plane with vertices $(1,1)$, $(-1,1)$, $(-1,-1)$, $(1,-1)$.

From our list of examples and from Exercise 1, it is clear that all the elements of one set may also be elements of a second set. For example, each element of the set P of positive integers is also an element of the set Z of all integers. Thus if A and B are sets and each element of B is also an element of A, we say that B is a *subset* of A (B is *included in* A, B is *contained in* A, A contains B) and we write $B \subseteq A$ or $A \supseteq B$.

We mentioned earlier that a set is determined by the objects of which it is composed. Thus, if $A \subseteq B$ and $B \subseteq A$, then each element of A is an element of B, and conversely, each element of B is an element of A; that is, A and B are composed of exactly the same elements, and so $A = B$. Indeed, when we wish to show that two sets, X and Y, are equal, we will usually accomplish this by showing that $X \subseteq Y$ and $Y \subseteq X$.

If $B \subseteq A$ but $B \neq A$, we call B a *proper subset* of A and write $B \subset A$.[3] Clearly, $B \subset A$ means that: (1) if $x \in B$, then $x \in A$; and (2) there exists $y \in A$ such that $y \notin B$.

Example 1.1.8 Let $A = \{1,2,3,4,5,6,7\}$ and $B = \{2,4,6\}$. Then $B \subseteq A$ and, moreover, $B \subset A$.

In the event that B has an element x such that $x \notin A$, then B is not a subset of A. This is indicated by $B \nsubseteq A$. Thus, $B \subset A$ if and only if $B \subseteq A$ and $A \nsubseteq B$.

EXERCISES

2. Let $A = \{1,2,3,4\}$. Show that $\{1\}$, $\{1,2\}$, and $\{2,3,4\}$ are subsets of A.
3. List all possible subsets of the set $\{1,2,4\}$.
4. Consider the sets described in Examples 1.1.1, 1.1.2, 1.1.4 and 1.1.7. Which are subsets of which?
5. Suppose that A is a set consisting of two elements: x, the set of all even integers, and y, the set of all odd integers. Let $S = \{2,4,6\}$. Is S a subset of A? Is S a subset of x? Is S a subset of y?
6. Let $S = \{1,2\}$, and denote by T the set with the single element S. That is $T = \{S\}$. Is $S = T$? Explain.
7. Let A, B, C be sets. Suppose $A \subseteq B$ and $B \subseteq C$. Prove that $A \subseteq C$. (This is called the transitive law for set inclusion.)

Some sets may be defined by a complete listing of their elements. This listing, however, need not be unique. For example, let $A = \{1,2,1,1,3,4,2\}$, $B = \{1,2,3,4\}$, and $C = \{1,3,4,2\}$. Then, using the criterion for equality of sets discussed above, we see that $A = B = C$. For, even though 1 appears three times in the listing for A, $1 \in A$ and $1 \in B$. Similarly, $2 \in B$, $3 \in B$, and $4 \in B$, and thus $A \subseteq B$. Conversely, $B \subseteq A$ and so $A = B$. Similar arguments establish that $B = C$ and $A = C$. Hence, we see that the number of times that an element appears in the listing is immaterial, as is the order of the listing.

1.2 Union and Intersection

Whether or not the reader has had any experience with the study of sets, he has undoubtedly manipulated sets in special circumstances. A locus of points in geometry is simply a set of points satisfying a certain property.

---

[3] Our use of the symbols $\subseteq$ and $\subset$ is exactly analogous to the use of the $\leq$ and $<$ symbols to denote weak and strict inequalities.

The solution to a set of two linear equations in two variables is the set of all points that satisfy both equations. Again, the set of points that satisfy the equation $x^2 + y^2 = 1$ is the same as the set of points satisfying $y = \sqrt{1 - x^2}$ together with points that satisfy $y = -\sqrt{1 - x^2}$. In this section we shall formalize some operations with sets that the examples above illustrate.

Definition 1.2.1 Let A and B be sets.

(a) $A \cup B$ (read either the *union* of A and B or A *union* B) is the set defined by $A \cup B = \{x \mid x \in A \text{ or } x \in B \text{ (or both)}\}$.

(b) $A \cap B$ (read the *intersection* of A and B or A *intersection* B or the *common part* of A and B) is the set defined by

$$A \cap B = \{x \mid x \in A \text{ and } x \in B\}.$$

Clearly, $x \in A \cup B$ if and only if x is a member of at least one of the sets A and B, whereas $x \in A \cap B$ if and only if x is a member of both of the sets A and B.

Example 1.2.2 Let $A = \{1,2,4,6,9\}$ and $B = \{2,6,10,13\}$. Then

$$A \cup B = \{1,2,4,6,9,10,13\} \text{ and } A \cap B = \{2,6\}.$$

Example 1.2.3 Let $A = \{1\}$. Then $A \cup A = \{1\} \cup \{1\} = \{1\} = A$, for, as we have seen previously, the sets $\{1\}$ and $\{1,1\}$ are identical, having the same members. It also follows that $A \cap A = A$. We will later show that for any set A, $A \cup A = A$ and $A \cap A = A$ (see Exercise 7).

Example 1.2.4 Let

$$A = \{\ldots,-4,-2,0,2,4,6,\ldots\} \quad \text{and} \quad B = \{\ldots,-6,-3,0,3,6,9,\ldots\}.$$

Clearly $A \cap B = \{\ldots,-12,-6,0,6,12,18,\ldots\}$, whereas

$$A \cup B = \{x \mid x \text{ is an integer and either } x/2 \text{ or } x/3 \text{ is an integer}\}.$$

Example 1.2.5 Let A be the set of points in the plane with abscissa greater than or equal to zero; that is, $A = \{(x,y) \mid x, y \text{ are real numbers and } x \geq 0\}$. Let $B = \{(x,y) \mid x, y \text{ are real numbers and } x \leq 0\}$. Then $A \cup B$ is the set of all points in the plane, while $A \cap B$ is just the y axis.

Example 1.2.6 Let A be the set of points in the plane satisfying $2x + y - 5 > 0$. Let B be the set of points satisfying $x - y + 2 < 0$. Then the sets corresponding to $A \cup B$ and $A \cap B$ are illustrated in Figure 1.1.

6   SET THEORY

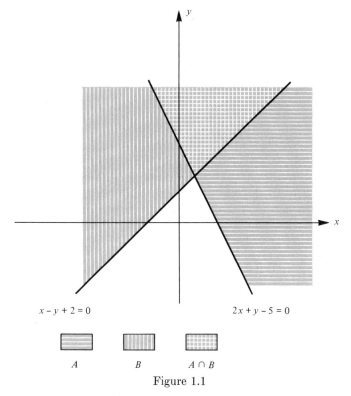

Figure 1.1

EXERCISES

1. Let $A = \{1,2,3,4\}$, $B = \{2,3,5\}$, and $C = \{4,5,1\}$. Find $A \cup B$, $B \cup C$, and $C \cup A$. Find $A \cap B$, $B \cap C$, $A \cap C$. What is $(A \cap B) \cap C$?
2. Let $A = \{1,2,3,4,5,6\}$, $B = \{2,3,4\}$, and $C = \{1,2,5\}$. Find $B \cup C$, $B \cap C$, and show that each of these sets is a subset of A.
3. Let A and B be sets of points in the plane defined by: $A = \{(x,y) \mid x$ is any real number and y is any nonnegative real number$\}$, $B = \{(x,y) \mid x$ is any nonnegative real number and y is any real number$\}$. Describe the sets of points $A \cup B$, $A \cap B$.
4. Let $A = \{(x,y) \mid x \geq y\}$ and $B = \{(x,y) \mid x$ is any real number and y is any nonnegative real number$\}$ be sets of points in the plane. Describe the sets $A \cap B$ and $A \cup B$ geometrically.
5. Let A and B be sets of points in the plane defined as follows: $A = \{(x,y) \mid y - x^2 \geq 0\}$ and $B = \{(x,y) \mid y + x - 1 \leq 0\}$. Sketch the region of the plane represented by $A \cap B$.
6. Let C be the set of points in the plane defined by
$$C = \{(x,y) \mid 1 \leq x^2 + y^2 \leq 4\}.$$
Find sets A and B such that $C = A \cap B$.

7. Let A, B, and C be sets. Prove the following:
   (a) $A \cup B = B \cup A$.
   (b) $A \cap B = B \cap A$.
   (c) $A \cup A = A$.
   (d) $A \cap A = A$.
8. Let A, B, and C be sets such that $B \subseteq A$ and $C \subseteq A$. Prove that $(B \cup C) \subseteq A$.

Often we can clearly illustrate important properties of $\cup$ and $\cap$ and relations between them by drawings called *Venn diagrams*. These diagrams will often help in checking the validity of equations involving sets.

Let a set be symbolized by the region interior to a closed curve. Then, if A and B are as shown in Figure 1.2, $A \cup B$ is represented by the

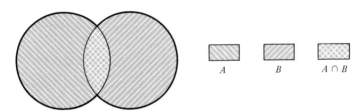

Figure 1.2

region within the heavy curve and $A \cap B$ is represented by the double-hatched area.

As an illustration of the use of Venn diagrams, we consider the following equation among sets A, B, and C:

$$(A \cup B) \cap C = (A \cap C) \cup (B \cap C).$$

We observe that in Figure 1.3 $(A \cup B) \cap C$ is the region containing both horizontal and slanting lines. We also notice that the region corre-

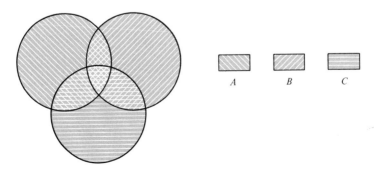

Figure 1.3

sponding to $A \cap C$ together with that corresponding to $B \cap C$ is the same as that of $(A \cup B) \cap C$. Thus

$$(A \cup B) \cap C = (A \cap C) \cup (B \cap C).$$

This discussion is based on the apparently plausible claim that unions and intersections of arbitrary sets can be simulated precisely by sets of points in the plane and, moreover, that such representations are the most general. Without pursuing this possibility any further, we now give a proof of the set equation that is based solely on our definitions.

THEOREM 1.2.7 If A, B, and C are sets, then

$$(A \cup B) \cap C = (A \cap C) \cup (B \cap C).$$

Proof We have mentioned earlier that given sets X and Y, $X = Y$ if and only if $X \subseteq Y$ and $Y \subseteq X$. We use this to achieve a proof by showing that

(i) $(A \cup B) \cap C \subseteq (A \cap C) \cup (B \cap C)$ and
(ii) $(A \cap C) \cup (B \cap C) \subseteq (A \cup B) \cap C$.

(i) Let $x \in (A \cup B) \cap C$. By the definition of intersection, $x \in A \cup B$ and $x \in C$. By the definition of union, $x \in A$ or $x \in B$, or both.

Hence, of the following possibilities, at least one must hold: (a) $x \in A$ and $x \in C$, or (b) $x \in B$ and $x \in C$.

If (a), then $x \in A \cap C$; if (b), $x \in B \cap C$. Since either (a) or (b) holds, we have $x \in A \cap C$ or $x \in B \cap C$, or both. By definition of union, $x \in (A \cap C) \cup (B \cap C)$. Thus, any x in $(A \cup B) \cap C$ is also in $(A \cap C) \cup (B \cap C)$, and therefore

$$(A \cup B) \cap C \subseteq (A \cap C) \cup (B \cap C).$$

(ii) Let $x \in (A \cap C) \cup (B \cap C)$. Then, either (a) $x \in A \cap C$ or (b) $x \in B \cap C$, or both.

(a) If $x \in A \cap C$, then $x \in A$ and $x \in C$. Since $x \in A$, $x \in A \cup B$, and, since $x \in C$, $x \in (A \cup B) \cap C$.
(b) If $x \in B \cap C$, then again we conclude that $x \in (A \cup B) \cap C$. (Why?)

Since (a) or (b) must be true, $x \in (A \cup B) \cap C$. Hence

$$(A \cap C) \cup (B \cap C) \subseteq (A \cup B) \cap C.$$

By our opening remark,

$$(A \cup B) \cap C = (A \cap C) \cup (B \cap C). \qquad \blacksquare$$

Even though our definitions of union and intersection were formulated for a pair of sets A, B, it is not difficult to see how these ideas can be extended to larger collections of sets. For example, we could define the union of three sets A, B, and C as that set which is obtained by first forming the union of A and B and then the union of $A \cup B$ with C. On the other hand we could also form the union of the set A with the set $B \cup C$. The statement that these two processes give rise to the same set is called the "associative law" for the union of sets; we prove it below.

THEOREM 1.2.8 Let A, B, and C be sets. Then

$$(A \cup B) \cup C = A \cup (B \cup C).$$

Proof As usual we will first show that $(A \cup B) \cup C \subseteq A \cup (B \cup C)$ and then that $(A \cup B) \cup C \supseteq A \cup (B \cup C)$.
(a) Let $x \in (A \cup B) \cup C$. Then $x \in A \cup B$ or $x \in C$. If $x \in A \cup B$, then either $x \in A$ or $x \in B$. If $x \in A$, then $x \in A \cup (B \cup C)$. If $x \in B$, then $x \in B \cup C$ and hence $x \in A \cup (B \cup C)$. Finally if $x \in C$, then again $x \in B \cup C$ and so $x \in A \cup (B \cup C)$. Thus we have shown that $(A \cup B) \cup C \subseteq A \cup (B \cup C)$.
(b) If $x \in A \cup (B \cup C)$, then $x \in A$ or $x \in B \cup C$. If $x \in A$, then $x \in A \cup B$ and so $x \in (A \cup B) \cup C$. If $x \in B \cup C$, then either $x \in B$ or $x \in C$. If $x \in B$, then $x \in A \cup B$ and so $x \in (A \cup B) \cup C$. If $x \in C$, then $x \in (A \cup B) \cup C$. Hence $A \cup (B \cup C) \subseteq (A \cup B) \cup C$, and the theorem is proved. ∎

Of course, a similar argument can be given for intersection, and we leave the proof of the associative law for the intersection of sets as an exercise.

Now if we are given a collection of sets $A_1, A_2, \ldots, A_n$, we could define the union of this collection by a process of iteration. That is, we could first form $A_1 \cup A_2$, then $(A_1 \cup A_2) \cup A_3$, and so on. It would then be necessary to prove that if the union of these n sets, in the given order, were formed in some other manner [for example, by first forming $A_1 \cup A_2$, $A_3 \cup A_4$, and then $(A_1 \cup A_2) \cup (A_3 \cup A_4)$ and so forth], we would always end up with the same set, which we will denote by $A_1 \cup A_2 \cup \cdots \cup A_n$ or by

$$\bigcup_{i=1}^{n} A_i.$$

This "generalized associative law" will be proved in another context in Chapter 4. At this stage we will simply assume this result for unions and the analogous result for intersections. It is clear that $\bigcup_{i=1}^{n} A_i$ is the set of all elements x such that $x \in A_i$ for at least one of $A_1, A_2, \ldots, A_n$. Hence

$$\bigcup_{i=1}^{n} A_i = \{x \mid x \in A_i \text{ for some } i = 1, \ldots, n\}.$$

Similarly

$$\bigcap_{i=1}^{n} A_i = \{x \mid x \in A_i \text{ for every } i = 1, \ldots, n\}.$$

We conclude this section with a theorem that shows the precise connection between the idea of the union of two sets and the notion of subset. The analogous theorem for intersection will be left for the exercises.

THEOREM 1.2.9 Let A and B be sets. Then $A \subseteq B$ if and only if

$$A \cup B = B.$$

Proof Suppose first that $A \subseteq B$. We must show that $A \cup B = B$. Let $x \in A \cup B$. Then either $x \in A$ or $x \in B$. If $x \in A$, then since $A \subseteq B$ it follows that $x \in B$. Thus $A \cup B \subseteq B$. If now $x \in B$, then clearly $x \in A \cup B$ whence $B \subseteq A \cup B$ and so $A \cup B = B$.

Conversely, suppose that $A \cup B = B$. To show that $A \subseteq B$, let $x \in A$. Then $x \in A \cup B = B$ and so $x \in B$, thus proving the theorem. ∎

EXERCISES

9. Let $A = \{1,2,3,4\}, B = \{2,3,4,5\}, C = \{3,4,5,6\}$. Find the following sets: (a) $A \cup B$; (b) $A \cup C$; (c) $B \cap C$; (d) $(A \cup B) \cup C$; (e) $(A \cup B) \cap C$; (f) $A \cup (B \cap C)$; (g) $B \cup (A \cap C)$; (h) $C \cup (A \cap B)$; (i) $A \cap (B \cup C)$.

10. Let
    $$A = \{x \mid x \text{ is an integer and } 1 \leq x \leq 100\},$$
    $$B = \{y \mid y \text{ is an integer and } 1 \leq y \leq 150\},$$
    $$C = \{z \mid z \text{ is an integer and } 0 \leq z \leq 200\}.$$

    Find $A \cup B, B \cup C, A \cup C, A \cup (B \cup C), (A \cup B) \cup C, A \cap B, B \cap C, A \cap C, A \cap (B \cup C), B \cup (A \cap C)$.

11. Prove: Let A, B, and C be sets. Then
    $$A \cap (B \cap C) = (A \cap B) \cap C.$$

12. Let A, B, and C be sets.
    (a) Use Venn diagrams to discover all equalities between sets of column I and column II.

I	II
(i) $A \cup (B \cap C)$	(1) $(A \cup B) \cap (A \cup C)$
(ii) $A \cap (B \cup C)$	(2) $(A \cap B) \cup (A \cap C)$
(iii) $(C \cup B) \cap A$	(3) $(B \cup A) \cup C$

(b) Prove each of the equalities you have discovered, in the manner of Theorem 1.2.7.

13. Let A and B be sets. Prove that $A \cap B = A$ if and only if $A \subseteq B$ (see Example 1.2.3).
14. Let A and B be sets. Prove: (a) $A \subseteq A \cup B$; (b) $A \cap B \subseteq A$; and (c) $A \cap B \subseteq A \cup B$.
15. Prove the following generalization of Theorem 1.2.7: Let A, B, C, and D be sets. Then

$$(A \cup B) \cap (C \cup D)$$
$$= (A \cap C) \cup (A \cap D) \cup (B \cap C) \cup (B \cap D).$$

16. Let A, B, C, and D be sets. Prove that

$$(A \cup B \cup C) \cap D = (A \cap D) \cup (B \cap D) \cup (C \cap D).$$

17. State the generalization of the result in Exercise 16 if we replace the sets A, B, C by the collection of sets $\{A_i \mid i = 1, \ldots, n\}$. Use the notation $\bigcup_{i=1}^{n} A_i$.
18. Let A and B be sets. Use Venn diagrams to examine the equation: $(A \cup B) \cap A = A$. Give a full proof of this fact. Do the same for $(A \cap B) \cup A = A$.

1.3 The Empty Set

Let $X = \{0,1,2\}$ and let $Y = \{4,5\}$. Then $X \cup Y = \{0,1,2,4,5\}$. But what is $X \cap Y$? Since there are no elements common to X and Y, we ask if it is even possible to speak of the set $X \cap Y$. But since for any object x, the statement $x \in X \cap Y$ is false and the statement $x \notin X \cap Y$ is true, it follows that $X \cap Y$ is a set, although one very unlike all the sets we have previously considered. $X \cap Y$ is the set with no elements.

Definition 1.3.1 The set with no elements is called the *empty set* (*null set*, *void set*) and is denoted by $\varnothing$.

Although we have worded the definition as though there were only one empty set, the possibility of distinct empty sets should be considered. For example, the set of United States Silver Certificates printed before A.D. 1503 has no elements. Also, the set of all integers that are integral multiples of π has no elements. Although these sets have different descriptions, we shall prove (Corollary 1.3.3) that they are exactly the same empty set.

THEOREM 1.3.2 Let A be any set and let $\emptyset'$ denote an empty set. Then $\emptyset' \subseteq A$.

Proof If $\emptyset'$ is not a subset of A, there must exist an element x such that $x \in \emptyset'$, $x \notin A$. Since $\emptyset'$ has no elements, clearly no such x can exist. Thus, $\emptyset' \subseteq A$. ∎

COROLLARY 1.3.3 The empty set $\emptyset$ is unique.

Proof Let $\emptyset$ and $\emptyset'$ be two empty sets. By Theorem 1.3.2, $\emptyset \subseteq \emptyset'$ and $\emptyset' \subseteq \emptyset$. As we have remarked earlier, $A \subseteq B$ and $B \subseteq A$ means $A = B$. Thus, $\emptyset = \emptyset'$. ∎

Definition 1.3.4 Let A and B be sets such that $A \cap B = \emptyset$. Then A and B are called *disjoint sets* and we say A and B are *disjoint*.

Thus $A = \{0,1,2,4\}$ and $B = \{5,6\}$ are disjoint.

EXERCISES

1. Let A and B be sets. Show that if $A \cup B = \emptyset$, then $A = \emptyset$ and $B = \emptyset$.
2. Let A be any set. Show that
   (a) $A \cup \emptyset = \emptyset \cup A = A$.
   (b) $A \cap \emptyset = \emptyset \cap A = \emptyset$ (hence A and $\emptyset$ are disjoint for every set A).
3. (a) Let $A = \{1,2,3\}$. Find all subsets of A, including $\emptyset$ and A.
   (b) Let $B = \{1,2,3,4\}$. Find all subsets of B, including $\emptyset$ and B.
   (c) Let $A = \{1,2,3,\ldots,n\}$. Can you suggest the number of subsets of A?
4. Show that if $A \subseteq B$ and $A \cap B = \emptyset$, then $A = \emptyset$.
5. Show that if $(A \cup B) \cap B = \emptyset$, then $B = \emptyset$.
6. Show that if $(A \cup B) \cap B = A$, then $A = B$.
7. Let A, B, and C be sets such that (a) $A \cup B = A \cup C$ and (b) $A \cap B = A \cap C = \emptyset$. Prove that $B = C$.
8. Let A, B, and C be sets such that
   (a) $A \cup B = A \cup C$,
   (b) $A \cap B = A \cap C$.
   Prove that $B = C$.

1.4 Universal Sets and Complements

In most of the examples of sets and subsets used to this point we have had a "large" or "master" set in mind from which we selected certain subsets. The set of all real numbers, the set of all integers, and the set of all points in the plane are among the master sets we have used. In a given discussion such a set is usually called the *universe* or *universal set*. Thus if P_2 is the set of all squares of integers and P_3 is the set of all cubes of integers, then we may take as our universe the set Z of all integers. Notice that, once this is done, we are assured that the union of any two sets and the intersection of any two sets are still sets contained in the same universe. This is true since if $A \subseteq C$ and $B \subseteq C$, then $A \cup B \subseteq C$ and $A \cap B \subseteq C$.

Now once a universe I is designated and a given set A is described, $A \subseteq I$, we have automatically also designated another set, namely, the set of all elements of the universe I *not* in A. For example, if the universe is the set of all real numbers and A is the set of all rational numbers, then the set of real numbers not in A is just the set of all irrational numbers. We will now formalize this concept.

Definition 1.4.1 Let I be a universal set. Let $A \subseteq I$. Then the *complement of A* is defined as the set

$$A^\sim = \{x \mid x \in I \text{ and } x \notin A\}.$$

To further illustrate, let I be the set of all points in the plane. Let $A = \{(0,0)\}$; that is, A is the set with only one element, the origin of the plane. Then $A^\sim$ is the set of all points in the plane whose distance from the origin is positive.

A number of important and useful relations among complements, unions, and intersections of sets are given in the following two theorems.

THEOREM 1.4.2 (*De Morgan's Laws*) Let I be a universal set. Let A and B be subsets of I. Then

(a) $(A \cup B)^\sim = A^\sim \cap B^\sim$,
(b) $(A \cap B)^\sim = A^\sim \cup B^\sim$.

Before we give the proof of this theorem, it is useful to illustrate part (a) by the Venn diagram in Figure 1.4. In the diagram we see that $(A \cup B)^\sim$ is simply the region outside of both circles. It is not hard to convince oneself that this region is the same as that given by $A^\sim \cap B^\sim$. We will now give a complete proof of the theorem.

Proof of Theorem 1.4.2 (a) As usual we will establish

$$(A \cup B)^\sim = A^\sim \cap B^\sim$$

by showing that

$$(A \cup B)^\sim \subseteq A^\sim \cap B^\sim \quad \text{and} \quad (A \cup B)^\sim \supseteq A^\sim \cap B^\sim.$$

Let $x \in (A \cup B)^\sim$. Then $x \in I$, but $x \notin A \cup B$. Hence $x \notin A$ and $x \notin B$. Thus $x \in A^\sim$ and $x \in B^\sim$. Therefore $x \in A^\sim \cap B^\sim$ and $(A \cup B)^\sim \subseteq A^\sim \cap B^\sim$.

14 SET THEORY

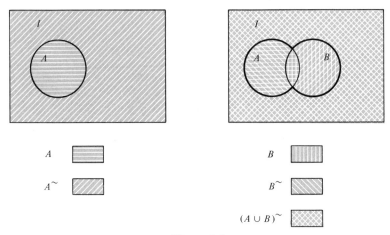

Figure 1.4

Now let $x \in A^\sim \cap B^\sim$. Then $x \in A^\sim$ and $x \in B^\sim$. Hence $x \notin A$ and $x \notin B$. Therefore $x \notin A \cup B$ and so $x \in (A \cup B)^\sim$. This shows that $A^\sim \cap B^\sim \subseteq (A \cup B)^\sim$ and hence

$$(A \cup B)^\sim = A^\sim \cap B^\sim.$$

(b) Let $x \in (A \cap B)^\sim$. Then $x \notin A \cap B$; hence either $x \notin A$ or $x \notin B$. Thus either $x \in A^\sim$ or $x \in B^\sim$, and so by definition of union $x \in A^\sim \cup B^\sim$. Thus $(A \cap B)^\sim \subseteq A^\sim \cup B^\sim$. Now if $x \in A^\sim \cup B^\sim$, then either $x \in A^\sim$ or $x \in B^\sim$. Thus $x \notin A$ or $x \notin B$. Hence $x \notin A \cap B$, and so $x \in (A \cap B)^\sim$. Hence

$$(A \cap B)^\sim = A^\sim \cup B^\sim. \qquad \blacksquare$$

The reader should note that the proof given above involves nothing more than the following notion from elementary logic: the negation of a statement of the form "either this *or* that" is simply "not this *and* not that." Another worthwhile point is that Theorem 1.4.2 can easily be generalized to a collection of sets as follows.

THEOREM 1.4.3 Let $A_1, A_2, \ldots, A_n$ be a collection of subsets of a universe I. Then

(a) $\left(\bigcup_{i=1}^{n} A_i \right)^\sim = \bigcap_{i=1}^{n} A_i^\sim,$

(b) $\left(\bigcap_{i=1}^{n} A_i \right)^\sim = \bigcup_{i=1}^{n} A_i^\sim.$

The proof of these more general theorems is very similar to the proof given and we omit it here.

THEOREM 1.4.4 Let I be a universal set and A a subset of I. Then

(a) $(A^\sim)^\sim = A$,
(b) $\varnothing^\sim = I$, and
(c) $I^\sim = \varnothing$.

Proof The proof is left as an exercise. ∎

EXERCISES

1. Let $I = \{1,2,3,4,5,6\}$, $A_1 = \{1,3,5\}$, $A_2 = \{2,3,4\}$. Verify directly that $(A_1 \cup A_2)^\sim = A_1^\sim \cap A_2^\sim$ and $(A_1 \cap A_2)^\sim = A_1^\sim \cup A_2^\sim$.
2. Let I be the set of all integers. Let $A = \{1,2,\ldots\}$ be the set of all positive integers and $B = \{0,1,2\}$. Find $A \cup B^\sim$, $A^\sim \cap B$, $A^\sim \cap A$, and $(A^\sim \cap B)^\sim$.
3. Let I be the plane. Let

$$A_1 = \{(x,y) \mid x^2 + y^2 \leq 4\}, \quad A_2 = \{(x,y) \mid x^2 + y^2 \leq 1\}.$$

Describe the sets $A_2^\sim \cap A_1$ and $A_1^\sim \cup A_2^\sim$ geometrically.
4. Show that if A is any subset of the universe I, then $A^\sim \cup A = I$ and $A^\sim \cap A = \varnothing$.
5. Let A and B be subsets of the universe I. Show that $A \subseteq B$ if and only if $A \cap B^\sim = \varnothing$. Conclude from this that $A \subseteq B$ if and only if $A^\sim \cup B = I$.
6. Prove Theorem 1.4.3.
7. Prove Theorem 1.4.4

The set $A^\sim$ has been defined with respect to the universal set I, which contains A as a subset. Another notion of "complement" can be defined for a set A relative to another set B, where B need not contain A as a subset.

Definition 1.4.5 Let A and B be sets. Then the *relative complement* of B in A is the set defined by

$$A - B = \{x \mid x \in A \text{ and } x \notin B\}.$$

If $A = \{1,2,3,4\}$ and $B = \{3,5\}$, then

$$A - B = \{1,2,4\} \quad \text{and} \quad B - A = \{5\}.$$

In terms of complements, if I is a universal set and $A \subseteq I$, then $A^\sim = I - A$. Thus the notion of complementation with respect to a universal set is a special case of relative complementation.

These concepts can be graphically illustrated by the Venn diagrams in Figure 1.5.

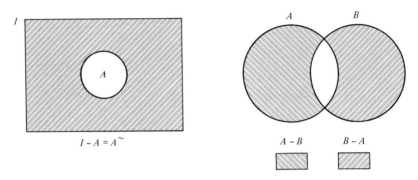

Figure 1.5

Example 1.4.6 Let $I = \{1,\ldots,10\}$ be a universe, and let $A = \{1,2,3,4,5\}$ and $B = \{2,4,6,8\}$. Then $A^\sim = \{6,7,8,9,10\}$, $B^\sim = \{1,3,5,7,9,10\}$, $A - B = \{1,3,5\}$, and $B - A = \{6,8\}$.

Example 1.4.7 Let A be the set of all circles in the plane with center at the origin and radius ≥ 1, and let B be the set of all circles in the plane with center at the origin and radius ≤ 1. Then $A - B$ is the set of all circles in the plane, with center at the origin, and radius > 1, and $B - A$ is the set of all circles in the plane with center at the origin and radius < 1.

EXERCISES

8. Let $A = \{1,2,3,4\}$ and $B = \{2,3,4,5\}$. Calculate $A \cap B$, $A \cup B$, $A - B$, $B - A$, $A - (A - B)$, and $B - (B - A)$. Which pairs of these are equal?
9. Let $A = \{1,2,\ldots,10\}$, $B = \{1,5,9,13,77\}$,
$$C = \{12,13,14,16,25,29,74\}.$$
Find $A - B$, $B - C$, $C - A$, $C - B$, $(A - B) \cup C$, $(C - A) \cup A$, $A - (A - C)$, $(A - B) - C$, $A - (B - C)$.
10. Given sets A and B, prove that
$$A \cap B = (A \cup B) - ((A - B) \cup (B - A))$$
$$= (A - (A - B) = B - (B - A).$$
Investigate these equations by means of Venn diagrams.

1.5 Cartesian Product

We have studied several ways in which pairs or collections of sets can be combined to produce new sets. One technique of great importance is already familiar to the reader from his study of analytic geometry. We recall that each point on a line can be represented by a real number, and that in a plane with rectangular axes each point can be denoted by two numbers x and y, its distances from the y axis and x axis, respectively. These numbers are grouped inside a pair of parentheses and the result (x,y) is called an *ordered pair* of numbers. For example, the ordered pair $(1, \sqrt{2})$ is the point whose abscissa is 1 and whose ordinate is $\sqrt{2}$. Of course, $(\sqrt{2},1)$ is a point different from $(1, \sqrt{2})$ (see Figure 1.6). Hence the order of the entries in the ordered pair is essential.

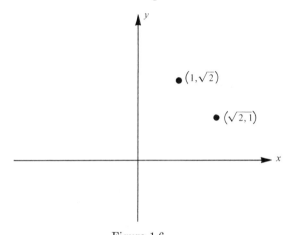

Figure 1.6

Definition 1.5.1 If a and b are elements in sets A and B, respectively, then (a,b) is called an *ordered pair*. Moreover, $(a,b) = (a',b')$ if and only if $a = a'$ and $b = b'$.

If the line L is defined to be the set of all real numbers, then the plane can be defined as the set of all ordered pairs (a,b) for all a and $b \in L$.

Definition 1.5.2 Let A and B be two sets. The *Cartesian product* of A and B is the set $A \times B = \{(a,b) \mid a \in A \text{ and } b \in B\}$.

Thus the plane is equal to $L \times L$.

Example 1.5.3 If $A = \{1,2,3\}$ and $B = \{3,4,5\}$, then

$$A \times B = \{(1,3),(2,3),(3,3),(1,4),(2,4),(3,4),(1,5),(2,5),(3,5)\}$$

and

$$B \times A = \{(3,1),(4,1),(5,1),(3,2),(4,2),(5,2),(3,3),(4,3),(5,3)\}.$$

Example 1.5.3 shows, of course, that except in special cases,

$$A \times B \neq B \times A.$$

The next example indicates that the sets A and B need not be of the same "type."

Example 1.5.4 Let $A = \{$red,green$\}$ and $B = \{$apple,lime$\}$. Then

$$A \times B = \{(\text{red,apple}),(\text{red,lime}),(\text{green,apple}),(\text{green,lime})\}.$$

Notice that on occasion we may wish to select a certain subset of the cartesian product $A \times B$ for some special purpose.

Example 1.5.5 Let P be the set of all positive integers and $A = \{0,1,2\}$. Let B be a subset of $P \times A$ defined as follows: $B = \{(x,y) \mid y$ is the smallest nonnegative remainder obtained when x is divided by 3$\}$. Then, for example, $(3,0) \in B$, $(4,1) \in B$, and $(-7,2) \in B$, but $(8,0) \notin B$ and $(10,2) \notin B$. We will have more to say about this particular example in Chapter 2.

EXERCISES

1. Let $A = \{0,1\}$ and $B = \{3,4,5\}$. Find $A \times B$.
2. Let A, B, and C be sets with $A \subseteq B$. Show that $A \times C \subseteq B \times C$.
3. When does $A \times B = B \times A$?
4. Let $A = \{x \mid x$ is an integer$\}$ and $B = \{y \mid y$ is an integer and $-5 \leq y \leq 5\}$. What are each of the following sets: $A \times B$, $B \times A$, $A \times A$, $B \times B$? Can you describe them geometrically?
5. Show that $(A \times C) \cup (B \times C) = (A \cup B) \times C$.
6. Is it true that $(A \times B) \cup (C \times D) = (A \cup C) \times (B \cup D)$? Explain.
7. Suppose $A = \emptyset$ and $B \neq \emptyset$. What are $A \times B$ and $B \times A$?
8. Let A, B, and C be sets. Is it true that $(A \times B) \times C = A \times (B \times C)$? Could you define $A \times B \times C$ in terms of "triples" instead of pairs? Do you know of any examples of sets like $A \times B \times C$?
9. Let $A_1, A_2, \ldots, A_n$ be a finite collection of sets. How could you define $A_1 \times A_2 \times \cdots \times A_n$?

1.6 Functions

One of the most important uses of ordered pairs is in the definition of functions. The reader has already dealt with functions at some length in analytic geometry and calculus. Briefly, the essential idea involved is that of a correspondence; that is, once an element of one set called the domain is specified, a unique corresponding element in another set called the codomain is determined. Since in this book we are going to work extensively with functions of various types, it is important to state very carefully exactly how we will use this term.

It seems that two possible procedures are available. In the first procedure, we could define a function to be the *rule of correspondence* that actually tells us how to choose the element of the codomain corresponding to a particular element of the domain. In some cases this rule is quite easy to describe. For example, the function that assigns the real number x^2 to the number x provides a correspondence from the domain of all real numbers to the codomain of nonnegative real numbers. Some rules of correspondence are not quite so easy to describe. More particularly, we will soon have the need to combine functions in various ways, and it is often inconvenient to carry along their descriptions as rules of correspondence. An alternate, second approach, in keeping with our general reliance on the set theory already introduced, is to define a function in terms of ordered pairs (x,y), where x is an arbitrary element of the domain and y is the element of the codomain that corresponds to x. This set of ordered pairs together with the domain and codomain will then completely specify the function under consideration. Notice that in the sense of analytic geometry, we are defining a function by its "graph." In particular, the set of ordered pairs associated with a function from calculus is precisely the set of points in the plane called the "graph" of the function.

Definition 1.6.1 A *function* from the set A, called the *domain*, to the set B, called the *codomain*, consists of the sets A and B and a subset f of $A \times B$ satisfying the following criteria:

(a) If $a \in A$, then there is a $b \in B$ such that $(a,b) \in f$.
(b) If $(a,b_1) \in f$ and $(a,b_2) \in f$, then $b_1 = b_2$.

Even though the symbol f denotes the set of ordered pairs we will usually designate the function by f, and we will write: $f: A \longrightarrow B$ to give the complete description of the function. We will indicate that $(a,b) \in f$ by the usual notation $f(a) = b$ or $fa = b$. If $A = B$ and $f: A \longrightarrow A$, we usually say that f is a *function on A*.

Before we proceed to give some examples, we make a few remarks about the definition. First, note that the definition of a function is not complete until the sets A and B are specified. It is not enough simply to specify the set of ordered pairs, for whereas the set of ordered pairs will determine the domain, it does not determine the codomain. The reason for this is seen in part (a) of the definition, which simply states that every element of the domain must occur as the first component of an ordered pair in f. But the definition does not state that every element of B must occur as the second component of an ordered pair in f. We do give a special name to the subset of B consisting of just those elements that occur as second components of ordered pairs in f. This set is called the *range* of f. For example, if A and B are both the set of all real numbers and f is the function from A to B with associated set of ordered pairs $\{(x,x^2) \mid x \in A\}$ [that is, $f(x) = x^2$], then while the codomain B is the set of real numbers, the range of f is the set of nonnegative real numbers, a proper subset of B. Thus if we again let A be the set of real numbers, if we let C be the set of nonnegative real numbers, and if we define $g\colon A \longrightarrow C$ by $g(x) = x^2$, then the set of ordered pairs so defined is the same as the set of ordered pairs for f, but the functions f and g are different. Note that for the function g, the range and codomain are identical. This distinction may seem a small one, but it will become more significant in the sequel.

The second remark concerning the definition is that we use the word "function" to mean "single-valued function." This is the content of (b) in the definition. For if $(a,b_1) \in f$ and $(a,b_2) \in f$, then $f(a) = b_1$ and $f(a) = b_2$. We then insist that $b_1 = b_2$.

Finally, it should be noted that functions are also referred to as "mappings" or "maps."

Example 1.6.2 Let $A = \{1,2,3,4\}$ and $B = \{0,1\}$. Then $f\colon A \longrightarrow B$ defined by $f = \{(1,1),(2,0),(3,1),(4,0)\}$ is a function with domain A and codomain B. In this case the range of f is also B.

Example 1.6.3 Let A be a set, $A \neq \varnothing$. Then $f = \{(a,a) \mid a \in A\}$ is called the *identity* function on A; that is, f is the function on A such that $f(a) = a$, for all a in A.

Example 1.6.4 Let $A = B$ be the set of all real numbers. Then $f\colon A \longrightarrow B$, $f = \{(x,x-1) \mid x \in A\}$ is a function whose graph in the plane is given by $y = x - 1$.

Example 1.6.5 Let $A = \{1,-1,i,-i\}$ with $i^2 = -1$. Then

$$f = \{(a,a^4) \mid a \in A\}$$

is the function defined by $f(a) = 1$, for all $a \in A$.

Example 1.6.6 Let A be the set of all first-class letters mailed inside the United States in 1969. Let B be the set of positive integers. Let f be the function equal to $\{(a,b) \mid b$ is the cost in cents of postage on a, for all $a \in A\}$. Another way of defining f is as follows: if a denotes a letter whose weight is greater than $n - 1$ ounces but less than or equal to n ounces, then $f(a) = 6n$.

Example 1.6.7 Let K denote the set of all real numbers and let F be a subset of K. We define the function $f\colon K \longrightarrow \{0,1\}$ as $f = \{(k,a) \mid a = 0$

if $k \notin F$; $a = 1$ if $k \in F\}$. This function is called the *characteristic function* of the set F (relative to K).

Example 1.6.8 Let Z be the set of integers. Let $f\colon (Z \times Z) \longrightarrow Z$ be defined by $f = \{((a,b), a + b) \mid a, b \in Z\}$. Put another way, $f(a,b) = a + b$. Notice that the domain is $Z \times Z$ and the codomain (and the range) is Z. Thus the operation of addition of integers can be thought of as a function from the set of ordered pairs of integers to the integers.

Example 1.6.9 As in Example 1.6.8 we may define multiplication of integers by the function $g\colon (Z \times Z) \longrightarrow Z$ with $g = \{((a,b), ab) \mid a, b \in Z\}$.

The functions of Examples 1.6.8 and 1.6.9 from $Z \times Z \longrightarrow Z$ are special cases of a type of function that may be defined in a more general setting. Since we will have later occasion to use this idea in a more general context, we give a formal definition here.

Definition 1.6.10 Let A be a nonempty set and let f be a function with domain $A \times A$ and codomain A, $f\colon A \times A \longrightarrow A$. Then f is called a *binary operation on* A.

EXERCISES

1. Let $f\colon Z \longrightarrow Z$ be defined by $f(x) = 2x + 1$. Express f as a set of ordered pairs of integers. What is the range of f?
2. Let $f = \{(x,x^3) \mid x \text{ is a real number}\}$. Assuming f is a function, what are the domain and range of f? Can you say anything about the codomain of f?
3. Let $f = \{(x,x^4) \mid x \text{ is a real number}\}$. Assuming that f is a function, what are the domain and range of f?
4. Let $A = \{1,2,3\}$ and $B = \{4,5,6,7\}$. Which of the following subsets of $A \times B$ are functions from A to B?
    (a) $\{(1,4),(2,5)\}$.
    (b) $\{(1,5),(1,4),(2,6),(3,7)\}$.
    (c) $\{(1,4),(2,5),(2,7),(1,6)\}$.
    (d) $\{(1,7),(2,6),(3,4)\}$.
    (e) $\{(1,7),(2,7),(3,7)\}$.
5. Find all functions on the set $A = \{1,2,3\}$.
6. Let $f\colon A \longrightarrow B$ be a function defined by $f = \{(x, x^2 + 1) \mid x \in A\}$, where $A = B$ is the set of all real numbers. Find a function g whose set of ordered pairs (graph) is the same as that for f, but where g is a different function from f.
7. Find a binary operation on the set of integers Z that is different from both addition and multiplication.
8. Suppose that $f\colon A \longrightarrow B$ and that A contains exactly n elements. Show that the range of f can contain at most n elements.

The roles played by the domain and the codomain in the definition of function (1.6.1) are clearly nonsymmetrical; that is, we see from (a) that the entire domain A must be "used up" as first components of ordered pairs in f, but some elements of B need not appear at all as second components.

On the other hand, it follows from (b) that a given element b may appear as a second component in more than one ordered pair of f, but that no element of A may appear twice as a first component. There are two especially important classes of functions whose defining properties restore the balance discussed above between domain and codomain.

Definition 1.6.11 Let $f: A \longrightarrow B$; that is, let f be a function with domain A and codomain B.

(a) If the range of f is B, then f is called an *onto* function, or simply *onto*.
(b) If $f(a_1) = f(a_2)$ implies that $a_1 = a_2$, then f is called a *one-to-one* function, or simply *1-1*.
(c) If f is both 1–1 and onto, then we say that f *is a 1–1 correspondence between A and B*.

If f is an onto function, clearly the range of f coincides with its codomain. And if f is 1–1, each element of its range occurs precisely once as the second component of an ordered pair of f.

We use some special notation to denote functions that are onto or 1–1 or 1–1 and onto:

$$f: A \xrightarrow{\text{onto}} B, \qquad g: A \xrightarrow{1-1} B, \qquad h = A \xrightarrow[\text{onto}]{1-1} B.$$

If A and B are two sets and there exists a 1–1 correspondence between them, then we can describe this by saying that there is a "pairing" of elements of A with elements of B in which every element of A has a unique partner and vice versa. As a consequence, if A contains, say, precisely nine elements, then B contains precisely nine elements. The existence of a 1–1 correspondence between A and B thus implies that the A and B have the same "size." This fact is of far-reaching significance in the theory of sets, since it allows one to compare the relative sizes of two sets even when one or both of the sets are not finite. It follows, for example, that the set of all integers has the same size as the set of all even integers, even though one of these is a proper subset of the other (see Exercise 12.)

In case there exists a 1–1 correspondence between A and the set P of positive integers, we shall call A *countably infinite*. We shall use this idea in Chapter 6.

Example 1.6.12 Let $f: A \longrightarrow B$, where $A = B$, the set of all real numbers, and $f = \{(x, x^3) \mid x \in A\}$. Then f is onto, since if u is a real number, so is

$\sqrt[3]{u}$, and so $(\sqrt[3]{u}, u) \in f$. Also f is 1-1, since if $f(x) = x^3$ and $f(y) = x^3$, then $x^3 = y^3$ and hence $x = y$.

Example 1.6.13 Let $A = B = K$, the set of real numbers, and let $f = \{(x,x^2) \mid x$ is a real number$\}$. In this example f is not 1-1, since $f(1) = f(-1) = 1$. Also, f is not onto, since, for example, there is no real number x such that $f(x) = x^2 = -1$. However, if we consider $f: A \longrightarrow K^+$, where K^+ is the set of all nonnegative real numbers, then f is onto K^+.

Similarly, if $g = \{(x,x^2) \mid x \in K^+\}$, then

$$g: K^+ \xrightarrow[\text{onto}]{1\text{-}1} K^+.$$

Example 1.6.14 Let $A = \{1, -1, i, -i\}$ with $i^2 = (-i)^2 = -1$, and let $B = \{0, 1\}$. Then again $f: A \longrightarrow B$, $f = \{(a, a^4) \mid a \in A\}$, is a function that is neither 1-1 nor onto.

Example 1.6.15 Let $A = B$, the set of all real numbers, and let $f: A \longrightarrow B$ be defined by $f = \{(x, x+1) \mid x \in A\}$. Then f is 1-1 and onto.

It is interesting to note that a function whose domain and codomain are the set of real numbers and that has an easily drawn graph in the plane can be tested for the properties of 1-1 and onto in a completely geometrical way. For the function will be onto if every horizontal line intersects the graph of the function at least once and it will be 1-1 if every horizontal line intersects the graph at most once. (What must be true about every vertical line?)

Example 1.6.16 Let $f: A \longrightarrow B$ with $A = B$, the set of all real numbers. If

$$f = \left\{ \left(x, \frac{1}{x}\right) \,\bigg|\, x \in A, x \neq 0 \right\} \cup \{(0,0)\},$$

then f is both 1-1 and onto. Note that $f(x) = 1/x$ if $x \neq 0$ and $f(0) = 0$.

Example 1.6.17 Let $f: A \longrightarrow B$ with $A = B$, the set of all real numbers. If

$$f = \left\{ \left(x, \frac{x^2}{1+x^2}\right) \,\bigg|\, x \in A \right\},$$

then it is not difficult to see that $x^2/(1+x^2) < 1$ for all x and so f is not onto. The function f is not 1-1 either since $f(x) = f(-x)$ for all x.

Example 1.6.18 Addition and multiplication on Z, the set of integers, are not 1-1 from $Z \times Z$ to Z, but they are onto.

Example 1.6.19 Let A be the set of all nondegenerate triangles in the plane, and let B be the set of all positive real numbers. Let $f: A \longrightarrow B$ be defined as follows:

If $\triangle \in A$, then $f(\triangle) = $ area $\triangle$.

Clearly, f is not 1–1 but f is onto.

EXERCISES

9. Let $A = \{1,2,3,4,5\}$, let $B = \{1,2,3\}$. Let $f: A \longrightarrow B$ be defined as follows:
   If a is odd, $f(a) = 1$.
   If a is even, $f(a) = 2$.

   Show that f is neither 1–1 nor onto.
10. Let $A = \{1,2,3,4,5\}$, $B = \{6,7,8,9,10\}$, and $C = \{11,12,13,14,15\}$. Let $f: A \longrightarrow B$ be defined by $f(a) = a + 5$ for all $a \in A$, and let $g: B \longrightarrow C$ be defined by $g(b) = b + 5$, for all $b \in B$. Show that both f and g are 1–1 and onto.
11. Let f be the identity function on a set A. Prove that f is both 1–1 and onto.
12. Show that there exists a one-to-one correspondence between the set of even integers and the set of all integers.
13. Let F be a set with 12 elements, and let $f: F \xrightarrow{1-1} F$. Show that f is onto.
14. Find a set A and a function $f: A \xrightarrow{1-1} A$ that is not onto.

1.7 Composition of Functions

If we think of a function as a rule that assigns to each member of a given set a corresponding element in another set, then it makes sense to consider the action of two successive functions. For example, if $f: Z \longrightarrow Z$ by $f(x) = x + 1$ and $g: Z \longrightarrow Z$ by $g(x) = x^2 - 1$, then we can determine a new function, denoted by $g \circ f$, which is defined by:

$$(g \circ f)(x) = g(f(x)) = g(x + 1) = (x + 1)^2 - 1 = x^2 + 2x.$$

The crucial fact here is that the range of f is contained in the domain of g, for otherwise $g \circ f$ would make no sense. For convenience, however, we will require that the codomain of f equal the domain of g.

Definition 1.7.1 Let $f: A \longrightarrow B$ and $g: B \longrightarrow C$. Then the *composite* $g \circ f$ (or *product* $g \circ f$) is the function $g \circ f: A \longrightarrow C$ defined by

$$(g \circ f)(x) = g(f(x)).$$

We could also define $g \circ f$ directly in terms of ordered pairs as follows: $(a,c) \in g \circ f$ with $a \in A$ and $c \in C$ if and only if there is an element $b \in B$ such that $(a,b) \in f$ and $(b,c) \in g$. Hence we can obtain all the ordered pairs of $g \circ f$ as follows: combine every pair in f with second entry b (if any) with that unique pair in g with first entry b for all $b \in B$.

Example 1.7.2 Let $A = B = C$ be the set of real numbers. Let $f\colon A \longrightarrow B$ and $g\colon B \longrightarrow C$ be defined by

$$f = \{(x, x+1) \mid x \in A\}, \qquad g = \{(x, x-2) \mid x \in B\}.$$

Since the domain of B and the codomain of A are the same, it makes sense to define $g \circ f$. If $(a, a+1)$ is in f, then $(a+1, a-1)$ is in g, since $a + 1 \in B = A$. Hence the pair $(a, a-1) \in g \circ f$ and in fact $g \circ f = \{(x, x-1) \mid x \in A\}$. We could also compute $g \circ f$ as follows:

$$(g \circ f)(x) = g(f(x)) = g(x+1) = (x+1) - 2 = x - 1.$$

It is clear from the definition that the order of writing $g \circ f$ is important, for it may happen that either $f \circ g$ makes no sense or that $f \circ g$ is not the same function as $g \circ f$.

Example 1.7.3 Let $f\colon A \longrightarrow B$ with $A = B$, the set of real numbers, and f defined by $f(x) = x^2$. Let $g\colon A \longrightarrow B$ with $g(x) = x + 1$. Now

$$(g \circ f)(x) = g(f(x)) = g(x^2) = x^2 + 1.$$

But

$$(f \circ g)(x) = f(g(x)) = f(x+1) = (x+1)^2 = x^2 + 2x + 1.$$

Clearly both $g \circ f$ and $f \circ g$ are defined, but they are not equal.

Example 1.7.4 Let $f\colon A \longrightarrow B$ with A the nonnegative real numbers, B the set of all real numbers, and $f(x) = \sqrt{x}$. Let $g\colon B \longrightarrow C$ with C the set of all real numbers and $g(x) = -x^2$. Then

$$(g \circ f)(x) = g(f(x)) = g(\sqrt{x}) = -x.$$

But since the codomain of g is not equal to the domain of f (in fact it is not even a subset), it follows that $f \circ g$ is not defined. [Note that $(f \circ g)(x) = f(-x^2)$, and $\sqrt{-x^2}$ is not a real number except for $x = 0$.]

Composition has been defined for a pair of functions, but it can easily be extended to larger numbers of functions, provided the appropriate conditions on domains and codomains are satisfied. For example, if we have three functions

$$f\colon A \longrightarrow B, \qquad g\colon B \longrightarrow C, \qquad h\colon C \longrightarrow D,$$

then we may form the composite of f, g, and h in that order in two different ways: $h \circ (g \circ f)$ and $(h \circ g) \circ f$, both of which make sense. Of course, the question that came up when we extended the definitions of union and intersection again arises: are these two compositions of functions the same?

THEOREM 1.7.5 Let $f: A \longrightarrow B$, $g: B \longrightarrow C$, and $h: C \longrightarrow D$. Then $h \circ (g \circ f) = (h \circ g) \circ f$.

Proof To show that $h \circ (g \circ f) = (h \circ g) \circ f$ we must show that the image of any element $x \in A$ is the same under both of these functions. This is easily seen, since

$$[h \circ (g \circ f)](x) = h[(g \circ f)(x)] = h[g(f(x))] = (h \circ g)(f(x))$$
$$= [(h \circ g) \circ f](x). \quad \blacksquare$$

In the special case of Theorem 1.7.5, in which $A = B = C$ and $f = g = h$, we have $f \circ (f \circ f) = (f \circ f) \circ f$. By defining $f^2 = f \circ f$, we may also define f^3 unambiguously by setting $f^3 = f^2 \circ f = f \circ f^2$. Given $f: A \longrightarrow A$, no matter how we insert parentheses into the expression

$$\underbrace{f \circ f \circ \cdots \circ f}_{n \text{ times}}$$

in a meaningful way, we always get the same result. (See the discussion of associative laws in Chapter 2. The proof mentioned in Chapter 2, also applicable here, will be given in Chapter 4.)

EXERCISES

In the following exercises K denotes the set of real numbers.

1. Let Q be the set of rational numbers. Let $f: Q \longrightarrow K$, where $f(x) = \sqrt[3]{x}$, $g: K \longrightarrow K$, where $g(x) = x^6$. What is the function $g \circ f$?
2. Let $A = \{(a,a') \mid a, a' \in K\}$. Let $f: A \longrightarrow K$, where $f(a,a') = a^2 + (a')^2$, and $g: K \longrightarrow K$, where $g(x) = \sqrt{x}$. What is $g \circ f$? Is $f \circ g$ a function?
3. Let $f: K \longrightarrow K$, where $f(x) = -x$, and let $g: K \longrightarrow K$, where $g(x) = 1/x$, for $x \neq 0$ and $g(x) = 0$, for $x = 0$. Show that $f^2 = I$ (the identity function), $g^2 = I$, and $f \circ g = g \circ f$.
4. Let $f: K \longrightarrow K$, where $f(x) = x + \pi$. For $n > 0$, show that f^n is defined by $f^n(x) = x + n\pi$.
5. Let $f: K \longrightarrow K$, where $f(x) = x^3$ and $g: K \longrightarrow K$, where $g(x) = 1 + x^2$. Determine $f \circ g$ and $g \circ f$. Are they equal?
6. Let $f: K \longrightarrow K$, where $f(x) = 2x - 1$. Find a function $g: K \longrightarrow K$ such that $(g \circ f)(x) = x$, all $x \in K$, and $(f \circ g)(x) = x$, all $x \in K$.

7. Let $f: A \xrightarrow[\text{onto}]{\text{1-1}} B$, $g: B \xrightarrow[\text{onto}]{\text{1-1}} C$. Show that

$$g \circ f: A \xrightarrow[\text{onto}]{\text{1-1}} C.$$

The reader will recall that a function that is 1-1 and onto introduces a symmetry into the roles of the domain and codomain. One of the most important consequences of this symmetry is that each such function can be "inverted" or "turned backwards"; that is, it is possible to define another function that when composed with the first gives the identity function.

THEOREM 1.7.6 Let $f: A \xrightarrow[\text{onto}]{\text{1-1}} B$. Then there exists a unique function

$$g: B \xrightarrow[\text{onto}]{\text{1-1}} A$$

such that $g \circ f = I_A$, the identity on A, and $f \circ g = I_B$, the identity on B. Indeed, the function g is defined as follows: for all $b \in B$, $g(b) = a$, where $f(a) = b$.

Proof Since f is onto, for each $b \in B$ there is an $a \in A$ such that $f(a) = b$. Furthermore this choice of a is unique, since f is 1-1. We define $g: B \longrightarrow A$ as follows:

$$g(b) = a, \quad \text{where } f(a) = b.$$

Since f is 1-1 and onto, g is a function from B to A. And g is in fact 1-1 and onto. (Why?) For any $a \in A$,

$$(g \circ f)(a) = g(f(a)) = g(b) = a.$$

Hence $g \circ f = I_A$. For any $b \in B$,

$$(f \circ g)(b) = f(g(b)) = f(a) = b,$$

and $f \circ g = I_B$.

To prove the uniqueness of g let $g': B \longrightarrow A$ satisfy $g' \circ f = I_A$ and $f \circ g' = I_B$. Then for each $b \in B$,

$$g'(b) = (g \circ f)(g'(b)) = g((f \circ g')(b)) = g(b). \quad \blacksquare$$

Definition 1.7.7 Let $f: A \xrightarrow[\text{onto}]{\text{1-1}} B$. Then the function g of Theorem 1.7.6 is called the *inverse function* of f. We denote g by f^{-1}.

For example, let K be the set of real numbers and let $f: K \longrightarrow K$ be defined by

$$f(x) = x^3.$$

It is easy to see that f is 1-1 and onto. By Theorem 1.7.6 the function

f^{-1} is defined by: $f^{-1}(a) = b$, where $f(b) = a$. But $f(b) = b^3$; hence $b = \sqrt[3]{a}$. This gives $f^{-1}(a) = \sqrt[3]{a}$, and f^{-1} is defined for all real numbers.

On the other hand, let $h: K \longrightarrow K$ be defined by

$$h(x) = x(x-1)(x+1).$$

Then h is onto but not 1-1, since $h(1) = h(-1) = 0$. Hence if there were a function g such that $g \circ h = I_K$, then $g(0) = 1$ and $g(0) = -1$, which contradicts the definition of a function. Hence we see that it would not be possible to define the inverse of a non one-to-one function.

Let $f: A \xrightarrow[\text{onto}]{1-1} B$. Then associated with f is the collection of ordered pairs such that every $a \in A$ occurs in precisely one ordered pair and every $b \in B$ occurs in precisely one ordered pair. Then the function $f^{-1} = \{(b,a) \mid (a,b) \in f\}$. For example, if K is the set of real numbers, then, if $f = \{(x,x^3) \mid x \in K\}$, we see that $f^{-1} = \{(x^3,x) \mid x \in K\}$. The reader should verify that

$$\{(x^3,x) \mid x \in K\} = \{(x, \sqrt[3]{x}) \mid x \in K\}.$$

EXERCISES

8. Show that the function g defined in Theorem 1.7.6 is 1-1 and onto.
9. Let $f: K \longrightarrow K$ and $g: K \longrightarrow K$, where $f(x) = -x$ and $g(x) = 1/x$, $x \neq 0$, and $g(0) = 0$. Show that f^{-1}, g^{-1}, and $(g \circ f)^{-1}$ exist. Moreover, show that $(g \circ f)^{-1} = g^{-1} \circ f^{-1}$.
10. Let $f: A \xrightarrow[\text{onto}]{1-1} B$ and $g: B \xrightarrow[\text{onto}]{1-1} A$. Show that $g = f^{-1}$ if and only if $f = g^{-1}$.
11. Let $f: A \longrightarrow B$ and $g: B \longrightarrow A$. If $g \circ f = I_A$, show that f is 1-1 and g is onto.
12. Give an example illustrating each of the following:
    (a) $f: A \xrightarrow[\text{onto}]{1-1} B$, $g: B \xrightarrow[\text{onto}]{1-1} A$, but $g \circ f \neq I_A$ and $f \circ g \neq I_B$.
    (b) $f: A \longrightarrow B$, $g: B \longrightarrow A$, $g \circ f = I_A$, but f is not onto and g is not 1-1.
    (c) $f: A \longrightarrow B$, $g: B \longrightarrow A$, $g \circ f = I_A$, but $f \circ g \neq I_B$.
13. Let $f: A \xrightarrow[\text{onto}]{1-1} A$. Show that f^2 and f^3 are both 1-1 and onto.
14. Let $f: A \xrightarrow[\text{onto}]{1-1} A$. Prove that $(f^2)^{-1} = (f^{-1})^2$ and $(f^3)^{-1} = (f^{-1})^3$.

1.8 Equivalence Relations and Partitions

Since a function f is determined by its domain A, codomain B, and the set of ordered pairs $\{(a,f(a)) \mid a \in A\}$, it can be thought of as a special subset of $A \times B$ once A and B are specified. Other notions of correspondence,

1.8 EQUIVALENCE RELATIONS AND PARTITIONS

not quite so restrictive as a function, can be expressed in a similar way. The word "relation" as it is used in everyday language is simply an expression that two objects (either from different sets or from the same set) are associated in some way. For example, we could define the relation "father of" between the set A of all members of the PTA of Leal School and the set B of all pupils of that school by the set of all ordered pairs $\{(a,b) \mid a \in A, b \in B, a \text{ is the father of } b\}$. Notice that this relation is not necessarily a function, since for a given father a there may be several children, say b_1, b_2, and b_3. Then (a,b_1), (a,b_2), and (a,b_3) would all be in the relation, with b_1, b_2, and b_3 all different, clearly violating the criteria for a function. In general a *relation* between two sets A and B is defined as a subset of the Cartesian product $A \times B$. We also say that a subset of $A \times A$ is a *relation on* A.

In this section we wish to focus on a very special relation defined on a set A, namely an equivalence relation. We will use the following notation: if $R \subseteq A \times A$ is a relation on A, we denote that $(a,b) \in R$ by aRb.

Definition 1.8.1 An *equivalence relation* R on a set A is a relation R on A satisfying

(a) aRa, for all $a \in A$. (reflexive law)
(b) If aRb, then bRa. (symmetric law)
(c) If aRb and bRc, then aRc. (transitive law)

Moreover, if aRb holds, we say that a is *equivalent* to b. Since aRb implies bRa, we see that b is equivalent to a as well. Thus if aRb, we also say a and b are *equivalent*.

It is easily seen that "$=$" is an equivalence relation on K, the real number system. However, "$<$" is not an equivalence relation on K since $2 < 3$ does not imply $3 < 2$.

Example 1.8.2 Let us consider the following relation defined for people: person a is equivalent to person b if a and b were born on the same day of the year (possibly different years). We write aRb if this is the case. Clearly, a was born on his own birthday, so aRa. Also, if aRb (say, for example, a was born on July 22, 1944, and b on July 22, 1931), then bRa. And further, if aRb and bRc, then aRc. This relation is, hence, an equivalence relation on the set of all people.

Example 1.8.3 As another important illustration, let R be the relation defined on the set of integers as follows: aRb if and only if $a - b$ is an integral multiple of 3; that is, $(a,b) \in R$ if and only if there is an integer d such that $a - b = 3d$. Now it is clear that $a - a = 0 = 3 \cdot 0$, and so aRa holds for any integer a. Second, if aRb, then there is an integer d

such that $a - b = 3 \cdot d$. But then $b - a = 3(-d)$. Since $(-d)$ is an integer, we have bRa. Thus, if aRb, then also bRa. Third, suppose aRb and bRc. Then there are integers d_1 and d_2 such that $a - b = 3d_1$ and $b - c = 3d_2$. But then $(a - b) + (b - c) = 3d_1 + 3d_2$; that is, $a - c = 3(d_1 + d_2)$. Since $d_1 + d_2$ is an integer, we have aRc. Thus, R is an equivalence relation on the set of all integers. We note, for example, that $3R0$, since $3 - 0 = 3 \cdot 1$; $5R2$, since $5 - 2 = 3 \cdot 1$, and $8R(-4)$, since $8 - (-4) = 3 \cdot 4$.

Example 1.8.4 Let A be the set of all triangles in the plane, and let R be the relation of congruence of triangles. It is not difficult to show that R is an equivalence relation (see Exercise 2).

Not all relations, of course, are equivalence relations. In fact, there exist relations that satisfy any two but not all of the properties of Definition 1.8.1. Examples of such relations are given in Exercises 8, 9, and 10.

EXERCISES

1. Let R be the relation of Example 1.8.3. Show that if a is any integer then exactly one of the following holds: (i) $aR0$, (ii) $aR1$, (iii) $aR2$.
2. Let $\cong$ denote the relation of congruence among triangles in the plane. Show that $\cong$ is an equivalence relation.
3. Let $\sim$ denote the relation of similarity among triangles in the plane. Show that $\sim$ is an equivalence relation.
4. Let R be the relation defined on the set of integers Z as follows: aRb if and only if $a + b$ is an integral multiple of 2. Show that R is an equivalence relation on Z.
5. Let $A = \{1,2,3,4,5,6,7,8,9,0\}$. Let $A_1 = \{1,2,3,4\}$, $A_2 = \{5,6,7\}$, $A_3 = \{8,9\}$, $A_4 = \{0\}$. Define the relation R on A by aRb if and only if $a \in A_i$ and $b \in A_j$ implies $i = j$. Show that R is an equivalence relation on A.
6. Let R be a relation defined on the set of all people as follows: aRb means that a is an ancestor of b. Is R an equivalence relation? Show that only (c) of Definition 1.8.1 is satisfied.
7. Let R be the relation defined on the set of integers Z as follows: aRb if and only if $a + b$ is an integral multiple of 3. Show that R is not an equivalence relation on Z. In particular, show that only (b) of Definition 1.8.1 is satisfied.
8. Let R be the relation defined on the set of integers Z as follows: aRb if and only if either a is an integral factor of b or b is an integral factor of a. Show that (a) and (b) of Definition 1.8.1 are satisfied, but that (c) is not.
9. Show that weak inequality ($\leq$) on the real number system satisfies (a) and (c) of Definition 1.8.1, but that it does not satisfy (b).

10. Let $S = Z \cup \{\sqrt{2}\}$ with Z the set of integers. Define aRb if $a + b \in Z$. Show that (b) and (c) of Definition 1.8.1 are satisfied, but not (a).

In each of our examples, the equivalence relation R splits the set into a number of subsets as follows: two elements a and b are in the same subset if and only if aRb.

Definition 1.8.5 Let R be an equivalence relation on a set A. We call the set of elements equivalent to a given element a the *equivalence class determined by* a. This will be denoted by $\bar{a}$. Thus, $\bar{a} = \{x \mid x \in A \text{ and } xRa\}$.

It is easy to see that in Example 1.8.3,

$$\bar{0} = \{\ldots, -6, -3, 0, 3, 6, 9, \ldots\}, \quad \bar{1} = \{\ldots, -5, -2, 1, 4, 7, 10, \ldots\},$$

and $\bar{2} = \{\ldots, -7, -4, -1, 2, 5, 8, \ldots\}$. Moreover, if $a \in Z$, then $\bar{a}$ is one of these three sets.

LEMMA 1.8.6 Let $\sim$ be an equivalence relation on a set A. Then

(a) $a \sim b$ if and only if $\bar{a} = \bar{b}$,
(b) $b \in \bar{a}$ if and only if $\bar{a} = \bar{b}$.

Proof (a) Let $a \sim b$. Let $x \in \bar{a}$. By definition of $\bar{a}$, $x \sim a$. Since $\sim$ is an equivalence relation, $x \sim a$ and $a \sim b$ implies $x \sim b$, and so $x \in \bar{b}$. Thus, $\bar{a} \subseteq \bar{b}$. Now, since $a \sim b$, by (b) of Definition 1.8.1, $b \sim a$. Interchanging the role of a and b in the argument above, we get $\bar{b} \subseteq \bar{a}$. Thus, $\bar{a} = \bar{b}$.

Conversely, suppose $\bar{a} = \bar{b}$. In particular, since $a \sim a$, $a \in \bar{a}$, and so $a \in \bar{b}$; that is, $a \sim b$.

(b) Let $b \in \bar{a}$. Then $b \sim a$, and by (a) $\bar{b} = \bar{a}$. Conversely, if $\bar{b} = \bar{a}$, then by (a) $b \sim a$ and so $b \in \bar{a}$. ∎

THEOREM 1.8.7 Let $\sim$ be an equivalence relation on a set A. Then

(a) For fixed a and b in A, either $\bar{a} \cap \bar{b} = \varnothing$ or $\bar{a} = \bar{b}$.
(b) If P is the collection of equivalence classes defined by $\sim$, then $A = \cup \, \bar{a}$, the union of all the equivalence classes $\bar{a}$ in P.

Proof (a) Either $\bar{a} \cap \bar{b} = \varnothing$ or $\bar{a} \cap \bar{b} \neq \varnothing$. If the latter is true, then there exists an element $c \in A$ such that $c \in \bar{a}$ and $c \in \bar{b}$. By (b) of Lemma 1.8.6, $\bar{c} = \bar{a}$ and $\bar{c} = \bar{b}$. Thus, $\bar{a} = \bar{b}$.

(b) For each $a \in A$, $a \in \bar{a}$. It is clear then that

$$A = \cup \, \bar{a}. \qquad \blacksquare$$

Theorem 1.8.7 proves that an equivalence relation splits a set into a collection of disjoint subsets. Such a collection is called a *partition*.

Definition 1.8.8 Let A be a set, $A \neq \emptyset$, and let P be a collection of subsets $\{A_\alpha\}$ of A such that

(a) $A_\alpha \neq \emptyset$, all α.
(b) $A = \cup A_\alpha$, the union of all the sets in the collection P.
(c) If $A_\alpha \neq A_\beta$, then $A_\alpha \cap A_\beta = \emptyset$.

Then we say that P is a *partition* of the set A.

Theorem 1.8.7 shows that every equivalence relation on a set A gives rise to a partition of A. We now show that the converse is also true.

THEOREM 1.8.9 Let P be a partition of a set A. Define the relation R on A as follows: aRb if a and b both belong to the same subset A_α of the partition P. Then R is an equivalence relation on A.

Proof Let P be a partition of A. Let R be defined as in the statement of the theorem.

(a) aRa, since a is in one and only one A_α.
(b) aRb implies bRa. (Why?)
(c) aRb and bRc imply aRc. (Why?) ∎

Definition 1.8.10 In Theorem 1.8.9, the equivalence relation R is called the equivalence relation *induced* by the partition P. Similarly, in Theorem 1.8.7, we call P the partition *induced* by the equivalence relation $\sim$.

EXERCISES

11. (a) Define a relation R for points on the surface of the earth as follows: pRq if p and q have the same latitude. Is R an equivalence relation? If so, determine the partition induced by R on the surface of the earth, as in Theorem 1.8.7.
    (b) In (a), replace R by the relation R': $pR'q$ if p and q have the same longitude.

12. Let $f: A \longrightarrow B$ and define the relation R on A as follows: $a_1 R a_2$ if $f(a_1) = f(a_2)$ for $a_1, a_2 \in A$. Show that R is an equivalence relation on A.

13. Let $f: A \longrightarrow A$. Show that f is an equivalence relation on A if and only if f is the identity function on A.

14. (a) Let $\sim$ be an equivalence relation on a set $A \neq \emptyset$. Let P be the partition of A induced by $\sim$. If R is the equivalence relation induced by this partition P according to Theorem 1.8.9, show that $a \sim b$ if and only if aRb.
    (b) Let P be a partition of a set $A \neq \emptyset$. Let R' be the equivalence relation induced by P according to Theorem 1.8.9. Let P' be the partition induced by R' according to Theorem 1.8.7. Show that $P = P'$; that is, show that $A_\alpha \in P$ if and only if $A_\alpha \in P'$.

2 Number Theory

Before one can begin to study almost any mathematical subject in an organized way, it is necessary to have some familiarity with properties of the integers. In fact, in Chapter 1 we have already made use of certain counting and ordering properties of the set of integers. A systematic study of the integers is itself an important branch of mathematics, known as the theory of numbers. Our purpose here, however, is neither to give a complete "definition" or "construction" of the integers, nor to present a comprehensive view of the results of the theory of numbers. Rather, our goal is to establish certain results about the integers that we shall derive from a relatively small but familiar set of basic properties. In addition, the development of the material in this chapter will serve as a springboard and model in our study of abstract algebraic systems.

2.1 The Arithmetic Properties of the Integers

We begin our study of number theory by first listing the basic arithmetic properties of the set of integers $Z = \{\ldots, -2, -1, 0, 1, 2, 3, \ldots\}$, and deriving some of their elementary consequences.

ARITHMETIC PROPERTIES

A_0: There exists a binary operation on the integers called *addition*. Thus, for $a, b \in Z$, there exists a unique integer $a + b$ called the *sum* of a and b.

M_0: There exists a binary operation on the integers called *multiplication*. Thus, for $a, b \in Z$, there exists a unique integer $a \cdot b$ (or ab) called the *product* of a and b.

A₁: If $a, b, c \in Z$, then $(a + b) + c = a + (b + c)$ (associative law of addition).

A₂: If $a, b \in Z$, then $a + b = b + a$ (commutative law of addition).

A₃: There exists a unique integer 0 (zero) such that $a + 0 = a$, for all $a \in Z$. (0 is called an *identity for addition*.)

A₄: If $a \in Z$, then there exists a unique integer denoted by $-a$ such that $a + (-a) = 0$. $(-a)$ is called the *negative* of a, or the *additive inverse* of a.

M₁: If $a, b, c \in Z$, then $(a \cdot b) \cdot c = a \cdot (b \cdot c)$ (associative law of multiplication).

M₂: If $a, b \in Z$, then $a \cdot b = b \cdot a$ (commutative law of multiplication).

M₃: There exists a unique integer $1 \neq 0$ such that $a \cdot 1 = a$, for all $a \in Z$. (1 is called an *identity for multiplication*, or simply an *identity*.)

M₄: Let $a, b, c \in Z$, $a \neq 0$. If $a \cdot b = a \cdot c$, then $b = c$ (cancellation law of multiplication).

D: If $a, b, c \in Z$, then $a \cdot (b + c) = (a \cdot b) + (a \cdot c) = a \cdot b + a \cdot c$ (distributive law).

In properties A_0 and M_0, that $a + b$ and $a \cdot b$ are integers is also expressed by saying that Z is closed with respect to addition and multiplication. That $a + b$ and $a \cdot b$ are unique is equivalent to saying that addition and multiplication are binary operations on Z. Thus if $a = b$ and $c = d$, then $a + c = b + d$ and $a \cdot c = b \cdot d$ are just consequences of the definition of a binary operation. Of course, these are simply the familiar properties "equals added to equals give equals" and "equals multiplied by equals give equals."

Before examining some of the elementary consequences of the properties, we take a closer look at the associative laws and A_4 and M_4.

The associative laws permit us to extend meaningfully the concepts of addition and multiplication of two integers to "addition" and "multiplication" of three integers. Given the integers a, b, and c, the sum $a + b$ is a well-defined integer; thus so is $(a + b) + c$. Similarly, $a + (b + c)$ is a well-defined integer, since we have formed it by adding the two integers a and $(b + c)$. By the associative law, $(a + b) + c = a + (b + c)$. We may therefore now define the "sum" $a + b + c$ as either the integer $(a + b) + c$ or the integer $a + (b + c)$, since these two integers are precisely the same. Of course, $a \cdot b \cdot c$ can be defined analogously.

Extending these notions, we can define the sum and product of n integers $a_1, a_2, \ldots, a_n$ in a specific order so that parentheses need not be used. That is, the sum $a_1 + a_2 + \cdots + a_n$ has only one interpretation as an integer and the product $a_1 \cdot a_2 \cdots a_n$ has only one interpretation as an integer. The uniqueness in interpretation of this sum (this product) is known as the *generalized associative law*. The proof of a generalized associative law in a broad setting appears in Chapter 4. The generalized associative laws for addition and multiplication are used freely in this chapter without specific reference.

Throughout this book, certain basic properties concerning equalities will be taken for granted. For example, such things as $a = b$ implies $b = a$ and $a = b$ and $b = c$ imply $a = c$ will be assumed without question. Moreover, if we have an expression such as $a + b = a + c$ and are given $b = d$, we may write $a + d = a + c$, without stopping to justify this substitution in terms of A_0 and the basic properties of equality.

We point out here that there exist well-known binary operations that are not associative. For example, the operation of subtraction: $(2 - 3) - 5 \neq 2 - (3 - 5)$. Subtraction is also an example of a binary operation that is not commutative. Clearly, $3 - 5 \neq 5 - 3$.

A_4 and M_4 are the first pair of properties in the lists that do not appear to be analogous. To have a complete parallelism between A_4 and M_4, since 1 is the multiplicative identity and 0 is the additive identity, we would hope to have in place of M_4 the following property: if $a \in Z$, then there exists an integer a^{-1} such that $a \cdot a^{-1} = 1$. This can happen in Z only when $a = 1$ or $a = -1$. M_4 is therefore a weakened version of a division property. For it says in effect that one can "divide out" a common nonzero factor that occurs on both sides of an equation. It is, in fact, the strongest possible "division" property that holds for the integers. The analogous "cancellation" law for addition is a consequence of the stronger property A_4.

THEOREM 2.1.1   Let $a, b, c \in Z$. If $a + b = a + c$, then $b = c$.

Proof   By A_4, there exists an integer $(-a)$ such that

$$a + (-a) = 0$$

and by A_2, $a + (-a) = (-a) + a$, whence $(-a) + a = 0$. By A_0, $(-a) + (a + b) = (-a) + (a + c)$, and by A_1,

$$((-a) + a) + b = (-a) + (a + b) = (-a) + (a + c)$$
$$= ((-a) + a) + c.$$

Thus $0 + b = 0 + c$, and by A_2, $b + 0 = 0 + b = 0 + c = c + 0$. Finally, by A_3, $b = b + 0 = c + 0 = c$; that is, $b = c$. ∎

EXERCISES

1. Show that $-0 = 0$.
2. Prove if $a, b, c \in Z$ and $a + b = c + b$, then $a = c$.

We shall now deduce several elementary and familiar theorems that follow from the arithmetic properties.

THEOREM 2.1.2   If $a \in Z$, then $a \cdot 0 = 0 \cdot a = 0$.

Proof   By A_3, $0 = 0 + 0$. Using D, $a \cdot 0 = a \cdot (0 + 0) = a \cdot 0 + a \cdot 0$. By A_3, $a \cdot 0 + 0 = a \cdot 0$. Thus $a \cdot 0 + 0 = a \cdot 0 + a \cdot 0$. Now, by Theorem 2.1.1, $0 = a \cdot 0$. Finally, by M_2, $0 \cdot a = 0$.

The reader should note where A_0 is used in this proof.   ∎

THEOREM 2.1.3   Let $a \in Z$. Then $-(-a) = a$.

Proof   By A_4, $a + (-a) = 0$ and $(-a) + [-(-a)] = 0$. By A_2, $(-a) + a = 0$. Thus, $(-a) + [-(-a)] = (-a) + a$. By Theorem 2.1.1, $-(-a) = a$.   ∎

THEOREM 2.1.4   Let $a, b \in Z$. Then $(-a)b = a(-b) = -(ab)$.

Proof $a \cdot 0 = a[b + (-b)] = ab + a(-b)$, by the distributive law D. By Theorem 2.1.2, $a \cdot 0 = 0$, whence $ab + a(-b) = 0$. (What properties were used to get this equation?) Also, by A_4, $ab + [-(ab)] = 0$. Thus $ab + a(-b) = ab + [-(ab)]$.

By the cancellation law for addition (Theorem 2.1.1), we now have

$$a(-b) = -(ab).$$

Similarly,

$$(-a)b = -(ab).$$ ∎

COROLLARY 2.1.5   Let $a, b \in Z$. Then $(-a)(-b) = ab$.

Proof

$$\begin{aligned}(-a)(-b) &= -[a(-b)] & &\text{by Theorem 2.1.4} \\ &= -[-(ab)] & &\text{by Theorem 2.1.4} \\ &= ab & &\text{by Theorem 2.1.3}\end{aligned}$$ ∎

To simplify our notation, we agree to write $a - b$ in place of $a + (-b)$. That is, we define the symbol $a - b$ to be $a + (-b)$.

The reader should note that Corollary 2.1.5 clears up the mystery often connected with the "law of signs." For the corollary follows from the basic arithmetic properties of the integers.

EXERCISES

3. Let a_1, a_2, a_3, a_4 be four integers. Explain in terms of A_0 and A_1 why each of the following expressions is a unique integer and why all of the expressions are equal to each other.
   (a) $a_1 + (a_2 + (a_3 + a_4))$.   (b) $(a_1 + a_2) + (a_3 + a_4)$.
   (c) $((a_1 + a_2) + a_3) + a_4$.   (d) $(a_1 + (a_2 + a_3)) + a_4$.
   (e) $a_1 + ((a_2 + a_3) + a_4)$.

4. In Exercise 3, replace each $+$ by $\cdot$ and explain why all the indicated expressions are unique integers and why they are all equal.
5. Prove that the integer $-a$ is unique; that is, show that if $a + (-a) = 0$ and also $a + a' = 0$, then $a' = -a$.
6. Let $a, b \in Z$. Prove that $-(a + b) = (-a) + (-b)$.
7. Let $a, b \in Z$. Prove that $a \cdot b = 0$ if and only if at least one of a and b is 0.
8. Let a, b, c be in Z, $a \neq 0$. Prove that $a(b - c) = 0$ if and only if $b = c$.
9. Prove that property M_4 can be replaced by either the property of Exercise 7 or the property of Exercise 8; that is, prove that any one of these three properties implies any of the others.
10. Let a be an integer. Prove that $a \cdot a = a$ if and only if $a = 0$ or $a = 1$.
11. (a) Is property D' true in Z? $D': a + (b \cdot c) = (a + b) \cdot (a + c)$.
    (b) Let $A, B,$ and C be sets. Using $\cup$ instead of $+$ and $\cap$ instead of $\cdot$, then the following are analogous to D and D' in the calculus of sets: $A \cap (B \cup C) = (A \cap B) \cup (A \cap C)$,
    $$A \cup (B \cap C) = (A \cup B) \cap (A \cup C).$$
    Prove that these equations hold.
12. Define binary operations $\oplus$ and $\otimes$ on Z by $a \oplus b = a + b$, $a \otimes b = a$. Which of the properties $A_0 - A_4$, $M_0 - M_4$, and D are true when $+$ is replaced by $\oplus$ and $\cdot$ is replaced by $\otimes$?
13. Let a, b be integers such that $0 \leq a < b$. Prove that $0 < b - a < b$.

2.2 The Order Properties of the Integers

As the reader knows from his earlier studies, the system of integers satisfies certain "order" properties in addition to the arithmetic properties. In this section we shall list two properties on order, from which a number of familiar properties are deducible.

The approach to be taken here will lead us to the usual facts that $1 < 2$, $2 < 3, \ldots, n < n + 1$; that any negative integer $k < 0$, and that $0 < m$, for any positive integer m. Furthermore, the usual rules intertwining the order and arithmetic properties will also follow.

ORDER PROPERTIES

There exists a subset P of Z, called the set of positive integers, such that

O_1: If $a \in Z$, then one and only one of the following is true:
(a) $a \in P$; (b) $a = 0$; (c) $-a \in P$ *(law of trichotomy)*.
O_2: If $a, b \in P$, then $a + b \in P$ and $a \cdot b \in P$.

A third order property is given in a later section. It is interesting to note that the arithmetic properties of O_1 and O_2 imply $1 \in P$.

THEOREM 2.2.1 $1 \in P$.

Proof   By M_3, $1 \neq 0$, and therefore either $1 \in P$ or $-1 \in P$. But if $-1 \in P$, $(-1)(-1) = 1 \in P$ by O_2 and Corollary 2.1.5, a contradiction. Hence $1 \in P$. ∎

Although O_1 and O_2 are very familiar, the use of the word "order" in connection with these properties may not be apparent. Since $1 \in P$, clearly $1 + 1 = 2 \in P$, $2 + 1 = 3 \in P$, $3 + 1 = 4 \in P$, and so forth. When we write 1, 2, 3, 4, we say that these four integers are in their "natural order." If the integer a is to the left of the integer b in this list, then we observe that $b - a$ is positive, and in fact we can define "natural order" in precisely this way. This is done below.

Definition 2.2.2   Let $a, b \in Z$. We say that a *is less than* b if $b - a$ is positive, and we write $a < b$. Alternatively, we say that b *is greater than* a and write $b > a$.

In the event that $a < b$ or $a = b$, we write $a \leq b$. Similarly, we write $a \geq b$ if either $a > b$ or $a = b$.

In view of the definition of $<$, it is now clear that $1 < 2$, $2 < 3$, ... Thus the "natural order" of the integers coincides with our concept of $<$, which we have defined in terms of positiveness.

We note that the "natural" order of the integers coincides with the order of the positive integers in terms of "magnitude." Without going into great detail at this point, suffice it to say that the positive integers 1, 2, 3, ... are used both to denote the order in which elements or objects appear and the magnitude or quantity that certain objects can be assigned. There will be more on this later; now we note only that whether we view the positive integers as being used to order elements or to measure magnitude, we get the same natural order in both cases.

Definition 2.2.3   Let $a \in Z$. If $(-a)$ is positive, we say that a is a *negative* integer, or that a is *negative*.

We observe that if a is negative, then a can be neither positive nor zero. For if a is both negative and positive, then both $-a$ and a are positive, contradicting O_1. For the case that a is both negative and zero, we see $-a \in P$ and $a = 0$, again contradicting O_1.

If a is positive, then $a - 0 = a + 0 = a$ is positive. Thus $0 < a$. Also, if $0 < a$, then a is positive. In light of this, O_1 and O_2 become

O_1':   If $a \in Z$, exactly one of the following is true:
(a) $a > 0$, (b) $a = 0$, (c) $(-a) > 0$.

O_2':   If $a, b \in Z$, and if $0 < a$ and $0 < b$, then $0 < a + b$ and $0 < ab$.

2.2 THE ORDER PROPERTIES OF THE INTEGERS

THEOREM 2.2.4 Let $a \in Z$. Then a is negative if and only if $a < 0$.

Proof Suppose that a is negative. By definition, this means that $-a$ is positive. But $-a = 0 + (-a) = 0 - a$, so $0 - a$ is positive, and by Definition 2.2.2 it follows that $a < 0$.

Conversely, suppose that $a < 0$. Then again by definition, $0 - a$ is positive, that is, $-a$ is positive, whence, by definition, a is negative. ∎

It is easy to give yet another formulation of O_1, namely O_1'': If $a \in Z$, then one and only one of the following is true: (a) $a < 0$, (b) $a = 0$, (c) $(-a) < 0$. The proof of O_1'' is left to the exercises.

EXERCISE

1. Prove the formulation O_1'' given immediately after Theorem 2.2.4.

As we have previously remarked, our point of departure in studying the integers is to assume that the integers satisfy certain basic properties, namely, A_0, A_1, A_2, A_3, A_4, M_0, M_1, M_2, M_3, M_4, D, O_1, and O_2, and an additional property O_3 to be given later. Any other properties of the integers that we shall use can be deduced from this basic set.

Although we have already given two statements of the law of trichotomy, we now give a third version, whose proof will again illustrate the use of our basic properties.

THEOREM 2.2.5 Let $a, b \in Z$. Then one and only one of the following is true:
$$a < b; \quad a = b; \quad a > b.$$

Proof Consider $b - a$. By O_1, exactly one of the following is true: (a) $b - a$ is positive, (b) $b - a = 0$, (c) $-(b - a)$ is positive. If (a), then $b - a$ is positive and so $a < b$, by definition. If (b), then $b - a = 0$, whence
$$b + (-a) = 0,$$
$$[b + (-a)] + a = 0 + a,$$
$$b + [(-a) + a] = a \quad \text{(why?)},$$
and
$$b = a \quad \text{(why?)}.$$

If (c), then $-(b - a)$ is positive. But
$$-(b - a) = (-b) + (-(-a)) \quad \text{(by Exercise 6, Section 2.1)}$$
$$= (-b) + a = a - b,$$
so $a - b$ is positive. Thus by definition, $a > b$. ∎

As the reader knows, a number of results involving inequalities are frequently used. We give several that follow readily from our discussion on order properties.

THEOREM 2.2.6 Let a, b, c be integers such that $a < b$ and $b < c$. Then $a < c$ (transitive law).

Proof By definition of $<$, $b - a$ and $c - b$ are both positive. But

$$\begin{aligned}(c - b) + (b - a) &= [c + (-b)] + [b + (-a)] \\ &= c + \{(-b) + [b + (-a)]\} \quad \text{(why?)} \\ &= c + \{[(-b) + b] + (-a)\} \\ &= c + \{0 + (-a)\} = c + (-a).\end{aligned}$$

Since $(c - b) + (b - a)$ is positive, $c - a$ is positive and $a < c$. ∎

THEOREM 2.2.7 Let $a, b, c \in Z$ and suppose $a < b$. Then

$$a + c < b + c.$$

Proof We must show that $(b + c) - (a + c)$ is positive. But $(b + c) - (a + c) = b - a$, which is positive, since $a < b$. ∎

THEOREM 2.2.8 Let $a, b, c \in Z$, $a < b$, and $c > 0$. Then $ac < bc$.

Proof From $a < b$ we have immediately that $b - a$ is positive. Since c is also positive, the product $(b - a)c$ is positive by O_2, but then $bc - ac = (b - a)c$ is also positive, whence $ac < bc$, by definition. ∎

Theorems 2.2.7 and 2.2.8 are generally known to every student of calculus as saying that the direction of an inequality is preserved when the same quantity is added to both sides of the inequality, and that the direction is preserved when both sides of the inequality are multiplied by the same positive quantity. On the other hand Theorem 2.2.9 states that the direction of an inequality is reversed when both sides of an inequality are multiplied by a negative quantity.

THEOREM 2.2.9 Let $a, b, c \in Z$, $a < b$, and $c < 0$. Then

$$bc < ac.$$

Proof We leave this proof as an exercise for the reader. ∎

EXERCISES

2. Show that $\leq$ is a relation on Z. Which properties of an equivalence relation does it satisfy?
3. Prove Theorem 2.2.9.
4. Let $a, b, c \in Z$. Suppose that $a + c < b + c$. Prove that $a < b$.
5. Let $a, b, c \in Z$, $c > 0$, and suppose $ac < bc$. Prove that $a < b$.
6. Let $a, b, c \in Z$, $c < 0$, and suppose $bc < ac$. Prove that $a < b$.
7. Let $a, b \in Z$, and suppose that $a < b$. Prove that $-b < -a$.

8. Knowing that $2 < 3$ and $3 < 4$, explain in two ways why $2 < 4$.
9. Prove that $3 \cdot 3 > 0$, without using the fact that the product is 9.

Also following directly from the list of properties of integers given to this point is the property that the square of any nonzero integer is positive.

THEOREM 2.2.10 Let a be an integer, $a \neq 0$. Then $a^2 = a \cdot a > 0$.

Proof Since $a \neq 0$, by the law of trichotomy either a is positive or $-a$ is positive. In the first instance, $a \cdot a$ is positive by O_2. In the second instance, $(-a)(-a)$ is positive by O_2, but $(-a)(-a) = a \cdot a$, so again a^2 is positive. ∎

We should mention that the proofs of the theorems in this section apply far more generally than just to the system of integers. Many of them recur later in the text.

Before concluding this section, we briefly discuss the role of the nonnegative integers in counting and ordering sets.

So far we have used the integers primarily as formal objects that we know how to manipulate relative to the operations of $+$ and $\cdot$. But we also know that the elements of the set $\{0,1,2,\ldots,n,\ldots\}$ can be used to count the number of objects in a set and to arrange these objects in some order. Indeed, our first experiences with numbers seem to be as counters and arrangers.

To illustrate, when we say that a set has n objects, $n > 0$, we mean that we can place the objects of the set in a 1–1 correspondence with the objects in the set $\{1,2,3,\ldots,n\}$. As is well known, any two ways of counting objects in a set turn out to give the same final answer. More precisely, this means it is possible to place the elements in the sets $\{1,2,\ldots,n\}$ and $\{1,2,\ldots,m\}$ in one-to-one correspondence if and only if $n = m$. The reader may feel this statement is trivial, but sophisticated arguments are needed in order to prove it in full. We do not do this here, but suggest that the reader consult P. R. Halmos, *Naive Set Theory* (cited in footnote 1, p. 1).

In addition to measuring magnitudes of sets, the nonnegative integers can sometimes be used to place the elements in a given set in a specific order. For example, when a store is crowded, customers are asked to take numbers and await their turns for service. Although the last customer to be served that day will hold the number that tells the total number of customers served during the day, earlier customers may very well not care how many customers have been served up to their turn, but only when their turn will come. The nonnegative integers are being used in this context to place the customers in a particular order.

It is clear that a set with more than one object can be ordered in a variety of ways, and so the assignment of natural numbers to the elements of the set is not at all unique. But again, when we are measuring the quantity of objects in the set, the final number used is unique. The ideas involved in counting magnitude and in ordering sets have been generalized to the case of arbitrary sets. While it is sometimes difficult to see a distinction between the counting and ordering properties of the nonnegative integers when applied to finite sets, the difficulty sometimes disappears in the infinite case. Again, we shall leave this for the reader to consult in Halmos. We simply state that when numbers are used to count quantities they are called *cardinal numbers*, and when they are used to order, or arrange objects, they are called *ordinal numbers*.

2.3 Absolute Value

A useful idea, especially for notational convenience, is that of absolute value. The definition we will use is the same as that for the real number system and is probably known to the reader.

Definition 2.3.1 Let a be any integer. The *absolute value* of a, denoted by $|a|$, is defined as follows:

$$|a| = \begin{cases} a & \text{if } a \geq 0, \\ -a & \text{if } a < 0. \end{cases}$$

Clearly, in terms of this definition, $|-3| = 3$, $|1| = 1$, $|-5| = 5$, $|0| = 0$, $|32| = 32$.

A number of very easy consequences follow from this definition. We shall prove some of them, leave others to the reader, and use other elementary consequences later in the text without proof.

THEOREM 2.3.2 Let $a, b \in Z$. Then $|ab| = |a| \, |b|$.

Proof The proof of this theorem illustrates a more or less general approach in that we consider several cases depending on the positiveness of the integers a and b. Thus we must consider the following cases:

  I. $a \geq 0, b \geq 0$;
 II. $a \geq 0, b < 0$;
III. $a < 0, b \geq 0$;
 IV. $a < 0, b < 0$.

  I. From $a \geq 0$ and $b \geq 0$, we get immediately that $ab \geq 0$. Thus, by definition, $|ab| = ab$, $|a| = a$, and $|b| = b$. Clearly, we now have $|ab| = |a| \, |b|$.
 II. From $a \geq 0$ and $b < 0$, we see that $ab \leq 0$. Thus $|ab| = -ab$, $|a| = a$, and $|b| = -b$. Thus, $|ab| = |a| \, |b|$.

The proofs in cases III and IV are left for the reader, since they have much the same flavor as the first two cases. ∎

A basic result, which the reader has already met in calculus, is the triangle inequality. We prove it here for integers only, although the proof can easily be extended to other number systems. We first need a preliminary result.

THEOREM 2.3.3 Let $a \in Z$. Then $-|a| \leq a \leq |a|$.

Proof If $a \geq 0$, then $|a| = a$. Thus, $|a| \geq 0$, and so multiplying through by -1, we have $-|a| \leq 0$, so that $-|a| \leq 0 \leq a = |a|$, and this immediately implies $-|a| \leq a \leq |a|$.

If $a < 0$, then $|a| = -a$, and of course $-a > 0$, whence $|a| > 0$. Then $-|a| < 0$, and so $-|a| = a < 0 < -a = |a|$, and so we can again conclude that $-|a| \leq a \leq |a|$.

Thus, whether $a \geq 0$ or $a < 0$, we can reach the desired conclusion. ∎

THEOREM 2.3.4 (*The Triangle Inequality*) Let $a, b \in Z$. Then

$$|a + b| \leq |a| + |b|.$$

Proof By Theorem 2.3.3, we get

$$-|a| \leq a \leq |a| \quad \text{and} \quad -|b| \leq b \leq |b|.$$

We add the above inequalities, and this results in

(*) $\qquad -(|a| + |b|) \leq a + b \leq |a| + |b|.$

There are now two possibilities. First, suppose that $a + b \geq 0$. Then by definition, $|a + b| = a + b$, and so $|a + b| = a + b \leq |a| + |b|$. Second, suppose that $a + b < 0$. Then $|a + b| = -(a + b)$. From the first part of (*), $-(|a| + |b|) \leq a + b$, so $-(a + b) \leq -(-(|a| + |b|))$, and finally $|a + b| = -(a + b) \leq |a| + |b|$. ∎

EXERCISES

1. Prove: Let $a \in Z$. Then $|a| = 0$ if and only if $a = 0$. If $a \neq 0$, then $|a| > 0$.
2. Prove: Let $a, b \in Z$. Then $|a| = |b|$ if and only if $a = \pm b$; that is, $a = b$ or $a = -b$.
3. Let a, b, c be integers such that $0 \leq a < c$, $0 \leq b < c$. Show that $|a - b| < c$.
4. Let b, c be integers, $c > 0$. Prove that $-c \leq b \leq c$ if and only if $|b| \leq c$. (*Hint:* To see the exact meaning of this theorem, try it out for various values of b and c.)
5. Let b and c be integers. Prove that $||b| + |c|| = |b| + |c|$.
6. Let $a, b, c \in Z$, $c \geq 0$. Prove that $|a - b| \leq c$ if and only if

$$a - c \leq b \leq a + c.$$

2.4 Well Ordering and Mathematical Induction

Let P denote the set of positive integers. Let S be a nonempty subset of P. We wish to establish that S has a smallest integer; that is, that there is an integer $m \in S$ such that $m \leq n$, for all integers $n \in S$. To do this, we first choose any integer n_1 in S. If, upon surveying the elements of S, we find $n_1 \leq n$ for all $n \in S$, then n_1 is the desired integer. If not, there is an integer $n_2 < n_1$. If $n_2 \leq n$, for all $n \in S$, n_2 is the integer we desire. If n_2 does not work, we pick n_3 from S, and so forth. In this way, we get a string of integers in S:

$$n_1 > n_2 > n_3 > \cdots > n_k > \cdots.$$

Since these integers are all positive, it is not difficult to be convinced that the string must stop at, say, $n_N = m$, and this will be our desired smallest integer.

Of course, this line of reasoning entails certain problems. First, in order to determine if n_1 is the smallest integer of an infinite set S of positive integers, we must test in turn all integers of the form $n - n_1$, $n \in S$, for positiveness. Since S is infinite, there is no effective way of doing this. Second, even assuming we can tell when n_1 is not the smallest and we find an $n_2 < n_1$, and then an $n_3 < n_2$, and so forth, how do we really know that the chain $n_1 > n_2 > \cdots > n_k > \cdots$ must eventually end? The crux of the matter is this: In assuming that this chain ends, we are assuming that a certain nonempty set of positive integers, namely $\{n_1, n_2, \ldots, n_k, \ldots\}$, has a smallest element. This is tantamount to assuming from the very outset that S has a smallest integer. This fact, however, cannot be derived from the arithmetic and order properties already stated.[1] Thus, we assume it to be a basic property of the integers, and we add this third order property to our list.

O_3: *Law of Well Ordering for Positive Integers (LWO)* Let P denote the set of all positive integers. Let $S \subseteq P$, $S \neq \varnothing$. Then there exists an integer $m \in S$ such that $m \leq n$, for all $n \in S$. We call m the *smallest* or *least* integer (element) of S, and we say that S has a least member.

Before we give a simple application of O_3, we prove two of its elementary consequences.

LEMMA 2.4.1 The law of well ordering for positive integers implies that 1 is the smallest positive integer.

---

[1] This can be shown by displaying a mathematical system that satisfies all the other properties, but not this one. The rational number system is such an example, since the set of positive rational numbers does not have a smallest element.

2.4 WELL ORDERING AND MATHEMATICAL INDUCTION 45

Proof By Theorem 2.2.1, 1 is a positive integer. Now we let $M = \{m \mid m \in P \text{ and } 0 < m < 1\}$. If $M = \emptyset$, 1 is the smallest positive integer in P. Otherwise, $M \neq \emptyset$, and by O_3 there is a smallest integer n in M. Then $0 < n < 1$ and also $0 < n \cdot n < 1 \cdot n < 1$. But then $n^2 \in M$ and $n^2 < n$, a contradiction. Thus, $M = \emptyset$, and 1 is the smallest positive integer. ∎

COROLLARY 2.4.2 If $n \in P$, then there does not exist an integer k such that $n < k < n + 1$.

Proof Suppose there exists an integer k such that $n < k < n + 1$. Then $0 < k - n < 1$, contradicting the theorem. ∎

We now give a simple application of O_3, proving the formula

$$(*) \qquad 1 + 2 + \cdots + n = \frac{n(n + 1)}{2}$$

is true for all positive integers n. We prove this by showing that the set E of positive integers for which $(*)$ is true is equal to P. Thus, we observe that when $n = 1$, the formula becomes $1 = 1 \cdot (1 + 1)/2 = 1$; this means that 1 is in E. For $n = 2$, we have

$$1 + 2 = \frac{2(2 + 1)}{2}, \qquad \text{so again } 2 \in E.$$

Continuing along these lines, it is clear that we could only verify $(*)$ for a finite collection of positive integers n. We observe, however, that if $E \neq P$, then there is a positive integer n for which

$$1 + 2 + \cdots + n \neq \frac{n(n + 1)}{2}.$$

Let

$$S = \left\{ n \mid n \in P \text{ and } 1 + 2 + \cdots + n \neq \frac{n(n + 1)}{2} \right\}$$

that is, $S = P - E$. By our assumption that $E \neq P$, $S \neq \emptyset$. By O_3, S has a least element, say m. Hence,

$$1 + 2 + \cdots + m \neq \frac{m(m + 1)}{2}.$$

On the other hand, $m \neq 1$, since $1 \in E$, and so by Lemma 2.4.1, $m > 1$. Thus, $m - 1 > 0$; that is, $m - 1 \in P$. But by the choice of m, $m - 1 \notin S$, since $m - 1 < m$. (Why is $m - 1 < m$?)
Thus

$$1 + 2 + \cdots + (m - 1) = \frac{(m - 1)(m - 1 + 1)}{2} = \frac{(m - 1)m}{2}.$$

Adding m to both sides of the equation,

$$1 + 2 + \cdots + (m-1) + m = \frac{(m-1)m}{2} + m$$

$$= \frac{(m-1)m}{2} + \frac{2m}{2}$$

$$= \frac{m^2 + m}{2}$$

$$= \frac{m(m+1)}{2},$$

and so the formula is true for m. Hence, $m \in E$, that is, $m \notin S$, a contradiction. To avoid this contradiction we must conclude that $S = \emptyset$, and hence $E = P$; that is, the formula is true for all positive integers. ∎

We see that the proof consists of two essential parts. First, we showed $1 \in E$. Second, we observed that if $m - 1 \in E$, with $m > 1$, then $m \in E$. Actually, as we shall see shortly, this result could have been proved by merely establishing the following about E:

(a) $1 \in E$,
(b) whenever $n \in E$, then $n + 1 \in E$.

For, from an intuitive point of view, assuming that $1 \in E$, then by (b), $1 + 1 = 2 \in E$. By (b) again, since $2 \in E$, $2 + 1 = 3 \in E$. And again, $3 + 1 = 4 \in E$. To proceed once more along these lines would be futile. We shall show that (a) and (b) together imply $E = P$.

THEOREM 2.4.3 *Principle of mathematical induction* (PMI) Let E be a subset of P such that

(a) $1 \in E$ and
(b) whenever $n \in E$, then also $n + 1 \in E$.

Then $E = P$.

Proof To show that $E = P$, let $S = P - E$. If $S = \emptyset$, then $E = P$ and we are finished. Hence, assume $S \neq \emptyset$. Then, by LWO, there exists a positive integer $m \in S$ such that $m \leq k$, for all $k \in S$. Since $1 \in E$, clearly $m \neq 1$, and so $m > 1$ by Lemma 2.4.1. Thus, $m - 1 > 0$. Since m is the least positive integer in S, $m - 1 \notin S$, that is, $m - 1 \in E$. Hence, by (b) $m \in E$, contradicting that $m \in S$. Therefore, $S = \emptyset$ and $E = P$. ∎

Although the formula (*) was proved by using the law of well ordering and motivated our formulation of mathematical induction, we now give a proof (basically the one given before) in terms of mathematical induction.

2.4 WELL ORDERING AND MATHEMATICAL INDUCTION

Proof of (∗) For the integer $n = 1$ the left-hand side of (∗) is simply 1, and the right-hand side is

$$\frac{1(1+1)}{2} = \frac{1 \cdot 2}{2} = \frac{2}{2} = 1,$$

and so (∗) is true for 1. [That is, 1 belongs to the set E of integers for which (∗) is true.]

Now assume that (∗) holds for the integer n; that is, assume that $1 + 2 + \cdots + n = n(n + 1)/2$, for the positive integer n. Adding $n + 1$ to both sides, the left-hand side becomes

$$1 + 2 + \cdots + n + (n + 1)$$

and the right-hand side becomes

$$\frac{n(n+1)}{2} + (n+1) = \frac{(n+1)(n+2)}{2},$$

which is the right-hand side of (∗) with n replaced by $n + 1$. The truth of (∗) in the case for n necessarily implies the truth of (∗) for the case of $n + 1$, since we have gone from the n case to the $n + 1$ case simply by addition of equals to equals. Thus, (∗) is true for $n + 1$ whenever it is true for n. (Whenever n is in E, then also $n + 1$ is in E.)

By induction, we can now conclude that (∗) is true for all positive integers. (Since 1 is in E, and since whenever n is in E, then also $n + 1$ is in E, we have that $E = P$.) ∎

Mathematical induction can be used in a variety of situations. The proof of (∗) is an application of induction in an arithmetic situation. To illustrate the use of induction in a geometric situation we prove the next theorem.

THEOREM 2.4.4 Let n lines be drawn in the Euclidean plane such that no two lines are parallel and no three lines have a common point. Then there are $(n^2 + n + 2)/2$ regions bordered by these lines and not having any of these lines in their interiors.

Proof For $n = 1$, the plane is clearly divided into two such regions by one line, and also

$$\frac{1^2 + 1 + 2}{2} = \frac{4}{2} = 2.$$

Thus, the formula holds for $n = 1$.

Assume now that the formula holds for the positive integer n. It is necessary to show that the formula also holds for the integer $n + 1$.

We will name the $n + 1$ lines under consideration $L_1, L_2, \ldots, L_n, L_{n+1}$.

Our induction assumption is that the lines $L_1, \ldots, L_n$ divide the plane into $(n^2 + n + 2)/2$ regions. Now under the hypotheses, the line L_{n+1} must cut each of the first n lines (in some order), and no two lines are cut at the same point. The line L_{n+1} passes through some region before it hits each of the lines $L_1, \ldots, L_n$, and this region is cut into two regions. Also, after L_{n+1} cuts the last of the lines, it divides one more region into two regions. Thus, the total number of regions is increased by $n + 1$.

From the assumption that there are $(n^2 + n + 2)/2$ regions formed by just the lines $L_1, \ldots, L_n$, we now have a total of $(n^2 + n + 2)/2 + (n + 1)$ regions, and this number is easily seen to be

$$\frac{(n+1)^2 + (n+1) + 2}{2}.$$

This completes the proof by induction. ∎

As another application of induction, we prove that for each positive integer n, $2^n > n$.

THEOREM 2.4.5 Let n be a positive integer. Then $2^n > n$.

Proof For $n = 1$, clearly $2^1 = 2 > 1$.

Assume now that for the positive integer n, $2^n > n$. Then, multiplying through by 2, $2^{n+1} > 2n$. But clearly $2n \geq (n + 1)$, since $n \geq 1$ implies $n + n \geq n + 1$ for all positive integers n. And so $2^{n+1} > n + 1$. By induction, the theorem is now proved. ∎

Our examples show that induction is often used in problems in which a statement involving positive integers is to be proved. Such a statement need not take the form of an equation as in the example following Corollary 2.4.2. As an illustration, consider the following statement, which we denote by $Q(n)$: Any nonempty set of positive integers containing an integer less than or equal to n has a least element.

To prove that this statement is true for all positive integers n, we need only establish

(a) that $Q(1)$ is true, that is, any nonempty set of positive integers containing an integer less than or equal to 1 has a least element, and
(b) that whenever $Q(k)$ is true, then $Q(k + 1)$ is true.

For if S is the set of positive integers for which $Q(n)$ is true, then (a) and (b) above are equivalent to (a') $1 \in S$ and (b') if $k \in S$, then $k + 1 \in S$, respectively. If (a) and (b) have been proved, then by PMI, $S = P$; that is, $Q(n)$ is true for all positive integers n.

Clearly, any equation or inequality involving n can be viewed as some statement $Q(n)$. In a sense, this idea has already been used in our illustrations in this section.

2.4 WELL ORDERING AND MATHEMATICAL INDUCTION

EXERCISES

1. Prove each of the following by mathematical induction:
   (a) $2 + 4 + 6 + \cdots + (2n) = n(n + 1)$, for all positive integers n.
   (b) $n < 3^n$, for each $n > 0$.
   (c) $1 + 3 + 5 + \cdots + (2n + 1) = (n + 1)^2$, for each integer $n > 0$.
   (d) $1 + 3 + 9 + \cdots + 3^n = (3^{n+1} - 1)/2$, for each integer $n > 0$.
   (e) For each positive integer n,
   $$1^3 + 2^3 + \cdots + n^3 = \frac{(n(n+1))^2}{4}.$$
   (f) For each positive integer n, there is a positive integer c_n such that $n^3 + 5n = 6 \cdot c_n$.
   (g) For each positive integer n, there is an integer d_n such that $4^n - 1 = 3 \cdot d_n$.
   (h) For each positive integer n, there is a positive integer k_n such that $5^{2n} - 1 = 24 \cdot k_n$.
   (i) Let a be a positive integer. For each positive integer n, $(1 + a)^n > 1 + na$.
   (j) For each positive integer n,
   $$\frac{1}{1 \cdot 2} + \frac{1}{2 \cdot 3} + \frac{1}{3 \cdot 4} + \cdots + \frac{1}{n(n+1)} = \frac{n}{n+1}.$$

2. Prove: If a and b are in Z and $a < b$, then $a + 1 \leq b$. (*Hint:* Use 2.4.1.)

3. Let a, b be positive integers. Show that there exists an integer m such that $ma > b$ (Archimedean property).

4. Let A be a nonempty subset of Z. A is said to be *well ordered* if whenever $B \subseteq A$, $B \neq \emptyset$, there exists an element $b \in B$ such that $b \leq c$, where c is any integer in B; that is, B has a least element b. Prove:
   (a) Z is not well ordered.
   (b) If A is a well ordered subset of Z and if $B \subseteq A$, $B \neq \emptyset$, then B is well ordered.

5. Let A be a nonempty subset of Z and let c be an integer satisfying $c \leq m$ for all $m \in A$. Show that A is well ordered. [*Hint:* If $c < 0$, consider the set $A' = \{m - c + 1 \mid m \in A\}$.] (This statement is equivalent to LWO.)

6. If A is well ordered and if C is a subset of Z such that for each element $c \in C$, there is an $a \in A$ such that $a \leq c$, show that C is well ordered and that $A \cup C$ is well ordered.

7. Let S be a nonempty set of negative integers. Suppose (i) $-1 \in S$

and (ii) whenever $-n \in S$, $n \geq 1$, then also $-n - 1 \in S$. Prove that S is the set of all negative integers.

8. What is wrong with the following proof?
Theorem   All horses have the same color.
Proof   Let $P(n)$ be the proposition that all horses in a set of n horses are the same color.

For $n = 1$, $P(n)$ is clearly true. Let n be an integer for which $P(n)$ is true. We want to prove that $P(n + 1)$ is true.

Let $H_1, \ldots, H_{n+1}$ be the $n + 1$ horses in a set of $n + 1$ horses. Then consider the subset of n horses, $S' = \{H_1, \ldots, H_n\}$. By assumption, these are all the same color. Now, replace H_n in S' by H_{n+1}. In the resulting set $S'' = \{H_1, \ldots, H_{n-1}, H_{n+1}\}$ of n horses all are the same color, since this is again a set of n horses. Since H_n and H_1 are the same color, and H_{n+1} and H_1 are the same color, all $n + 1$ horses are the same color. This concludes the "proof."

9. For $n \geq 0$ define $n!$ as follows:

$$0! = 1$$

and $n! = n(n - 1)!$ for $n \geq 1$; that is

$$n! = n \cdot (n - 1) \cdot (n - 2) \cdots 2 \cdot 1.$$

For $n \geq 0$ define the symbol

$$\binom{n}{k} = \frac{n!}{k!(n - k)!} \quad \text{if } 0 \leq k \leq n$$

and

$$\binom{n}{k} = 0 \quad \text{if } k < 0 \text{ or } k > n.$$

(a) Show that $\binom{n}{0} = 1$, $\binom{n}{n} = 1$, $\binom{n}{1} = n$, $\binom{n}{n-1} = n$ for $n \geq 0$, and

$$\binom{n}{n-k} = \binom{n}{k}.$$

(b) Prove that

$$\binom{n+1}{k} = \binom{n}{k} + \binom{n}{k-1}$$

for all $n \geq 0$ and all integers k. (This can be proved without PMI.)

(c) Prove by induction on n that $\binom{n}{k}$ is an integer for all $n \geq 0$ and all integers k.

(d) Prove that
$$\binom{n}{0} + \binom{n}{1} + \binom{n}{2} + \cdots + \binom{n}{n} = 2^n.$$

10. A well-known puzzle involves a board having three pegs and a set of n round discs, each disc having a hole in its center and the discs being of n different radii. All of the discs are arranged on one peg, in order of increasing radius, the largest disc on the bottom and the smallest at the top. We wish to move all of the discs from their original peg to one of the other two pegs so that, when we are done, the discs have the same arrangement on the new peg as on the original peg. The rules of moving discs stipulate that only one disc can be moved at a time, and when moved, a disc must be placed on some peg. Also, no disc can be placed on top of a smaller one at any time before the task is completed. *Challenge:* Use induction to show that the discs can be moved in $2^n - 1$ moves. Actually, this is the smallest number of moves that will work. (*Hint:* Clearly, with only one disc, one move $= 2^1 - 1$ move is enough. With 2 discs, suppose the pegs are labeled P_1, P_2, P_3. Suppose all discs are on P_1 and we wish to move them to P_2. Step 1: Move the smallest disc to P_3. Move the other disc to P_2. Now move the smallest disc from P_3 to P_2. Thus in $2^2 - 1 = 3$ moves, the discs have been moved in the appropriate manner. To complete the problem, try a few more special cases, and then proceed to do the problem by induction.)
11. Let S denote a sphere. Suppose that n planes pass through the center of S in such a way that no three planes have a line in common. Each plane must intersect the surface of S. Use induction to prove that the n planes divide the surface of S into $n^2 - n + 2$ regions.

2.5 Other Forms of Induction

In Section 2.4 we chose LWO as our last basic property of the integers. Theorem 2.4.3 (PMI) was proved using LWO with the other basic properties. In this section we shall show that LWO can be proved as a theorem under the assumption of PMI and the other properties; that is, we could just as well have chosen the principle of mathematical induction for this last property in place of LWO. In other words, the two mathematical statements—LWO and PMI—are logically equivalent.

Besides showing that PMI and LWO are equivalent, we shall discuss some other forms of mathematical induction. These will range from the course-of-values formulation of induction, to very simple modifications of PMI, LWO, and course-of-values induction. All of the various forms that we shall give are logically equivalent. We will leave many of the arguments to the reader. We begin by showing that PMI implies LWO in the presence of the arithmetic properties and O_1 and O_2.

In the proof of Theorem 2.4.3, the fact that 1 is the least positive integer (Lemma 2.4.1) is strongly used. This fact is also needed in the proof that PMI implies LWO. The reader will recall that LWO was used to prove Lemma 2.4.1. Hence, in order to avoid circularity, we now give a proof of this fact using only PMI.

LEMMA 2.5.1 The principle of mathematical induction implies that 1 is the least positive integer.

Proof Let $E = \{m \mid m \in P \text{ and } m \geq 1\}$. Clearly, $1 \in E$, since $1 \geq 1$. Next, assume that $n \in E$, that is, that $n \geq 1$. Then $n + 1 > 1$, whence $n + 1 \in E$. By PMI, $E = P$; that is, for any positive integer m, $m \geq 1$. ∎

THEOREM 2.5.2 The principle of mathematical induction implies the law of well ordering.

Proof We must prove that if $S \neq \varnothing$, $S \subseteq P$, then S has a least element. To achieve this, let $Q(n)$ be the statement: any set of positive integers containing an integer $\leq n$ has a least element. Let $E = \{m \mid m \in P$ and $Q(m)$ is true$\}$. Now, $Q(1)$ is true, since if a set of positive integers has an integer ≤ 1, it must, by Lemma 2.5.1, contain 1, which is its least element. Thus, $1 \in E$. Assume now that $n \in E$. That is, if B is a set of positive integers, and if B contains an integer $\leq n$, B has a least integer. Now let C be a set of positive integers containing an integer $\leq n + 1$. If C has no integer less than $n + 1$, then clearly $n + 1$ is the least integer of C. If C has an integer less than $n + 1$, it follows from Lemma 2.5.1 that C has an integer $\leq n$. (Why?) Thus, since $n \in E$, C has a least integer. Thus, under any circumstances, C has a least integer, and $n + 1 \in E$. By PMI, $E = P$.

Now, for $S \subseteq P$, $S \neq \varnothing$, S has some integer $m > 0$. Since $Q(m)$ is true for all $m > 0$, S has a least integer. This completes the proof. ∎

Theorems 2.4.3 and 2.5.2 show the logical equivalence of the law of well ordering and the principle of mathematical induction.

Another formulation of induction, called course-of-values induction, has been found very useful in applications. It is equivalent to either (and hence both) of PMI and LWO.

Course-of-values formulation of mathematical induction (CVI). Let E be a set of positive integers such that

(a) $1 \in E$ and
(b) the integer $n > 1$ is in E, whenever k is in E for all k satisfying $1 \leq k < n$.

Then $E = P$.

THEOREM 2.5.3 LWO implies course-of-values induction.

Proof We are required to prove that in the presence of the usual arithmetic properties of the integers, and in the presence of the first two order properties, if LWO holds, then CVI holds.
Suppose that $E \neq P$. Then E is a proper subset of P, and so there is an integer n in P that is not in E. Let m be the least such integer in P but not in E (such an m exists by LWO). By hypothesis (a), $1 \in E$, and so $m > 1$. Also, for all k such that $1 \leq k < m$, k is in E, for otherwise we contradict the selection of m. By (b), m must now be in E. This is a contradiction, and so we must conclude that $E = P$. ∎

Before we can prove that CVI implies LWO, we remind ourselves that we wish to assume at this stage only the arithmetic properties of the integers, the order properties O_1 and O_2, and the statement of CVI. We cannot assume that 1 is the least positive integer at this point, so as before, we must give a proof of this fact using only those properties.

THEOREM 2.5.4. CVI implies that 1 is the least positive integer.

Proof Let $E = \{m \mid m \in P \text{ and } m \geq 1\}$. Clearly, $1 \in E$, since $1 \geq 1$. Next, assume that n is an integer satisfying the following: for all k such that $1 \leq k < n$, k is in E. Trivially, n must also be in E. By CVI $E = P$, and so any integer m satisfies $m \geq 1$. Thus, 1 is the least positive integer. ∎

THEOREM 2.5.5 CVI implies LWO.

Proof We must prove that if $S \neq \varnothing$, $S \subseteq P$, then S has a least positive integer. To achieve this, let $Q(n)$ be the statement: any set of positive integers containing an integer $\leq n$ has a least element. Let $E = \{m \mid m \in P \text{ and } Q(m) \text{ is true}\}$. Now $Q(1)$ is true, since if a set of positive integers has an integer ≤ 1, it must by Theorem 2.5.4 contain 1, which is then the least element of such a set. Thus, $1 \in E$. Assume now that n is an integer, $n > 1$, such that for any k, $1 \leq k < n$, the integer k is in E. Now let C be a set of positive integers containing an integer $\leq n$. If C has no integer less than n, then clearly n is the least integer of C. If C has an integer $<n$, it follows from Theorem 2.5.4 that C has an integer $\leq n - 1$ (since 2.5.4 implies that there are no integers between $n - 1$ and n). Since $n - 1$ is in E, $Q(n - 1)$ is true, and so applying $Q(n - 1)$ to C, we see that C has a least element. But this says then that $Q(n)$ is true, and so $n \in E$. Now by CVI, $E = P$.
To conclude the proof, let $S \subseteq P$, $S \neq \varnothing$. Then S has some integer $m > 0$. Since $Q(m)$ holds, S has a least integer. ∎

The reader has undoubtedly observed the very close similarity in proofs given in this and the last section. The proofs that LWO imply PMI and

CVI are basically the same, and the proofs of the two converses are also essentially the same.

Theorems 2.4.3 and 2.5.2 show the equivalence of PMI and LWO. Theorems 2.5.3 and 2.5.5 show the equivalence of LWO and CVI. Since CVI and PMI are both equivalent to LWO, they are clearly equivalent to each other. Thus, we have concluded the proof that these three formulations of induction are equivalent to each other in the presence of the arithmetic properties and properties O_1 and O_2.

There are times when it is not convenient to start a proof by induction at 1. For example, a statement might be false for 1, but might be true for all integers, say, ≥ 6, or, on the other hand a given statement might be true for the set P and some nonpositive integers. It would be convenient to have a theorem that would let us start with 6, say, use some type of induction, and get the result. To illustrate, we start with the following example, where -3 is the initial integer.

Example 2.5.6 Let S be a set of integers $k \geq -3$ such that

(a) -3 is in S,
(b) whenever $n \geq -3$ is in S, then so is $n + 1$ in S.

Then S is the set of all integers $n \geq -3$.

Proof Let E be the set of integers defined as follows: $n \in E$ if and only if $n - 4$ is in S. Now 1 is in E, for $1 - 4 = -3$ is in S. Also, for any n in E, $n - 4$ in S implies that $n - 4 \geq -3$, whence $n \geq 1$. Thus, E is a set of positive integers. Suppose that n is in E. Then $n - 4$ is in S; that is, $n - 4 \geq -3$. But then $(n + 1) - 4 \geq -2 \geq -3$, so also $n + 1$ is in E. By PMI, $E = P$. Thus, n is in E for all positive integers; that is, $n - 4$ is in S for all positive integers. But any $k \geq -3$ has the form $n - 4$, for some positive integer. Thus, all $k \geq -3$ are in S, and S is the set of all such integers. ∎

It is not difficult to see exactly how we have used PMI in doing this example. What is important is that the argument can be switched around ever so slightly to show that the statement of 2.5.6 also implies PMI.

Example 2.5.7 The statement of 2.5.6 implies PMI.

Proof Let E be a subset of P satisfying the hypotheses of the statement of PMI. Thus, 1 is in E, and whenever n is in E, then also $n + 1$ is in E. Let S be the set defined as follows: k is in S if and only if $k + 4$ is in E. Now since 1 is in E, $k = -3$ is in S. Also, it is easy to see that $k + 4$ is in $E(k + 4 \geq 1)$, if and only if $k \geq -3$. Now let k be in S. Then $k + 4$ is in E. By assumption, $k + 4$ in E implies $(k + 4) + 1$ is in E; that is, $(k + 1) + 4$ is in E. So, by definition, $k + 1$ is in S. By assuming the truth of the statement of 2.5.6, S is now the set of *all* integers $k \geq -3$. Since any positive integer t has the form $t = s + 4$, for an appropriate integer $s \geq -3$, all integers $t \geq 1$ are in E. Thus, $E = P$, and we have shown that 2.5.6 implies PMI. ∎

It should be clear from these arguments that a host of statements are equivalent to PMI, and their proofs of equivalency are easily given through mimicking the two arguments above.

THEOREM 2.5.8 The following statement is equivalent to PMI: Let $c \in Z$. Let S be a subset of $\{k \mid k \in Z, k \geq c\}$ such that

(a) $c \in S$ and
(b) whenever n is in S, then also $n + 1$ is in S.

Then $S = \{k \mid k \in Z \text{ and } k \geq c\}$.

Proof This proof is left as an exercise. ∎

Theorem 2.5.8 is often referred to as the PMI.
The course-of-values version of 2.5.8 is given below.

THEOREM 2.5.9 The following statement is equivalent to CVI: Let $c \in Z$. Let S be a subset of the set of integers $k \geq c$ such that

(a) $c \in S$ and
(b) whenever n is an integer $> c$ such that all k satisfying $c \leq k < n$ are in S, then also $n \in S$.

Then $S = \{k \mid k \in Z, k \geq c\}$.

Proof The proof is left as an exercise. ∎

EXERCISES

1. Prove Theorem 2.5.8.
2. Prove Theorem 2.5.9.
3. Without making use of LWO, prove that PMI implies CVI and that CVI implies PMI. Note that one of these is easy, the other more difficult.
4. (a) Show that $(\frac{4}{3})^n > n$, for $n = 1$ and for $n \geq 7$.
   (b) Show directly that $(\frac{4}{3})^n < n$, for $n = 2, 3, 4, 5, 6$.
5. Prove that for $n \geq 4$, $2^n < n!$. (See Exercise 9, Section 2.4.)
6. Let n lines in the plane be in general position; that is, there is a set of n lines such that no two are parallel and such that no three are concurrent (intersect in the same point). Show that if $n \geq 2$, there are $(n^2 - n)/2$ points of intersection.
7. Show that for $n \geq 5$, $2^n \geq n^2$.
8. Show that for integers $n \geq -1$, $2n^3 - 9n^2 + 13n + 25 > 0$.

2.6 Elementary Concepts: Divisibility and the Division Algorithm

Although, as we have seen, the binary operation of division cannot be defined on the integers, some of the multiplicative properties of pairs of integers are most conveniently described by the use of words such as "division," "divisible," and "quotient." These terms are traditionally used in number theory, since division is possible for certain pairs of integers. We say, for example, that 18 is "exactly" divisible by 3, but not by 4. That is, the equation $18 = 3x$ has a solution in integers, while $18 = 4x$ does not.

Definition 2.6.1 Let b be an integer. We say the integer a is a *factor* of b or a *divisor* of b if there exists an integer c such that $b = ac$. We denote this by $a \mid b$. If a is not a divisor of b, we write $a \nmid b$. If $a \mid b$, we also say that b is a *multiple* of a, that b is *divisible* by a, or that a divides b.

As a consequence of this definition, we see that any integer a divides 0. Indeed, even $0 \mid 0$. We also observe that if $a \neq 0$, $a \in Z$, then $0 \nmid a$.

It also follows directly from our definition of divisibility that every integer a is divisible by 1, -1, a, and $-a$. For certain integers, such as 3 and 7, this list exhausts all possibilities. For other integers, such as 6 and 8, this list does not exhaust all divisors. Those integers that have only the obvious divisors play a special role in number theory and will be discussed in Section 2.8. An important result obtained in this chapter will be that $n > 1$ can be written as a product of integers with only the obvious divisors.

A number of very elementary results follow immediately from the definition of divisibility. Although exceedingly easy to prove, they are used frequently in the sequel, usually without reference. We shall prove some of them here, leaving others for the exercises.

THEOREM 2.6.2 Let $a, b, c \in Z$ and suppose $a \mid b$ and $a \mid c$. Let u and v be any integers. Then $a \mid (bu + cv)$.

Proof Since $a \mid b$, by definition there exists an integer d such that $b = ad$. Similarly, there exists an integer e such that $c = ae$. Multiplying these two equations through by u and v, respectively, we have $bu = adu$, and $cv = aev$. Adding, $(bu + cv) = adu + aev = a(du + ev)$. Thus, there is an integer $du + ev$ which satisfies the definition that $a \mid (bu + cv)$. ∎

It follows from Theorem 2.6.2 that if one has a relationship of the form $x = y + z$ and if a divides any two of the three quantities, x, y, and z, then a divides the third quantity. This fact, which could be proven directly without resort to the theorem, will be used repeatedly.

2.6 DIVISIBILITY AND THE DIVISION ALGORITHM

THEOREM 2.6.3 If $a \in Z$, then

$$1 \mid a, \quad -1 \mid a, \quad a \mid a, \quad -a \mid a, \quad |a| \mid a, \quad a \mid |a|.$$

Proof To see that $1 \mid a$, simply observe that $a = a \cdot 1$. To see that $-a \mid a$, observe that $a = a(-1)$. The other parts are equally easy to prove. ∎

A somewhat more difficult theorem is the following.

THEOREM 2.6.4 Let $a, b \in Z$, $b \neq 0$. If $a \mid b$, then $|b| \geq |a|$.

Proof Since $a \mid b$, there is an integer c such that $b = ac$. Since $b \neq 0$, then necessarily $a \neq 0$ and $c \neq 0$, for if either a or c is 0, then the product $ac = 0$.

By Theorem 2.3.2, $|b| = |a| |c|$. Since $c \neq 0$, $|c| > 0$. There are two possibilities concerning $|c|$.

(a) $|c| = 1$. In this case $|b| = |a|$.
(b) $|c| > 1$. Then $|b| = |a| |c| > |a| \cdot 1 = |a|$, so $|b| > |a|$. ∎

Another result that is used repeatedly is that divisibility is transitive.

THEOREM 2.6.5 Let a, b, c be integers such that $a \mid b$ and $b \mid c$. Then $a \mid c$.

Proof Since $b \mid c$, there is an integer d such that $c = bd$. Since $a \mid b$, there is an integer e such that $b = ae$. Substituting ae for b in $c = bd$, we find that $c = aed = a(ed)$. Thus, there is an integer ed such that $c = a(ed)$, and so $a \mid c$. ∎

EXERCISES

1. Let a, x, y, z be integers such that $x = y + z$ and such that a divides two of x, y, and z. Prove that a divides the third of x, y, and z, also.
2. Let a, b be integers such that $a \mid b$. Prove that $a \mid (-b)$, $(-a) \mid b$, $(-a) \mid (-b)$, $|a| \mid |b|$, and $a \mid bc$, for any integer c.
3. Let a, b, c be integers, with $c \neq 0$. Prove that $(ca) \mid (cb)$ if and only if $a \mid b$; that is, prove that if $a \mid b$, then $(ca) \mid (cb)$, and prove that if $(ca) \mid (cb)$, then $a \mid b$.
4. Let m and n be integers. Suppose $m \mid n$ and $n \mid m$. Prove that $m = n$ or $m = -n$. (The conclusion can be written $m = \pm n$.)
5. Find all factors of 24.
6. Find all factors of 212.
7. Find all factors of 89.

In the development of elementary number theory, one theorem plays a key role. This theorem is known as the division algorithm, and it establishes the elementary fact, known to all school children, that an integer b may be "divided" by an integer a in such a way that the remainder is less than $|a|$. An important exercise for the reader, as he progresses through this book, is to examine the proofs carefully and see how the line of reasoning of many of them can be traced back to the division algorithm. Our proof of the division algorithm will rest on the law of well ordering for positive integers.

THEOREM 2.6.6 (*Division Algorithm*) If a and b are in Z, $a \neq 0$, then there exist unique integers q and r such that

$$b = aq + r, \quad 0 \leq r < |a|.$$

Proof I. *Existence of q and r.*

CASE 1 $a > 0$. Let $S = \{b - ax \mid x \in Z, b - ax \geq 0\}$. We first show that $S \neq \varnothing$. If $b \geq 0$, then for $x = 0$, $b - a \cdot 0 = b$, whence $b - a0 \in S$ and $S \neq \varnothing$. If $b < 0$, then, since $a > 0$, $a \geq 1$ and $ab \leq b$. Thus $b - ab \in S$ and $S \neq \varnothing$.

By the law of well ordering S has a least element r. Thus there exists an integer q such that

$$b - aq = r, \quad 0 \leq r.$$

To complete the proof of existence for the case $a > 0$, we need to show that $r < |a|$. Hence assume that $r \geq |a| = a$. Then $r = a + c$ with $0 \leq c < r$ and $b - aq = a + c$. Hence $c = b - a(q + 1) \in S$, and, since $c < r$, this contradicts the choice of r. Thus, $r < |a|$.

CASE 2 $a < 0$. Since $-a > 0$, by Case 1 there exist q', r such that

$$b = (-a)q' + r, \quad 0 \leq r < |a|.$$

Setting $q = -q'$,

$$b = aq + r, \quad 0 \leq r < |a|.$$

II. *Uniqueness of q and r.* Suppose $b = aq + r = aq' + r'$ with

$$0 \leq r < |a| \quad \text{and} \quad 0 \leq r' < |a|.$$

Then $a(q - q') = r' - r$. Hence, $a \mid (r - r')$ and, if $r - r' \neq 0$, $|a| \leq |r - r'|$ by Theorem 2.6.4. But, since $0 \leq r < |a|$ and $0 \leq r' < |a|$, by Exercise 3, Section 2.3, we have $|r - r'| < |a|$. This is a contradiction, hence $r - r' = 0$. Therefore $r = r'$. Since $a \neq 0$, $q - q' = 0$, and so $q = q'$. ∎

The integer r of Theorem 2.6.6 is called the *remainder* in the division of b by a, and q is called the *quotient*.

To illustrate Theorem 2.6.6, in the case that $a = 7$ and $b = 47$, we see that $47 = 6 \cdot 7 + 5$, so that in this example the quotient is $q = 7$ and the remainder is $r = 5$.

As a second example, let $a = -31$ and let $b = -453$. Then $-453 = (-31)(15) + 12$, so $q = 15$ and $r = 12$. If a is again -31 and now $b = 453$, then $453 = (-31)(-14) + 19$, so now $q = -14$ and $r = 19$.

EXERCISES

8. Prove Theorem 2.6.6 by induction on b.
9. Find the remainders when 73 and 96 are each divided by 11 in terms of the division algorithm. Add the integers 73 and 96, and find the remainder of this sum upon division by 11. Show that this last remainder is the remainder obtained upon dividing the sum of the first two remainders by 11.
10. Let $b_1, b_2 \in Z$, $a \neq 0$. Let

$$b_1 = aq_1 + r_1, \quad 0 \leq r_1 < |a|,$$
$$b_2 = aq_2 + r_2, \quad 0 \leq r_2 < |a|.$$

Prove that if

$$b_1 + b_2 = aq_3 + r_3, \quad 0 \leq r_3 < |a|,$$

and if

$$r_1 + r_2 = aq_4 + r_4, \quad 0 \leq r_4 < |a|,$$

then $r_3 = r_4$.
11. Let $b_1, b_2 \in Z$, $a \neq 0$. Let

$$b_1 = aq_1 + r_1, \quad 0 \leq r_1 < |a|,$$
$$b_2 = aq_2 + r_2, \quad 0 \leq r_2 < |a|.$$

Prove that $a \mid (b_1 - b_2)$ if and only if $r_1 = r_2$.

2.7 Greatest Common Divisor and Euclidean Algorithm

Definition 2.7.1 Let a, b be integers. If $n \in Z$ and $n \mid a$ and $n \mid b$, then n is called a *common divisor* of a and b.

If a and b are integers and $a \neq 0$, then the set of common divisors of a and b is finite. The fundamental theorem of this section is the existence of a unique nonnegative common divisor that is divisible by all common divisors.

Definition 2.7.2 Let a and b be integers. The *greatest common divisor* (gcd) of a and b, denoted by (a,b), is defined as the nonnegative integer d such that

(a) $d \mid a$ and $d \mid b$
and (b) if $e \mid a$ and $e \mid b$, then $e \mid d$.

It is easy to show that $(132,630) = 6$, $(14,21) = 7$, $(380,95) = 95$, $(-72,18) = 6$, and $(-32,-12) = 4$. Moreover, we see that $(0,0) = 0$. The use of the word "greatest" is not a misnomer, for if either $a \neq 0$ or $b \neq 0$, then since any common divisor e of a and b divides d, $d \geq |e|$.

There is, of course, the problem of showing that any two integers a and b have a gcd, and that the gcd is indeed unique.

THEOREM 2.7.3 Let $a, b \in Z$. Then (a,b) exists and is unique. Moreover, there exist integers s and t such that

$$(a,b) = as + bt.$$

Proof Since $(0,0) = 0 = 0 \cdot s + 0 \cdot t$ (any s and t), the theorem is true if $a = b = 0$. Thus, we will assume that at least $a \neq 0$.
Let

$$S = \{ax + by \mid x, y \in Z \text{ and } ax + by > 0\}.$$

Since $0 < a \cdot a + b \cdot 0 \in Z$, $S \neq \varnothing$. By the law of well ordering S has a smallest positive integer d, whence there exist integers s and t such that $d = as + bt$. We shall show that d is (a,b).

By the division algorithm, there exist integers q and r such that $a = dq + r$, $0 \leq r < d$. Then

$$r = a - dq = a - (as + bt)q$$
$$= a(1 - sq) + b(-tq).$$

If $0 < r$, then $r \in S$. But this would contradict the fact that d is the smallest integer in S. Hence $r = 0$, and $a = dq$, that is, $d \mid a$. Similarly, $d \mid b$.

Now, from the equation $d = as + bt$, we see that if $e \mid a$ and $e \mid b$, then $e \mid d$, completing the proof of the existence of (a,b).

We leave the proof of uniqueness of (a,b) as an exercise. ∎

EXERCISES

1. Let $a, b \in Z$. Prove the uniqueness of (a,b). (*Hint:* Let d_1, d_2 both satisfy Definition 2.7.2. We must show $d_1 = d_2$. First show that $d_1 \mid d_2$ and $d_2 \mid d_1$, whence $d_1 = \pm d_2$.)
2. Find each of the following:
   (a) $(32,96)$.   (b) $(8,-18)$.
   (c) $(4,21)$.   (d) $(7,-24)$.
   (e) $(0,27)$.   (f) $(-16,-21)$.
3. For each part of 2, can you find integers s and t so that

$$(a,b) = as + bt.$$

2.7 GREATEST COMMON DIVISOR AND EUCLIDEAN ALGORITHM

4. Let $S \subseteq Z$, $S \neq \emptyset$, and assume that $a + b$ and $a - b$ are in S whenever $a, b \in S$. Prove there exists an integer $c \in S$ such that
$$S = \{nc \mid n \in Z\}.$$

Definition 2.7.4 If a, b, x, and y are integers, then $ax + by$ is called a *linear combination* of a and b.

Although Theorem 2.7.3 shows the existence of (a,b), the proof of the theorem does not give us a systematic way of finding (a,b). An efficient method is available, however, and we present it here. This method, the *Euclidean algorithm*, also gives another proof of the existence of (a,b). In the exercises of Section 2.9 we give a fairly simple way of finding (a,b), which, however, unlike the Euclidean algorithm, does not show how to find (a,b) as a linear combination of a and b.

Euclidean algorithm Let a and b be integers, $a \neq 0$. Then by the division algorithm there exist integers q_1 and r_1 such that

(1) $\qquad b = aq_1 + r_1, \qquad 0 \leq r_1 < |a|.$

If $r_1 = 0$, then $(a,b) = |a|$, and we are finished. If, however, $r_1 > 0$, then there are integers q_2 and r_2 such that

(2) $\qquad a = r_1q_2 + r_2, \qquad 0 \leq r_2 < r_1.$

If $r_2 = 0$, then $r_1 \mid a$; hence, by (1), $r_1 \mid b$. Also, if $c \mid a$ and $c \mid b$, then by (1) $c \mid r_1$. By Definition 2.7.2, $r_1 = (a,b)$ and $r_1 = b \cdot 1 + a(-q_1)$. If $r_2 > 0$, we continue, getting the complete chain:

(1) $\qquad b = aq_1 + r_1, \qquad 0 \leq r_1 < |a|,$
(2) $\qquad a = r_1q_2 + r_2, \qquad 0 \leq r_2 < r_1,$
(3) $\qquad r_1 = r_2q_3 + r_3, \qquad 0 \leq r_3 < r_2,$

$\qquad \cdot$
$\qquad \cdot$

$(n-1) \qquad r_{n-3} = r_{n-2}q_{n-1} + r_{n-1}, \qquad 0 \leq r_{n-1} < r_{n-2},$
$(n) \qquad r_{n-2} = r_{n-1}q_n + r_n, \qquad 0 \leq r_n < r_{n-1},$
$(n+1) \qquad r_{n-1} = r_nq_{n+1}.$

Since $0 \leq r_i < r_{i-1}$ at the ith step, we see that some remainder eventually must be zero. (Why?) This is our $(n+1)$st step.

Now, starting with $(n+1)$ and working back to (1), we see that $r_n \mid r_{n-1}$ by $(n+1)$; by (n), $r_n \mid r_{n-2}$; by $(n-1)$, $r_n \mid r_{n-3}$; and so, by a simple induction argument, $r_n \mid a$ and $r_n \mid b$. Now, if $c \mid a$ and $c \mid b$, then by (1), $c \mid r_1$; by (2), $c \mid r_2$; by (3), $c \mid r_3$; and, continuing, at (n), $c \mid r_n$. Since $r_n > 0$, $r_n = (a,b)$.

We next show how to find s and t such that $(a,b) = r_n = as + bt$. Starting at (n),

$$r_n = r_{n-2} - r_{n-1}q_n$$
$$= r_{n-2} - q_n(r_{n-3} - q_{n-1}r_{n-2}) \quad [\text{by } (n-1)]$$
$$= -q_n r_{n-3} + (1 + q_n q_{n-1})r_{n-2} = \cdots$$
$$= as + bt. \quad \blacksquare$$

Example 2.7.5 Find $(132,630)$.

$$630 = 132 \cdot 4 + 102,$$
$$132 = 102 \cdot 1 + 30,$$
$$102 = 30 \cdot 3 + 12,$$
$$30 = 12 \cdot 2 + 6,$$
$$12 = 6 \cdot 2 + 0.$$

Then, once again, $6 = (132,630)$. Moreover,

$$6 = 30 + 12(-2)$$
$$= 30 + (102 + 30(-3))(-2)$$
$$= 102(-2) + 30(7)$$
$$= 102(-2) + (132 + 102(-1))(7)$$
$$= 132 \cdot 7 + 102(-9)$$
$$= 132 \cdot 7 + (630 + 132(-4))(-9)$$
$$= 630(-9) + 132 \cdot 43.$$

Example 2.7.6 Find $(-14,73)$.

$$73 = (-14)(-5) + 3,$$
$$-14 = 3(-5) + 1,$$
$$3 = 1 \cdot 3 + 0.$$

Thus $(-14,73) = 1$. Also

$$1 = (-14) + 3(5)$$
$$= (-14) + [73 + (-14)5]5$$
$$= 73 \cdot 5 + (-14)[1 + 5 \cdot 5]$$
$$= 73 \cdot 5 + (-14)(26).$$

Although (a,b) is unique, the integers s and t of Theorem 2.7.3 are not unique. The reader should find other sets of values for s and t in the examples above.

THEOREM 2.7.7 Let a, b, s, and t be integers such that $as + bt = 1$. Then

$$(a,b) = (a,t) = (s,b) = (s,t) = 1.$$

Proof Let $d = (a,b)$. Then $d \mid a$ and $d \mid b$, whence $d \mid (as + bt)$, that is, $d \mid 1$. Since d is nonnegative and $d \mid 1$, clearly $d = 1$.
The other three parts are done similarly. $\blacksquare$

2.7 GREATEST COMMON DIVISOR AND EUCLIDEAN ALGORITHM

THEOREM 2.7.8 Let $a, b, m \in Z$. Suppose $a \neq 0$ and $m \geq 0$. Then

$$(am, bm) = m(a,b).$$

Proof It suffices to show that if $d = (am, bm)$, then $d = m(a,b)$. We know that there are integers s and t such that $(a,b) = as + bt$. Multiplying this through by m, we have $m(a,b) = mas + mbt$. Since $d \mid am$ and $d \mid bm$, clearly, $d \mid (mas + mbt)$, whence $d \mid m(a,b)$. Conversely, since $(a,b) \mid a$ and $(a,b) \mid b$, there are integers u and v such that $a = u(a,b)$ and $b = v(a,b)$. Thus, $am = um(a,b)$ and $bm = vm(a,b)$, whence $m(a,b) \mid am$ and $m(a,b) \mid bm$, so that $m(a,b) \mid (am, bm)$, that is, $m(a,b) \mid d$. We now conclude that $m(a,b) = (am, bm)$. ∎

THEOREM 2.7.9 Let a, b, m be integers. Then $(a,m) = (b,m) = 1$ if and only if $(ab, m) = 1$.

Proof First, suppose that $(a,m) = (b,m) = 1$. Then there are integers s and t such that $as + mt = 1$ and there are integers u and v such that $bu + mv = 1$. Then, multiplying the equations together, we get

$$ab(su) + m(asv + but + mtv) = 1.$$

By Theorem 2.7.7, $(ab, m) = 1$.

Conversely, from $(ab, m) = 1$, there are integers s and t such that $abs + mt = 1$. Writing this as $a(bs) + mt = 1$, we get $(a,m) = 1$, by 2.7.7. Similarly, $(b,m) = 1$. ∎

Since the Euclidean algorithm implies the existence of (a,b) and shows how to express (a,b) as a linear combination of a and b, we may ask why we first gave an existence proof in Theorem 2.7.3. The answer is that the first proof is easily applied to more general algebraic systems.

EXERCISES

5. Express $(24,63)$ as a linear combination of 24 and 63, $(15,23)$ as a linear combination of 15 and 23.
6. Let $(c,m) = d \neq 0$. Putting $c = dc_1$ and $m = dm_1$, prove that $(c_1, m_1) = 1$.
7. Let $a, b \in Z$. Define the *least common multiple* (lcm) of a and b, denoted by $[a,b]$, to be the nonnegative integer l such that (a) $a \mid l$ and $b \mid l$ and (b) if $a \mid n$ and $b \mid n$, then $l \mid n$. Prove that any two integers a and b have a least common multiple.
8. Let $a_1, a_2, \ldots, a_n \in Z$. Define the *greatest common divisor* of $a_1, a_2, \ldots, a_n$, denoted by $(a_1, a_2, \ldots, a_n)$, as the nonnegative integer d

satisfying (a) $d \mid a_i, i = 1, \ldots, n$, (b) if $e \mid a_i, i = 1, \ldots, n$, then $e \mid d$. Prove that $(a_1, \ldots, a_n)$ exists and can be expressed in the form

$$(a_1, a_2, \ldots, a_n) = a_1 b_1 + a_2 b_2 + \cdots + a_n b_n,$$

where the b_i are integers. (*Hint:* Mimic the proof of Theorem 2.7.3.)

9. Let $a, b, c \in Z$ satisfy $(a,b,c) = 1$. Show that $(a + b, b, c) = 1$.
10. Let $a_1, a_2, \ldots, a_n \in Z$. Define the least common multiple of $a_1, a_2, \ldots, a_n$ and prove its existence.

2.8 Prime Numbers

The numbers mentioned in Section 2.6 as having only the obvious divisors play a central role in the theory of numbers.

Definition 2.8.1 A number $p > 1$ is called a *prime number* if $n \mid p$ implies that $n = 1, -1, p$, or $-p$. An integer $m > 1$ is called a *composite number* if it is not a prime number.

Clearly the integers 2, 3, 5, 7, 11, 13 are prime numbers, since they have only themselves and 1 for positive divisors. The numbers 4, 6, 9, 10, 12 are composite numbers.

It is easy to see that if n is composite, then there exist integers $n_1 > 1$ and $n_2 > 1$ such that $n = n_1 \cdot n_2$.

It is interesting to make some elementary remarks about the list of prime numbers. First, there are infinitely many prime numbers. This will be seen later in Section 2.9.

Second, if we try to find all the prime numbers up to a certain integer, then the "sieve of Eratosthenes" gives us a systematic, albeit time-consuming, method for doing so. To see how the sieve works, we write down all the positive integers greater than 1 and $\leq n$. To illustrate, suppose $n = 90$.

```
 2  3  4  5  6  7  8  9 10 11 12 13 14 15 16 17 18 19
20 21 22 23 24 25 26 27 28 29 30 31 32 33 34 35 36 37
38 39 40 41 42 43 44 45 46 47 48 49 50 51 52 53 54 55
56 57 58 59 60 61 62 63 64 65 66 67 68 69 70 71 72 73
74 75 76 77 78 79 80 81 82 83 84 85 86 87 88 89 90
```

We then circle 2 and cross out all its subsequent multiples. Since 3 is the first number not crossed out in this list, 3 must be a prime number, for it has no divisors smaller than itself (2 is the only possibility at this time), other than 1. We circle 3, then cross out all subsequent multiples of 3. The next number in the list not crossed out is 5, and this must be a prime,

2.8 PRIME NUMBERS

for the only possible nontrivial divisors of 5 would be 2 and 3, and 5 is not a multiple of either of these. We circle 5 and then cross out all subsequent multiples of 5; we find that 7 is the next number not crossed out. We continue this method until we have circled 11 and crossed out all subsequent multiples of 11. Now the numbers that have not yet been crossed out must all be prime numbers, for if one of them is composite, it must be of the form $m = m_1 \cdot m_2$, and certainly one of these factors must be smaller than $\sqrt{90}$. Say m_1 is smaller than $\sqrt{90}$. Then it is less than 10. If m_1 is a prime, we would have crossed out m as being a multiple of m_1. If m_1 is not a prime, it was crossed out as we crossed out multiples of earlier prime divisors of m_1. This implies that m is also a multiple of that prime divisor of m_1 and must have been crossed out.

We show the list of all numbers up to 90 after all crossings-out have been made. Thus the list of all primes up to 90 is: 2, 3, 5, 7, 11, 13, 17, 19, 23, 29, 31, 37, 41, 43, 47, 53, 59, 61, 67, 71, 73, 79, 83, 89.

(2) (3) 4̸ (5) 6̸ (7) 8̸ 9̸ 1̸0̸ (11) 1̸2̸ 13 1̸4̸ 1̸5̸ 1̸6̸ 17 1̸8̸ 19
2̸0̸ 2̸1̸ 22 23 2̸4̸ 2̸5̸ 2̸6̸ 27 2̸8̸ 29 3̸0̸ 31 3̸2̸ 3̸3̸ 3̸4̸ 3̸5̸ 3̸6̸ 37
3̸8̸ 3̸9̸ 4̸0̸ 41 4̸2̸ 43 4̸4̸ 4̸5̸ 4̸6̸ 47 4̸8̸ 4̸9̸ 5̸0̸ 5̸1̸ 5̸2̸ 53 5̸4̸ 5̸5̸
5̸6̸ 5̸7̸ 5̸8̸ 59 6̸0̸ 61 6̸2̸ 6̸3̸ 6̸4̸ 6̸5̸ 6̸6̸ 67 6̸8̸ 6̸9̸ 7̸0̸ 71 7̸2̸ 73
7̸4̸ 7̸5̸ 7̸6̸ 77 7̸8̸ 79 8̸0̸ 8̸1̸ 8̸2̸ 83 8̸4̸ 8̸5̸ 8̸6̸ 8̸7̸ 8̸8̸ 89 9̸0̸

To use the sieve method for any positive integer $n > 1$, we need only cross out multiples of all those prime numbers up to $\sqrt{n}$. The reader should note that the sieve method lets us find primes up to a number n; it tells nothing about the existence of primes beyond n.

An old and as yet unanswered question is the "twin-prime conjecture." The reader may notice that in the list of primes there appear pairs of primes of the form p, $p + 2$; for example, 5,7; 11,13; 17,19; ...; 71,73. The "twin-prime conjecture" is that there are infinitely many such pairs of twin primes.

Another interesting question is whether for each even number $n > 4$ there are primes p_1 and p_2 such that $n = p_1 + p_2$. This conjecture, Goldbach's conjecture, is easily seen to be true for all cases of reasonable size. The answer is not yet known in general.

It is possible to go on and give many other simply worded questions about the prime numbers that are unanswered. In part, this is what has made number theory so appealing to so many nonmathematicians (as to so many mathematicians). Many questions are easily understood, even though the proofs may be highly difficult.

Closely allied to the concept of prime number is the idea of two numbers being relatively prime.

Definition 2.8.2 The integers a and b are *relatively prime* if $(a,b) = 1$. We also say that a *is prime to* b.

We note that if p is a prime number and if $p \nmid a$, then a and p are relatively prime. For if $d \mid p$ and $d \mid a$, $d > 0$, then either $d = p$ or $d = 1$. Since $d \neq p$, $d = 1$ and so $(a,p) = 1$.

It is not the case, however, that if a and b are relatively prime then one of them is a prime number. For $(4,9) = 1$ and neither 4 nor 9 is a prime number.

THEOREM 2.8.3 *(Euclid's Lemma)* Let p be a prime number and let $p \mid ab$, $a, b \in Z$. Then $p \mid a$ or $p \mid b$.

Proof Suppose $p \nmid a$. Then $(a,p) = 1$, whence, by Theorem 2.7.3, there exist integers s and t such that

$$1 = as + pt.$$

Then $b = b(as + pt) = (ab)s + p(bt)$. Since $p \mid ab$ and $p \mid p$, clearly $p \mid (ab)s + p(bt)$, whence $p \mid b$. ∎

COROLLARY 2.8.4 Let n be a positive integer and let $a_1, a_2, \ldots, a_n$ be integers such that $p \mid (a_1 \cdots a_n)$, p a prime. Then $p \mid a_i$, for some i such that $1 \leq i \leq n$.

Proof We prove this corollary by using mathematical induction.

If $n = 1$, the theorem is trivially true.

Assume the theorem for $n - 1 \geq 1$. Let $p \mid (a_1 \cdots a_{n-1} \cdot a_n)$. By Euclid's lemma, either $p \mid a_n$ or $p \mid (a_1 \cdots a_{n-1})$. If $p \mid a_n$, we are finished. If $p \nmid a_n$, then $p \mid (a_1 \cdots a_{n-1})$. By our induction hypothesis, $p \mid a_i$ for some i satisfying $1 \leq i \leq n - 1$. In either case $p \mid a_i$ for some i such that $1 \leq i \leq n$. Thus, the theorem is true for n if it is true for $n - 1$. The theorem then follows from PMI. ∎

An interesting corollary of Corollary 2.8.4 is the next result. It is of importance in that it is the main idea used by Euclid in proving the existence of infinitely many prime numbers. That proof itself must wait until the next section.

THEOREM 2.8.5 Let $p_1, p_2, \ldots, p_n$ be n distinct prime numbers and let $a_1, \ldots, a_n$ be n integers such that $p_i \mid a_i$, for each $i = 1, 2, \ldots, n$. Then $p_i \nmid (a_1 a_2 \cdots a_n + 1)$ for each i.

Proof Suppose $p_i \mid (a_1 a_2 \cdots a_n + 1)$. Since $p_i \mid a_i$, clearly $p_i \mid a_1 a_2 \cdots a_n$. Then $p_i \mid 1$, which is absurd. ∎

COROLLARY 2.8.6 Let p be a prime number and let $a_1, a_2 \in Z$ satisfy $0 < a_i < p$, for $i = 1, 2$. Then

$$p \nmid a_1 a_2.$$

Proof The proof is left as an exercise. ∎

Euclid's lemma can be easily generalized to the case where $(a,m) = 1$, with m dividing ab.

THEOREM 2.8.7 Let a, b, m be integers such that $(a,m) = 1$ and $m \mid ab$. Then $m \mid b$.

Proof This is almost identical to the proof of Euclid's lemma. From $(a,m) = 1$, there exist integers s and t such that $1 = as + mt$. Multiplying through by b, we have $b = abs + mbt$. Clearly, the hypotheses imply that $m \mid b$. ∎

EXERCISES

1. (a) Find three integers a, b, and c such that $a \mid bc$, but $a \nmid b$ and $a \nmid c$.
   (b) Find three integers a, b, and c, a not prime, such that $a \mid bc$ and $a \mid b$.
2. Prove Corollary 2.8.6.
3. Let $a, b, c, d \in Z$ and suppose $ab = cd$. Show that any prime divisor of a is a prime divisor of c or d.
4. Let p and q be prime numbers such that $p \mid q$. Show that $p = q$.
5. Prove that p is prime if and only if whenever $p = n_1 n_2$, then $|n_1| = 1$ or $|n_2| = 1$.
6. Let $p > 1$ and suppose that whenever $p \mid ab$, then $p \mid a$ or $p \mid b$. Prove that p is a prime.
7. Let $p_1, p_2, \ldots, p_n$ be n distinct prime numbers. Let $a_1, a_2, \ldots, a_n$ be n integers such that for each $i = 1, 2, \ldots, n$, $p_i \nmid a_j$, for $j \neq i$. Prove that if $p_i \mid (a_1 a_2 \cdots a_n)$, then $p_i \mid a_i$. Also prove that if $p_i \nmid (a_1 a_2 \cdots a_n)$, then $p_i \nmid a_j$, $j = 1, 2, \ldots, n$.
8. Use the "sieve" to find all prime numbers up to 180; 300.
9. Show that $M_n = 2^n - 1$ is prime for $n = 2, 3, 5, 7$. Show that M_{11} is composite.
10. Let $a \mid m$, $b \mid m$, and $(a, b) = 1$. Show that $ab \mid m$. [*Hint:* There are integers s and t such that $as + bt = 1$. Multiply through by m. Does $ab \mid am$? Does $ab \mid bm$?]
11. Let $a, b, c, d \in Z$. If $ab \mid cd$ and $(a,d) = 1$, show that $a \mid c$.

2.9 The Fundamental Theorem of Arithmetic

The fundamental theorem of arithmetic, proved in this section, states the familiar fact that any integer $n > 1$ can be expressed as the product of prime numbers in essentially one way. This theorem, an example of an algebraic structure theorem, shows the importance of prime numbers, since they "generate" the set of all positive integers greater than one. As we progress, we shall see other examples of mathematical systems where certain elements act as the "building blocks" for the whole system.

Definition 2.9.1 Let n be a positive integer that is either itself a prime number or that can be expressed as a product of prime numbers. Then we say that n has a *factorization into prime numbers*, or that n has a *prime factorization*. If $n = p_1 p_2 \cdots p_s$, $s \geq 1$, and each p_i is a prime, we call this expression a *prime factorization* for n.

THEOREM 2.9.2 (*The Fundamental Theorem of Arithmetic*) Let n be a positive integer, $n > 1$. Then n has a prime factorization, and if

$$n = p_1 p_2 \cdots p_r \quad \text{and} \quad n = q_1 q_2 \cdots q_s$$

are two prime factorizations for n, then $r = s$ and the two factorizations differ only in the order of the factors.

Proof I. *Existence.* Let $S = \{k \mid k$ is an integer, $k > 1$, k does not have a prime factorization$\}$. If $S = \emptyset$, then the theorem is proved. Thus, assume $S \neq \emptyset$. By the law of well ordering, S has a least member, say m. If the only positive factors of m are m and 1, then m is a prime and $m \notin S$. Thus, m can be written $m = m_1 \cdot m_2$, where $1 < m_1 < m$, $1 < m_2 < m$. By our choice of m, $m_1 \notin S$ and $m_2 \notin S$. Hence,

$$m_1 = u_1 u_2 \cdots u_g, \quad g \geq 1,$$

where each u_i is prime, and

$$m_2 = v_1 v_2 \cdots v_h, \quad h \geq 1,$$

where each v_i is prime. But

$$m = m_1 m_2 = u_1 \cdots u_g v_1 \cdots v_h$$

is a prime factorization for m. Thus, $m \notin S$, a contradiction. Therefore $S = \emptyset$ and we have shown that every integer $n > 1$ has a prime factorization.

II. *Uniqueness.* The uniqueness is certainly true for $n = 2$. Thus, we shall assume that the theorem holds for all integers k such that $2 \leq k < n$ and prove that the theorem then holds for n. Then, by course-of-values induction, the theorem will follow for all integers $n > 1$.

Suppose, therefore, that

$$n = p_1 p_2 \cdots p_r = q_1 q_2 \cdots q_s$$

are two prime factorizations for n. If $r = 1$, then n is prime, whence we must have $s = 1$ and also $p_1 = q_1$. Thus, we may assume $r > 1$ and $s > 1$.

2.9 THE FUNDAMENTAL THEOREM OF ARITHMETIC 69

Now it is clear that $p_1 \mid q_1 q_2 \cdots q_s$, and so, by Corollary 2.8.4, $p_1 \mid q_t$ for some t. But since q_t is a prime, $p_1 = q_t$. We may assume that the q_i's are so arranged that $t = 1$. Thus

$$p_1 p_2 \cdots p_r = p_1 q_2 \cdots q_s.$$

Since $p_1 \neq 0$, we may cancel and get $p_2 \cdots p_r = q_2 \cdots q_s = n'$. But $1 < n' < n$, and by our induction hypothesis we may conclude (a) that $r - 1 = s - 1$ and (b) that the factorization $p_2 \cdots p_r$ is just a rearrangement of the q_i's, $i = 2, \ldots, r$. Thus, $r = s$, and since $p_1 = q_1$, we have proved the theorem for n. Hence, by induction, the theorem is true for all integers $n > 1$. ∎

The fundamental theorem of arithmetic is often called *the unique factorization theorem* for positive integers. It is now clear that we excluded 1 from the set of prime numbers so that the fundamental theorem of arithmetic could be stated in the form of Theorem 2.9.2.

We remark that there exist mathematical systems very similar to the integers in which the analogue of the fundamental theorem of arithmetic is false. For example, let E be the set of even integers. E satisfies all the arithmetic and order properties of Z, except that E does not have a multiplicative identity. If in E we define a prime number n to be one that cannot be written as the product of two other numbers in E, then we see that 2, 6, and 18 are prime in E. There are, then, two distinct prime factorizations of 36 in E, since $36 = 2 \cdot 18 = 6 \cdot 6$.

Among the many applications of the fundamental theorem of arithmetic, we present two below. Our first application is Euclid's proof that there exist infinitely many prime numbers.

THEOREM 2.9.3   There exist infinitely many prime numbers.

Proof   The main idea here is the idea used in Theorem 2.8.5. Thus, suppose that there is only a finite number of prime numbers. These can be listed: $p_1 = 2, p_2 = 3, p_3 = 5, \ldots, p_N$, where p_N is the last of the finite list of primes. We form the integer $K = p_1 p_2 \cdots p_N + 1$. Now by the fundamental theorem of arithmetic, K must have a prime factor, say p, and by assumption that there is only a finite number of prime numbers, p must be one of those prime numbers we have listed, so let $p = p_i$, for appropriate i. Then $p_i \mid (p_1 p_2 \cdots p_N)$, so that $p_i \mid K - (p_1 p_2 \cdots p_N)$, whence $p_i \mid 1$. This is absurd, so we conclude that there are infinitely many prime numbers. ∎

The second result we prove is known to every student of calculus. In order to state the result, we assume that the reader is familiar with numbers of the form p/q, where p and q are integers, $q \neq 0$. Such numbers are called *rational numbers*, and we will use them for illustrations from time

70   NUMBER THEORY

to time, developing them more fully in Chapter 5. Real numbers that are not rational numbers are called *irrational numbers*. We will prove here that $\sqrt{2}$ is irrational. A full discussion of the real number system will appear in Chapter 6.

THEOREM 2.9.4   The real number $\sqrt{2}$ is irrational.

Proof Suppose that $\sqrt{2}$ is rational. Then there exist integers p and q, $q \neq 0$, such that $\sqrt{2} = p/q$. Then squaring both sides, $2 = p^2/q^2$, whence $2q^2 = p^2$. Now we factor p and q into their prime factorizations. This will induce a prime factorization for each of p^2 and q^2.

Each time that the prime number 2 appears in a factorization for p, it appears twice in the prime factorization for p^2. Thus, 2 appears an even number of times when p^2 is factored into prime numbers. Similarly, 2 appears an even number of times when q^2 is factored into prime numbers. But in the equation $2q^2 = p^2$, this would mean that 2 appears an odd number of times in $2q^2$ and an even number of times in p^2, a contradiction. This contradiction means that $\sqrt{2}$ is not rational, that is, $\sqrt{2}$ is irrational. ∎

The proof of this theorem can be generalized to prove that $\sqrt{p}$, for any prime p, is irrational, and to show that if n is not a perfect square, then $\sqrt{n}$ is irrational.

We have used exponents previously in examples and exercises but we have not yet formally defined what is meant by x^n.

Definition 2.9.5   Let x be a nonzero integer. Define x^n, $n \geq 0$, as follows:

(a) $x^0 = 1$,
(b)   for $n \geq 1$, $x^n = x^{n-1} \cdot x$.

This definition is an example of a *recursive* or *inductive definition*. It is not difficult to convince oneself that this provides a method for computing x^n for integers $x \neq 0$, $n \geq 0$. The complete proof that x^n is uniquely defined for all $x \neq 0$ and $n \geq 0$ utilizes the principle of mathematical induction and is rather subtle. We refer the interested reader to P. R. Halmos, *Naive Set Theory*.

To look at a specific example, let $x = 3$. Then

$$3^0 = 1, \quad 3^1 = 3^{1-1} \cdot 3 = 3^0 \cdot 3 = 3,$$
$$3^2 = 3^{2-1} \cdot 3 = 3^1 \cdot 3 = 3 \cdot 3 = 9,$$
$$3^3 = 3^{3-1} \cdot 3 = 3^2 \cdot 3 = 9 \cdot 3 = 27,$$
$$3^4 = 3^{4-1} \cdot 3 = 3^3 \cdot 3 = 27 \cdot 3 = 81, \text{ ad infinitum.}$$

With our definition, we are now in a position to derive the usual laws of exponents. The derivations are left as exercises.

2.9 THE FUNDAMENTAL THEOREM OF ARITHMETIC

EXERCISES

1. If $n \geq 0$ and $m \geq 0$ and if $x \neq 0$, prove that $x^n \cdot x^m = x^{n+m}$.
2. Prove: $(xy)^n = x^n y^n$, if $n \geq 0$, all $x, y \in Z$, $x \neq 0$, $y \neq 0$.

Our exponential notation allows the following useful restatement of Theorem 2.9.2.

COROLLARY 2.9.6 If $n > 1$ is a positive integer, then n has a unique representation in the form

$$n = p_1^{e_1} p_2^{e_2} \cdots p_r^{e_r},$$

where p_i is prime, $i = 1, 2, \ldots, r$, $p_i < p_j$, for $i < j$, and $e_i > 0$, for all i.

Proof The proof is left as an exercise. ∎

Definition 2.9.7 The representation of n in Corollary 2.9.6 is called the *prime power factorization of n*.

Among the basic theorems concerning exponents are the familiar formulas $x^n x^m = x^{n+m}$ of Exercise 1 and $(xy)^n = x^n y^n$ of Exercise 2. We now prove the important binomial theorem.

THEOREM 2.9.8 (*Binomial Theorem*) Let $a, b \in Z$, and let n be a positive integer. Then

$$(a + b)^n = \binom{n}{0} a^n + \binom{n}{1} a^{n-1}b + \binom{n}{2} a^{n-2}b^2 + \cdots + \binom{n}{n} b^n,$$

where $\binom{n}{i}$ is the *binomial coefficient* defined by

$$\binom{n}{i} = \frac{n!}{i!(n-i)!},$$

with

$$k! = \begin{cases} 1, & k = 0, \\ k(k-1)(k-2) \cdots 3 \cdot 2 \cdot 1, & k > 0. \end{cases}$$

Proof We prove this by induction on n. For the case $n = 1$ the result is immediate. Assume then that the result has been proven for $n - 1$. We must prove it is true for n. Thus, assume

$$(a + b)^{n-1} = \binom{n-1}{0} a^{n-1} + \binom{n-1}{1} a^{n-2}b + \binom{n-1}{2} a^{n-3}b^2$$
$$+ \cdots + \binom{n-1}{n-1} b^{n-1}.$$

72 NUMBER THEORY

Multiplying both sides through by $(a + b)$, and using the basic properties of the integers, we get

$$(a + b)^n = \left[\binom{n-1}{0}a^{n-1} + \binom{n-1}{1}a^{n-2}b + \cdots + \binom{n-1}{i}a^{n-(i+1)}b^i \right.$$
$$\left. + \cdots + \binom{n-1}{n-1}b^{n-1}\right](a + b)$$
$$= \binom{n-1}{0}a^n + \left[\binom{n-1}{0} + \binom{n-1}{1}\right]a^{n-1}b$$
$$+ \left[\binom{n-1}{1} + \binom{n-1}{2}\right]a^{n-2}b^2$$
$$+ \cdots + \left[\binom{n-1}{i} + \binom{n-1}{i+1}\right]a^{n-(i+1)}b^{i+1}$$
$$+ \cdots + \binom{n-1}{n-1}b^n.$$

Since $\binom{n-1}{0} = \binom{n}{0} = 1$ and $\binom{n-1}{n-1} = \binom{n}{n} = 1$, the result will follow if we can show

$$\binom{n-1}{i} + \binom{n-1}{i+1} = \binom{n}{i+1} \qquad \text{for } 0 \leq i \leq n - 2.$$

Now

$$\binom{n-1}{i} + \binom{n-1}{i+1} = \frac{(n-1)!}{i!(n-(i+1))!} + \frac{(n-1)!}{(i+1)!(n-(i+2))!}$$
$$= n!\left[\frac{1}{i!n \cdot (n-(i+1))!}\right.$$
$$\left. + \frac{1}{(i+1)!(n-(i+2))!n}\right]$$
$$= n!\left[\frac{i+1+n-(i+1)}{n \cdot (i+1)!(n-(i+1))!}\right]$$
$$= \frac{n! \cdot n}{n(i+1)!(n-(i+1))!}$$
$$= \frac{n!}{(i+1)!(n-(i+1))!} = \binom{n}{i+1}.$$

Thus, the theorem is proved by induction. ∎

EXERCISES

3. Let m and n be integers such that $m \mid n$. Show that if
$$n = p_1^{e_1}p_2^{e_2}\cdots p_r^{e_r}, \qquad e_i > 0, \quad p_i \text{ prime, for each } i,$$

2.9 THE FUNDAMENTAL THEOREM OF ARITHMETIC

then
$$m = p_1^{f_1} p_2^{f_2} \cdots p_r^{f_r}, \quad \text{with } 0 \leq f_i \leq e_i, \text{ for each } i.$$

4. Let $p_1, p_2, \ldots, p_s$ be distinct prime numbers. Let $j_1, j_2, \ldots, j_s, k_1, k_2, \ldots, k_s$ be integers ≥ 0. Let $m = p_1^{j_1} \cdots p_s^{j_s}$ and $n = p_1^{k_1} \cdots p_s^{k_s}$. Then $m \mid n$ if and only if $j_i \leq k_i$, $i = 1, 2, \ldots, s$.
5. Let $m = p_1^{j_1} \cdots p_s^{j_s}$ and $n = p_1^{k_1} \cdots p_s^{k_s}$, where the p_i are distinct primes, and $j_i \geq 0$, $k_i \geq 0$, all i. Show that
$$(m,n) = p_1^{e_1} \cdots p_s^{e_s}$$
and
$$[m,n] = p_1^{f_1} \cdots p_s^{f_s},$$
where $e_i = \min(j_i, k_i)$ (the smaller of j_i and k_i), for each i, and $f_i = \max(j_i, k_i)$ (the larger of j_i and k_i), for each i.
6. Let $a, b \in Z$. Show that $ab = (a,b) \cdot [a,b]$.
7. Use the fundamental theorem of arithmetic to prove the following: Let $a \mid m$, $b \mid m$ and $(a,b) = 1$. Then $ab \mid m$. Compare Exercise 10, Section 2.8.
8. Use the fundamental theorem of arithmetic to prove the following: Let a, b, c and $d \in Z$. If $ab \mid cd$ and $(a,d) = 1$, then $a \mid c$. Compare Exercise 11, Section 2.8.
9. Let $d, m, n \in Z$, with $(m,n) = 1$. Suppose $d \mid mn$. Show that there exist integers r and s such that $d = rs$, $r \mid m$, $s \mid n$, and $(r,s) = 1$. (*Hint:* Factor mn into prime factors and d into prime factors.)
10. Let $n \geq 1$ and suppose $2^n + 1$ is prime. Show that $n = 2^j$, for some $j \geq 0$.
11. Let $p \geq 2$. Show that if $2^p - 1$ is prime, then p is also prime.
12. Let p and q be distinct primes. Show that
$$(1 + p + p^2)(1 + q + q^2 + q^3)$$
is the sum of all the divisors of $p^2 q^3$.
13. Let $m = p_1^{e_1} \cdots p_s^{e_s}$ be the prime power factorization for m. How many positive divisors of m are there, and what are they? Show that the sum of these divisors is
$$(1 + p_1 + \cdots + p_1^{e_1}) \cdots (1 + p_s + \cdots + p_s^{e_s}).$$
14. Prove that there exist infinitely many primes of the form $6n - 1$. (*Hint:* Consider $p_1 p_2 \cdots p_k - 1$, where $p_1 = 2$, $p_2 = 3$, $\ldots$, p_k are the first k primes.)
15. Prove that there are infinitely many primes of the form $4n - 1$.
16. Define $F_n = 2^{2^n} + 1$ (the nth *Fermat number*), $n = 0, 1, 2, \ldots$. Prove that $(F_n, F_m) = 1$, $n \neq m$. (*Hint:* First prove by mathemati-

cal induction that

$$F_0 F_1 \cdots F_{n-1} + 2 = F_n, \quad n = 1, 2, \ldots.$$

Then observe that if $p \neq 1$ and if $p \mid F_n$ and $p \mid F_m$, $n \neq m$, then p must be odd and even. Show that this exercise gives another proof of Theorem 2.9.3.)

17. Show that an integer $n > 1$ is a square if and only if all exponents in its prime power factorization are even integers.
18. Prove: If n is a positive integer that is not a perfect square, then $\sqrt{n}$ is irrational.

2.10 Congruences

If a and b are integers, then in view of the division algorithm, it makes sense to talk about the quotient and remainder obtained upon dividing a by b. If b is kept fixed and we look at the remainders obtained by allowing a to take on successive integral values, then a periodic sequence of remainders always occurs. For example, if $b = 3$ and a assumes the values 0, 1, 2, 3, 4, 5, 6, 7, then the remainders obtained are, respectively, 0, 1, 2, 0, 1, 2, 0, 1. (What does the sequence of quotients look like?)

This property provides a way of classifying integers according to the remainder obtained upon division by a fixed integer. In fact, often this remainder is the only thing of interest. For example, in counting the hours of the day, we begin again after reaching the number 12. With the sole exception of using 12 instead of 0, this is equivalent to counting the hours sequentially and naming each hour with the remainder obtained upon division by 12. In this section we shall study a relation on the integers that is defined in terms of remainders.

Definition 2.10.1 Let a, b, and m be integers, $m \geq 0$. Then we say that a *is congruent to* b modulo m if there is an integer k such that $a = b + km$. We denote this by $a \equiv b \pmod{m}$. If it is not the case that $a \equiv b \pmod{m}$, then we write $a \not\equiv b \pmod{m}$.

In the event that $m = 0$, it is immediate that $a \equiv b \pmod{m}$ if and only if $a = b$. Thus, the study of the integers modulo 0 will give us no useful new information concerning the integers. Although many of the theorems we shall study are true even in the case of $m = 0$, we shall assume henceforth that $m > 0$. This will simplify many of the statements we must make.

For $m > 0$, there are two other ways of determining whether a and b are congruent modulo m, which we will use freely when working with congruences.

2.10 CONGRUENCES

THEOREM 2.10.2 Let a, b, m be integers, $m > 0$. Then the following three statements are equivalent:

(a) $a \equiv b \pmod{m}$.
(b) $m \mid (a - b)$.
(c) The remainders of a and b in the division algorithm upon division by m are the same.

Proof (a) implies (b). If $a = b + km$, then $a - b = km$, whence $m \mid (a - b)$.

(b) implies (c). Suppose that $a = k_1 m + r_1$ and $b = k_2 m + r_2$, where $0 \leq r_i < m$, for $i = 1, 2$. If $r_1 \neq r_2$, then without loss of generality we may assume that $r_1 > r_2$. Then $a - b = m(k_1 - k_2) + (r_1 - r_2)$, and since $m \mid (a - b)$ and $m \mid m$, we have that $m \mid (r_1 - r_2)$. But from the conditions on r_1 and r_2 imposed by the division algorithm, $0 < r_1 - r_2 < m$, so certainly we cannot have $m \mid (r_1 - r_2)$. Thus, it is untenable that $r_1 \neq r_2$, so $r_1 = r_2$.

(c) implies (a). Express a and b in terms of the division algorithm as in the paragraph above. Then, assuming $r_1 = r_2$, we get $a - k_1 m = b - k_2 m$, so $a = b + (k_1 - k_2)m$, so $a \equiv b \pmod{m}$. ∎

Example 2.10.3 We illustrate with some examples. Let $m = 5$. Then $23 \equiv 3 \pmod{5}$, because $23 = 3 + 5 \cdot 4$, or because $5 \mid (23 - 3)$. Similarly, using (c), $-8 \equiv 20 \pmod{7}$, for $-8 = 7(-2) + 6$ and $20 = 7 \cdot 2 + 6$. Or we see that $-8 \equiv 20 \pmod{7}$, since $7 \mid (-8 - 20)$, or because $-8 = 20 + 7(-4)$. It should be easy for the reader to give a host of examples.

For a fixed integer m, congruence modulo m is a relation on the set Z. Before proceeding, the reader should study Example 1.8.3.

THEOREM 2.10.4 For a fixed integer $m \geq 0$, the relation $a \equiv b \pmod{m}$ defines an equivalence relation on the set of integers Z.

Proof By the definition of equivalence relation, we must show that the set $\{(a,b) \mid a \equiv b \pmod{m}\}$ defines a relation that is reflexive, symmetric, and transitive.

(a) Let $a \in Z$. Since $m \mid 0$, $m \mid (a - a)$. And so by definition

$$a \equiv a \pmod{m}. \quad \text{(reflexive)}$$

(b) Let $a \equiv b \pmod{m}$. Then $m \mid (a - b)$ and $m \mid (-1)(a - b)$. Hence $m \mid (b - a)$ and $b \equiv a \pmod{m}$. (symmetric)

(c) Let $a \equiv b \pmod{m}$ and $b \equiv c \pmod{m}$. Then $m \mid (a - b)$ and $m \mid (b - c)$. Hence $m \mid [(a - b) + (b - c)]$ and $m \mid (a - c)$. Therefore $a \equiv c \pmod{m}$. (transitive) ∎

Since by Theorem 2.10.4 congruence modulo m is an equivalence relation on Z, it is interesting to determine the equivalence classes. Since a and b are congruent modulo m if and only if a and b differ by a multiple of m, then an equivalence class is the set of all those integers differing from a fixed element of the equivalence class by a multiple of m. When $m = 7$, say, the equivalence class containing 4 will then also contain 11, 18, 25, 32, -3, -10, and so forth. It is also clear that an equivalence class under this equivalence relation consists of all those integers that have the same remainder under division by m. Thus if we think of the clock example, all those hours during the course of a year that are, say, 3:00 P.M. would be in the same equivalence class. More will be said about the equivalence classes in the next section.

THEOREM 2.10.5 Let $a \equiv b \pmod{m}$ and $c \equiv d \pmod{m}$. Then

(a) $a + c \equiv b + d \pmod{m}$.
(b) $a - c \equiv b - d \pmod{m}$.
(c) $ac \equiv bd \pmod{m}$.

Proof We leave (a) and (b) as exercises. To prove (c), since $a \equiv b \pmod{m}$, there is an integer k_1 such that $a = b + mk_1$. Similarly from $c \equiv d \pmod{m}$, $c = d + mk_2$ for some integer k_2. Then

$$ac = (b + mk_1)(d + mk_2) = bd + m(k_1 d + k_2 b + mk_1 k_2).$$

Hence $ac \equiv bd \pmod{m}$. ∎

Theorems 2.10.4 and 2.10.5 indicate that in many respects the relation $a \equiv b \pmod{m}$ behaves very much like the relation of equality. On the other hand, the cancellation law of multiplication holds only in a restricted sense. For example, if $ab \equiv ac \pmod{m}$ and $a \not\equiv 0 \pmod{m}$, the perfect analogue to the cancellation law for integers would imply that $b \equiv c \pmod{m}$. But this is not in general true. For if $a = 2$, $b = 3$, $c = 1$, $m = 4$, then $2 \cdot 3 \equiv 2 \cdot 1 \pmod 4$, but $3 \not\equiv 1 \pmod 4$. We give one particularly useful analogue of the cancellation law. We first need a lemma.

LEMMA 2.10.6 Let a, b, c, d be integers such that $ab \mid cd$ and such that $(a,d) = 1$. Then $a \mid c$.

Proof This lemma is Exercise 11, Section 2.8, and Exercise 8, Section 2.9. We give a proof here appropriate to Section 2.9. Since $ab \mid cd$, there exists an integer r such that $cd = abr$. In the prime power factorizations for $cd = abr$, we must get the same prime powers appearing on both sides. Since $(a,d) = 1$, no prime appearing in the factorization for a appears in the factorization for d, so each prime power of a must appear in the factorization for c, whence a must be a factor of c. ∎

2.10 CONGRUENCES   77

THEOREM 2.10.7   Let a, b, c, m be integers, $m > 0$, and suppose that $ac \equiv bc \pmod{m}$ and $(c,m) = 1$. Then $a \equiv b \pmod{m}$.

Proof From $ac \equiv bc \pmod{m}$, we have $m \mid (ac - bc)$, that is, $m \mid (a - b)c$. Since $(m,c) = 1$, it follows that $m \mid (a - b)$ by Lemma 2.10.6. ∎

Example 2.10.8 Let p be a prime number and let a be an integer such that $p \nmid a$. Then, of course, $(a,p) = 1$. Now consider the set of integers $a \cdot 0, a \cdot 1, a \cdot 2, \ldots, a \cdot (p - 1)$. No two of these integers are congruent modulo p. For if $a \cdot t \equiv a \cdot s \pmod{p}$, then by Theorem 2.10.7, $t \equiv s \pmod{p}$. But for t and s in the range 0, 1, 2, …, $p - 1$, the only way that $t \equiv s \pmod{p}$ is for $t = s$. For $p = 5$, and $a = 3$, we observe that no two of 0, 3, 6, 9, 12 are congruent modulo 5.

Example 2.10.9 The integers 2^i and 2^j, $i \geq 0$, $j \geq 0$, have the same remainder upon division by 5 if and only if $i \equiv j \pmod{4}$. To see this, first observe that for any integer $k \geq 0$,

$$(2)^{4k} \equiv (2^4)^k \equiv (16)^k \equiv 1 \pmod{5}, \quad \text{since } 16 \equiv 1 \pmod{5}.$$

Next, there are integers k_1 and k_2 such that $i = r + 4k_1$ and $j = s + 4k_2$, where r and s satisfy $0 \leq r \leq 3$ and $0 \leq s \leq 3$. Then we have

$2^i \equiv 2^j \pmod{5}$ if and only if $2^{r+4k_1} \equiv 2^{s+4k_2} \pmod{5}$
                if and only if $2^r \cdot 2^{4k_1} \equiv 2^s \cdot 2^{4k_2} \pmod{5}$
                if and only if $2^r \cdot 1 \equiv 2^s \cdot 1 \pmod{5}$
                if and only if $2^r \equiv 2^s \pmod{5}$.

Since the remainders of 2^0, 2^1, 2^2, and 2^3 are 1, 2, 4, and 3, respectively, upon division by 5, we see that $2^r \equiv 2^s \pmod{5}$ if and only if $r = s$, that is, if and only if i and j have the same remainder upon division by 4, that is, if and only if $i \equiv j \pmod{4}$.

Example 2.10.10 Congruences can help us decide that an equation does not have any integer solutions. To illustrate, consider the equation $x^3 + 3x^2 - 1 = 0$. If there is any integer x_0 that satisfies the equation, then $x_0^3 + 3x_0^2 - 1 = 0$. Thus, for every integer m, $x_0^3 + 3x_0^2 - 1 \equiv 0 \pmod{m}$. If we can exhibit one integer $m > 0$ such that $x_0^3 + 3x_0^2 - 1 \not\equiv 0 \pmod{m}$, then the integer x_0 satisfying the original equation does not exist. In this particular case, let $m = 4$. Then for any integer w, $w = 4t + s$, where $s = 0, 1, 2,$ or 3. By Theorem 2.10.5, $w^3 + 3w^2 - 1 \equiv s^3 + 3s^2 - 1 \pmod{4}$ so if w is a solution to the equation, then s has the property that $s^3 + 3s^2 - 1 \equiv 0 \pmod{4}$. But $s^3 + 3s^2 - 1$ is congruent modulo 4 to 3, 3, 3, and 1, for $s = 0, 1, 2,$ and 3, respectively. Thus, the original equation has no solution in integers. Note that for $m = 3$, this technique fails in this equation. There may also be methods of getting at the answer other than through congruences.

EXERCISES

1. Show that each of the following polynomial equations does not have a solution in the system of integers. (*Note:* If one choice for m fails to work, try other choices. You need not choose very large m to succeed.)
   (a) $2x^2 - 2x + 5 = 0$.
   (b) $5x^5 - 4x^4 + 3x^3 - 2x^2 + x - 1 = 0$.
   (c) $2x^3 - x + 3 = 0$.
2. Find the remainders of each of the following powers of 3 under division by 5: 3, 3^2, 3^3,..., 3^{11}. [*Hint:* Use Theorem 2.10.5, part (c).]
3. Determine the units digit of the integer $7^{25} + 21^4 + 16$.
4. Use Theorem 2.10.5 to show that $6! + 1$ is divisible by 7.
5. Show that $12! + 1$ is divisible by 13.
6. Show that there is an integer x such that $x^2 \equiv -1 \pmod{13}$. Can you determine all possible integers satisfying this condition?
7. Let $f: Z \longrightarrow Z$ be defined by

    $$f(x) = a_n x^n + a_{n-1} x^{n-1} + \cdots + a_1 x + a_0, \ a_i \in Z.$$

    Show that if $a \equiv b \pmod{m}$, then $f(a) \equiv f(b) \pmod{m}$.
8. Let n be an integer with decimal representation:

    $$n = a_k 10^k + a_{k-1} 10^{k-1} + \cdots + a_1 \cdot 10 + a_0, \quad 0 \leq a_i \leq 9.$$

    (a) Prove that the remainder, when n is divided by 3, equals the remainder when $a_0 + a_1 + \cdots + a_k$ is divided by 3. Hence prove that $3 \mid n$ if and only if $3 \mid a_0 + a_1 + \cdots + a_k$.
    (b) Replace 3 in part (a) by 9.
    (c) Prove that $11 \mid n$ if and only if

    $$11 \mid (a_0 - a_1 + a_2 - \cdots + (-1)^k a_k).$$

9. Since $10 \equiv 3 \pmod 7$, $100 \equiv 2 \pmod 7$, and $1000 \equiv 6 \pmod 7$, prove that $7 \mid (1000 a_3 + 100 a_2 + 10 a_1 + a_0)$ if and only if

    $$7 \mid (a_0 + 3a_1 + 2a_2 + 6a_3).$$

10. For $x \in Z$, let

    $$f(x) = a_n x^n + \cdots + a_1 x + a_0,$$
    $$g(x) = b_n x^n + \cdots + b_1 x + b_0, \quad a_i, b_i \in Z.$$

    Show that if $c \equiv d \pmod m$ and $a_i \equiv b_i \pmod m$ for $i = 0, 1, \ldots, n$, then $f(c) \equiv g(d) \pmod m$.

2.11 Residue Classes

Theorem 2.10.4 states that the relation $a \equiv b \pmod{m}$ defines an equivalence relation on the set Z of integers. Then, by Theorem 1.8.7, this relation partitions Z into a collection of disjoint subsets. In the special case of Example 1.8.3, congruence modulo 3 determined three equivalence classes. In this section we shall prove in general that congruence modulo m, $m > 0$, determines m equivalence classes.

THEOREM 2.11.1 Let $Z_m = \{0, 1, 2, \ldots, m - 1\}$. If a and $b \in Z_m$, $a \neq b$, then $a \not\equiv b \pmod{m}$.

Proof Since $a \neq b$, we may assume without loss of generality that $a < b$. Since $0 \leq a < b \leq m - 1$, we get $0 < b - a \leq m - 1 - a < m$. Hence $b - a$ is not a multiple of m. Thus $b \not\equiv a \pmod{m}$. ∎

THEOREM 2.11.2 Let z be any integer. Then there exists one and only one integer $r \in Z_m$ such that $z \equiv r \pmod{m}$.

Proof By the division algorithm, there exist unique integers q and r such that $z = qm + r$, $0 \leq r < m$. By the definition of congruence, $z \equiv r \pmod{m}$. Clearly, $r \in Z_m$. To show that r is the only integer in Z_m such that $z \equiv r \pmod{m}$, assume $z \equiv r_1 \pmod{m}$, where $r_1 \in Z_m$. Then $z \equiv r \pmod{m}$ and $z \equiv r_1 \pmod{m}$ imply that $r \equiv r_1 \pmod{m}$. By Theorem 2.11.1, $r = r_1$. ∎

Definition 2.11.3 The set of integers $Z_m = \{0, 1, \ldots, m - 1\}$ is called the set of *least positive residues modulo m*.

THEOREM 2.11.4 If m is a positive integer, then there exist exactly m equivalence classes for the equivalence relation "congruence modulo m." These equivalence classes are $\bar{0}, \bar{1}, \ldots, \overline{m-1}$. [The equivalence class $\bar{z}$ is simply that set $\{x \mid x \in Z \text{ and } x \equiv z \pmod{m}\}$.]

Proof By Theorem 2.11.2, any $z \in Z$ is congruent to precisely one least positive residue and so $z \in \bar{r}$ for some $r \in Z_m$. Since no two distinct least positive residues are congruent modulo m, $\bar{r}_i \neq \bar{r}_j$ for $r_i \neq r_j$ and the theorem follows. ∎

Definition 2.11.5 The equivalence classes of Theorem 2.11.4 are called *residue classes modulo m*.

If x belongs to the residue class $\bar{z}$, then by Lemma 1.8.6 $\bar{x} = \bar{z}$. Thus, if $x_0 \in \bar{0}, x_1 \in \bar{1}, \ldots, x_{m-1} \in \overline{m-1}$, we see that $\{\bar{x}_0, \bar{x}_1, \ldots, \bar{x}_{m-1}\}$ consists of all the residue classes.

Definition 2.11.6 The *complete set of residue classes modulo m* is the collection $\{\bar{0}, \bar{1}, \ldots, \overline{m-1}\}$. A set of integers $\{x_0, x_1, \ldots, x_{m-1}\}$ is called a *com-*

plete set of residues if $\{\bar{x}_0, \bar{x}_1, \ldots, \bar{x}_{m-1}\}$ is the complete set of residue classes. The integer x_i is called a *representative* of the residue class $\bar{x}_i$.

To illustrate, let $m = 6$. Then $\{0,1,2,3,4,5\}$ is the set of least positive residues modulo 6. This set is also a complete set of residues modulo 6, as is the set $\{12, -5, 20, -9, 10, 35\}$, since $\overline{12} = \bar{0}$, $\overline{-5} = \bar{1}$, $\overline{20} = \bar{2}$, $\overline{-9} = \bar{3}$, $\overline{10} = \bar{4}$, and $\overline{35} = \bar{5}$.

The residue classes modulo 6 are easily seen to be:

$$\bar{0} = \{\ldots, -12, -6, 0, 6, 12, \ldots\},$$
$$\bar{1} = \{\ldots, -11, -5, 1, 7, 13, \ldots\},$$
$$\bar{2} = \{\ldots, -10, -4, 2, 8, 14, \ldots\},$$
$$\bar{3} = \{\ldots, -9, -3, 3, 9, 15, \ldots\},$$
$$\bar{4} = \{\ldots, -8, -2, 4, 10, 16, \ldots\},$$
$$\bar{5} = \{\ldots, -7, -1, 5, 11, 17, \ldots\}.$$

An important characterization of a complete set of residues is given in the next theorem.

THEOREM 2.11.7 The set $\{a_0, a_1, \ldots, a_{m-1}\}$ is a complete set of residues modulo m if and only if $a_i \not\equiv a_j \pmod{m}$, for $i \neq j$.

Proof Suppose $\{a_0, \ldots, a_{m-1}\}$ is a complete set of residues. Then if $a_i \equiv a_j \pmod{m}$, we get that $\bar{a}_i = \bar{a}_j$, and this would imply fewer than m residue classes determined by the a_i's. Thus, $a_i \not\equiv a_j \pmod{m}$, for $i \neq j$. Conversely, if $a_i \not\equiv a_j \pmod{m}$, for $i \neq j$, then reversing the above argument shows that we get m distinct residue classes; thus $\{a_0, a_1, \ldots, a_{m-1}\}$ is a complete set of residues. ∎

Another way of stating Theorem 2.11.7 is that a set of m integers is a complete set of residues if and only if no two of the integers are congruent modulo m. It is also immediate that given a complete set of residues, any integer whatsoever is congruent modulo m to one and only one of the integers in the complete set of residues.

Example 2.11.8 An interesting application of some of the ideas of this section is a proof of the fact that for all integers n, $30 \mid (n^5 - n)$. By Exercise 10, Section 2.8 (also by Exercise 7, Section 2.9), it will suffice to show that $n^5 - n$ is divisible by 2, 3, and 5. We begin by factoring as follows:

$$n^5 - n = n(n + 1)(n - 1)(n^2 + 1).$$

For any n, the integers $n - 1$, n, and $n + 1$ are consecutive, and since no two of these are congruent modulo 3, they constitute a complete set of residues, whence (exactly) one of them must be divisible by 3. For similar reasons one of these is divisible by 2. If it should happen that one of n,

$(n - 1)$, and $n + 1$ is divisible by 5, then we are finished. If none of these is divisible by 5, then either $n + 2 \equiv 0$ (mod 5) or $n + 3 \equiv 0$ (mod 5), since these two integers together with the previous three are a set of five consecutive integers. But if $n + 2 \equiv 0$ (mod 5), then $n \equiv -2$ (mod 5) and $n^2 + 1 \equiv 0$ (mod 5). If $n + 3 \equiv 0$ (mod 5), then again $n^2 + 1 \equiv 0$ (mod 5). Hence, in any event $5 \mid n^5 - n$, and we have proved that $30 \mid n^5 - n$.

Example 2.11.9 As a second application of the results of this section, we prove that the integer 1007 is not the sum of three or fewer squares; that is, there are no nonnegative integers x, y, z such that $1007 = x^2 + y^2 + z^2$. To do this, observe that each of $x, y,$ and z is congruent modulo 8 to one of 0, 1, 2, 3, 4, 5, 6, 7. Thus, the square of each must be congruent modulo 8 to one of $0^2, 1^2, 2^2, 3^2, 4^2, 5^2, 6^2, 7^2$, which in turn are congruent to 0, 1, 4, 1, 0, 1, 4, 1, respectively. Now $1007 \equiv 7$ (mod 8), so if $1007 = x^2 + y^2 + z^2$, then $x^2 + y^2 + z^2 \equiv 7$ (mod 8). But given that the only possible choices modulo 8 for x^2, y^2, z^2 come from the list 0, 1, 4, it is impossible to choose $x, y,$ and z such that $x^2 + y^2 + z^2$ is congruent to 7 modulo 8. Thus, 1007 is not the sum of three squares.

EXERCISES

1. Let $a, b \in Z$. Prove that $a \equiv b$ (mod m) if and only if each is congruent to the same least positive residue modulo m.
2. Let $m = 7$. Let x be any integer. (a) Show that $x, x + 3, x + 3^2, x + 3^3, x + 3^4, x + 3^5,$ and $x + 3^6$ is a complete set of residues modulo 7. (b) Show that $x, x + 2, x + 2^2, x + 2^3, x + 2^4, x + 2^5,$ and $x + 2^6$ is not a complete set of residues modulo 7.
3. Show that for every integer n, $11 \nmid 4(n^2 + 1)$.
4. Find all integers n such that $13 \mid 4(n^2 + 1)$.
5. Let r_n be the product of a set of n consecutive integers. Show that $n \mid r_n$.
6. Show that no integer of the form $8k + 7$ is the sum of three or fewer squares.
7. Show that the set $\{0, 1 \cdot 5, 2 \cdot 5, 3 \cdot 5, 4 \cdot 5, 5 \cdot 5, 6 \cdot 5\}$ is a complete set of residues modulo 7.
8. Let p be a prime number, and let a be an integer such that $p \nmid a$. Show that the set $\{0, 1 \cdot a, 2 \cdot a, 3 \cdot a, \ldots, (p - 1) \cdot a\}$ is a complete set of residues modulo p.
9. Let $m > 0$ and suppose $(a,m) = 1$. Show that the set

$$\{0, 1 \cdot a, 2 \cdot a, \ldots, (m - 1) \cdot a\}$$

is a complete set of residues modulo m.

2.12 Residue-Class Arithmetic

In Section 2.10 we saw that if $a \equiv b \pmod{m}$ and $c \equiv d \pmod{m}$, then $a + c \equiv b + d \pmod{m}$ and $ac \equiv bd \pmod{m}$. Using the concept of residue classes, we can rewrite this last statement as follows: if $\bar{a} = \bar{b}$ and $\bar{c} = \bar{d}$, then

$$\overline{a + c} = \overline{b + d} \quad \text{and} \quad \overline{ac} = \overline{bd}.$$

In view of the statement above we can make a very natural definition for the "addition" and "multiplication" of two residue classes:

Definition 2.12.1 Let $\bar{a}$ and $\bar{c}$ be residue classes modulo m. Define $\bar{a} + \bar{c} = \overline{a + c}$ and $\bar{a} \cdot \bar{c} = \overline{a \cdot c}$; that is, the *sum* of $\bar{a}$ and $\bar{c}$ is the residue class containing the integer $a + c$, and the *product* of $\bar{a}$ and $\bar{c}$ is the residue class containing the integer ac.

Note that we have used the symbols "$+$" and "$\cdot$" in two different ways. In $\bar{a} + \bar{c}$, the $+$ is used to represent a binary operation between residue classes, whereas in $a + c$, the $+$ is the usual addition of integers. We use $\cdot$ similarly.

From the discussion preceding Definition 2.12.1, we see that these operations are *well defined;* that is, if $\bar{a} = \bar{b}$ and $\bar{c} = \bar{d}$, then

$$\bar{a} + \bar{c} = \bar{b} + \bar{d} \quad \text{and} \quad \bar{a}\bar{c} = \bar{b}\bar{d}.$$

We now let $\bar{Z}_m$ denote the set of all residue classes of integers modulo m, with $m > 0$. Then, by Definition 2.12.1, there exist binary operations $+$ and $\cdot$ on $\bar{Z}_m$ that, we shall show, satisfy perfect analogues of all the arithmetic properties with the exception of M_4. We will formulate a modified version of M_4 that does hold. Thus, a kind of "miniature arithmetic" very similar in nature to the arithmetic of the integers exists for $\bar{Z}_m$.

THEOREM 2.12.2 Let $\bar{Z}_m$ be the set of residue classes modulo m. Then

$\bar{A}_0$: There exists a binary operation on $\bar{Z}_m$ called addition. Thus, for $\bar{a}, \bar{b} \in \bar{Z}_m$, there exists a unique residue class $\bar{a} + \bar{b} = \overline{a + b}$ called the sum of $\bar{a}$ and $\bar{b}$.

$\bar{A}_1$: If $\bar{a}, \bar{b}, \bar{c} \in \bar{Z}_m$, then $(\bar{a} + \bar{b}) + \bar{c} = \bar{a} + (\bar{b} + \bar{c})$.

$\bar{A}_2$: If $\bar{a}, \bar{b} \in \bar{Z}_m$, then $\bar{a} + \bar{b} = \bar{b} + \bar{a}$.

$\bar{M}_0$: There exists a binary operation $\cdot$ on $\bar{Z}_m$ called multiplication. Thus for $\bar{a}, \bar{b} \in \bar{Z}_m$, there exists a unique residue class $\bar{a} \cdot \bar{b} = \overline{ab}$ called the product of $\bar{a}$ and $\bar{b}$.

$\bar{M}_1$: If $\bar{a}, \bar{b}, \bar{c} \in \bar{Z}_m$, then $(\bar{a}\bar{b})\bar{c} = \bar{a}(\bar{b}\bar{c})$.

$\bar{M}_2$: If $\bar{a}, \bar{b} \in \bar{Z}_m$, then $\bar{a} \cdot \bar{b} = \bar{b} \cdot \bar{a}$.

$\bar{A}_3$: There exists an element $\bar{0}$ in $\bar{Z}_m$ such that $\bar{a} + \bar{0} = \bar{a}$, for all $\bar{a} \in \bar{Z}_m$. ($\bar{0}$ is called the *zero residue class* modulo m, or simply the *zero element* of $\bar{Z}_m$.)

$\bar{M}_3$: If $m > 1$, there exists an element $\bar{1} \in \bar{Z}_m$, $\bar{1} \neq \bar{0}$ such that $\bar{a} \cdot \bar{1} = \bar{a}$, for all $\bar{a} \in \bar{Z}_m$. ($\bar{1}$ is called the *multiplicative identity* of $\bar{Z}_m$.)

$\bar{A}_4$: Given $\bar{a} \in \bar{Z}_m$, there exists $\bar{b} \in \bar{Z}_m$ such that $\bar{a} + \bar{b} = \bar{0}$.

$\bar{M}_4$: Given $\bar{a}, \bar{b}, \bar{c} \in \bar{Z}_m$, if $(a,m) = 1$, then $\bar{a} \cdot \bar{b} = \bar{a} \cdot \bar{c}$ implies $\bar{b} = \bar{c}$.

$\bar{D}$: Given $\bar{a}, \bar{b}, \bar{c}$ in $\bar{Z}_m$, then $\bar{a}(\bar{b} + \bar{c}) = \bar{a}\bar{b} + \bar{a}\bar{c}$.

Proof $\bar{A}_0$ and $\bar{M}_0$ are consequences of Definition 2.12.1 and the discussion of that definition.

To prove $\bar{A}_1$, the associative law of addition in $\bar{Z}_m$, we see that

$$(\bar{a} + \bar{b}) + \bar{c} = \overline{(a+b)} + \bar{c} = \overline{(a+b) + c}$$
$$= \overline{a + (b+c)} \quad \text{(by the associative law for addition on } Z\text{)}$$
$$= \bar{a} + \overline{(b+c)} = \bar{a} + (\bar{b} + \bar{c}).$$

The proof of $\bar{M}_1$ is similar.

We leave the proofs of $\bar{A}_2, \bar{M}_2, \bar{A}_3, \bar{M}_3$ to the reader.

To prove $\bar{A}_4$, we observe that $\bar{a} + \overline{(-a)} = \bar{0}$. To prove $\bar{M}_4$, we note that $\bar{a} \cdot \bar{b} = \bar{a} \cdot \bar{c}$ implies $\overline{ab} = \overline{ac}$, that is, $ab \equiv ac \pmod{m}$. By Theorem 2.10.7, $ab \equiv ac \pmod{m}$ and $(a,m) = 1$ implies $b \equiv c \pmod{m}$, that is, $\bar{b} = \bar{c}$.

The proof of $\bar{D}$ is left as an exercise. ∎

Although $\bar{M}_4$ is not as strong as M_4 (see Section 2.2), it is, however, the best possible analogue. For example, if $m = 6$, then $\bar{2} \cdot \bar{3} = \bar{6} = \bar{0}$. Thus $\bar{2} \cdot \bar{3} = \bar{2} \cdot \bar{0}$, yet $\bar{3} \neq \bar{0}$. In addition, in this example we see that the product of two nonzero elements may be zero.

In case m is a prime, then $\bar{M}_4$ is the exact analogue of M_4. For if m is a prime, $\bar{M}_4$ becomes: given $\bar{a}, \bar{b}, \bar{c} \in \bar{Z}_m$, $\bar{a} \neq \bar{0}$, then $\bar{a} \cdot \bar{b} = \bar{a} \cdot \bar{c}$ implies $\bar{b} = \bar{c}$.

EXERCISES

1. Complete the proof of Theorem 2.12.2.
2. Let $\bar{a}, \bar{b} \in \bar{Z}_m$. Let $S = \{a_i + b_j \mid a_i \in \bar{a}, b_j \in \bar{b}\}$. Show that $S = \bar{a} + \bar{b}$.
3. Let $\bar{2} \in \bar{Z}_6$. Show there does not exist $\bar{x}$ in $\bar{Z}_6$ such that $\bar{2}\bar{x} = \bar{1}$.
4. Let $\bar{a} \in \bar{Z}_5$, $\bar{a} \neq \bar{0}$. Show that there exists $\bar{x} \in \bar{Z}_5$, such that $\bar{a}\bar{x} = \bar{1}$.
5. Show that it is impossible to find a subset of $\bar{Z}_m$ satisfying properties O_1 and O_2 of the integers.

84 NUMBER THEORY

2.13 Linear Congruences and Chinese Remainder Theorem

We have already noted that an equation of the form $ax = b$, $a \in Z$, $b \in Z$, has a solution in Z if and only if $a \mid b$. In this section we will study those conditions under which $ax \equiv b \pmod{m}$ has solutions.

Definition 2.13.1 Any integer x_0 satisfying the *linear congruence* $ax \equiv b \pmod{m}$ is called a *solution* of the linear congruence.

For example, 3 and 7 are solutions of $3x \equiv 1 \pmod 4$.

THEOREM 2.13.2 If $(a,m) = 1$, then the linear congruence $ax \equiv b \pmod m$ has a solution. Further, if x_0 is a solution, then the set of all solutions is precisely the equivalence class $\bar{x}_0$. Thus we say the solution is unique modulo m.

Proof Since $(a,m) = 1$, there are integers s and t such that $as + mt = 1$. Thus $asb = b - bmt$. Hence, $a(sb) \equiv b \pmod m$, or $x_0 = sb$ satisfies the linear congruence. Now if $y \in \bar{x}_0$, then $ax_0 \equiv ay \pmod m$, whence $ay \equiv b \pmod m$ and y is a solution. Conversely, if $ay \equiv b \pmod m$, then $ax_0 \equiv ay \pmod m$ and, by Theorem 2.10.7, $x_0 \equiv y \pmod m$, that is, $y \in \bar{x}_0$. ∎

COROLLARY 2.13.3 Let a, b, and p be integers, p a prime, $p \nmid a$. Then $ax \equiv b \pmod p$ always has a solution, which is unique modulo p.

Proof Since $(a,p) = 1$, apply Theorem 2.13.2. ∎

We see from Corollary 2.13.3, that for a prime modulus p, a linear congruence $ax \equiv b \pmod p$ has a solution as long as $a \not\equiv 0 \pmod p$. This is the direct analogue to the condition that $rx = s$, r, s rational numbers, has a solution in rational numbers if $r \neq 0$. Thus, "division" by any "nonzero" element is possible modulo a prime. Hence, as in the rational number system, any "nonzero" element has an "inverse."

Definition 2.13.4 a and a' are *inverses modulo* m if $aa' \equiv 1 \pmod m$.

COROLLARY 2.13.5 A number a has an inverse modulo m if and only if $(a,m) = 1$. If a has an inverse, say a', a' is unique modulo m.

Proof (a) If $(a,m) = 1$, then by Theorem 2.13.2 $ax \equiv 1 \pmod m$ has a solution, and this is an inverse of a. (b) Now suppose a has an inverse a'. Then $aa' \equiv 1 \pmod m$ and $aa' - 1 = km$, for some integer k, or $aa' - km = 1$. By Theorem 2.7.7, $(a,m) = 1$. (c) Uniqueness of a'. If a' and a'' are both inverses of a, then a' and a'' both satisfy $ax \equiv 1 \pmod m$, hence $aa' \equiv aa'' \pmod m$. Since $(a,m) = 1$, $a' \equiv a'' \pmod m$ by Theorem 2.10.7. ∎

2.13 LINEAR CONGRUENCES AND CHINESE REMAINDER THEOREM

Our next result deals with the solution of a special class of simultaneous congruences in one unknown. The solution to the more general problem (see Exercise 5) was known by the Chinese Mathematician Sun-Tsu in the first century A.D..

THEOREM 2.13.6 (*Chinese Remainder Theorem*) If $(m_1,m_2) = 1$, then the congruences (1) $x \equiv a_1 \pmod{m_1}$ and (2) $x \equiv a_2 \pmod{m_2}$ have a common solution, which is unique modulo $m_1 m_2$.

Proof (1) has a solution, since $a_1 + km_1$ satisfies (1) for all integers k. All that remains is to prove there is an integer k_1 such that

$$a_1 + k_1 m_1 \equiv a_2 \pmod{m_2}.$$

This is equivalent to showing that $m_1 k \equiv a_2 - a_1 \pmod{m_2}$ has a solution for k. But, since $(m_1,m_2) = 1$, this last congruence has a solution by Theorem 2.13.2.

Now suppose x_0 and x_1 are both common solutions to (1) and (2). Then $x_0 \equiv a_1 \pmod{m_1}$, and $x_1 \equiv a_1 \pmod{m_1}$. This implies that

$$x_0 - x_1 \equiv 0 \pmod{m_1},$$

or that $m_1 \mid (x_0 - x_1)$. Similarly, $m_2 \mid (x_0 - x_1)$. Thus $m_1 m_2 \mid (x_0 - x_1)$, since $(m_1,m_2) = 1$, by Exercise 10, Section 2.8, or Exercise 7, Section 2.9. This means simply that $x_0 \equiv x_1 \pmod{m_1 \cdot m_2}$, and we have proven the uniqueness of the solution modulo $m_1 m_2$. ∎

We illustrate the Chinese remainder theorem by solving a system of two simultaneous linear congruences.

Solve the congruences

$$x \equiv 7 \pmod{5} \quad \text{and} \quad x \equiv 3 \pmod{6}.$$

Since $(5,6) = 1$, we know a solution exists. Call it x_0. Then $x_0 \equiv 7 \pmod 5$, so $x_0 = 7 + 5t$, for an appropriate t. Substituting into the second congruence, $7 + 5t \equiv 3 \pmod 6$. Thus, $5t \equiv -4 \equiv 2 \pmod 6$. Now $5 \equiv -1 \pmod 6$, so this becomes $-t \equiv 2 \pmod 6$, that is, $t \equiv -2 \pmod 6$, or $t \equiv 4 \pmod 6$. Then $x_0 = 7 + 5 \cdot 4 = 27$ satisfies both congruences. This is easily verified.

Solve $x \equiv 3 \pmod 4$ and $x \equiv 7 \pmod 9$. Again, a solution exists, since $(4,9) = 1$. The solution must be of the form $3 + 4t$, from the first congruence. From the second congruence we get

$$3 + 4t \equiv 7 \pmod 9, \quad \text{whence } 4t \equiv 4 \pmod 9.$$

Since $(4,9) = 1$, we can divide out the coefficient 4 of $4t$ in this congruence, getting $t \equiv 1 \pmod 9$, so with $t = 1$, we see that $3 + 4 \cdot 1 = 7$ satisfies both congruences.

These ideas can be extended to solve three (or more) linear congruences. Solve $x \equiv 2 \pmod 3$, $x \equiv 4 \pmod 5$, and $x \equiv 4 \pmod 8$. Note that the moduli are relatively prime in pairs.

The solution must be of the form $2 + 3t$ from the first congruence. Substituting into the second, $2 + 3t \equiv 4 \pmod 5$, whence $3t \equiv 2 \pmod 5$. Multiplying both sides by 2 (you can multiply through only by integers a such that $(a,5) = 1$ in order not to introduce extraneous solutions), we have $6t \equiv 4 \pmod 5$, that is $t \equiv 4 \pmod 5$. Thus t is of the form $4 + 5s$, so the solution is of the form $2 + 3(4 + 5s) = 14 + 15s$. Substituting into the last congruence, $14 + 15s \equiv 4 \pmod 8$, whence $15s \equiv -10 \equiv 6 \pmod 8$. Since $15 \equiv -1 \pmod 8$, we have $-s \equiv 6 \pmod 8$, and so $s \equiv -6 \equiv 2 \pmod 8$. With $s = 2$, the solution now is $14 + 15 \cdot 2 = 44$. It is easy to see this satisfies all the congruences.

EXERCISES

1. Solve the simultaneous congruences:
   (a) $x \equiv 7 \pmod{21}$, $x \equiv 3 \pmod 8$.
   (b) $x \equiv 4 \pmod{15}$, $x \equiv 7 \pmod{16}$.
   (c) $x \equiv 2 \pmod{61}$, $x \equiv 29 \pmod{50}$.
2. Solve the simultaneous congruences:
   (a) $x \equiv 1 \pmod{11}$, $x \equiv 1 \pmod 6$, $x \equiv 3 \pmod 7$.
   (b) $x \equiv 3 \pmod{11}$, $x \equiv 2 \pmod{12}$, $x \equiv 7 \pmod{17}$.
3. Solve each of the following congruences:
   (a) $3x \equiv 4 \pmod 5$.
   (b) $8x \equiv 3 \pmod{27}$.
4. Let $\bar{a}, \bar{b} \in \bar{Z}_p$, $\bar{b} \neq \bar{0}$, and p a prime. Show that $\bar{b}\bar{x} = \bar{a}$ has a solution in $\bar{Z}_p$.
5. Prove the following generalization of Theorem 2.13.2: The linear congruence $ax \equiv b \pmod m$ has a solution if and only if $d \mid b$, where $d = (a,m)$. Moreover, if a solution exists, there is a unique solution (mod m_1), where $m_1 = m/d$, and thus there are exactly d solutions modulo m, that is, there are exactly d solutions x_i, $0 \le x_i < m$, no two of which are congruent modulo m.
6. Prove the following more general case of the Chinese remainder theorem: If $(m_i, m_j) = 1$, $i \neq j$, $i, j = 1, 2, \ldots, n$, then $x \equiv a_1 \pmod{m_1}$, $x \equiv a_2 \pmod{m_2}$, \ldots, $x \equiv a_n \pmod{m_n}$ have a common solution, and this solution is unique modulo $m_1 \cdot m_2 \cdots m_n$.

2.14 Euler φ-Function

The least positive residues modulo m that have inverses modulo m are those relatively prime to m. An important function that counts the number of these integers is called the *Euler φ-function*, or the *totient*. It will occur in several applications.

Definition 2.14.1 Let $m \geq 1$. The *Euler φ-function* is the function φ with domain P, the set of positive integers, defined as follows: $\varphi(m)$ equals the number of integers in Z_m that are prime to m.

To illustrate, $\varphi(1) = 1$, $\varphi(2) = 1$, $\varphi(4) = 2$, and if p is any prime $\varphi(p) = p - 1$.

Definition 2.14.2 The residue class $\bar{r}$ modulo m is *prime to m* if $(r,m) = 1$.

We note that if $x \in \bar{r}$, then $(x,m) = (r,m)$. Hence Definition 2.14.2 is independent of the choice of the residue class representative.

As a consequence of Definitions 2.14.1 and 2.14.2, $\varphi(m)$ is equal to the number of residue classes prime to m.

Definition 2.14.3 A *reduced set of residues modulo m* is a set of integers $\{r_1, r_2, \ldots, r_{\varphi(m)}\}$ such that exactly one of them lies in each residue class prime to m. If each r_i satisfies $0 \leq r_i < m$, we call $\{r_1, r_2, \ldots, r_{\varphi(m)}\}$ the *reduced set of least positive residues modulo m*.

As an example, suppose $m = 6$. Then $\{0,1,2,3,4,5\}$ is a complete set of residues modulo 6, and $\{1,5\}$ is a reduced set of residues module 6. In addition, $\{6,13,26,39,10,17\}$ is also a complete set of residues, while $\{13,17\}$ is a reduced set of residues modulo 6.

EXERCISES

1. (a) Prove that if A and B are two complete sets of residues modulo m, then there is a 1–1 correspondence between A and B such that corresponding pairs are congruent modulo m.
   (b) In (a) let A and B be reduced sets of residues.
2. Prove that a necessary and sufficient condition that a set of integers is a reduced set of residues modulo m is that each be prime to m, that no two be congruent to each other, and that they be $\varphi(m)$ in number.

In deriving an explicit formula for $\varphi(m)$, we shall proceed in two steps. First, we shall determine $\varphi(p^k)$ explicitly where p is a prime and k a positive integer. Second, we shall show that if $(m_1, m_2) = 1$, then $\varphi(m_1 \cdot m_2) = \varphi(m_1) \cdot \varphi(m_2)$. Then combining these two steps with the fundamental theorem of arithmetic, we shall derive the complete formula for $\varphi(m)$.

THEOREM 2.14.4 Let p be a prime number, $k \in Z$, $k > 0$. Then

$$\varphi(p^k) = p^k \left(1 - \frac{1}{p}\right).$$

Proof We need count only those integers in the set

$$S' = \{0, 1, 2, \ldots, p^k - 1\}$$

that are prime to p^k, or equivalently in the set $S = \{1, 2, \ldots, p^k - 1, p^k\}$ those that are prime to p^k. We shall count the number of integers in S that are not prime to p^k and subtract that number from p^k. Now, if $a \in S$ and $(a,p^k) \neq 1$, then $p \mid (a,p^k)$ and also $p \mid a$, since p is a prime. Thus, we may write $a = np$. But then $p \mid a$ for any value of n satisfying $1 \leq n \leq p^{k-1}$ and so there are p^{k-1} integers in S not prime to p^k. Thus

$$\varphi(p^k) = p^k - p^{k-1} = p^k \left(1 - \frac{1}{p}\right). \quad \blacksquare$$

THEOREM 2.14.5 If $(m_1,m_2) = 1$, m_1, m_2 positive integers, then

$$\varphi(m_1 \cdot m_2) = \varphi(m_1) \cdot \varphi(m_2).$$

Proof If either m_1 or m_2 is 1, the proof is trivial. Thus, we may assume $m_1 > 1$, $m_2 > 1$.

Now let $X = \{x_1, x_2, \ldots, x_{\varphi(m_1)}\}$ be the reduced set of least positive residues modulo m_1 and let $Y = \{y_1, y_2, \ldots, y_{\varphi(m_2)}\}$ be the reduced set of least positive residues modulo m_2. Finally, let $W = \{w_1, w_2, \ldots, w_{\varphi(m_1 m_2)}\}$ be the reduced set of least positive residues modulo $m_1 m_2$. We shall show that the number of elements in $X \times Y$ is equal to the number of elements in W, which will establish the result. This will be done by showing that to each (x_i, y_j) in $X \times Y$ there exists a unique w_k in W, and that to each w_k in W there exists an element (x_i, y_j) in $X \times Y$. Moreover, we shall see that if we start with (x_i, y_j) and get w_k, then the pair (x_m, y_n) determined by w_k will be precisely the original (x_i, y_j). That is, we shall establish that there exists a 1–1 correspondence between $X \times Y$ and W.

(a) Thus let $x_i \in X$, $y_j \in Y$. Then the congruences

$$w \equiv x_i \pmod{m_1} \quad \text{and} \quad w \equiv y_j \pmod{m_2}$$

have a common solution w_k, since $(m_1, m_2) = 1$, and this solution is unique modulo $m_1 m_2$ (Chinese remainder theorem). Moreover, $(w_k, m_1 m_2) = 1$, by Theorem 2.7.9. Thus, $w_k \in W$; that is, the pair (x_i, y_j), $x_i \in X$, $y_j \in Y$ determines a unique element w_k of W.

(b) Conversely, suppose that $w_k \in W$. Then clearly, $(w_k, m_1 m_2) = 1$, and so again by Theorem 2.7.9 $(w_k, m_1) = (w_k, m_2) = 1$. By Theorem 2.13.2 there is a unique element $x_i \in X$ such that $x_i \equiv w_k \pmod{m_1}$, and there

2.15 THE THEOREMS OF FERMAT AND EULER

is a unique element $y_j \in Y$ such that $y_j \equiv w_k \pmod{m_2}$. Thus, any $w_k \in W$ gives rise to a unique pair (x_i, y_j), where $x_i \in X$, $y_j \in Y$.

Now let w_k be the element that corresponds to (x_i, y_j) in (a), and let (x_m, y_n) be the ordered pair that corresponds to w_k in (b). We see that by (a), $w_k \equiv x_i \pmod{m_1}$, and in part (b) we have that $x_m \equiv w_k \pmod{m_1}$. Thus, $x_i \equiv x_m \pmod{m_1}$, that is, $x_i = x_m$. Similarly, $y_j = y_n$, and the proof is complete. ∎

THEOREM 2.14.6 If $m = p_1^{e_1} p_2^{e_2} \cdots p_s^{e_s}$ is the prime power factorization of m, then

$$\varphi(m) = m\left(1 - \frac{1}{p_1}\right)\left(1 - \frac{1}{p_2}\right) \cdots \left(1 - \frac{1}{p_s}\right).$$

Proof The proof is left as an exercise ∎

EXERCISES

3. Find $\varphi(36)$, $\varphi(81)$, $\varphi(101)$.
4. Prove Theorem 2.14.6.
5. Show that $\varphi(n)$ is even for all $n > 2$.
6. Find all integers n such that $\varphi(n) = 2$.
7. Let p^α be a power of the prime p. Show that

$$\varphi(1) + \varphi(p) + \cdots + \varphi(p^\alpha) = p^\alpha.$$

8. Let m be a positive integer. Show that

$$\sum_{d|m} \varphi(d) = m, \quad \text{where} \sum_{d|m} \varphi(d) = \varphi(d_1) + \varphi(d_2) + \cdots + \varphi(d_r)$$

with $\{d_1, d_2, \ldots, d_r\}$ the set of all positive divisors of m, and $d_i \neq d_j$, for $i \neq j$.

9. Find all pairs of positive integers m, p such that $\varphi(m) = m/p$, where p is a prime.
10. Let $n > 1$ and let k_i, $i = 1, \ldots, \varphi(n)$, be the integers satisfying $0 < k_i < n$ and $(k_i, n) = 1$. Show that

$$k_1 + k_2 + \cdots + k_{\varphi(n)} = \left(\frac{n}{2}\right) \cdot \varphi(n).$$

2.15 The Theorems of Fermat and Euler

The theorems of Fermat and Euler proved in this section will be used for illustrative purposes in Chapter 4, although different proofs will be given there.

THEOREM 2.15.1 (*Euler*) Let a and m be integers, $m > 0$. If $(a,m) = 1$, then $a^{\varphi(m)} \equiv 1 \pmod{m}$.

Proof Let $r_1, r_2, \ldots, r_{\varphi(m)}$ be a reduced set of residues modulo m. Now if $ar_i \equiv ar_j \pmod{m}$, for some i and j, $i \neq j$, then since $(a,m) = 1$, we have $r_i \equiv r_j \pmod{m}$ by Theorem 2.10.7, contradicting the statement that $r_1, \ldots, r_{\varphi(m)}$ is a reduced set of residues modulo m. Hence $ar_i \not\equiv ar_j \pmod{m}$, for all i, j, $i \neq j$. Moreover for each i, by Theorem 2.7.9, we have $(ar_i, m) = (a,m)(r_i,m) = 1 \cdot 1 = 1$. Thus, since the integers $ar_1, \ldots, ar_{\varphi(m)}$ are $\varphi(m)$ in number, each prime to m, and since no two are congruent modulo m, they form a reduced set of residues modulo m by Exercise 2, Section 2.14. By Exercise 1, Section 2.14, the ar_i's and the r_j's can be paired off into congruent pairs modulo m. Thus, by Theorem 2.10.5,

$$ar_1 \cdot ar_2 \cdots ar_{\varphi(m)} \equiv r_1 r_2 \cdots r_{\varphi(m)} \pmod{m},$$

whence

$$a^{\varphi(m)} r_1 r_2 \cdots r_{\varphi(m)} \equiv r_1 r_2 \cdots r_{\varphi(m)} \pmod{m}.$$

Since $(r_i, m) = 1$, for all i, clearly $(r_1 r_2 \cdots r_{\varphi(m)}, m) = 1$, and thus $a^{\varphi(m)} \equiv 1 \pmod{m}$ by Theorem 2.10.7. ∎

COROLLARY 2.15.2 If p is a prime and if a is an integer such that $p \nmid a$, then $a^{p-1} \equiv 1 \pmod{p}$.

Proof Let $m = p$ in Theorem 2.15.1. ∎

THEOREM 2.15.3 (*Fermat*) If p is a prime, then $a^p \equiv a \pmod{p}$, for all integers a.

Proof If $a \equiv 0 \pmod{p}$, then $a^p \equiv 0 \equiv a \pmod{p}$. If $a \not\equiv 0 \pmod{p}$, then $(a,p) = 1$. By Corollary 2.15.2, $a^{p-1} \equiv 1 \pmod{p}$. Then

$$a^p \equiv a \pmod{p}. \qquad ∎$$

EXERCISES

1. Let $a, u, m \in Z$ be such that $(a,m) = 1$ and u is the smallest positive integer such that $a^u \equiv 1 \pmod{m}$. Prove that $u \mid \varphi(m)$.
2. Let a and b be relatively prime integers. Show that there exist integers m and n such that $a^m + b^n \equiv 1 \pmod{ab}$.

3 Permutations

In this chapter we will study a special class of functions called permutations, and we will see that composition of functions defines a binary operation on sets of permutations. Our main interest lies in the examples permutations provide for our study of groups, and for a very important role they play in group theory.

3.1 Permutations: Introduction

Definition 3.1.1 Let A be a set, $A \neq \emptyset$. A *permutation* T on the set A is a function $T: A \xrightarrow[\text{onto}]{1-1} A$.

THEOREM 3.1.2 Let T_1, T_2, T_3 be permutations on a set A. Then

(a) $T_1 \circ T_2$ is a permutation on A,
(b) $T_1 \circ (T_2 \circ T_3) = (T_1 \circ T_2) \circ T_3$.

Proof (a) See Exercise 7, Section 1.7; (b) see Theorem 1.7.5. ∎

Of special interest are the permutations on the set $X_n = \{1,2,3,\ldots,n\}$. We denote the set of all permutations on X_n by S_n.

Let $T \in S_n$. For each integer $j \in X_n$, $T(j) = Tj$ is an integer in X_n. Moreover, since T is 1-1, $T(j) = T(h)$ if and only if $j = h$. Since T is also onto, we now easily observe that the set

$$\{T(1), T(2), \ldots, T(n)\} = X_n.$$

Therefore T can be thought of as effecting a rearrangement of the integers $1, 2, \ldots, n$.

92 PERMUTATIONS

One simple way to display the permutation T on X_n is to write

(1) $$T = \begin{pmatrix} 1 & 2 & \cdots & n \\ T(1) & T(2) & \cdots & T(n) \end{pmatrix},$$

where the image Tj of j under T is written directly below j. For example,

$$T = \begin{pmatrix} 1 & 2 & 3 \\ 3 & 1 & 2 \end{pmatrix}$$

is the permutation on X_3 defined by $T(1) = 3$, $T(2) = 1$, $T(3) = 2$. Of course, any array of the form

$$\begin{pmatrix} 1 & 2 & \cdots & n \\ j_1 & j_2 & \cdots & j_n \end{pmatrix},$$

where $j_1, \ldots, j_n$ is just a rearrangement of X_n, defines a permutation.

THEOREM 3.1.3 There exist $n!$ permutations on X_n; that is, S_n has $n!$ elements.

Proof There are n choices for $j_1 = T(1)$, then $(n-1)$ choices for $j_2 = T(2)$, $(n-2)$ for $j_3 = T(3)$, and so forth, yielding $n!$ choices altogether. ∎

The notation (1) provides us with a ready means of finding the composite, or product, of two permutations. Suppose

$$T = \begin{pmatrix} 1 & 2 & \cdots & n \\ T(1) & T(2) & \cdots & T(n) \end{pmatrix} \text{ and } U = \begin{pmatrix} 1 & 2 & \cdots & n \\ U(1) & U(2) & \cdots & U(n) \end{pmatrix}.$$

To find $U \circ T$ we must determine $(U \circ T)(j)$ for each j in X_n. Since $(U \circ T)(j) = U(T(j))$, we find $T(j)$ under j in the display for T, and then we find $U(Tj)$ under $T(j)$ in the display for U.

For example, let

$$T = \begin{pmatrix} 1 & 2 & 3 & 4 \\ 1 & 4 & 2 & 3 \end{pmatrix} \text{ and } U = \begin{pmatrix} 1 & 2 & 3 & 4 \\ 3 & 4 & 1 & 2 \end{pmatrix}.$$

Then to find $U \circ T$ we trace out the successive images of each $j \in X_n$ as follows:

$$T = \begin{pmatrix} 1 & 2 & 3 & 4 \\ 1 & 4 & 2 & 3 \end{pmatrix}$$

$$U = \begin{pmatrix} 1 & 2 & 3 & 4 \\ 3 & 4 & 1 & 2 \end{pmatrix}.$$

Thus, $U \circ T = \begin{pmatrix} 1 & 2 & 3 & 4 \\ 3 & 2 & 4 & 1 \end{pmatrix}$

Moreover,

$$T \circ U = \begin{pmatrix} 1 & 2 & 3 & 4 \\ 1 & 4 & 2 & 3 \end{pmatrix} \circ \begin{pmatrix} 1 & 2 & 3 & 4 \\ 3 & 4 & 1 & 2 \end{pmatrix} = \begin{pmatrix} 1 & 2 & 3 & 4 \\ 2 & 3 & 1 & 4 \end{pmatrix}.$$

We note that $U \circ T \neq T \circ U$.

EXERCISES

1. Let

$$T = \begin{pmatrix} 1 & 2 & 3 & 4 & 5 \\ 2 & 1 & 4 & 3 & 5 \end{pmatrix}, \quad U = \begin{pmatrix} 1 & 2 & 3 & 4 & 5 \\ 3 & 2 & 1 & 5 & 4 \end{pmatrix}, \quad V = \begin{pmatrix} 1 & 2 & 3 & 4 & 5 \\ 5 & 4 & 2 & 3 & 1 \end{pmatrix}$$

be in S_5. Find each of the following: $T \circ U$, $T \circ V$, $V \circ U$, $U \circ V$, $T \circ (U \circ V)$, $(T \circ V) \circ V$, T^2, V^2.

2. List the elements of S_3 in the form

$$\begin{pmatrix} 1 & 2 & 3 \\ T_i(1) & T_i(2) & T_i(3) \end{pmatrix}, \quad i = 1, \ldots, 6,$$

and find T_1 and T_2 such that $T_1 \circ T_2 \neq T_2 \circ T_1$.

Exercises 3 through 10 foreshadow events of Chapter 4. The reader should use induction, where necessary, or approach these exercises from an intuitive point of view. Special cases may clarify the more general situation in some of these exercises.

3. Let A be a set, $A \neq \emptyset$. Let T be a permutation on A. Show that T^{-1} exists and that T^{-1} is also a permutation on A.
4. Let $T \in S_n$. For $m \in Z$, $m > 0$, define T^m by

$$\underbrace{T \circ T \circ \cdots \circ T}_{m \ T\text{'s}}.$$

Show $T^m \in S_n$.

5. Let $T \in S_n$ and let $m \in Z$, $m > 0$. Show that $(T^m)^{-1}$ is also in S_n.
6. Let $i \in X_n$ and let $T \in S_n$. Suppose that $T(i) = i$. Prove that for all integers m, $T^m(i) = i$.
7. Let $T \in S_n$. For $m \geq 0$, define $T^{-m} = (T^m)^{-1}$. Show that

$$T^{-m} = (T^{-1})^m.$$

[Thus $T^{-m} = (T^{-1})^m = (T^m)^{-1}$, for all integers m.]

8. Let k, l be integers, and let $T \in S_n$. Show that $T^k \circ T^l = T^{k+l}$.
9. Let T be a permutation. Show that $(T^k)^l = T^{kl}$, for all integers k and l.

3.2 Cycles and Cyclic Decomposition

In this section we will study a class of permutations called cycles. The importance of cycles lies in the fact that any permutation can be written uniquely, up to order of factors, as a product of disjoint cycles.

Definition 3.2.1 A permutation T in S_n is called a *cycle* if those integers x in X_n such that $Tx \neq x$ can be arranged in an order $x_1, x_2, \ldots, x_k$ so that $T(x_1) = x_2$, $T(x_2) = x_3$, $\ldots$, $T(x_{k-1}) = x_k$, $T(x_k) = x_1$. An integer x such that $Tx \neq x$ is said to be *moved by* T, and an integer x such that $Tx = x$ is said to be *left fixed by* T.

The permutation

$$T_1 = \begin{pmatrix} 1 & 2 & 3 & 4 & 5 & 6 \\ 1 & 3 & 5 & 4 & 2 & 6 \end{pmatrix} \in S_6$$

is an example of a cycle, for the elements in X_6 moved by T_1 are 2, 3, and 5, and clearly $T_1(2) = 3$, $T_1(3) = 5$, and $T_1(5) = 2$. Also, the permutation

$$T_2 = \begin{pmatrix} 1 & 2 & 3 & 4 & 5 & 6 \\ 3 & 1 & 4 & 2 & 5 & 6 \end{pmatrix} \in S_6$$

is a cycle, for we write the integers moved by T_2 in the order 1, 3, 4, 2 and observe that $T_2(1) = 3$, $T_2(3) = 4$, $T_2(4) = 2$, and $T_2(2) = 1$.

The permutation

$$T_3 = \begin{pmatrix} 1 & 2 & 3 & 4 & 5 & 6 \\ 2 & 3 & 4 & 1 & 6 & 5 \end{pmatrix}$$

is not a cycle, for no matter how we try to arrange the elements moved by T_3, we cannot meet the conditions of the definition. Note, however, that T_3 is the product of the two cycles U and V,

$$U = \begin{pmatrix} 1 & 2 & 3 & 4 & 5 & 6 \\ 2 & 3 & 4 & 1 & 5 & 6 \end{pmatrix}, \quad V = \begin{pmatrix} 1 & 2 & 3 & 4 & 5 & 6 \\ 1 & 2 & 3 & 4 & 6 & 5 \end{pmatrix}.$$

What occurs in T_3 is that basically two subsets of X_6 go into "loops" under T_3, namely, $\{1,2,3,4\}$ and $\{5,6\}$; that is, T_3 acts cyclically on these two sets simultaneously.

If T is a cycle and if $T(x_1) = x_2$, $T(x_2) = x_3$, $\ldots$, $T(x_k) = x_1$, then it follows that if $Ti \neq i$ and $Tj \neq j$, there is an integer m such that $T^m(i) = j$. (See Section 1.7 and Exercises 4-9 of Section 3.1.) We observe that $T^2(x_1) = x_3$, $T^3(x_2) = x_5$, and, in general, $T^m(x_i) = x_{i \oplus m}$, where $i \oplus m$ is taken to be that residue modulo k between 1 and k congruent to $i + m$. The reader should verify these claims for the examples given above.

The definition of a cycle implies that the identity permutation is also a cycle. For there is no subset of integers x such that $Tx \neq x$ and hence the set of all such integers (the empty set $\varnothing$) can be appropriately ordered.

Definition 3.2.2 Let T be a cycle in S_n.

(a) If T is not the identity permutation, then T is said to be a cycle of *length* r if T moves precisely r elements of X_n.

(b) If T is the identity permutation, then T is said to be a cycle of *length* 1.

We note that if the cycle T is not the identity, then the length r of T must be greater than 1, for no permutation moves precisely one element of X_n.

Let T be a cycle of length r. If $r = 1$, then T is the identity, and we shall often denote T by $T = (1)$ or $T = (i)$ for any $i = 1, \ldots, n$. If $r \geq 2$, the set A of elements moved by T is $\{x_1, Tx_1, T^2x_1, \ldots, T^{r-1}x_1\}$, for any $x_1 \in A$. In view of this, we shall frequently denote the cycle T of length r by $T = (x_1 \ Tx_1 \ \ldots \ T^{r-1}x_1)$. In this notation, we see that the image of an element y moved by T is the element immediately to the right of y, or if y is at the extreme right of the list, Ty is the element at the extreme left of the list. Given any subset $\{y_1, \ldots, y_k\} \subseteq X_n - A$ and given any $x \in A$, we agree we may also denote T by

$$(x \ Tx \ \cdots \ T^{r-1}x) \ (y_1) \ \cdots \ (y_k),$$

since each (y_i) denotes the identity. To illustrate, let

$$T = \begin{pmatrix} 1 & 2 & 3 & 4 & 5 & 6 & 7 \\ 3 & 4 & 2 & 5 & 1 & 6 & 7 \end{pmatrix} \in S_7.$$

Then T is a cycle of length 5, which may be denoted equally well by any of the following:

(1 3 2 4 5), (2 4 5 1 3), (5 1 3 2 4),
(1 3 2 4 5)(7), (6)(1 3 2 4 5), (6)(7)(3 2 4 5 1).

Note that cycles of length 1 can be added or deleted without affecting the permutation. Although it is unclear to which S_n the cycle T belongs without the addition of the missing letters, we shall usually rely on context to suggest the proper S_n; that is (1 3 2 4 5) may be in S_5 or S_{18}, say, depending on context.

Definition 3.2.3 Let $T, U \in S_n$ be cycles of lengths r and s, respectively. If $r > 1$ and $s > 1$, we say that T and U are *disjoint* cycles if $A \cap B = \varnothing$, where A and B are the sets of elements moved by T and U, respectively. If either $r = 1$ or $s = 1$, we also say T and U are *disjoint*.

With T, U, A, and B as in the definition above, we observe that if $r > 1$ and if $x \in A$, then (a) x is moved by T, that is, $T(x) \neq x$, and (b) $x \notin B$, whence U leaves x fixed, that is $Ux = x$. Thus, it will be easy for the reader to verify (Exercise 2 below) that if T and U are disjoint cycles, then $T \circ U = U \circ T$. We see, for example, that if

$$T = \begin{pmatrix} 1 & 2 & 3 & 4 & 5 & 6 \\ 3 & 2 & 5 & 4 & 1 & 6 \end{pmatrix} \text{ and } U = \begin{pmatrix} 1 & 2 & 3 & 4 & 5 & 6 \\ 1 & 6 & 3 & 4 & 5 & 2 \end{pmatrix},$$

then $A = \{1,3,5\}$ and $B = \{2,6\}$ and T and U are disjoint. Thus we know that $T \circ U = U \circ T$, without even finding these products.

EXERCISES

1. Prove that the cycles T and U are disjoint if and only if whenever $Ti \neq i$, then $Ui = i$, and whenever $Ui \neq i$, then $Ti = i$.
2. Let T and $U \in S_n$, T and U disjoint cycles. Show that $T \circ U = U \circ T$.
3. Let V_1, V_2 be disjoint cycles in S_n such that $V_1 \circ V_2 = (1)$. Show that $V_1 = V_2 = (1)$. Extend this to a product of l cycles, $l \geq 2$, disjoint in pairs.
4. Let T be a cycle of length $r > 1$. Show that if $Ti \neq i$, then $T^n i \neq T^{n-1} i$, for all integers $n > 0$.

If T is a cycle of length $r > 1$ and U a cycle of length $s > 1$, with $T = (y_1 \cdots y_r)$, $U = (z_1 \cdots z_s)$, we write

$$T \circ U = (y_1 \cdots y_r)(z_1 \cdots z_s).$$

Let

$$T = \begin{pmatrix} 1 & 2 & 3 & 4 & 5 \\ 2 & 3 & 1 & 4 & 5 \end{pmatrix} \text{ and } U = \begin{pmatrix} 1 & 2 & 3 & 4 & 5 \\ 1 & 4 & 3 & 5 & 2 \end{pmatrix}$$

be in S_5. Then T and U are cycles, $T = (1\ 2\ 3)$ and $U = (2\ 4\ 5)$. Using our original notation, we see that

$$T \circ U = \begin{pmatrix} 1 & 2 & 3 & 4 & 5 \\ 2 & 3 & 1 & 4 & 5 \end{pmatrix}\begin{pmatrix} 1 & 2 & 3 & 4 & 5 \\ 1 & 4 & 3 & 5 & 2 \end{pmatrix} = \begin{pmatrix} 1 & 2 & 3 & 4 & 5 \\ 2 & 4 & 1 & 5 & 3 \end{pmatrix}$$
$$= (1\ 2\ 4\ 5\ 3).$$

Using cycle notation, we find the product as follows:

$$T \circ U = (1\ 2\ 3)(2\ 4\ 5).$$

The cycle U takes 1 to 1 and T takes 1 to 2, whence the product $T \circ U$ takes 1 to 2. U takes 2 to 4 and T leaves 4 fixed, whence $T \circ U$ takes 2 to 4. Similarly, U takes 4 to 5 and T leaves 5 fixed. Then U takes 5 to 2 and T takes 2 to 3. Finally, U leaves 3 fixed and T takes 3 to 1. Thus, we see that the product $T \circ U$ takes 1 to 2, 2 to 4, 4 to 5, 5 to 3, and 3 to 1; and so $T \circ U = (1\ 2\ 4\ 5\ 3)$. Despite this long description, the reader will find it more convenient to multiply permutations using cycle notation. As another example: (1 2 3)(1 4)(1 3) = (1)(2 3 4).

A product of cycles is clearly a single permutation. Conversely, we shall prove that any permutation can be written as a product of cycles. This will provide a very convenient method for expressing and multiplying permutations.

Example 3.2.4 Let

$$T = \begin{pmatrix} 1 & 2 & 3 & 4 & 5 & 6 & 7 & 8 \\ 2 & 4 & 5 & 1 & 3 & 7 & 8 & 6 \end{pmatrix}.$$

Then we may write T as a product of disjoint cycles as follows:

$$T1 = 2, \quad T2 = 4 \quad \text{and} \quad T4 = 1.$$

Hence one of the cycles will be (1 2 4).

$$T3 = 5 \quad \text{and} \quad T5 = 3.$$

Therefore a second cycle will be (3 5).

$$T6 = 7, \quad T7 = 8, \quad \text{and} \quad T8 = 6.$$

The final cycle will be (6 7 8). Thus T = (1 2 4)(3 5)(6 7 8).

THEOREM 3.2.5 Let $T \in S_n$, $T \neq (1)$, the identity permutation. Then T can be uniquely written as a product of disjoint cycles. That is, there exist cycles $T_1, T_2, \ldots, T_s \in S_n$, T_i and T_j disjoint for $i \neq j$, each of length at least 2, such that $T = T_1 \circ T_2 \circ \cdots \circ T_s$. Moreover, if $U_1, U_2, \ldots, U_m$ is another set of disjoint cycles each of length at least 2 such that

$$T = U_1 \circ U_2 \circ \cdots \circ U_m,$$

then $m = s$ and, for some arrangement $U_{j_1}, \ldots, U_{j_s}$ of the U_j's, $U_{j_i} = T_i$, $i = 1, \ldots, s$.

Proof I. *Existence.* Let $T \in S_n$, $T \neq (1)$. Let x_1 be the smallest integer in X_n moved by T. Then with $x_2 = T(x_1)$, $x_3 = T(x_2)$, ..., we eventually get a repetition. Say that $T^k(x_1) = T^j(x_1)$, $j < k$, is the first repetition. Then $T^j(T^{k-j}(x_1)) = T^j(x_1)$. Since T^j is a permutation, $T^{k-j}(x_1) = x_1$. Thus, the first integer repeated is x_1, and we have

$$T^{k_1}(x_1) = x_1,$$

for a least positive integer k_1. We put T_1 equal to the cycle

$$(x_1 \quad T(x_1) \quad T^2(x_1) \quad \cdots \quad T^{k_1-1}(x_1)).$$

Let x_2 be the smallest integer in X_n moved by T and not moved by T_1. Using the same type of argument just used, we get another cycle, say T_2, disjoint from T_1, with

$$T_2 = (x_2 \quad T(x_2) \quad T^2(x_2) \quad \cdots \quad T^{k_2-1}(x_2)).$$

To see that T_2 and T_1 are disjoint, we need only observe that if $T^k(x_1) = T^j(x_2)$, for some k and j, then by applying powers of T to this equality, we get that $x_2 = T^p(x_1)$, for some integer p, contradicting the way we chose x_2.

We continue in this manner until we have a factorization of T into disjoint cycles, say $T = T_1 \circ T_2 \circ \cdots \circ T_s$, where

$$T_i = (x_i \quad Tx_i \quad T^2(x_i) \quad \cdots \quad T^{k_i-1}(x_i)), \quad \text{for } i = 1, \ldots, s.$$

II. *Uniqueness.* Let $T = T_1 \circ \cdots \circ T_s$ be the factorization obtained above and let $T = U_1 \circ \cdots \circ U_m$ be another factorization of T into disjoint cycles, each of length greater than 1. Let x_1 be defined as in I. Hence for some i, $U_i(x_1) \neq x_1$. Then, since U_i is disjoint from U_j, $j \neq i$, it follows from Exercise 1 in this section that $U_j(x_1) = x_1$, for all $j \neq i$. Since by Exercise 2 $U_i \circ U_j = U_j \circ U_i$, we may renumber the U_p's so that U_i becomes U_1. Since x_1 is moved by U_1 and $U_1(x_1) = T(x_1)$, it follows that

$$U_1 = (x_1 \quad T(x_1) \quad T^2(x_1) \quad \cdots \quad T^{k_1-1}(x_1)) = T_1,$$

and therefore

$$U_2 \circ \cdots \circ U_m = U_1^{-1} \circ T = T_1^{-1} \circ T = T_2 \circ \cdots \circ T_s.$$

We now complete the proof by induction on s. If $s = 1$, then

$$T_1^{-1} \circ T = (1) = U_2 \circ \cdots \circ U_m.$$

By Exercise 3, $U_i = (1)$ for $i = 2, \ldots, m$; hence $m = 1$ and $U_1 = T_1 = T$.

If $s > 1$, assuming the theorem for $s - 1$, we see that $m - 1 = s - 1$ and for some rearrangement $U_2 = T_2$, $U_3 = T_3$, $\ldots$, $U_s = T_s$. This completes the proof. ∎

EXERCISES

5. Write

$$\begin{pmatrix} 1 & 2 & 3 & 4 & 5 & 6 & 7 & 8 \\ 2 & 4 & 1 & 3 & 6 & 8 & 7 & 5 \end{pmatrix}$$

as a product of disjoint cycles.

6. Express each of the following products in terms of disjoint cycles (assume all permutations are in S_7):
   (a) $(123)(467)(345)(146)$
   (b) $(43)(156)(712)$
   (c) $(1234)(123)(12)$

Verify each of your answers by writing all the permutations in the form
$$\begin{pmatrix} 1 & 2 & 3 & 4 & 5 & 6 & 7 \\ T1 & T2 & T3 & T4 & T5 & T6 & T7 \end{pmatrix}.$$

7. Let T be a permutation in S_n. Define the *order* of T to be the smallest positive integer m such that T^m is the identity permutation.
   (a) Prove that for any $T \in S_n$, the order of T exists.
   (b) Show that if T is a cycle of length r, then T has order r.
   (c) Show that if T and S are disjoint cycles of orders t, s, respectively, then the order of $T \circ S$ is lcm $[t,s]$.
8. If p is a prime, describe all elements of S_n of order p.
9. Find an element of largest order in S_n, for

$$n = 2, 3, 4, 5, 6, 7, 8, 9, 10, 17.$$

3.3 Parity of Permutations

Although the factorization of a permutation into disjoint cycles is unique up to the arrangement of the factors, there is another important way of factoring a permutation in which the uniqueness of the factors is no longer present. This method is factorization into 2-cycles (*transpositions*). We shall see, however, that the parity (evenness or oddness) of the number of factors is an invariant of the permutation.

Definition 3.3.1 A cycle of length 2 is called a *transposition*. (1 2) is a transposition and (1 2)(3) is a transposition.

LEMMA 3.3.2 For $n \geq 2$, any cycle $C \in S_n$ can be factored into a product of transpositions.

Proof The proof is by induction on the length of the cycle. If C is of length 1, then $C = (1)$. Hence $C = (1\ 2)(1\ 2)$. If C is a cycle of length $r \geq 2$, it is easily verified that

$$(i_1 i_2 \cdots i_r) = (i_1 i_3 \cdots i_r)(i_1 i_2).$$

By induction, $(i_1 i_3 \cdots i_r)$ can be factored into transpositions; thus, so can $(i_1 i_2 i_3 \cdots i_r)$. ∎

THEOREM 3.3.3 Any permutation can be factored into a product of transpositions.

Proof If T is the identity, the result follows from Lemma 3.3.2. Otherwise, T can be written as a product of disjoint cycles, and each cycle can be factored into a product of transpositions by Lemma 3.3.2. ∎

It is clear that the decomposition of a permutation into a product of transpositions is not at all unique. For example,

$$(1\ 2\ 3\ 4) = (1\ 4)(1\ 3)(1\ 2)$$
$$= (2\ 1)(2\ 4)(2\ 3)$$
$$= (1\ 2)(2\ 4)(2\ 4)(1\ 2)(2\ 3)(2\ 3)(1\ 4)(1\ 3)(1\ 2).$$

Although the first two factorizations require 3 transpositions each, and the last requires 9, both 3 and 9 are odd.

THEOREM 3.3.4 Let $T \in S_n$. If

$$T = C_1 \circ C_2 \circ \cdots C_r \quad \text{and} \quad T = D_1 \circ D_2 \circ \cdots \circ D_s$$

are two factorizations of T into transpositions, then r and s are either both even or both odd.

Before proving Theorem 3.3.4, we make a brief comment. Although polynomials will not be introduced until Chapter 5, we are familiar with them from our experiences in algebra and calculus. Thus, here we will assume the reader has some knowledge of polynomials (sometimes called polynomial forms). The particular polynomial we will use will be denoted $f(X)$, defined by

$$f(X) = (x_1 - x_2)(x_1 - x_3) \cdots$$
$$(x_1 - x_n)(x_2 - x_3)(x_2 - x_4) \cdots (x_2 - x_n) \cdots (x_{n-1} - x_n).$$

For a permutation T in S_n, we define the operation of T on $f(X)$ by putting

$$T[f(X)] = (x_{T1} - x_{T2})(x_{T1} - x_{T3}) \cdots (x_{T1} - x_{Tn})(x_{T2} - x_{T3}) \cdots$$
$$(x_{T2} - x_{Tn}) \cdots (x_{T(n-1)} - x_{Tn}).$$

Since T is a permutation (1–1 and onto) on X_n, each factor of $f(X)$ or its negative appears in $T[f(X)]$. Thus, $T[f(X)] = f(X)$ or $-f(X)$. For example, if $n = 3$ and $T = (1\ 3)$, then $f(X) = (x_1 - x_2)(x_1 - x_3)(x_2 - x_3)$ and

$$T[f(X)] = (x_{T1} - x_{T2})(x_{T1} - x_{T3})(x_{T2} - x_{T3})$$
$$= (x_3 - x_2)(x_3 - x_1)(x_2 - x_1)$$
$$= -(x_1 - x_2)(x_1 - x_3)(x_2 - x_3)$$
$$= -f(X).$$

Proof of Theorem 3.3.4 Let $T = (ij)$ be a transposition in S_n, with $i < j$. Then

$$T[f(X)] = (x_{T1} - x_{T2}) \cdots (x_{T(n-1)} - x_{Tn}).$$

Under T, the factor $x_l - x_k$ becomes $x_{Tl} - x_{Tk}$, and we note that $x_{Tl} - x_{Tk} = x_l - x_k$ if $l \neq i, j$ and $k \neq i, j$. Thus, $x_{Tl} - x_{Tk} \neq x_l - x_k$ if either $l = i$ or j, or $k = i$ or j. We may place into pairs all but one of the factors of $f(X)$ changed by T, using the following scheme.

(1) Pair factors $x_i - x_k$ and $x_k - x_j$, where $i < k < j$ (if any such factors exist; that is, there may not be a k such that $i < k < j$).
(2) Pair factors $(x_k - x_i)$ and $(x_k - x_j)$, for $k < i$ (if any: i may be equal to 1).
(3) Pair factors $(x_i - x_k)$ and $(x_j - x_k)$, for $j < k$ (if any: j may be equal to n).

These three ways of pairing factors will give us all the factors of $f(X)$ that are changed by T with the exception of $x_i - x_j$.

Using (1), the product $(x_i - x_k)(x_k - x_j)$ becomes under T the product

$$(x_{Ti} - x_{Tk})(x_{Tk} - x_{Tj}) = (x_j - x_k)(x_k - x_i) = (x_i - x_k)(x_k - x_j).$$

In (2) the product $(x_k - x_i)(x_k - x_j)$, which is a factor of $f(X)$, will also go into itself under T.

Similarly, using the pairing of (3), the product $(x_i - x_k)(x_j - x_k)$, also a factor of $f(X)$, will be transformed into itself under application of T.

It is now clear that the only lasting change effected on $f(X)$ by T is through the action of T on $(x_i - x_j)$, which becomes $(x_{Ti} - x_{Tj}) = (x_j - x_i) = -(x_i - x_j)$. Thus, putting all this information together, we conclude that $T[f(X)] = -f(X)$, where T is the transposition (ij).

Now if $T = (i_1j_1)(i_2j_2)\cdots(i_rj_r)$, then $T[f(X)]$ is the polynomial obtained by applying T all at once to $f(X)$ or by applying the transpositions one at a time in order. Thus $T[f(X)] = (-1)^r f(X)$, as each $(i_l j_l)$ effects one change of sign.

Thus $T[f(X)] = f(X)$ if T is written as an even number of transpositions and $T[f(X)] = -f(X)$ if T is written as an odd number of transpositions. Since the sign must be always $+$, or always $-$, we see that we can write T either as an even number of transpositions and never as an odd number, or as an odd number but never as an even number. This completes the proof of the theorem. ∎

Definition 3.3.5 A permutation $T \in S_n$, $n \geq 2$, is *even* (*odd*) if it can be written as a product of an even (odd) number of transpositions. We say that the *parity* of T is even (odd), accordingly.

It is easy to determine the parity of a product of two permutations from their respective parities. For if

$$T = (i_1j_1)\cdots(i_rj_r) \quad \text{and} \quad S = (m_1n_1)\cdots(m_sn_s),$$

then $T \circ S$ is a product of $r + s$ 2-cycles. Hence the parity of $T \circ S$ is given by the following table:

T	S	T ∘ S
Even	Even	Even
Even	Odd	Odd
Odd	Odd	Even
Odd	Even	Odd

The set of permutations S_n on X_n is partitioned into two classes, the even and the odd permutations. We conclude this section by showing that both these classes have the same number of elements.

THEOREM 3.3.6 If $n \geq 2$, and A_n is the set of all even permutations on X_n, then A_n contains $n!/2$ elements.

Proof Let B_n be the set of all odd permutations in S_n. Then $S_n = A_n \cup B_n$, $A_n \cap B_n = \varnothing$. Let $\alpha\colon A_n \longrightarrow B_n$ be defined as follows: for $T \in A_n$, $\alpha(T) = T \circ (1\ 2)$. We claim that α is 1–1 and onto. For, if $\alpha(T) = \alpha(U)$, we have $T \circ (1\ 2) = U \circ (1\ 2)$. Then

$$T \circ (1\ 2) \circ (1\ 2) = U \circ (1\ 2) \circ (1\ 2).$$

But $(1\ 2) \circ (1\ 2) = (1)$, whence $T = U$, and α is 1–1. Next, let $U \in B_n$. Then U is odd and $U \circ (1\ 2)$ is even. Hence

$$\alpha(U \circ (1\ 2)) = U \circ (1\ 2) \circ (1\ 2) = U,$$

and α is onto. This shows that A_n and B_n have the same number of elements, say m. Clearly, $m + m = n!$ and $m = n!/2$. ∎

EXERCISES

1. Find the parity of each of the following permutations:
   (a) $\begin{pmatrix} 1 & 2 & 3 & 4 & 5 & 6 & 7 \\ 2 & 3 & 4 & 1 & 6 & 7 & 5 \end{pmatrix}$.
   (b) $(1\ 2\ 3)(2\ 4\ 5)(1\ 6\ 7)$.
   (c) $(1\ 4\ 5\ 6\ 3)(1\ 4\ 6\ 5\ 3)(1\ 4\ 3\ 6\ 5)$.
2. Show that, for $r \geq 2$,
   $$(i_1 i_2 \cdots i_r) = (i_1 i_r) \cdots (i_1 i_4)(i_1 i_3)(i_1 i_2).$$
3. Show that the cycle $(1\ 2\ \ldots\ n)$ is even if and only if n is odd.
4. Let $T \in A_n$, $U \in S_n$. Show that $UTU^{-1} \in A_n$.
5. Let $T \in A_n$. Show that $T^{-1} \in A_n$.
6. Show that the set A_n is "closed" under permutation multiplication.

3.3 PARITY OF PERMUTATIONS

That is, if $T, S \in A_n$, then $T \circ S \in A_n$. Is a similar statement true about B_n?

7. (a) Let $T \in S_n$, $n \geq 4$, and $U = (1\ 2\ 3\ 4)$. Show that
$$TUT^{-1} = (T1\ T2\ T3\ T4).$$
   (b) Let $T \in S_n$, $n \geq 5$, and $U = (1\ 2\ 3)(4\ 5)$. Find TUT^{-1} as a product of disjoint cycles.

8. Let $T, U \in S_n$. Let $U = C_1 \cdots C_k$ be a factorization of U into disjoint cycles, where $C_i = (j_1 j_2 \cdots j_{t_i})$. Show that $TUT^{-1} = D_1 \cdots D_k$, where $D_i = (Tj_1\ Tj_2\ \ldots\ Tj_{t_i})$.

9. Let $T, U \in S_n$. $T = C_1 \circ \cdots \circ C_k$ is a factorization of T into disjoint cycles, and $U = D_1 \circ \cdots \circ D_k$ is a factorization of U into disjoint cycles. Suppose C_i and D_i have the same length, $i = 1, 2, \ldots, k$. Prove that there exists $V \in S_n$ such that $VTV^{-1} = U$.

4
The Theory of Groups

4.1 Definition of Group and Examples

In Chapter 2 we saw that a great many theorems about the integers are deducible from the relatively short list of properties stated in Section 2.1. These properties could have been stated in the "abstract," using the terminology and notation of set theory.

For example, if we replace the set Z of integers by an arbitrary nonempty set S and the operation of addition by a function $f: S \times S \longrightarrow S$, then properties A_0, A_1, and A_2 become

B_0: There is a function $f: S \times S \longrightarrow S$.
B_1: If $a, b, c \in S$, then $f(f(a,b),c) = f(a,f(b,c))$.
B_2: If $a, b \in S$, then $f(a,b) = f(b,a)$.

Notice that in this form no direct mention is made of integers. We merely define a function and list some of its properties. In a similar way we can define an abstract "multiplication" and state the distributive law. Such a procedure has the obvious advantage of simultaneously including all mathematical systems that share these abstract properties.

Historically, the important abstractions developed (often in very slow stages) from specific problem areas. The theory of groups, which we will study in this chapter, arose from the theory of equations, more specifically from the attempt to find roots of a polynomial in terms of its coefficients. Considerations far too involved to explain here led to a systematic study of sets of permutations, which in turn led to the formulation of the concept of a group. The study of groups, although very much influenced by its historical origins, has developed in many different directions. The richness and breadth of this development has given the theory of groups a central position in mathematics.

4.1 DEFINITION OF GROUP AND EXAMPLES

Definition 4.1.1 Let G be a nonempty set and f a binary operation on G, that is, $f: G \times G \longrightarrow G$. We denote the element $f(a,b)$ by $a \circ b$. The pair (G,f) will be called a *group* if

(a) f is associative; that is, $a \circ (b \circ c) = (a \circ b) \circ c$ for all $a, b, c \in G$.
(b) There is an element $e \in G$ such that $a \circ e = e \circ a = a$ for all $a \in G$. We call e an *identity element* for (G,f).
(c) There exists an identity e of (G,f) with the property that for each $a \in G$ there is an element $b \in G$ such that $a \circ b = b \circ a = e$. We call b an *inverse* of a.

For purposes of brevity, when no ambiguity of meaning exists, we shall simply denote the group (G,f) by G. The context should clearly indicate whether we are concerned with the abstract set or with a set together with an associated binary operation.

When the function denoting the binary operation is not indicated, we shall usually use juxtaposition, as in the multiplication of numbers, to denote the group operation. That is, the element $f(a,b) = a \circ b$ will be denoted by ab. Following this convention, if b is an inverse of a, we write $b = a^{-1}$, and $ab = aa^{-1} = e$. On occasion we shall use $a + b$ or $a * b$ in place of $a \circ b$. With the notation $a + b$, an inverse of a will be denoted by $(-a)$ and a group identity by 0. We will write $a - b$ for $a + (-b)$.

Definition 4.1.2 Let G be a group. If $a \circ b = b \circ a$, for all $a, b \in G$, we say that G is *commutative* or *abelian*.[1] If there exists a pair of elements $a, b \in G$ such that $a \circ b \neq b \circ a$, we say that G is *noncommutative* or *nonabelian*. When $a \circ b = b \circ a$, we say that the elements a and b *commute*.

Definition 4.1.3 A group with a finite number of elements is called a *finite group;* otherwise, it is called an *infinite group*. If a group G is finite, and G consists of exactly n elements, we say that the *order of G* is n and we write $|G| = n$. If G is infinite, we write $|G| = \infty$.

The following list of examples will illustrate the diversity of mathematical systems that satisfy the axioms given above and that are, therefore, groups.

Example 4.1.4 Let G be the set of all integers and f the operation of addition; that is, $f(a,b) = a + b$. Property A_1 in Section 2.1 is the statement that f is associative. The integer 0 satisfies the requirement for an identity and $-a$ is clearly an inverse of a, since $a + (-a) = 0$. Thus $(G,+)$ is a group.

Example 4.1.5 Let G be the set of all rational numbers excluding 0. That is, G is the set of all real numbers that can be written in the form a/b, with

[1] After the Norwegian mathematician N. H. Abel (1802–1829).

a and b integers, $a \neq 0$, $b \neq 0$. For $u, v \in G$, let $u \circ v = uv$, the usual product of two real numbers. The associative property of multiplication of rational numbers follows from the associative law of multiplication for integers (see Section 5.7). The rational number $1/1 = 1$ is an identity and an inverse of a/b is simply b/a. Hence $(G, \circ)$ is a group.

EXERCISES

1. Let G be the set of all rational numbers. Show that $(G, +)$ is a group.
2. Let Z be the set of integers. Is the operation of subtraction a binary operation? Is $(Z, -)$ a group?
3. Why is it necessary in Example 4.1.5 above to exclude 0 from the set G?
4. Let p be a prime number and let $R_p = \{a/p^\alpha \mid a, \alpha \in Z$, with $\alpha \geq 0\}$. Show that $(R_p, +)$ is a group. Is it necessary to assume that p is a prime number?
5. Let H be the set of even integers. Show that $(H, +)$ is a group.

Example 4.1.6 Let G be the set of residue classes of integers modulo m, that is $G = \bar{Z}_m$, and let f be the addition of residue classes defined in Section 2.12. We showed there that f satisfies the associative law. Also $\bar{0}$ is an identity and the residue class $\overline{(m - a)}$ is an inverse of $\bar{a}$. Thus $(\bar{Z}_m, +)$ is a group.

Example 4.1.7 Let R be a nonsquare rectangle in the plane whose vertices are labeled as in Figure 4.1. Consider the following rigid motions

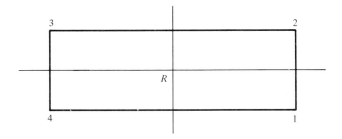

Figure 4.1

of R (called the *symmetries* of R):

r: rotation of 180° about its center.
h: reflection in the horizontal axis.
v: reflection in the vertical axis.
e: no movement at all (equivalent to a rotation of 360°).

Each of these movements can be thought of either as a transformation of R or as a permutation on its vertices. In either case these movements can be composed or "multiplied" as functions (Section 1.7), and it is easy to verify that the product of any pair of these is again one of the four. For example, $h \circ r = v$, as illustrated in Figure 4.2.

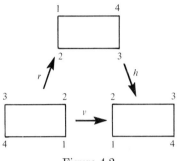

Figure 4.2

The element e is the identity function, and each element is its own inverse. Since we have already shown that composition of functions is an associative operation (Section 1.7), it follows that the set of symmetries of R is a group with respect to the operation of function composition. This group is sometimes called the Klein 4-group.

EXERCISE

6. Compute all products of pairs of elements of the Klein group above. Is the group commutative? What is its order?

Example 4.1.8 Let E_2 denote the points of the plane with Cartesian coordinates; that is, $E_2 = \{(x,y) \mid x, y \text{ are real numbers}\}$. Let D_4 represent the following set of permutations of E_2:

e (identity function): $(x,y) \longrightarrow (x,y)$,
r_1 (counterclockwise rotation of 90°): $(x,y) \longrightarrow (-y,x)$,
r_2 (rotation of 180°): $(x,y) \longrightarrow (-x,-y)$,
r_3 (rotation of 270°): $(x,y) \longrightarrow (y,-x)$,
v (reflection in the y axis): $(x,y) \longrightarrow (-x,y)$,
h (reflection in the x axis): $(x,y) \longrightarrow (x,-y)$,
d_1 (reflection through the line $y = x$): $(x,y) \longrightarrow (y,x)$,
d_{-1} (reflection through the line $y = -x$): $(x,y) \longrightarrow (-y,-x)$.

This set of transformations can also be thought of as the set of all symmetric movements of a square with center at the origin (0,0) and vertical and horizontal edges.

The set D_4 together with the composition of functions satisfies all of the requirements for a group. The associativity of this product follows from Theorem 1.7.5. The fact that the product of any two of these transformations is again in D_4 can be verified directly. For example,

$$(d_1 \circ r_1)(x,y) = d_1(r_1(x,y)) = d_1(-y,x) = (x,-y).$$

But

$$h(x,y) = (x,-y).$$

Hence

$$d_1 \circ r_1 = h.$$

The reader is asked to verify that e is an identity and that for each element of D_4 there exists an inverse. This group is called the *group of symmetries of a square*. Note that $r_1 \circ d_1 = v$ and $d_1 \circ r_1 = h$, so that D_4 is the first example of a nonabelian group we have given.

EXERCISES

7. Verify that the product of any two elements of D_4 is again an element of D_4. (Can you do this without enumerating all such products?)
8. Show that the subset $R = \{e, r_1, r_2, r_3\}$ together with the operation $\circ$ of D_4 restricted to R is a group. Can you find any other subsets of D_4 that form a group with respect to $\circ$?
9. Find all pairs of elements of D_4 that do not commute.
10. Show that in D_4 there is a unique identity and each element has a unique inverse.
11. Show that $(a^{-1})^{-1} = a$ for all elements a of the group D_4.
12. Show that if a, b are elements of the group D_4, then $(ab)^{-1} = b^{-1}a^{-1}$.

Example 4.1.9 Let X_n represent a regular n-gon in the plane; that is, X_n is a polygon all of whose sides and angles are equal. Thus each internal angle is $\theta = \pi(n-2)/n$ radians. The symmetries of X_n are those rigid movements of X_n that cause it to occupy the same space in the plane as it originally occupied. For example, if an equilateral triangle is placed thus on the coordinate plane:

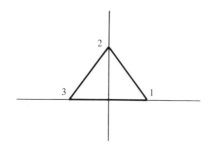

then the movement whose effect is

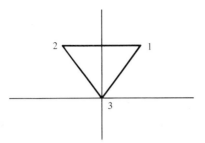

is not a symmetric movement. But the movements whose effects are

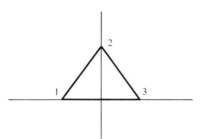

or

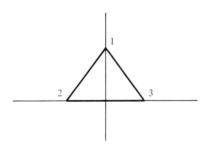

are symmetric movements.

Now for the arbitrary regular n-gon the symmetric movements may be described as follows.

Rotations These are the counterclockwise motions $r_0, r_1, r_2, \ldots, r_{n-1}$ of $2\pi i/n$ radians for $i = 0, \ldots, n - 1$. Thus r_0 represents no movement at all, r_1 represents a rotation of $2\pi/n$ radians, r_2 represents a rotation of $4\pi/n$ radians, and so forth. Notice that the restriction that rotations be counterclockwise is no restriction at all, since a clockwise rotation of $(2\pi/n)j$ radians is equivalent to a counterclockwise rotation of $(2\pi/n)(n - j)$ radians. It is easy to see how rotations may be multiplied.

The product $r_i \circ r_j$ is defined as $r_{(i+j)'}$, where $(i+j)'$ is the least positive residue of $i+j$ modulo n.

Reflections Here we must distinguish two cases. If n is even, then there are two classes of reflections to consider: a reflection through the line joining two opposite vertices, and a reflection through the line joining the midpoints of two opposite sides. There are precisely n such reflections. In the case $n = 4$, namely that of the square, d_1 and d_{-1} are reflections through lines joining opposite vertices, while v and h are reflections through lines joining the midpoints of opposite sides. (See Figure 4.3.)

If n is odd, then it makes no sense to talk about opposite sides and opposite vertices. Rather we consider reflections through those lines that join a vertex with the midpoint of the opposite side. Thus, for example, if $n = 3$, we have an equilateral triangle and we obtain three reflections, which are listed in Figure 4.4. Again there are precisely n of these.

Returning now to the general case, it is not difficult to verify that these two classes of symmetries exhaust all possible symmetries and hence that the product of any two such symmetries is again equal to one of them. (One symmetry followed by another is equivalent to a single symmetry.)

To show that this set of $2n$ motions of X_n is a group under composition of motions, we note that since the product of any two of these is still in the set we have a binary operation. Associativity follows from the fact that these are functions, and r_0 is clearly an identity. The rotation r_{n-i} clearly an inverse of r_i, and every reflection is its own inverse, since the application of a reflection twice yields the identity r_0. For each positive integer $n \geq 3$ this group is called the dihedral group of order $2n$, usually denoted by D_n. The group D_4 of Example 4.1.8 is, of course, a special case. We will have occasion to return to this class of examples later in the chapter.

Example 4.1.10 Let $G = \{1, -1, i, -i\}$, where $i^2 = -1$. If f is the usual multiplication of complex numbers, it is easy to see that (G, f) is a group of four elements.

Example 4.1.11 Let S be an arbitrary set and let G be the set of all 1–1 mappings from S onto itself, that is, the set of all permutations on the set S. Let f be the binary operation of composition of functions. Then, by Theorem 1.7.5, f is associative on G. The identity function satisfies the requirements for a group identity, and if $g \in G$, then the function g^{-1} defined in Definition 1.7.7 is a group inverse for g. The group G is usually called the *transformation group* for S. If the set S consists of n elements, say $\{1, 2, \ldots, n\}$, then G is called the *symmetric group on n letters*, denoted by S_n.

4.1 DEFINITION OF GROUP AND EXAMPLES

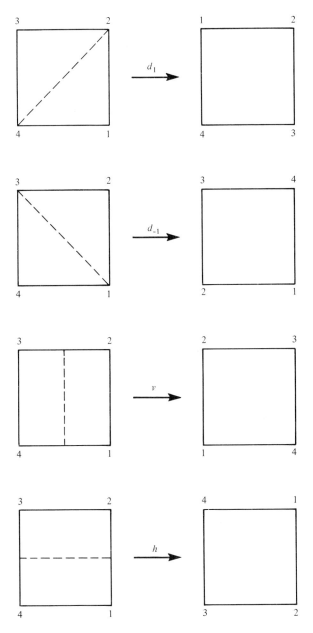

Figure 4.3

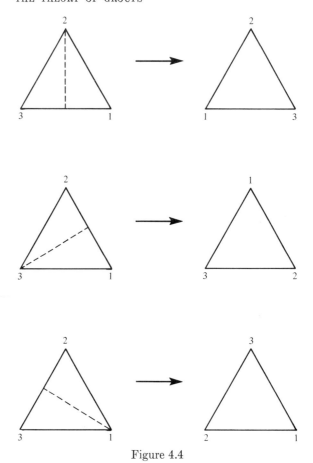

Figure 4.4

EXERCISES

13. Show that the set of rotations of the dihedral group D_n form a group of order n with respect to function composition.
14. Calculate the following elements of D_4: r_1v, r_2v, r_3v, ev.
15. Let v be some reflection in D_n. Show that any reflection can be obtained as the product vr_i with r_i one of the rotations in D_n.
16. Let v be a reflection in D_n. Describe the reflections vr_1 and r_1v.
17. Show that D_n is not an abelian group for $n \geq 3$.
18. Find the orders of S_2, S_3, S_n.
19. Let $G = \{\bar{1},\bar{3},\bar{5},\bar{7}\} \subseteq \bar{Z}_8$. Show that G is a group with respect to residue class multiplication modulo 8.
20. Let $G = \{\bar{1},\bar{5},\bar{7},\overline{11}\} \subseteq \bar{Z}_{12}$. Show that G is a group with respect to residue class multiplication modulo 12.

4.2 Elementary Consequences of the Definition of a Group

The examples listed above are a fairly wide sample of infinite and finite groups. They all share some rather obvious properties that do not appear explicitly in the definition of a group. In particular, in each group there appears to be only one choice for an identity element and for the inverse of an element. We shall show in this section that many of the properties that appear common to all of these examples do indeed follow from the group axioms and thus hold for all groups.

THEOREM 4.2.1 (*Uniqueness of Identity*) Let e_1 and e_2 be identity elements of the group G; that is, $e_1 a = a e_1 = a$ and $e_2 a = a e_2 = a$, for all $a \in G$. Then $e_1 = e_2$.

Proof Consider the element $e_1 e_2$. Since e_2 is an identity $e_1 e_2 = e_1$ and since e_1 is an identity, $e_1 e_2 = e_2$. Thus $e_1 = e_2$. ∎

THEOREM 4.2.2 Let a, b, and c be elements of a group G. If $ab = ac$, then $b = c$, and if $ba = ca$, then $b = c$; that is, cancellation on the left and right is always permissible.

Proof If $ab = ac$, then $a^{-1}(ab) = a^{-1}(ac)$. Thus $(a^{-1}a)b = (a^{-1}a)c$ and $eb = ec$, which implies $b = c$. Similarly if $ba = ca$, then $(ba)a^{-1} = (ca)a^{-1}$ and $be = ce$, which implies that $b = c$. ∎

THEOREM 4.2.3 (*Uniqueness of Inverse*) If a is an element of the group G, then there is precisely one element $a^{-1} \in G$ such that $aa^{-1} = a^{-1}a = e$.

Proof Suppose that $ba = e$ and $ca = e$ for some $b, c \in G$. Then $ba = ca$, whence $b = c$ by Theorem 4.2.2. ∎

COROLLARY 4.2.4 If $a \in G$, then $(a^{-1})^{-1} = a$.

Proof The element $(a^{-1})^{-1}$ is the unique element satisfying $a^{-1}x = e$. But $a^{-1}a = e$; hence $(a^{-1})^{-1} = a$. ∎

THEOREM 4.2.5 Let a, b be elements of a group G. Then $(ab)^{-1} = b^{-1}a^{-1}$.

Proof Since we have proved the uniqueness of the inverse of any group element, it will suffice to show that $(ab)(b^{-1}a^{-1}) = e$. Thus
$$(ab)(b^{-1}a^{-1}) = a(b(b^{-1}a^{-1})) = a((bb^{-1})a^{-1})$$
$$= a(ea^{-1}) = aa^{-1} = e,$$
and so $(ab)^{-1} = b^{-1}a^{-1}$. ∎

EXERCISES

1. Verify that the associative law holds in Example 4.1.10.
2. Let c be an element of the group G such that $cc = c$. Show that c is the identity of G.

3. Show that if a group G contains an element c with the property that $ca = c$ for all $a \in G$, then G consists of a single element.
4. Show that the set $\bar{Z}_p{}^* = \{\bar{1}, \bar{2}, \ldots, \overline{(p-1)}\}$ and the operation of multiplication of residue classes form a group if and only if p is a prime.
5. Let G be a group. Show that G is abelian if and only if $(ab)^2 = a^2b^2$, for all $a, b \in G$.
6. Show that if $a^2 = e$ for all elements a of a group G, then G is abelian.
7. Let $gG = \{gh \mid h \in G\}$ for some fixed g in the group G. Show that $G = gG$.
8. Let H be a nonempty finite set and let $\circ$ be an associative binary operation on H. Suppose that $a \circ b = a \circ c$ implies $b = c$ and $a \circ c = b \circ c$ implies $a = b$. Show that H is a group. Show that the hypothesis that H is finite cannot be omitted. [*Hint:* Let $\{a_1, \ldots, a_n\} = H$ and show that $\{a_1a_1, a_1a_2, \ldots, a_1a_n\} = H$.]
9. Let G be a nonempty set and $\circ$ an associative binary operation on G. Assume that there exists $e \in G$ such that $a \circ e = e \circ a = a$ for all $a \in G$, and further, that for each $a \in G$ there exists $b \in G$ such that either $a \circ b = e$ or $b \circ a = e$. Show that G is a group.
10. Let G be a nonempty set and $\circ$ an associative binary operation on G. Assume that there is an element e, called a right identity, such that $ae = a$ for all $a \in G$. Assume further that there is a right identity, say e, so that for each $a \in G$, there is an element b, called a right inverse of a, such that $ab = e$. Show that $(G, \circ)$ is a group.
11. Let A be the set of reduced residue classes modulo m, $m > 0$. Show that A is a group with respect to residue class multiplication.
12. Let $(G, \circ)$ be a group. Show that the equations $a \circ x = b$ and $y \circ a = b$ have unique solutions, and that these solutions are $x = a^{-1} \circ b$ and $y = b \circ a^{-1}$, respectively.
13. Prove the following: Let G be a nonempty set and $\circ$ a binary operation on G satisfying:
    (a) $a \circ (b \circ c) = (a \circ b) \circ c$, for all $a, b, c \in G$.
    (b) The equation $a \circ x = b$ has a solution in G for all $a, b \in G$.
    (c) The equation $y \circ a = b$ has a solution in G for all $a, b \in G$.
    Then $(G, \circ)$ is a group.

4.3 Multiplication Table for a Finite Group

Let G be a finite group of order n. Then it is possible to define its binary operation $\circ$ simply by listing the values of the n^2 products $a \circ b$ for all $a, b \in G$. One particularly convenient way to do this is by means of a "multiplication table," a table analogous to the ones used in grade school to illustrate addition or multiplication of numbers. In the case of a finite group, many of the group axioms can be translated into "geometrical" statements about the multiplication table. We illustrate this with the group S_3 of all permutations on the three numbers 1, 2, 3. Here the binary operation is permutation multiplication, defined in Section 1.7. The elements of S_3 are:

$$a_1 = (1), \quad a_2 = (1\ 2), \quad a_3 = (2\ 3), \quad a_4 = (1\ 3),$$
$$a_5 = (1\ 2\ 3), \quad \text{and} \quad a_6 = (1\ 3\ 2).$$

The multiplication table is constructed by listing the elements of G along the top and left side in the same order. We refer to the column headed by a_j as the jth column and to the row with a_i on the left as the ith row. We then fill in the boxes according to the rule: the contents of the box in the ith row and the jth column is the element $a_i a_j$.

∘	a_1	a_2	a_3	a_4	a_5	a_6
a_1	a_1	a_2	a_3	a_4	a_5	a_6
a_2	a_2	a_1	a_5	a_6	a_3	a_4
a_3	a_3	a_6	a_1	a_5	a_4	a_2
a_4	a_4	a_5	a_6	a_1	a_2	a_3
a_5	a_5	a_4	a_2	a_3	a_6	a_1
a_6	a_6	a_3	a_4	a_2	a_1	a_5

EXERCISES

1. Verify that all entries in the table above are correct.
2. Construct the multiplication table for Example 4.1.10.
3. Construct the multiplication table for Example 4.1.7.
4. State and prove a criterion for commutativity based on the multiplication table of a group.

It is now possible, by examining the formal properties of the table, to decide whether certain of the axioms are fulfilled. One of the elements is an identity if and only if its column and row contain elements in exactly the same order as in the labeling column and row. Notice that if there are two such columns, then either the two elements are equal or else the set is not a group. This follows from the fact that the identity element is unique. In the illustration above, a_1 is the identity element.

The presence of a_1 in each row guarantees that each element has an inverse. More generally, every element must appear in every row and every column; otherwise one element would have to appear twice in a row (or a column), and this would contradict the fact that the elements are distinct, since $uv = uw$ implies that $v = w$. The only group property that is difficult to check directly from properties of the multiplication table is associativity.[2]

EXERCISES

5. Let $G = \{a,b,c,d\}$ with binary operation defined by the following table. Is G a group? Why?

	a	b	c	d
a	a	b	d	c
b	b	a	c	d
c	d	c	a	b
d	c	d	b	a

6. In Exercise 5 replace G by $H = \{a,b,c,d,e\}$ and the table by:

	a	b	c	d	e
a	a	b	c	d	e
b	b	a	d	e	c
c	c	d	e	b	a
d	d	e	a	c	b
e	e	c	b	a	d

4.4 Isomorphism

It should be evident from the examples already given that a great many different mathematical systems are groups. In order to obtain a clear picture of how many different types of groups there are, we must be able to distinguish between those properties of a group that are "relevant" and those that are not.

As a trivial example of such a distinction, consider the Klein group described in Example 4.1.7 and suppose that the rectangle in question has sides of length 2 inches and 4 inches. If now we form the group of symmetries of a rectangle with sides of length 1 inch and 3 inches, then we may say that the two groups are "essentially" the same.

---

[2] See, for example, H. Zassenhaus, *The Theory of Groups*, 2nd ed., New York: Chelsea Publishing Company, 1958.

A slightly less trivial example is given by the groups $G = \{e, r_1, r_2, r_3\}$ described in Exercise 8, Section 4.1, and H, the group $\bar{Z}_4$ of Example 4.1.6 ($m = 4$). To demonstrate the similarity between these two groups we define a mapping $f: G \longrightarrow \bar{Z}_4$ as follows:

$$f(e) = \bar{0},$$
$$f(r_1) = \bar{1},$$
$$f(r_2) = \bar{2},$$
$$f(r_3) = \bar{3}.$$

Any product of elements of G corresponds to the sum of corresponding elements in $\bar{Z}_4$. For example, $f(r_1 r_2) = \bar{1} + \bar{2} = f(r_1) + f(r_2)$. We see, then, that the operations of $\bar{Z}_4$ and G behave in "essentially" the same way. This can be most easily demonstrated by examining the tables of G and $\bar{Z}_4$.

TABLE FOR G

∘	e	r_1	r_2	r_3
e	e	r_1	r_2	r_3
r_1	r_1	r_2	r_3	e
r_2	r_2	r_3	e	r_1
r_3	r_3	e	r_1	r_2

TABLE FOR $\bar{Z}_4$

+	$\bar{0}$	$\bar{1}$	$\bar{2}$	$\bar{3}$
$\bar{0}$	$\bar{0}$	$\bar{1}$	$\bar{2}$	$\bar{3}$
$\bar{1}$	$\bar{1}$	$\bar{2}$	$\bar{3}$	$\bar{0}$
$\bar{2}$	$\bar{2}$	$\bar{3}$	$\bar{0}$	$\bar{1}$
$\bar{3}$	$\bar{3}$	$\bar{0}$	$\bar{1}$	$\bar{2}$

The statement above can now be completely checked by noting that if all of the entries in the table for G are replaced by elements of $\bar{Z}_4$ according to the function f, then the two tables are identical. Alternatively, if the table for G is superimposed on the table for $\bar{Z}_4$, the set of pairs of elements occurring in the same box are the ordered pairs in the function f.

We now formalize this discussion in the following definition.

Definition 4.4.1 Two groups $(G, \circ)$ and $(H, *)$ will be called *isomorphic* if there exists a 1-1 onto function $f: G \longrightarrow H$ such that for all $a, b \in G$, $f(a \circ b) = f(a) * f(b)$. We shall also say that G is *isomorphic to* H and use the notation $G \cong H$. The function f is called an *isomorphism*.

We see then that two groups are isomorphic if their elements can be put into 1-1 correspondence in such a way that their respective binary operations correspond.

Example 4.4.2 Let $(Z, +)$ and $(W, +)$ be the groups of integers and even integers respectively under addition. Let $f: Z \longrightarrow W$ be defined by $f(n) = 2n$ for all $n \in Z$. Clearly f is 1-1 and onto. Since $f(a + b) = 2(a + b) = 2a + 2b = f(a) + f(b)$, f is an isomorphism and $Z \cong W$.

118   THE THEORY OF GROUPS

Example 4.4.3   Let D_3 be the dihedral group of order 6. Thus D_3 is the group of all symmetries of the equilateral triangle, say

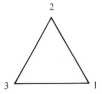

and hence consists of rotations r_0, r_1, and r_2 plus the reflections v_1, v_2, and v_3 corresponding to the reflections through the angle bisectors of vertices 1, 2, and 3, respectively.

Now let S_3 be the group of all permutations on the integers 1, 2, 3. Let $f: D_3 \longrightarrow S_3$ be defined as follows:

$f(r_0) = (1)$, the identity of permutation of S_3,
$f(r_1) = (1\ 2\ 3)$,
$f(r_2) = (1\ 3\ 2)$,
$f(v_1) = (2\ 3)$,
$f(v_2) = (1\ 3)$,
$f(v_3) = (1\ 2)$.

It is not difficult to see that these two groups are isomorphic, since the effect of applying one of the elements of D_3 to the triangle

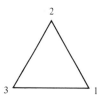

is simply to cause a permutation of the vertices. In order to completely verify the isomorphism it would suffice to construct the multiplication tables of the groups, listing the elements in accord with the mapping f defined above, and then checking these tables as we did with the example earlier in this section.

The following properties of an isomorphism are very useful in testing whether or not two groups are isomorphic.

THEOREM 4.4.4   Let f be an isomorphism from the group G onto the group H.

(a) If e is the identity of G, then $f(e)$ is the identity of H.
(b) If $a \in G$, then $(f(a))^{-1} = f(a^{-1})$.

Proof (a) Let e be the identity of G and let e' be the identity of H. Since $e = ee$, $f(e) = f(ee) = f(e)f(e)$. Also, $f(e) = e'f(e)$, so $e'f(e) = f(e)f(e)$. Right multiplying in H by $[f(e)]^{-1}$, we have $e' = f(e)$.

(b) Let $a \in G$. Then $e' = f(e) = f(aa^{-1}) = f(a)f(a^{-1})$. Thus, $f(a^{-1})$ is a right inverse in H of $f(a)$, and so is the inverse of $f(a)$, that is

$$[f(a)]^{-1} = f(a^{-1}). \qquad \blacksquare$$

The relation of isomorphism partitions any set of groups into equivalence classes, as we will prove below. All of the groups in each class are "essentially" the same.

THEOREM 4.4.5 If S is a set of groups, then the relation of isomorphism is an equivalence relation on S.

Proof (a) $A \cong A$ for any group $A \in S$. We simply define the map $f: A \longrightarrow A$ to be the identity map. Thus $f(a) = a$ for all $a \in A$. Clearly, f is 1–1 and onto and has the property $f(ab) = ab = f(a)f(b)$.

(b) Let $A, B \in S$. If $A \cong B$, then $B \cong A$. For since $A \cong B$, there is a 1–1 onto function $f: A \longrightarrow B$ with the property $f(xy) = f(x)f(y)$, for all $x, y \in A$. Since f is 1–1 and onto, f has an inverse function $g: B \longrightarrow A$ defined as follows: if $b \in B$, then there is $a \in A$ such that $f(a) = b$. We put $g(b) = a$. Thus g is 1–1 and onto. (See Section 1.7.) Now if $u, v \in B$, then there are $x, y \in A$ such that $f(x) = u$ and $f(y) = v$; hence $g(u) = x$ and $g(v) = y$. Thus

$$g(uv) = g(f(x)f(y)) = g(f(xy))$$
$$= xy = g(u)g(v),$$

which shows that g is an isomorphism. Hence $B \cong A$.

(c) Let $A, B, C \in S$. If $A \cong B$ and $B \cong C$, then $A \cong C$. For since $A \cong B$, there is a function

$$f: A \xrightarrow[\text{onto}]{1\text{-}1} B$$

satisfying $f(xy) = f(x)f(y)$ for all $x, y \in A$. Further since $B \cong C$, there is a function

$$g: B \xrightarrow[\text{onto}]{1\text{-}1} C$$

satisfying $g(uv) = g(u)g(v)$ for all $u, v \in B$. Now consider the composite $g \circ f$. It is clearly 1–1 and onto (see Exercise 7, Section 1.7), so

$$g \circ f: A \xrightarrow[\text{onto}]{1\text{-}1} C$$

and if $x, y \in A$, then

$$(g \circ f)(xy) = g(f(xy)) = g(f(x)f(y))$$
$$= g(f(x))g(f(y)) = ((g \circ f)(x))((g \circ f)(y)).$$

Thus $g \circ f$ is an isomorphism and $A \cong C$. $\qquad \blacksquare$

120   THE THEORY OF GROUPS

COROLLARY 4.4.6   The relation of isomorphism partitions any set of groups.

Proof   We have proved above that isomorphism is an equivalence relation, and so by Theorem 1.8.7 the corollary follows. ∎

EXERCISES

1. Show that if H is the additive group $(\bar{Z}_4,+)$ and G is the group of Example 4.1.10, then $G \cong H$.
2. Show that the groups of Exercises 19 and 20, Section 4.1, are both isomorphic to the Klein group (Example 4.1.7).
3. Let G be the group of all permutations on the set of three symbols $\{a,b,c\}$. Show that $G \cong S_3$.
4. Show that the Klein group is not isomorphic to $(\bar{Z}_4,+)$.
5. Show that S_3 is not isomorphic to $(\bar{Z}_6,+)$.
6. Let G be an abelian group and H a nonabelian group. Prove that G and H are not isomorphic.
7. Let $f: G \longrightarrow G$ be defined as follows: $f(g) = a^{-1}ga$ for a fixed in G, and all $g \in G$. Show that f is an isomorphism.

4.5   The Generalized Associative Law and the Law of Exponents

Let $(G,\circ)$ be a group. Since $\circ$ is a *binary* operation, it makes sense to form products of *pairs* of elements—that is, to combine elements of G two at a time. Hence, given three elements a_1, a_2, a_3 we may form a single product in this order in two distinct ways, namely, $(a_1 \circ a_2) \circ a_3$ and $a_1 \circ (a_2 \circ a_3)$. The associative law states that both of these products represent the same element of G. Hence we may write $a_1 \circ a_2 \circ a_3$ with no ambiguity. In the case of four elements, the number of products that can be formed is five, and it is not difficult to show that they are all equal in G.

EXERCISES

1. Form the five products of a_1, a_2, a_3, a_4 in that order and, using the associative law, show that they are all equal.
2. Let the binary operation be subtraction and compute the five different elements produced from the integers 1, 2, 3, 4 in that order as in Exercise 1. That is, compute $1 - (2 - (3 - 4))$, $(1 - 2) - (3 - 4)$, and so forth. Are these five elements equal?

THEOREM 4.5.1   (*Generalized Associative Law*)   Let G be a group and let $a_1, \ldots, a_n$ be n elements of G. Then any two products of $a_1, \ldots, a_n$ in

4.5 GENERALIZED ASSOCIATIVE LAW AND LAW OF EXPONENTS

that order are equal. Hence we may speak of *the product* of $a_1, \ldots, a_n$ in that order and denote it simply $a_1 a_2 \cdots a_n$.

Proof Let g and h represent two products of $a_1, \ldots, a_n$ in that order. Then $g = uv$, where u is a product of $a_1, \ldots, a_r$ in that order and v is a product of $a_{r+1}, \ldots, a_n$ in that order for some r, $1 \leq r < n$, and $h = xy$, where x is a product of $a_1, \ldots, a_s$ in that order and y is a product of $a_{s+1}, \ldots, a_n$ in that order for some s, $1 \leq s < n$. This is true because a product of $a_1, \ldots, a_n$ in the given order is formed by a repeated application of the binary operation, and so the last step must be the formation of a product from two elements of G. We now use induction, assuming the theorem is true for all integers less than n. Clearly, we may assume $n \geq 3$, $1 \leq r < n$, and $1 \leq s < n$. The theorem is true for $n = 3$ by the associative law. Now suppose that $r = s$. Since each product u, x, v, and y involves fewer than n of the a_i's, by induction $u = x$ and $v = y$, and so $g = h$. If $r > s$, then $u = xz$, where x is defined above and z is a product of $x_{s+1}, \ldots, x_r$ in that order. Thus $g = (xz)v$, and by the associative law $g = x(zv)$. But now zv is a product of $x_{s+1}, \ldots, x_n$ in that order, and by induction $zv = y$. Hence $g = x(zv) = xy = h$ and the theorem follows. Of course if $r < s$, then a similar argument may be used. ∎

The reader should note that the argument given above has very little to do with groups. Any set with an associative binary operation will do. Thus the proof of Theorem 4.5.1 also provides a proof for the generalized associative law for unions and intersections of finitely many sets, as mentioned in Chapter 1, as well as for sums and products of integers, as mentioned in Chapter 2.

We have been careful to include the phrase "in that order" in the arguments above, since we wish our results to be valid in the general noncommutative case. If G is an abelian group, however, we can state our main result in the following form.

THEOREM 4.5.2 Let G be an abelian group, and let $a_1, \ldots, a_n$ be elements of G. Let $a_{i_1}, a_{i_2}, \ldots, a_{i_n}$ be the same elements listed in some order. Then the product of $a_1, \ldots, a_n$ is equal to the product of $a_{i_1}, \ldots, a_{i_n}$ in their respective orders; that is, the order of $a_1, \ldots, a_n$ may be altered without changing the product.

Proof The proof is by induction on n. As in the proof of the previous theorem let $g = uv$ be the product of $a_{i_1}, \ldots, a_{i_n}$, with u the product of $a_{i_1}, \ldots, a_{i_r}$ in that order and v the product of $a_{i_{r+1}}, \ldots, a_{i_n}$ in that order. Now one of the a_{i_k}'s is equal to a_n, say $a_{i_j} = a_n$. First assume that $j > r$; that is, a_n occurs in v. Then by the induction assumption, $v = xa_n$, where x is the product of all of $a_{i_{r+1}}, \ldots, a_{i_n}$ except a_{i_j}, and $g = u(xa_n) = (ux)a_n$. But now by induction ux equals the product of $a_1, \ldots, a_{n-1}$ in that order and hence g equals the product of $a_1, \ldots, a_n$ in the given order. On the other hand if $j \leq r$, that is, if a_n occurs in u, then since G is abelian, $g = uv = vu$, and we now proceed as above. ∎

We are now in a position to develop the basic facts about exponents.

Definition 4.5.3 Let a be an element of a group G. We define a^n to be the product of $a_1, \ldots, a_n$ with $a_i = a$ for $i = 1, \ldots, n$, $n \geq 1$. For $n = 0$ we define $a^0 = e$, the identity of G.

It is useful to think of a^n as the product of a "n times."

LEMMA 4.5.4 Let G be a group and let $a \in G$. Then $a^n = a^{n-1}a = aa^{n-1}$ for all $n \in Z$, $n \geq 1$.

Proof If $n = 1$, then the lemma clearly follows. If $n > 1$, then a^{n-1} is by definition the product of $a_1, \ldots, a_{n-1}$ with $a_i = a$ for $i = 1, \ldots, n - 1$. Hence $a^{n-1}a$ is the product of $a_1, \ldots, a_{n-1}, a$ with $a_i = a$, $i = 1, \ldots, n - 1$. Thus, $a^{n-1}a = a^n$. Similarly $aa^{n-1} = a^n$. ∎

THEOREM 4.5.5 Let G be a group and let $a \in G$. Then $(a^n)^{-1} = (a^{-1})^n$ for $n \in Z$, $n \geq 0$.

Proof The proof is by induction on n. If $n = 0$, then $(a^0)^{-1} = e^{-1} = e = (a^{-1})^0$, since any element to the zeroth power is e. Assume the result is true for k, $k \geq 0$; that is, assume $(a^k)^{-1} = (a^{-1})^k$. We will prove the result is true for $k + 1$. Thus $a^{-1}(a^k)^{-1} = a^{-1}(a^{-1})^k$, and by Theorem 4.2.5, $a^{-1}(a^k)^{-1} = (a^ka)^{-1}$. But $a^{-1}(a^{-1})^k = (a^{-1})^{k+1}$ and $a^ka = a^{k+1}$ by Lemma 4.5.4, and so $(a^{k+1})^{-1} = (a^{-1})^{k+1}$. Hence the theorem follows by induction. ∎

Definition 4.5.6 Let G be a group and let $a \in G$. We define $a^{-n} = (a^{-1})^n$, where $n \in Z$, $n > 0$.

LEMMA 4.5.7 Let G be a group and let $a \in G$. Then $a^{-n} = (a^{-1})^n$ for all integers n.

Proof If $n \geq 0$, this follows from Definitions 4.5.6 and 4.5.3. If $n < 0$, then $a^{-n} = ((a^{-1})^{-1})^{-n} = (a^{-1})^n$ by Corollary 4.2.4 and Definition 4.5.6. ∎

THEOREM 4.5. (*Laws of Exponents*) Let a be an element of a group G. Then

(a) $a^n a^m = a^{n+m}$

and

(b) $(a^n)^m = a^{nm}$ for all $n, m \in Z$.

Proof (a) We first consider the case $n \geq 0$ and $m \geq 0$. It follows from Definition 4.5.3 and from Theorem 4.5.1 that $a^n a^m$ is the product of $a_1, a_2, \ldots, a_{n+m}$ with $a_i = a$ for all $i = 1, \ldots, n + m$ and so $a^n a^m = a^{n+m}$.

4.5 GENERALIZED ASSOCIATIVE LAW AND LAW OF EXPONENTS

If $n < 0$ and $m < 0$, then we can repeat the argument above by replacing $a^n a^m$ by $(a^{-1})^{-n}(a^{-1})^{-m}$.

Finally let $n \geq 0$ and $m < 0$. If $n = 0$, then the theorem follows trivially. Thus assume $n > 0$ and that the result holds for all k such that $0 \leq k < n$ and all $m < 0$. Then

$$a^n a^m = a^n (a^{-1})^{-m} = a^n(a^{-1}(a^{-1})^{-m-1}) = (a^n a^{-1})(a^{-1})^{-m-1}$$
$$= ((a^{n-1}a)a^{-1})(a^{-1})^{-m-1} = (a^{n-1}(aa^{-1}))(a^{-1})^{-m-1}$$
$$= (a^{n-1})(a^{-1})^{-m-1} = a^{n-1}a^{m+1} = a^{n-1+m+1}$$

by the induction assumption since $n - 1 < n$. Hence

$$a^n a^m = a^{n-1+m+1} = a^{n+m}$$

by induction.

The case $n < 0$, $m \geq 0$ can be handled similarly. This completes the proof of (a).

(b) If $m = 0$, then $(a^n)^0 = e = a^0 = a^{n \cdot 0}$. We now use induction on m. Thus, assume $m > 0$ and that the theorem holds for all k such that $0 \leq k < m$ and all n. Then

$$(a^n)^m = (a^n)^{m-1} a^n \text{ [by part (a)]} = a^{n(m-1)} a^n,$$

by the induction assumption. Hence by (a),

$$a^{n(m-1)} a^n = a^{n(m-1)+n} = a^{nm}$$

and the result follows by induction for nonnegative m.

If $m < 0$, then

$$(a^n)^m = ((a^n)^{-1})^{-m} = ((a^{-1})^n)^{-m}$$
$$= (a^{-1})^{n(-m)} = (a^{-1})^{-nm}$$

by the previous case and Theorem 4.5.5. But by Theorem 4.5.5 again,

$$(a^{-1})^{-nm} = a^{nm} \quad \text{and so} \quad (a^n)^m = a^{nm} \quad \text{for } m < 0 \text{ and all } n.$$

This completes the proof of the theorem. ∎

We recall that when a group is written additively, that is, when $+$ is used to represent the group operation, then the inverse of an element a is written as $-a$, and 0 is used to represent the group identity. The element we have written as a^n above becomes na, and the theorem just proved becomes

(a) $na + ma = (n + m)a$,
(b) $n(ma) = (nm)a$ for $n, m \in \mathbb{Z}$.

EXERCISES

3. Let a be an element of a group G. Prove that $aa^n = a^n a$ for all integers n.

124   THE THEORY OF GROUPS

4. Let a be an element of the group G, and let m, n be integers. Show that $a^n a^m = a^m a^n$.
5. Let a be an element of the group G. Prove that $a^n a^m a^p = a^{n+m+p}$ for integers n, m, p.
6. Let G be a group. Show that G is abelian if and only if $(ab)^n = a^n b^n$ for all $a, b \in G$ and all $n \in Z$.
7. Suppose that $a^{-1}ba = c$ for a, b, c in a group G. Prove $a^{-1}b^i a = c^i$ for all $i \in Z$.
8. Suppose that $a^{-1}ba = b^2$ for a, b in a group G. Show that $a^{-3}ba^3 = b^8$.
9. Let a, b be elements of a group G. And let $a^{-1}ba = b^j$ for some integer $j \geq 0$. Prove that $a^{-r}ba^r = b^{j^r}$ for all integers $r \geq 0$. (Note that if $j = 1$, then $a^{-1}ba = b$ implies that $a^{-r}ba^r = b$, for all integers $r \geq 0$.)

4.6   Subgroups

The reader may have noticed from some of the examples in Section 4.1 that certain subsets of a group $(G,\circ)$ are themselves groups with respect to the binary operation $\circ$ of G. That is, if the binary operation defined for G is restricted to a subset H of G, then it may happen that this subset with the restricted operation is a group. When this occurs, we say that the group $(H,\circ)$ is a *subgroup* of $(G,\circ)$.

Definition 4.6.1   Let $(G,\circ)$ be a group. Let $H \subseteq G$, $H \neq \emptyset$. Let $*$ be the restriction of $\circ$ to H, that is, $a * b = a \circ b$, for all $a, b \in H$. If $(H,*)$ is a group, we call H a *subgroup* of G.

Since $*$ is simply the operation $\circ$ restricted to the elements of H, we shall ordinarily write $(H,\circ)$ or simply H in place of $(H,*)$.

Example 4.6.2   Let $(Z,+)$ be the additive group of integers. Then the set of even integers forms a group with respect to addition. This follows from the fact that associativity of addition will hold on any subset of integers, that the sum of a pair of even integers is even and the negative of an even integer is even.

Example 4.6.3   Let G be the Klein 4-group—that is, the group of all symmetries of a nonsquare rectangle. Its elements consist of e, r, v, h (see Example 4.1.7). The set $\{e,v\}$ is a subgroup of G, since the products ee, ev, ve, and vv are all in $\{e,v\}$, and v is its own inverse.

It is easy to see that every group G has at least two subgroups: the group G itself and the subgroup that has only one element, namely the identity element. If $(H,\circ)$ is a subgroup of $(G,\circ)$, but $H \neq G$, that is,

$H \subset G$, we say that H is a *proper subgroup* of G. The subgroup consisting of only the identity element is called the *trivial subgroup* of G, and will be denoted by E.

Since the binary operation of a subgroup H is essentially the same as in the full group G, we should expect that the elements that act as inverse and identity in H will be the same as in G.

THEOREM 4.6.4 Let G be a group and let H be a subgroup of G. Then

(a) the identity element of H is the identity of G,
(b) if $a \in H$ and a^{-1} is the inverse of a in G, then $a^{-1} \in H$ and a^{-1} is the inverse of a in H.

Proof (a) Let e' be the identity element of the group H. Then $e'e' = e'$ in H and hence in G. Therefore, by Exercise 2 Section 4.2, $e' = e$, the identity of G.

(b) Let $a \in H$. Let b be the inverse in H of a. By (a), $ab = e$, the identity of both H and G. Hence b is the inverse of a in G. ∎

Often it becomes necessary to determine when a subset of elements of a group constitutes a subgroup with respect to the group operation. The determination can be made, of course, by directly testing each of the group axioms for the set in question. The task is simplified, however, when we realize that since the elements of the subset are elements of the group, certain of the group axioms are automatically true. (Name one.) It is necessary only to verify closure, that is, $a \circ b \in H$, for all $a, b \in H$, and to verify that $a^{-1} \in H$ for all $a \in H$. This can be accomplished in a single step, as we prove next.

THEOREM 4.6.5 Let H be a nonempty subset of a group G. Then H is a subgroup of G if and only if for all $a, b \in H$, the element $ab^{-1} \in H$.

Proof Suppose H is a subgroup of G. Then, if $a, b \in H$, also $b^{-1} \in H$, whence $ab^{-1} \in H$.

Suppose, conversely, whenever $a, b \in H$, also $ab^{-1} \in H$. We verify that all of the group axioms hold for H.

(a) H contains e, the identity of G. For, since $H \neq \emptyset$, there is an $a \in H$. By hypothesis, with $b = a$, $aa^{-1} = e \in H$. (Note that we need not assume that $a^{-1} \in H$.)

(b) For each $a \in H$, $a^{-1} \in H$. For, by (a), $e \in H$, whence for each $a \in H$, $ea^{-1} = a^{-1} \in H$, by hypothesis.

(c) H is closed with respect to the binary operation on G; that is, the binary operation of G restricted to H is a binary operation on H. To show this, let $a, b \in H$. By (b), $b^{-1} \in H$. Then, by hypothesis, $a(b^{-1})^{-1} \in H$; that is, $ab \in H$.

(d) The associative law holds in H. Let $a, b, c \in H$. Since $H \subseteq G$,

$$a, b, c \in G,$$

whence $a(bc) = (ab)c$.

Properties (a), (b), (c), and (d), taken together, show that H is a group, and the theorem is proved. ∎

We can use this criterion to show quickly that the set of even integers does form a subgroup of the group of all integers under addition. For, if a and b are even integers, then $a - b$ is an even integer.

If H is a finite subset of the group G, then we can state another criterion for a subset H of G to be a subgroup.

THEOREM 4.6.6 Let G be a group and H a finite nonempty subset of G. Then H is a subgroup of G if and only if $ab \in H$ whenever $a, b \in H$.

Proof If H is a subgroup of G, then clearly for each pair $a, b \in H$, the element $ab \in H$.

Now suppose that H is a nonempty subset of G satisfying the criterion: If $a, b \in H$, then $ab \in H$. Since H is finite, let $H = \{a_1, \ldots, a_r\}$, where all the a_i's are distinct. Now consider the set

$$a_1 H = \{a_1 a_1, a_1 a_2, \ldots, a_1 a_r\}.$$

If $a_1 a_i = a_1 a_j$, then by cancellation $a_i = a_j$, a contradiction. Hence the r elements of the set $a_1 H$ are distinct elements of H and hence $a_1 H = H$. Thus $a_1 \in a_1 H$, and so, for some i, $a_1 = a_1 a_i$. Hence $a_i = e$, and H contains the identity of G. Therefore $e \in a_1 H$, and so for some j, $a_1 a_j = e$. Thus $a_j = a_1^{-1} \in H$. By now considering $a_i H = H$, it follows that H contains the inverse a_i^{-1} for each of its members a_i. Thus, since H is closed with respect to the group operation, it follows that H is a subgroup of H. ∎

The theorem above has an interesting interpretation in terms of the multiplication table of G. For if the elements of H are the headings for the first r rows and columns, then the upper lefthand $r \times r$ square corner of the table will contain only elements of H if and only if H is a subgroup of G.

Example 4.6.7 Let $G = (\bar{Z}_6, +)$. Then the subgroups of $\bar{Z}_6$ are $E = \{\bar{0}\}$, $A_1 = \{\bar{0}, \bar{2}, \bar{4}\}$, $A_2 = \{\bar{0}, \bar{3}\}$, $A_3 = \bar{Z}_6$. Note that $\bar{0}$ is contained in each of these. Why is $\{\bar{0}, \bar{2}, \bar{3}\}$ not a subgroup of $\bar{Z}_6$?

Example 4.6.8 In this example we determine all subgroups of the dihedral group D_4. We have already noted that D_4 can be thought of either as the specific group of transformations of the plane given in Example 4.1.8 or as the special case $n = 4$ of Example 4.1.9. We list the elements of D_4 as in Example 4.1.8:

$$D_4 = \{e,r_1,r_2,r_3,v,h,d_1,d_{-1}\}.$$

The two subgroups D_4 and E are immediate. Now, since $v^2 = h^2 = d_1{}^2 = d_{-1}{}^2 = e$, it follows that $V = \{e,v\}$, $H = \{e,h\}$, $D_1 = \{e,d_1\}$, and $D_{-1} = \{e,d_{-1}\}$ are subgroups. We have already remarked earlier that a product of rotations is again a rotation, and so $R = \{e,r_1,r_2,r_3\}$ is a subgroup. Now $r_2{}^2 = e$ and so $R_2 = \{e,r_2\}$ is a subgroup. There are no other subgroups composed of rotations alone, since $r_1{}^2 = r_2$, $r_1{}^3 = r_3$, and $r_3{}^3 = r_1$. The remaining subgroups must contain some reflections, and

$$M = \{e,r_2,v,h\} \quad \text{and} \quad N = \{e,r_2,d_1,d_{-1}\}$$

are subgroups, as can be seen by calculating $M \cdot M$ and $N \cdot N$ and using Theorem 4.6.6 ($T \cdot T = \{ab \mid a, b \in T\}$). No subgroup other than V, H, D_1, and D_{-1} can contain reflections alone, since the product of any two distinct reflections (different from e) is a rotation, as can be verified directly. Now any other subgroup K of D_4 would have to contain a rotation and reflection. The only possibilities remaining for K would have to contain r_1 and a reflection, or r_3 and a reflection, or r_2 together with two reflections different from those already listed. In all of these cases K would turn out to be the entire group D_4. Again direct calculations will show these assertions.

We will shortly develop some elementary but powerful results that will allow us to analyze situations like that above with a minimum of effort.

EXERCISES

1. Find all subgroups of the groups S_3 and the Klein group. Find a nontrivial subgroup of S_3 isomorphic to a subgroup of the Klein group.
2. Let A be a subgroup of a group G and B a subgroup of A. Show that B is a subgroup of G.
3. Show directly that the set $M = \{e,r_2,v,h\}$ is a subgroup of D_4 (see Example 4.6.8).
4. Show directly that the only subgroup of D_4 that contains r_1 and v is the entire group D_4 (see Example 4.6.8).
5. Let $L_m = \{x \mid x \in Z \text{ and } m \mid x\}$. Show that L_m is a subgroup of the additive group of integers.
6. Show that if A is a subgroup of the additive group of integers, then $A = L_m$ for some integer m (see Exercise 5 above).
7. Let A be the set of all 3-cycles of S_4; that is,

$$A = \{(1\ 2\ 3),(1\ 2\ 4),(1\ 3\ 4),(2\ 3\ 4),(1\ 3\ 2),(1\ 4\ 2),(1\ 4\ 3),(2\ 4\ 3)\}.$$

Find the smallest subgroup of S_4 containing A. This subgroup is called the subgroup generated by A. What is its order? Find all of its subgroups.

8. Let A be the set of all 3-cycles of S_n. Show that A generates A_n. [*Hint:* Show that the product of two disjoint transpositions is equal to the product of two 3-cycles.]

9. Let $N = \{e, r_2, d_1, d_{-1}\}$ be the subgroup of D_4 of Example 4.6.8. Calculate the following sets:

$$d_1 N = \{d_1 e, d_1 r_2, d_1 d_1, d_1 d_{-1}\}, \quad vN = \{ve, vr_2, vd_1, vd_{-1}\}.$$

Is either of these sets equal to N?

10. Let H be a subgroup of the group G. Suppose that

$$gH = \{gh \mid h \in H\}.$$

Show that $g \in H$ if and only if $gH = H$.

11. Show by example that Theorem 4.6.6 is false if the hypothesis that H is finite is removed. [Consider the group $(Z, +)$. What is an appropriate choice for H?]

4.7 Cyclic Groups

As the reader may have already observed, it is comparatively easy to recognize certain subgroups of a group G. One need only choose an element a and the set of all powers (including negative powers) of a. Since every element in such a set has the form a^m for some integer m, and since $a^n \circ (a^m)^{-1} = a^n \circ a^{-m} = a^{n-m}$, it follows that such a set forms a subgroup of G. We have therefore proved the following theorem.

THEOREM 4.7.1 Let a be an element of a group G. Then the set $A = \{a^n \mid \text{all } n \in Z\}$ is a subgroup of G. ∎

If it should occur that for some choice of $a \in G$, the subgroup A is actually all of G, then G is called *cyclic* and has some very special properties that we investigate in this section.

Definition 4.7.2 Let G be a group. If there is an element $a \in G$ such that $G = \{a^m \mid m \in Z\}$, then G is called a *cyclic group* (*generated by the element* a), and a is called a *generator* of G. We denote this by $G = (a)$.

We observe that if $G = (a)$, every element $g \in G$ can be expressed in the form a^m, for some integer m. As we shall see by later examples, the choice of m need not be unique.

In additive notation, the definition becomes: there exists an element $a \in G$ such that for each $g \in G$, there exists an integer m such that $g = ma$. It is immediate that $(Z, +)$ is cyclic, since 1 is a generator.

4.7 CYCLIC GROUPS

Example 4.7.3 Let G be the Klein group. If we let a be the rotation of 180 degrees, then $a^2 = e$, $a^{-1} = a$, and hence the set of powers of a forms a cyclic subgroup of order 2. The reader can easily check that each nonidentity element of the group generates a cyclic subgroup of order 2. Therefore, G is not a cyclic group.

Example 4.7.4 Let $G = (\bar{Z}_6, +)$. Then the cyclic subgroup generated by $\bar{2}$ is the set $\{\bar{0}, \bar{2}, \bar{4}\}$, a proper subgroup of G. The set of multiples of $\bar{1}$, however, is the whole group; that is, $(\bar{1}) = G$. The reader should also verify that $(\bar{5}) = G$.

Example 4.7.5 Let $G = S_4$. Then the element (1 2 3 4) generates the cyclic group $\{(1\ 2\ 3\ 4), (1\ 3)(2\ 4), (1\ 4\ 3\ 2), (1)\}$.

EXERCISES

1. Show that all proper subgroups of S_3 are cyclic.
2. Find all cyclic subgroups of S_4. What are their orders?
3. Show that S_4 contains cyclic subgroups of orders 1, 2, 3, and 4, but of no larger order.
4. Show that the subgroup of D_4 consisting of rotations is a cyclic group.
5. Let G be a cyclic group of order 8. Find all subgroups of G and show that they are all cyclic.
6. Show that $(\bar{Z}_m, +)$ is cyclic for all positive integers m.
7. Show that every cyclic group is abelian.

Since a cyclic group consists of the powers of a single element, and since by the law of exponents $a^n a^m = a^{n+m}$, the binary operation of a cyclic group behaves similarly to addition in Z. Exactly how far this analogy goes is investigated in the next several theorems.

LEMMA 4.7.6 Let G be a cyclic group generated by a. Let e be the identity of G. Then $a^n = a^m$ for $n \neq m$ if and only if $a^x = e$ has a least positive solution, say d, and $n \equiv m \pmod{d}$.

Proof Suppose $a^n = a^m$, $m \neq n$. Assuming $n < m$, we see that $a^{m-n} = e$ implies that $a^x = e$ has a positive solution and hence a least positive solution d. By the division algorithm, $m - n = qd + r$, $0 \leq r < d$. Then

$$e = a^{m-n} = a^{qd}a^r = (a^d)^q a^r = e^q a^r = a^r.$$

Thus $a^r = e$. But since $0 \leq r < d$, we must have $r = 0$, for otherwise the choice of d as the smallest positive solution of $a^x = e$ is contradicted. Thus $m - n = qd$; that is, $m \equiv n \pmod{d}$.

Conversely, suppose $m \equiv n \pmod{d}$, where d is the least positive solu-

tion to $a^x = e$. Then $m - n = kd$, for some integer k, and
$$a^{m-n} = a^{kd} = (a^d)^k = e^k = e,$$
whence
$$a^m = a^n. \qquad \blacksquare$$

THEOREM 4.7.7 Let G be a cyclic group generated by a, and let e denote the identity of G.

If $|G| = d$, then d is the least positive solution of $a^x = e$ and
$$G = \{a, a^2, \ldots, a^{d-1}, a^d = e\}.$$

Conversely, if d is the least positive solution to $a^x = e$, then G is finite and $|G| = d$.

Proof If G is finite, then there must be a repetition in the list a, a^2, a^3, ... Thus, there are integers m and n, $m \neq n$, such that $a^m = a^n$. By Lemma 4.7.6, there is a least positive integer f satisfying $a^f = e$. The set of integers $\{0, 1, \ldots, f - 1\}$ is a complete set of least positive residues modulo f. We may thus conclude:

(a) If i and j are two distinct least positive residues, then $a^i \neq a^j$, for otherwise $i \equiv j \pmod{f}$ by Lemma 4.7.6. Hence the elements $e, a, a^2, \ldots, a^{f-1}$ are distinct.

(b) If m is any integer, then there exists a least positive residue $k \equiv m \pmod{f}$, and hence $a^k = a^m$. Thus the set $\{a, a^2, \ldots, a^{f-1}, a^f = e\}$ contains all the elements of G.

Therefore $|G| = f$ and so $f = d$; and d is the least positive solution of $a^x = e$.

Conversely, if d is the least positive solution of $a^x = e$, then by Lemma 4.7.6, $a^n = a^m$ if and only if $n \equiv m \pmod{d}$. It is now easy to see that $G = \{a, a^2, \ldots, a^{d-1}, a^d = e\}$ and $|G| = d$. $\qquad \blacksquare$

The close relation between cyclic groups and the arithmetic residue classes is more than incidental.

THEOREM 4.7.8 Let G be a cyclic group.

(a) If G is finite of order d, then G is isomorphic to $(\bar{Z}_d, +)$, the additive group of residue classes modulo d.

(b) If G is infinite, then G is isomorphic to $(Z, +)$, the additive group of integers.

Proof Let G be a cyclic group generated by the element a.

(a) If $|G| = d$ a positive integer, then
$$G = \{a^0, a, a^2, \ldots, a^{d-1}\}, \qquad \text{where } a^0 = a^d = e.$$

4.7 CYCLIC GROUPS 131

Let $f: G \longrightarrow \bar{Z}_d$ be defined by $f(a^i) = \bar{\imath}$ for $i = 0, \ldots, d - 1$. Clearly f is onto. To show that f is 1–1, suppose that $f(a^i) = \bar{\imath} = f(a^j) = \bar{\jmath}$. Then since $\bar{\imath} = \bar{\jmath}$, we have $i \equiv j \pmod{d}$ and $a^i = a^j$. Now consider $f(a^i a^j) = f(a^{i+j})$. Since, by the division algorithm, there are integers q and r such that $i + j = qd + r$, with $0 \leq r < d$, it follows that $i + j \equiv r \pmod{d}$ and hence $a^{i+j} = a^r$. Thus, in $\bar{Z}_d$,

$$f(a^i a^j) = f(a^{i+j}) = f(a^r) = \bar{r} = \overline{i+j} = \bar{\imath} + \bar{\jmath} = f(a^i) + f(a^j),$$

whence f is an isomorphism and $G \cong \bar{Z}_d$.
(b) If G is infinite and generated by the element a, then

$$G = \{a^i \mid i \in Z\}.$$

Now consider $f: Z \longrightarrow G$ defined by $f(i) = a^i$. Clearly f is onto. To show that f is 1–1, suppose that $f(i) = f(j)$, that is, $a^i = a^j$. Then $a^{i-j} = e$, the identity of G, whence $i - j = 0$, for if $i - j \neq 0$, then G would be finite. Finally, to show that f is an isomorphism, observe that

$$f(i + j) = a^{i+j} = a^i a^j = f(i)f(j) \quad \text{in } G.$$

Thus $G \cong Z$. ∎

COROLLARY 4.7.9 Let G and H be two cyclic groups of the same order. Then $G \cong H$.

Proof The proof is left as an exercise. ∎

By virtue of Corollary 4.7.9, we may now speak of *the* cyclic group of order n, C_n, for any positive integer n, with no ambiguity up to isomorphism.

Definition 4.7.10 Let a be an element of the group G. The *order of a*, denoted $o(a)$, is defined to be the least positive integer satisfying $a^x = e$. If no such integer exists, we say that a is an element of infinite order and write $o(a) = \infty$.

If a is an element of order n, then it follows that the cyclic subgroup generated by a has order n. Thus $o(a) = n$ if and only if $|(a)| = n$.

Perhaps the reader has noticed that, in the definition of cyclic group, care was taken to avoid talking about *the* generator of G. First, it is easy to see that if a is a generator of G, then a^{-1} is also a generator. Second, it can happen that G is generated by yet other elements. For instance, if G is the additive group $(\bar{Z}_8, +)$ of residue classes modulo 8, then G is generated not only by $\bar{1}$ but also by $\bar{3}$ or $\bar{5}$ or $\bar{7}$. Those elements of a cyclic group that can act as generators are completely described by the next theorem.

THEOREM 4.7.11 Let G be a cyclic group generated by a.

(a) If G is finite of order d, then a^k is a generator of G if and only if $(k,d) = 1$.

(b) If G is infinite, then a and a^{-1} are the only generators of G.

Proof (a) a^k is a generator of G if and only if $(a^k)^l = a$, for some integer l. By Lemma 4.7.6, however, $(a^k)^l = a$ if and only if $kl \equiv 1 \pmod{d}$. But this equation, by Corollary 2.13.5, has a solution if and only if $(k,d) = 1$. Hence (a) is proved.

(b) If a^k is a generator of G, then for some l, $a^{kl} = a$. Hence, either $kl = 1$ or the equation $a^x = e$ has a positive solution. In the second case, G would be finite, a contradiction. Hence $kl = 1$ and $k = \pm 1$. ∎

If we ask how many different elements of a cyclic group may be selected as generators, the theorem shows that this is equivalent to asking how many positive integers there are less than d and relatively prime to d. This was answered in Theorem 2.14.6, since $\varphi(d)$ (φ is the Euler φ-function) gives precisely this information.

EXERCISES

8. Prove Corollary 4.7.9.
9. Let G be the multiplicative group of reduced residue classes modulo 7. (See Exercise 11, Section 4.2.)
    (a) Show that G is cyclic.
    (b) To what additive group of residue classes of integers is this group isomorphic?

THEOREM 4.7.12 If G is a cyclic group and if H is a subgroup of G, then H is cyclic.

Proof Let G be generated by the element a. The elements of H form a subset of the powers of a. If $H = E$, then H is cyclic. Otherwise H contains a positive power of a. Let k be the least positive integer such that $a^k \in H$. Let $K = \{a^{kl} \mid l \in Z\}$. Then K is the cyclic subgroup of G generated by a^k, and $K \subseteq H$. We will show that $H = K$. Let a^s be an arbitrary element of H. By the division algorithm, there exist unique integers q and r such that
$$s = qk + r, \quad 0 \leq r < k.$$

Then $a^s = a^{kq+r} = a^{kq}a^r$, whence $a^r = a^s(a^k)^{-q}$. Since a^s and a^k are in H, we have $a^r \in H$. But $0 \leq r < k$, and so, by the choice of k, it follows that $r = 0$. Thus $a^s = (a^k)^q \in K$. ∎

THEOREM 4.7.13 Let G be a cyclic group of order n. If $d \mid n$, $d > 0$, then G contains one and only one subgroup of order d.

Proof Let a be a generator of G. If $d \mid n$, then $n = dr$, for some integer r. Since n is the smallest positive integer such that $a^n = e$, then d is the smallest positive integer such that $(a^r)^d = e$. Hence $o(a^r) = d$, and the cyclic subgroup generated by a^r has order d. Thus, G contains a subgroup of order d.

We must now show that (a^r) is the only subgroup of G of order d. Suppose there is another subgroup, say L, of order d. Then, by Theorem 4.7.12, L is cyclic, whence L is generated by a^s for some $s \geq 0$. Then $(a^s)^d = e = a^0$, and so $sd \equiv 0 \pmod{n}$, that is, $sd = kn$ for some integer k. But $n = rd$ and so $sd = krd$, which implies that $s = kr$. Therefore $a^s = a^{kr} = (a^r)^k$, and so $a^s \in (a^r)$, whence $L \subseteq (a^r)$. But L and (a^r) have the same finite order, whence $L = (a^r)$. Thus, G contains precisely one subgroup of order d, namely $(a^{n/d})$. ∎

THEOREM 4.7.14 Let a be an element of a group G and suppose that $o(a) = d$. Then $a^n = e$ if and only if $d \mid n$.

Proof If $d \mid n$, then $n = dk$ for some integer k. Then $a^n = a^{dk} = (a^d)^k = e^k = e$, since $o(a) = d$.

Conversely, if $a^n = e$, then $a^n = a^0$, and so $n \equiv 0 \pmod{d}$ by Lemma 4.7.6. Thus $n = dr$ for some integer r, and so $d \mid n$. ∎

Given two elements a and b in a group G, usually one cannot give the order of the element ab in terms of the orders of a and b. For example, the permutations (1 2 3) and (1 2) have orders 3 and 2, respectively, but their product (1 2 3)(1 2) = (1 3) has order 2. In D_4, the product of v and d_1, both elements of order 2, has order 4. There is an important case, however, in which it is possible to make a general statement about the order of a product.

THEOREM 4.7.15 Let G be an abelian group and let $a, b \in G$ be elements of finite relatively prime orders—that is, $(o(a), o(b)) = 1$. Then

$$o(ab) = o(a)o(b).$$

Before giving the proof of Theorem 4.7.15, we need two lemmas, which also have independent interest.

LEMMA 4.7.16 If a is an element of the group G, then $o(a) = o(a^{-1})$.

Proof The proof is left as an exercise. ∎

LEMMA 4.7.17 If a is an element of the group G and $o(a) = n$, then $o(a^d) \mid n$, for any integer d.

Proof Let a^d have order s; that is, s is the least positive integer such that $(a^d)^s = e$. By the division algorithm, there are integers q and r such that
$$n = sq + r, \quad 0 \leq r < s.$$
Then
$$(a^d)^n = (a^d)^{sq+r} = (a^d)^{sq}(a^d)^r,$$
$$(a^n)^d = ((a^d)^s)^q(a^d)^r,$$
$$e^d = e^q(a^d)^r,$$
$$e = (a^d)^r.$$

Thus, since $o(a^d) = s$, and $0 \leq r < s$, we have $r = 0$, so $n = sq$; that is, $o(a^d) \mid n$. ∎

Proof of 4.7.15 Let $o(a) = n$ and $o(b) = m$ with $(m,n) = 1$. Then $(ab)^{mn} = a^{mn}b^{mn}$ by Exercise 6, Section 4.5. But $a^{mn} = a^{nm} = (a^n)^m = e^m = e$ and $b^{mn} = (b^m)^n = e^n = e$, and so $(ab)^{mn} = e$. Thus, by Theorem 4.7.14, the order d of ab is a divisor of mn.

Now $e = (ab)^d = a^d b^d$ and so $a^d = b^{-d}$. But $o(a^d)$ is a divisor of $o(a) = n$ and $o(b^{-d})$ is a divisor of $o(b^{-1}) = o(b) = m$. Hence $o(a^d)$ divides both m and n, since $a^d = b^{-d}$. But $(n,m) = 1$, whence $o(a^d) = 1$, and so $a^d = b^{-d} = e$. Therefore $o(a) = n$ divides d, and $o(b^{-1}) = o(b) = m$ divides d. Since $(n,m) = 1$, we have $mn \mid d$. We now have $d \mid mn$ and $mn \mid d$, and so $mn = d = o(ab)$. ∎

EXERCISES

10. Prove Lemma 4.7.16.
11. Let $G = S_3$, the symmetric group on $\{1,2,3\}$. Show that
    (a) (123) and $(23)(123)(23)$ have the same order.
    (b) (13) and $(123)(13)(132)$ have the same order.
    (c) $(12)(123)$ and $(123)(12)$ have the same order.
    (These are special cases of Exercise 12.)
12. Let a and b be elements of a group G. Show that
    $$o(a) = o(g^{-1}ag) \quad \text{for all } g \in G;$$
    and that $o(ab) = o(ba)$.
13. Find the orders of all elements of $(\bar{Z}_{12}, +)$.
14. Let G be an abelian group of order pq, with $(p,q) = 1$. Show that if G contains elements a and b of orders p and q, respectively, then G is cyclic.
15. Let a be an element of a group G. Let $o(a) = mn$, where m and n are integers. Show that if $b = a^m$, then $o(b) = n$.
16. Let G be an abelian group containing elements a and b of orders 12

and 21, respectively. Show that there is an element c in G whose order is lcm [12,21].

17. Let G be an abelian group containing elements a and b of order m and n, respectively. Show that there is a $c \in G$ of order lcm $[m,n]$. [*Hint:* Let $m = p_1^{e_1} \cdots p_s^{e_s}$, $n = p_1^{f_1} \cdots p_s^{f_s}$, the p_i distinct primes, and

$$p_k^{e_k} < p_k^{f_k}, \quad 1 \leq k \leq t; \quad p_k^{f_k} \leq p_k^{e_k}, \quad t+1 \leq k \leq s.$$

Let

$$m_1 = p_1^{e_1} \cdots p_t^{e_t}, \quad n_1 = p_{t+1}^{f_{t+1}} \cdots p_s^{f_s}.$$

Consider a^{m_1} and b^{n_1}, and apply Theorem 4.7.15.]

18. Show that the order of any subgroup of a finite cyclic group divides the order of the group.

Thus far in this section we have discussed groups generated by one element. We wish to extend this notion to sets of elements that generate a group.

THEOREM 4.7.18 Let G be a group. Let $\{H_\alpha\}$ be a collection of subgroups of G. Then $\cap\, H_\alpha$ is a subgroup of G, where $\cap\, H_\alpha$ is the intersection of all those subgroups of G in the collection $\{H_\alpha\}$.

Proof Let $a, b \in \cap\, H_\alpha$. Then a and b are in each H_α. By Theorem 4.6.5, ab^{-1} is an element of each H_α, and so $ab^{-1} \in \cap\, H_\alpha$. Hence again by Theorem 4.6.5, $\cap\, H_\alpha$ is a subgroup of G. ∎

Now let G be a group and let A be a subset of G. Clearly there is at least one subgroup of G, namely G itself, that contains A. Thus the following definition will define a group according to Theorem 4.7.18.

Definition 4.7.19 Let G be a group and let A be a subset of G. Let $\{H_\alpha\}$ be the collection of all subgroups of G such that $A \subseteq H_\alpha$. Then $\cap\, H_\alpha$ is called the subgroup of G *generated* by A. A is called a set of *generators* of this subgroup. We denote this subgroup by (A).

Since any product of elements and inverses of elements in A is contained in any H_α containing A, and since this set of products is a subgroup of G, clearly (A) is simply the set of all such products. (See Exercise 29.) Thus (A) is the "smallest" subgroup of G containing the set A.

In case $A = \{a\}$, that is, A has exactly one element, then $(A) = (a)$, that is, (A) is the cyclic group generated by a.

It is not hard to see that a given group G may have different sets of generators. For example, if $G = (Z, +)$, then $A = \{1\}$ and $B = \{2, 3\}$ will both be sets of generators of G. Also, any group G will have at least one set of generators, namely, the set G itself. Although this latter set of generators is usually of little interest, certain sets of generators will be of importance.

Example 4.7.20 Let S_3 be the group of all permutations on $\{1,2,3\}$. Then S_3 is generated by (1 2) and (1 2 3). To see this, observe that any subgroup of S_3 that contains (1 2) and (1 2 3) also contains all possible products composed from these elements. Thus

(1 2 3)(1 2 3) = (1 3 2), (1 2)(1 2 3) = (2 3), (2 3)(1 2 3) = (1 3),

and (1 2)(1 2) = (1) are in such a subgroup. But these elements make up all of S_3. Hence S_3 is generated by (1 2) and (1 2 3). It is also easy to show that S_3 is generated by (1 3) and (2 3), since (2 3)(1 3) = (1 2 3).

Example 4.7.21 The group D_4 of Example 4.1.8 can be generated by r_1 and v. This is easily seen from Example 4.6.8, for in that example we determined all subgroups of D_4, and the only subgroup that contains r_1 and v is the whole group D_4.

EXERCISES

19. Prove that S_3 is generated by (1 3) and (1 2). Is S_3 generated by (1 2 3) and (1 3 2)? Explain.
20. Find all pairs of elements that generate D_4.
21. Let $A = \{a,b\}$ be a set of generators for the group G. Show that each of the following is also a set of generators of G:

    $\{a^{-1},b\}$; $\{a,ab\}$; $\{ab,ab^{-1}a^{-1}\}$.

22. Find a pair of elements that generates the Klein 4-group.
23. Show that for any positive integer n there exists a set of n integers that generates $(Z,+)$ and such that no subset of $n-1$ of these integers generates all of $(Z,+)$.
24. Show that (1 2) and (1 2 3 4) generate S_4.
25. Show that (1 2) and (1 2 $\cdots$ n) generate S_n.
26. Show that every element of D_4 can be written in the form $v^m r_1^n$ with $0 \leq m \leq 1$ and $0 \leq n \leq 3$.
27. Show that every element of D_n can be written in the form $v^\alpha r_1^\beta$, v an arbitrary reflection, r_1 the rotation of $2\pi/n$ radians, $0 \leq \alpha \leq 1$, $0 \leq \beta \leq n-1$.
28. Show that in D_n, $(v^{\alpha_1} r_1^{\beta_1})(v^{\alpha_2} r_1^{\beta_2}) = v_1^{\alpha_1+\alpha_2} r_1^{\beta_2+(-1)^{\alpha_2}\beta_1}$.
29. Let G be a group, $G = (A)$. Show that if $g \in G$, there exist elements $a_{i_1}, a_{i_2}, \ldots, a_{i_k}$ (with possible repetitions) of A such that

    $$g = a_{i_1}^{\epsilon_1} a_{i_2}^{\epsilon_2} \cdots a_{i_k}^{\epsilon_k},$$

    where each $\epsilon_j = 0, 1,$ or -1.

30. Let G be an infinite group, and suppose that $G = (A)$, where A is a finite subset of G. Show that if $G = (B)$, where B is an infinite subset of G, then there is a finite subset B' of B such that $G = (B')$.
31. Let a, b be elements of a group G and suppose that $ab = ba$. Show that the subgroup generated by $\{a,b\}$ is an abelian group.

4.8 Cosets and Lagrange's Theorem

In this section we prove the famous theorem of Lagrange that asserts that the order of a subgroup A of a finite group G must divide the order of G. Aside from its own usefulness, this theorem can be taken as the starting point for a great many significant results in the theory of finite groups.

The binary operation that is defined on the elements of a group can be extended to subsets of elements in the following way: if A and B are nonempty subsets of the group G, then the product AB is defined to be $\{ab \mid a \in A, b \in B\}$.

In general, little can be said about the set AB. For example, even if A and B are both subgroups of G, it still may not be true that the product AB is a subgroup of G.

In many special cases, however, the notion of a product of two sets is extremely important—especially when A consists of a single element and B is a subgroup.

Definition 4.8.1 Let G be a group. If $a \in G$ and H is a subgroup of G, we call the set $aH = \{ab \mid b \in H\}$ a *left coset of H* and we call the element a a *representative* of aH. The set $Ha = \{ba \mid b \in H\}$ is called a *right coset of H*.

Example 4.8.2 Let G be the Klein 4-group, $G = \{e,r,h,v\}$, of Example 4.1.7, and let $H = \{e,h\}$. Then the set $rH = \{re,rh\} = \{r,v\}$ is a left coset of H. Since G is an abelian group, the right coset $Hr = \{er,hr\} = \{r,v\}$ is equal to rH.

Example 4.8.3 Let $G = S_3$, the group of all permutations on the set $\{1,2,3\}$. Let $H = \{(1),(1\ 2)\}$. Then the left coset

$$(1\ 2\ 3)H = \{(1\ 2\ 3),(1\ 3)\}$$

and the right coset

$$H(1\ 2\ 3) = \{(2\ 3),(1\ 2\ 3)\}.$$

Thus $(1\ 2\ 3)H \neq H(1\ 2\ 3)$. Moreover, for any $g \in G$,

$$(1\ 2\ 3)H \neq Hg.$$

Thus, this example shows that the left coset gH need not be equal to the right coset Hg, and also that the left coset gH need not be the same as any right coset Hg'.

In the event $gH = Hg$, we will refer to gH simply as a *coset* of H.

While the right coset Hg need not be equal to the left coset gH, they do have the same "size." This notion, and more, is made precise in the next theorem.

THEOREM 4.8.4   Let H be a subgroup of a group G, and let g_1 and g_2 be elements of G. Then there exist functions f_1, f_2, and f_3 such that

(a) $f_1: g_1H \xrightarrow[\text{onto}]{1\text{-}1} H$,

(b) $f_2: g_1H \xrightarrow[\text{onto}]{1\text{-}1} g_2H$, and

(c) $f_3: g_1H \xrightarrow[\text{onto}]{1\text{-}1} Hg_2$.

Proof   (a) Let f_1 be defined by $f_1(g_1h) = h$. Clearly f_1 is a function, since if $g_1h = g_1h'$, then $h = h'$.

If $f_1(g_1h) = f_1(g_1h')$, then $h = h'$ and $g_1h = g_1h'$. Hence f_1 is 1-1. If $h \in H$, then $g_1h \in g_1H$ and $f(g_1h) = h$. Hence f_1 is onto.

(b) Define f_2 by $f_2(g_1h) = g_2h$. As above, it is easy to show that f_2 is 1-1 and onto.

(c) This is left as an exercise.   ∎

Note that the proof above does not require that H be a subgroup of G. The theorem is true if we replace subgroup by subset.

COROLLARY 4.8.5   Let G be a finite group and H a subgroup of G. If g is an element of G, then the cosets gH, Hg, and H all have the same number of elements.

Proof   It follows from Theorem 4.8.4 that there is a one-to-one correspondence between any pair of these sets. Hence, since they are finite, they have the same number of elements.   ∎

EXERCISES

1. Prove Theorem 4.8.4, part (c).
2. Let G be the symmetric group S_3. Choose subgroups A of order two and B of order three and calculate the product AB. Is it a subgroup of G? Do the same for two subgroups of order two.
3. Let A and B be subgroups of a group G. Show that $AB = BA$ if and only if AB is a subgroup of G.
4. Let A, B, C be subsets of a group G. Prove that $(AB)C = A(BC)$.
5. In Example 4.8.3, show that $(1\ 2\ 3)H \neq Hg$, for all $g \in S_3$.

6. Let G be a finite group and let A and B be subsets of G with m and n elements, respectively. Prove that if $|G| < m + n$, $G = AB$. [*Hint*: If $c \notin AB$, then consider $cB^{-1} \cap A$, where

$$B^{-1} = \{b^{-1} \mid b \in B\}.]$$

In Exercise 5 the reader should observe that a left coset may have more than one representative. The next theorem completely answers the question as to what elements are representative of bH.

THEOREM 4.8.6 Let H be a subgroup of G and bH be a left coset of H. Then, if $g \in bH$, $gH = bH$; that is, every element of bH is a representative of bH.

Proof Let $g \in bH$. Then $g = bh$, for some $h \in H$. Then, for any $h' \in H$, $gh' = bhh' = bh''$, where $h'' = hh' \in H$, since H is a subgroup of G. Thus, $gH \subseteq bH$. Similarly, since $b = gh^{-1}$, we can show that $bH \subseteq gH$. Thus $bH = gH$. ∎

It is easy to see that if H is a subgroup of a group G, then

$$G = \bigcup_{g \in G} gH.$$

We will now show that the set of cosets of H is a partition of G.

THEOREM 4.8.7 Let H be a subgroup of the group G, and let aH and bH be two left cosets of H. Then, either $aH = bH$ or $aH \cap bH = \varnothing$. That is, the set of left cosets of H is a partition of G.

Proof Suppose $aH \cap bH \neq \varnothing$. Then there exists an element $g \in G$ such that $g \in aH$ and $g \in bH$. By Theorem 4.8.6, $aH = gH = bH$. Since $G = \bigcup_{a \in G} aH$, the set of left cosets of H is a partition of G. ∎

THEOREM 4.8.8 (*Lagrange*) If G is a finite group of order n and if H is a subgroup of order m, then $m \mid n$.

Proof It follows from Corollary 4.8.5 that any left coset gH of H has the same number of elements as H. Thus, gH has m elements. Since G is finite, there are k distinct left cosets of H that are pairwise disjoint and that partition G, by Theorem 4.8.7. Then, since G is of order n, $n = |G| = km$ and $m \mid n$. ∎

The number k that appears in the proof of Lagrange's theorem is just the number of distinct left cosets of H in G. We call this integer the *index* of H in G and we write $[G : H] = k$.

A parallel development of the theory for right cosets would also show that there are k right cosets of H in G, and hence the index could also be defined as the number of right cosets. If G is an infinite group and H is a subgroup of G, then we say that H is of *infinite index* in G if the set of cosets $\{gH \mid g \in G\}$ is an infinite set.

COROLLARY 4.8.9 Let H be a subgroup of a finite group G. Then

$$|G| = [G : H] \cdot |H|.$$

Proof If the number of distinct left cosets of H is k and the order of H is m, then we have shown in the proof of Theorem 4.8.8 that $|G| = km$. But, by definition, the number of distinct left cosets of H is $[G : H]$ and so $|G| = [G : H] \cdot |H|$. ∎

EXERCISES

7. Let $H = \{(1),(23)\}$, a subgroup of S_3. Enumerate the left cosets of H and the right cosets of H.
8. Let G be a finite group, H a subgroup of G, and K a subgroup of H. Prove that $[G : K] = [G : H][H : K]$.

COROLLARY 4.8.10 The order of every element of a finite group G divides the order of G.

Proof By the remarks following Definition 4.7.10, the order of a is equal to the number of elements in the cyclic subgroup generated by a. By Theorem 4.8.8, $o(a) \mid |G|$. ∎

COROLLARY 4.8.11 A group G of order p, a prime, is cyclic and has no proper nontrivial subgroups.

Proof Let a be an element of G different from e. Then $o(a)$ is equal to the order of the cyclic subgroup generated by a, so by Lagrange's theorem $o(a) \mid |G|$. But $|G| = p$, a prime, and $o(a)$ is a positive integer greater than 1, whence $o(a) = p$. Thus, $(a) = G$, and so G is cyclic and has no proper nontrivial subgroup. ∎

Some important theorems in number theory can be proved by a simple application of Lagrange's theorem. We cite two of them here.

COROLLARY 4.8.12 Let a be an integer and p a prime such that $p \nmid a$. Then $a^{p-1} \equiv 1 \pmod{p}$.

Proof Let G be the multiplicative group of nonzero residue classes of integers modulo p. $|G| = p - 1$, and since $p \nmid a$, $\bar{a} \in G$. Thus, the order of $\bar{a}$ divides $p - 1$. By Theorem 4.7.14, $(\bar{a})^{p-1} = \bar{1}$. But $(\bar{a})^{p-1} = \overline{a^{p-1}}$, whence $a^{p-1} \equiv 1 \pmod{p}$. ∎

4.8 COSETS AND LAGRANGE'S THEOREM

COROLLARY 4.8.13 If a is any integer and p is a prime, then
$$a^p \equiv a \pmod{p}.$$
Proof The proof is left as an exercise. ∎

EXERCISES

9. Let A and B be subgroups of a group G. Let $|A| = 5$ and let $|B| = 2$. Show that $A \cap B = E$.
10. Let A and B be subgroups of a group G, $|A| = p$, a prime, and suppose $A \cap B \neq E$. Show that $A \subseteq B$.
11. Let H and K be two distinct subgroups of G with $|H| = |K| = p$, a prime. Show that the total number of elements in $H \cup K$ is $2p - 1$.
12. Let G be a group of order p^2, p a prime.
    (a) Show that G has at least one subgroup of order p.
    (b) Show that G can have at most $p + 1$ subgroups of order p.
13. Using Lagrange's theorem, prove Theorem 2.15.1.
14. Use Lagrange's theorem to prove that the table given in Exercise 6, Section 4.3, is not the multiplication table of a group. (*Hint:* Find the orders of the elements.)

As an application of Lagrange's theorem we prove that there are precisely two nonisomorphic groups of order 4.

THEOREM 4.8.14 Every group of order 4 is isomorphic either to the cyclic group of order 4 or to the group of symmetries of a nonsquare rectangle, the Klein group.

Proof Let G be a group of order 4, $G = \{e,a,b,c\}$, where e is the identity. By Lagrange's theorem, the elements a, b, c have either order 2 or order 4. If one of them has order 4, then G is cyclic of order 4.

Thus suppose none of a, b, c has order 4, that is, each has order 2. Then $a^2 = b^2 = c^2 = e$.

Consider the product ab. If $ab = e$, then $b = a^{-1} = a$, since a has order 2, contradicting the fact that $b \neq a$. Thus, $ab \neq e$. If $ab = a$, then $b = e$, a contradiction. If $ab = b$, then $a = e$, a contradiction. Therefore, the only possibility left is $ab = c$, which must hold, since $ab \in G$.

At this state, the multiplication table for G must appear as follows:

	e	a	b	c
e	e	a	b	c
a	a	e	c	
b	b		e	
c	c			e

Since every element of the group must appear in every row and column of its multiplication table, we must have $ac = b$. By filling in the remaining entries according to this property, we find in order that $bc = a$, $ba = c$, $ca = b$, and $cb = a$.

It is easy to establish that a group G with multiplication described above is isomorphic to the Klein group. ∎

We have mentioned earlier that even though A and B are subgroups of a group G, the set $AB = \{ab \mid a \in A, b \in B\}$ need not be a subgroup of G. Moreover, the number of elements in AB is not obvious. The next result lets us compute this number.

THEOREM 4.8.15 Let A and B be finite subgroups of the group G. Then

$$|AB| = \frac{|A| \cdot |B|}{|A \cap B|},$$

where $|AB|$ is the number of elements in AB.

Proof Since A and B are subgroups, $A \cap B = C$ is a subgroup of B (and of A). Since B is a finite group, let n be the number of distinct right cosets of C in B. Thus

$$B = Cb_1 \cup Cb_2 \cup \cdots \cup Cb_n$$

with $Cb_i \cap Cb_j = \emptyset$ as long as $i \neq j$. Now

$$AB = ACb_1 \cup ACb_2 \cup \cdots \cup ACb_n$$

by definition of the products AB and AC. But $C \subseteq A$ and so $AC = A$. Hence

$$AB = Ab_1 \cup \cdots \cup Ab_n.$$

Now if $Ab_i \cap Ab_j \neq \emptyset$ for some $i \neq j$, then there exist a, a' in A such that $ab_i = a'b_j$. Then

$$b_j b_i^{-1} = (a')^{-1} a,$$

and this element is in both A and B, whence

$$b_j b_i^{-1} \in A \cap B = C.$$

Hence, $Cb_i = Cb_j$, which is a contradiction. Thus, it follows that

$$Ab_i \cap Ab_j = \emptyset \quad \text{for all } i, j, i \neq j.$$

We now see that

$$|AB| = |A| \cdot n, \quad \text{where } n = \frac{|B|}{|C|} = \frac{|B|}{|A \cap B|},$$

and so

$$|AB| = \frac{|A| \cdot |B|}{|A \cap B|}. \quad ∎$$

EXERCISES

15. Let G be a group and H a subgroup. Define the relation $\sim$ on G by: $a \sim b$ if and only if $a^{-1}b \in H$. Show that this is an equivalence relation on G.
16. Prove that the equivalence classes determined in Exercise 15 are precisely the left cosets of H, and hence the left cosets of H partition G.
17. Let S be a subset of the group G. Let the relation $\sim$ on G be defined as follows: $a \sim b$ if and only if $ab^{-1} \in S$. Show that $\sim$ is an equivalence relation if and only if S is a subgroup of G.
18. Let G be a group with no nontrivial proper subgroups. Show that $|G| = p$, a prime, or $|G| = 1$.
19. Let G be a group and H a subgroup of G. Show that there is a 1–1 map from the set $\{gH\}$ of left cosets of H onto the set $\{Hg\}$ of right cosets of H. [*Hint:* Consider $gH \longrightarrow Hg^{-1}$.]
20. Let B be a group of order 20 and A and B subgroups of orders 4 and 5, respectively. Show that $G = AB$.
21. Let G be a group of order $p^\alpha m$ with p a prime and $(p,m) = 1$. Let A be a subgroup of order p^α and B a subgroup of order p^β, $0 < \beta \leq \alpha$, $B \not\subseteq A$. Show that AB is not a subgroup of G. (*Hint:* Compute the order of AB and use Lagrange's theorem.)
22. Let G be a group of order 6. Show that G is isomorphic to either C_6 or S_3.

4.9 Homomorphisms and Normal Subgroups

If G and H are two groups that are isomorphic, then, as abstract algebraic systems, they are related in the strongest possible way; that is, there exists a function $f\colon G \longrightarrow H$ that is 1–1, onto, and preserves products:

$$f(g_1 g_2) = f(g_1)f(g_2) \qquad \text{for all } g_1, g_2 \in G.$$

We have, however, already seen examples of groups that are related in a somewhat weaker manner. For example, there exists a mapping from the additive group of integers Z to the additive group $\bar{Z}_3$ of residue classes of integers modulo 3. The mapping here is defined by $f(a) = \bar{a}$, where $\bar{a} \in \bar{Z}_3$, for all $a \in Z$. It is easy to show that this function preserves sums, is onto, but is not 1–1. There are, nevertheless, advantages in examining the range of such a function, for we may be able to discover certain aspects of the behavior of the domain group. In the case above, we gain some insight into the additive behavior of the remainders of integers upon division by 3.

144   THE THEORY OF GROUPS

Definition 4.9.1   Let f be a function from the group $(G,\circ)$ to the group $(H,*)$ such that

$$f(a \circ b) = f(a) * f(b), \quad \text{for all } a, b \in G.$$

Then f is called a *homomorphism* from G to H. The range of f in H is called a *homomorphic image* of G, or the *homomorphic image of G under f*.

In the example above, f is an onto function and $\bar{Z}_3$ is therefore a homomorphic image of Z.

The reader should note that an isomorphism is a special type of homomorphism.

Example 4.9.2   Let S_n be the symmetric group on n letters. Let $f\colon S_n \longrightarrow \bar{Z}_2$ be defined by

$$f(g) = \begin{cases} \bar{0}, & \text{if } g \text{ is an even permutation,} \\ \bar{1}, & \text{if } g \text{ is an odd permutation.} \end{cases}$$

Using the table in Section 3.3, we can easily verify that f is a homomorphism.

Example 4.9.3   A homomorphism need not, of course, be an onto function. If Z is the additive group of integers and $\bar{Z}_4$ the additive group of residue classes modulo 4, then the mapping $f\colon Z \longrightarrow \bar{Z}_4$ defined by

$$f(u) = \begin{cases} \bar{2}, & \text{if } u \text{ is odd,} \\ \bar{0}, & \text{if } u \text{ is even,} \end{cases}$$

is a homomorphism from Z to $\bar{Z}_4$, but it is not onto.

Example 4.9.4   Let f be a function with domain D_4 and codomain $G = \{e,v,r,h\}$, the Klein 4-group (see Example 4.1.7). We define f as follows:

$$f(e) = f(r_2) = e$$

(e is used for the identity in both D_4 and G),

$$f(r_1) = f(r_3) = r,$$
$$f(v) = f(h) = v,$$
$$f(d_1) = f(d_{-1}) = h.$$

The function f is clearly onto G. To show that f is a homomorphism we must check the equation $f(ab) = f(a)f(b)$ for all $a, b \in D_4$. Some cases are quite easy; for example, if $a = e$, then, since $f(e) = e$,

$$f(eb) = f(e)f(b) = f(b).$$

Also, if $a = b = e, r_2, v, h, d_1,$ or d_{-1}, then

$$f(ab) = f(a^2) = f(e) = e.$$

If $a = r_1$, then
$$f(r_1{}^2) = f(r_2) = e = rr = f(r_1)f(r_1) = f(r_1)^2,$$
and similarly $f(r_3{}^2) = f(r_3)f(r_3)$. We leave the remainder of the argument to the reader.

EXERCISES

1. Show that the function of Example 4.9.3 is a homomorphism.
2. Complete the verification that the function of Example 4.9.4 is a homomorphism.
3. Find two homomorphisms from Z to $\bar{Z}_6$ that are onto. Find four homomorphisms that are not onto.
4. Find a homomorphism from D_4 to C_4, the cyclic group of order 4.
5. Find a homomorphism from D_4 to itself.
6. Find a homomorphism from the dihedral group D_n onto the cyclic group C_2.

THEOREM 4.9.5 Let f be a homomorphism from the group G to the group H. Then the range of $f(\text{Im } f)$ is a subgroup of H. That is, the homomorphic image of a group is a group.

The proof of this theorem is facilitated by the following two lemmas.

LEMMA 4.9.6 Let f be a homomorphism from the group G to the group H. Let e be the identity of G and e' the identity of H. Then $f(e) = e'$.

Proof $f(e)f(e) = f(ee) = f(e)$ by the definition of a homomorphism. By Exercise 2 of Section 4.2, $f(e)$ is the identity of H, that is, $f(e) = e'$. ∎

LEMMA 4.9.7 Let f be a homomorphism from the group G to the group H. Then, if $a \in G$,
$$f(a^{-1}) = [f(a)]^{-1},$$
that is, the image of the inverse of a is the inverse of the image of a.

Proof Let e be the identity of G. Then
$$f(a^{-1})f(a) = f(a^{-1}a) = f(e),$$
the identity of H by the previous lemma. Since inverses are unique,
$$f(a^{-1}) = [f(a)]^{-1}. \qquad ∎$$

Proof of Theorem 4.9.5 By Theorem 4.6.5, to show that Im f is a subgroup of H we need only show that if $h_1, h_2 \in \text{Im } f$, then $h_1 h_2{}^{-1} \in \text{Im } f$. Let $a, b \in G$ have the property that
$$f(a) = h_1, \qquad f(b) = h_2.$$

Then
$$h_1 h_2^{-1} = f(a)[f(b)]^{-1} = f(a)f(b^{-1}) = f(ab^{-1}) \in \text{Im } f. \quad \blacksquare$$

We note that since the image of a group G under a homomorphism f is again a group, then by suitably restricting the group H to its subgroup Im f, we may always modify f so that it is an onto mapping.

A homomorphism f from a group G can be restricted to a subgroup of G. It is natural to ask what is the image of a subgroup under the action of f.

THEOREM 4.9.8 Let f be a homomorphism from G to H. Let A be a subgroup of G, and f' the restriction of f to A. Then f' is a homomorphism from A to H, and the image of A under f' is a subgroup of H.

Proof If x and y are in the image of A under f', then there are $g, h \in A$ such that $f'(g) = x$ and $f'(h) = y$. Since f' is just the restriction of f to A, then
$$(f'(h))^{-1} = f'(h^{-1})$$
by Lemma 4.9.7. Hence
$$xy^{-1} = f'(g)(f'(h))^{-1} = f'(g)f'(h^{-1}).$$
Again, since f' is the restriction of f,
$$f'(g)f'(h^{-1}) = f'(gh^{-1}).$$
Hence $f'(g)(f'(h))^{-1} = xy^{-1}$ is in the image of A under f', and so the image of A under f' is a subgroup of H by Theorem 4.6.5. $\quad \blacksquare$

Next we give a useful lemma, which we will use in the sequel.

THEOREM 4.9.9 Let $f \colon G \longrightarrow H$ be a homomorphism from the group G to the group H. Let $a \in G$. Then
$$f(a^n) = (f(a))^n \quad \text{for any integer } n.$$

Proof If $n = 0$, then
$$f(a^0) = f(e) = e = (f(a))^0.$$
We can now prove the result for the case $n \geq 0$ by induction. For if $f(a^k) = (f(a))^k$ for $k \geq 0$, then
$$f(a^{k+1}) = f(a^k a) = f(a^k)f(a),$$
since f is a homomorphism. But by the induction assumption
$$f(a^k) = (f(a))^k,$$
and so
$$f(a^{k+1}) = (f(a))^k f(a) = (f(a))^{k+1}.$$

This gives the result for all $n \geq 0$. Now if $n < 0$, we consider
$$f(a^n) = f((a^{-1})^{-n}) = (f(a^{-1}))^{-n}$$
by the previous result. But since f is a homomorphism, $f(a^{-1}) = (f(a))^{-1}$, and so
$$f(a^n) = (f(a^{-1}))^{-n} = ((f(a))^{-1})^{-n} = (f(a))^n. \quad \blacksquare$$

When a homomorphism f of the group G to the group H is not an isomorphism, f has the effect of "identifying" the elements of certain subsets of G. Thus those elements of G that have the same image under f can be thought of as being "identified" by the mapping. A particularly important set of such elements are those that are mapped by f onto the identity of H.

Definition 4.9.10 Let f be a homomorphism from the group G to the group H. Let e' be the identity of H. Then the set $K = \{x \mid x \in G$ and $f(x) = e'\}$ is called the *kernel* of f. We often denote K by ker f.

Since $f(e) = e'$, $x \in$ ker f if and only if $f(x) = f(e)$.

Example 4.9.11 Let $G = S_3$, the symmetric group on three letters, and let $H = (\bar{Z}_2, +)$.

Define a map f from G onto H by
$$f(1) = \bar{0}, \quad f(1\ 2) = \bar{1}, \quad f(1\ 3) = \bar{1}, \quad f(2\ 3) = \bar{1},$$
$$f(1\ 2\ 3) = \bar{0}, \quad f(1\ 3\ 2) = \bar{0}.$$

To show that f is a homomorphism, we need to verify that
$$f(xy) = f(x) + f(y) \quad \text{for all } x, y \in S_3.$$
We leave this to the reader.

Note that in this example ker $f = \{(1),(1\ 2\ 3),(1\ 3\ 2)\}$. Moreover, the remaining elements of S_3 are "identified" in the sense that they all have the same image.

Example 4.9.12 In Example 4.9.4 the kernel of the homomorphism is simply the set $\{e, r_2\}$. In Example 4.9.3, ker f is the set of all even integers.

EXERCISES

7. Let f be a homomorphism from $(\bar{Z}_2, +)$ to $(\bar{Z}_3, +)$. Calculate the various possibilities for $f(\bar{1})$. What can you conclude about the kernel of such a homomorphism?
8. Repeat Exercise 7 replacing $(\bar{Z}_3, +)$ by $(\bar{Z}_4, +)$.
9. Prove that if f is a homomorphism from S_3 to $(\bar{Z}_3, +)$, then $f(x) = \bar{0}$ for all $x \in S_3$.

10. What is the kernel of the homomorphism that you defined in your answer to Exercise 6 of this section?

THEOREM 4.9.13 Let f be a homomorphism from the group G to the group H. Then ker f is a subgroup of G.

Proof It follows from Lemma 4.9.6 that ker $f \neq \emptyset$. Let $a, b \in $ ker f. It is enough to show that $ab^{-1} \in $ ker f. Since

$$f(ab^{-1}) = (f(a))(f(b^{-1})) = f(a)[f(b)]^{-1},$$

and $f(a) = f(b) = e'$, the identity of H,

$$f(ab^{-1}) = e'(e')^{-1} = e'e' = e'.$$

Thus $ab^{-1} \in $ ker f, and ker f is a subgroup of G. ∎

Now suppose J is a set of elements of G all of which have the same image under f. Assume that this image is not e'. It is natural to ask if this set J of "identified" elements is also a subgroup of G. By examining Example 4.9.11, the reader may quickly convince himself that this is not the case. Sets such as J, however, bear a strong relation to ker f.

THEOREM 4.9.14 Let f be a homomorphism from the group G to the group H with ker f denoted by K. For any $r \in $ Im f, let

$$T_r = \{u \mid u \in G, f(u) = r\}.$$

Then

(a) $vK = T_r$ for any $v \in T_r$,
(b) $Kv = T_r$ for any $v \in T_r$,
(c) $Kv = vK$ for $v \in G$,
(d) $f(u) = f(v)$ if and only if $uK = vK$.

Proof (a) Let $u \in T_r$. Then $r = f(u) = f(v)$. Hence

$$f(v)^{-1}f(u) = f(v^{-1})f(u) = f(v^{-1}u) = e',$$

the identity of H. Therefore $v^{-1}u \in K$ and $u \in vK$. Now let $vk \in vK$. Then

$$f(vk) = f(v)f(k) = f(v) = r,$$

which means that $vk \in T_r$. Therefore $T_r = vK$.

(b) In a similar manner to (a) we can prove that $T_r = Kv$.

(c) Clearly, since for any $v \in G$, $v \in T_r$, for some $r \in $ Im f, it follows from (a) and (b) that $vK = Kv$.

(d) If $f(u) = f(v)$, then $u, v \in T_r$ for some $r \in $ Im f. By (a), $vK = T_r = uK$. Let $v \in T_r$. Then $vK = uK$ implies by (a) that $vK = uK = T_r$ and hence $r = f(v) = f(u)$. ∎

4.9 HOMOMORPHISMS AND NORMAL SUBGROUPS

We see now that the "identification" that takes place in G as a result of the homomorphism f is precisely the decomposition of G into left cosets of the kernel K of f. In this case, however, we need not insist on left cosets, since, as we have shown, left cosets and right cosets coincide.

It is conceivable that a subgroup H of G other than K has the property $Hg = gH$ for all $g \in G$. Such subgroups are extremely important, as we shall soon see.

Definition 4.9.15 A subgroup N of the group G is called a *normal subgroup of G* (N *is normal in* G) if $gN = Ng$, for all $g \in G$. We denote this by $N \triangleleft G$.

Theorem 4.9.14 asserts, in part, that a kernel K of a homomorphism is normal. In the next section we prove that every normal subgroup is the kernel of some homomorphism.

It is convenient to have available alternate descriptions of normal subgroups.

THEOREM 4.9.16 Let G be a group and N a subgroup of G. Then the following are equivalent:
(a) $N \triangleleft G$,
(b) $g^{-1}Ng = N$ for all $g \in G$,
(c) $g^{-1}Ng \subseteq N$ for all $g \in G$.

Proof We will prove the theorem by showing that (a) implies (b), (b) implies (c), and (c) implies (a).

First assume that $N \triangleleft G$. If $g \in G$, then $gN = Ng$. Therefore

$$g^{-1}(gN) = g^{-1}Ng, \quad \text{whence } (g^{-1}g)N = N = g^{-1}Ng,$$

and so (a) implies (b).

Next suppose $N = g^{-1}Ng$, for all $g \in G$. Then clearly

$$g^{-1}Ng \subseteq N \quad \text{for all } g \in G$$

whence (b) implies (c).

Finally, if $g^{-1}Ng \subseteq N$ for all $g \in G$, then if $n \in N$,

$$g^{-1}ng = n_1 \quad \text{for some } n_1 \in N.$$

Hence $ng = gn_1$. Thus the set $Ng \subseteq gN$. Similarly if $g^{-1}ng = n_1$, then $g^{-1}n = n_1 g^{-1}$ and the set $g^{-1}N \subseteq Ng^{-1}$. But as g ranges over all the elements of G, then so does g^{-1}. Hence

$$g^{-1}N \subseteq Ng^{-1} \quad \text{implies that} \quad gN \subseteq Ng.$$

Thus $gN = Ng$ and $N \triangleleft G$ by definition. This completes the cycle of implications. ∎

150   THE THEORY OF GROUPS

If f is a homomorphism from G onto H that is also 1-1, then f is an isomorphism from G to H as defined in Section 4.4. A simple criterion for when a homomorphism is an isomorphism is found in the following theorem.

THEOREM 4.9.17   Let $f: G \longrightarrow H$ be a homomorphism from the group G onto the group H. Then f is an isomorphism from G onto H if and only if ker $f = E$.

Proof   Let f be a homomorphism from G onto H and suppose that ker $f = E$. We must show that f is 1-1. If there are elements $a, b \in G$ such that $f(a) = f(b)$, then

$$f(a)(f(b))^{-1} = e \in H.$$

But $f(a)(f(b))^{-1} = f(a)f(b^{-1}) = f(ab^{-1})$, and so

$$ab^{-1} \in \text{ker } f.$$

Since ker $f = E$,

$$ab^{-1} = e \quad \text{and} \quad a = b.$$

Thus f is 1-1, and therefore f is an isomorphism onto H.

If, on the other hand, f is an isomorphism onto H, then f is 1-1. Hence, if $a \in \text{ker } f$,

$$f(a) = f(e) = e \in H \quad \text{and so} \quad a = e.$$

Thus ker $f = E$, and the theorem is proved.   ∎

Example 4.9.18   Let G be the Klein 4-group, $G = \{e,r,h,v\}$, and let $\bar{Z}_2 = \{\bar{0},\bar{1}\}$ be the additive group of residue classes mod 2. If $f: G \longrightarrow \bar{Z}_2$ is defined by

$$f(e) = f(h) = \bar{0} \quad \text{and} \quad f(r) = f(v) = \bar{1},$$

then it is easy to verify that f is a homomorphism onto $\bar{Z}_2$, with ker $f = \{e,h\}$. We can directly determine the sets T_s of Theorem 4.9.14. For if we let $s = \bar{0}$ and $\bar{1}$, respectively, then

$$T_{\bar{0}} = \{e,h\} \quad \text{and} \quad T_{\bar{1}} = \{r,v\}, \text{ respectively.}$$

To verify that these are cosets of the kernel, we observe that

$$\{e,h\} = \text{ker } f \quad \text{and} \quad T_{\bar{1}} = r\{e,h\},$$

since $rh = v$. Also $T_{\bar{1}} = \{e,h\}r$, since in this special case G is an abelian group.

Example 4.9.19   The property of normality has been defined by the condition $gN = Ng$, for all g. This does not, of course, imply that $gn = ng$ for all $g \in G$ and all $n \in N$. As an illustration, consider D_4 and the normal subgroup $N = \{e,r_2,h,v\}$. In order to show that this subgroup is normal, it suffices to calculate the sets gN and Ng for all $g \in G$, $g \notin N$. For if $g \in N$, then we already know that $gN = Ng = N$. Now

$$r_1\{e, r_2, h, v\} = \{r_1, r_3, d_1, d_{-1}\}$$

and

$$\{e, r_2, h, v\} r_1 = \{r_1, r_3, d_{-1}, d_1\}.$$

Thus, even though $r_1 h \neq h r_1$, the two sets $r_1 N$ and $N r_1$ are equal. Similarly, we can show that $gN = Ng$, for all $g \in G$, by computing these sets for $g = r_3$, d_1, and d_{-1}. (Exercise 14 provides an easy, alternative proof that $N \triangleleft G$.)

Example 4.9.20 The property that the subgroup H of G is normal in G is sometimes called a *relative* property, since it depends on the relationship of a subgroup H to a group G. More precisely, if H and N are two subgroups of G with $H \triangleleft N$ and $N \triangleleft G$, it does not follow that H will be normal *relative to* G. For example, let $G = D_4$ and $N = \{e, r_2, h, v\}$ as in the previous example. We have shown that $N \triangleleft D_4$. Let $H = \{e, h\}$. Then it is easy to show that $H \triangleleft N$. But the element $r_1^{-1} h r_1 = r_3 h r_1 = v$, and so $r_1^{-1} H r_1 \not\subseteq H$ whence $H \ntriangleleft D_4$.

EXERCISES

11. Let $f: G \longrightarrow H$ be a homomorphism with kernel K. Prove without using Theorem 4.9.14 that $uK = vK$ implies $f(u) = f(v)$.
12. Let N be a subgroup of G. Prove $N \triangleleft G$ if and only if for each $a \in G$ and $n \in N$, there exists $n' \in N$ such that $an = n'a$.
13. Show that every subgroup of an abelian group G is normal in G.
14. Let H be a subgroup of a group G. Suppose $[G : H] = 2$. Show that $H \triangleleft G$. (*Hint:* For $g \notin H$, consider the cosets H, gH, and Hg.)
15. Show that A_n is a normal subgroup of S_n.
16. Complete the proof in Example 4.9.19, that $N \triangleleft D_4$. (See Exercise 14.)
17. Complete the proof in Example 4.9.20 that $H \triangleleft N$.
18. Let G be a cyclic group and let f be a homomorphism from G onto the group H. Show that H is cyclic.
19. Let G be a group and let A and B be normal subgroups of G. Show that $(A \cap B) \triangleleft G$.
20. Let G be a group, with $A \triangleleft G$ and B a subgroup of G. Show that $AB = \{ab \mid a \in A, b \in B\}$ is a subgroup of G, and $A \triangleleft AB$.
21. Let G be a group, with $A \triangleleft G$ and B a subgroup of G. Show that $(A \cap B) \triangleleft B$.
22. Let G and H be groups, and let $f: G \longrightarrow H$ be a homomorphism. Let $a \in G$. Show that $o(f(a)) \mid o(a)$.

4.10 Factor Groups and the First Isomorphism Theorem

If N is a normal subgroup of a group G, then by definition $aN = Na$, for all $a \in G$. This partial commutativity property allows us to define a binary operation on the set of cosets $\{aN \mid a \in G\}$ so that the collection of cosets and this binary operation form a group.

Definition 4.10.1 Let N be a normal subgroup of the group G. We define a binary operation on the set of cosets of N as follows:

$$aN \cdot bN = (ab)N.$$

To show that $\cdot$ is actually a binary operation, we must prove that the definition of multiplication is independent of the choice of coset representative—that is, if $aN = a'N$ and $bN = b'N$, then

$$(ab)N = aN \cdot bN = a'N \cdot b'N = (a'b')N.$$

Now if $aN = a'N$, then $a = a'n_1$, and if $bN = b'N$, then $b = b'n_2$ for some $n_1, n_2 \in N$. Therefore

$$(ab)N = (a'n_1 b' n_2)N = (a' n_1 b')N,$$

since $n_2 N = N$. Now by Exercise 12, Section 4.9, $n_1 b' = b' n_1'$ for some $n_1' \in N$. Hence

$$(a' n_1 b')N = (a' b' n_1')N = (a'b')N,$$

and the operation is well defined.

THEOREM 4.10.2 Let N be a normal subgroup of the group G. Then the set of cosets $\{aN \mid a \in G\}$ together with the operation $aN \cdot bN = (ab)N$ is a group, and its order is $[G : N]$.

Proof We have shown above that this operation is well defined. To show that it is associative, consider the product:

$$aN \cdot (bN \cdot cN) = aN \cdot (bc)N = a(bc)N = (ab)cN$$
$$= (abN) \cdot cN = (aN \cdot bN) \cdot cN.$$

Since $aN \cdot eN = (ae)N = aN = eN \cdot aN$, the coset $eN = N$ is an identity.

Finally, since

$$(aN) \cdot (a^{-1}N) = (aa^{-1})N = eN = (a^{-1}N)(aN),$$

$(aN)^{-1} = a^{-1}N$, and the set of cosets of N is a group.

The order of the group is clearly the number of cosets of N in G, that is, $[G : N]$. ∎

Definition 4.10.3 Let N be a normal subgroup of G. The set of cosets of N together with the binary operation of Definition 4.10.1 is called the

4.10 FACTOR GROUPS AND THE FIRST ISOMORPHISM THEOREM

quotient group of G with respect to the normal subgroup N, or the *factor group of G with respect to N*, and is denoted by G/N. Frequently, "with respect to N" is replaced by "modulo N."

The technique of treating a set of elements as a new single element and of defining a binary operation on the set of the new elements is not unfamiliar.

Example 4.10.4 The additive group of residue classes modulo m is an example of the procedure described above. For the set

$$(m) = \{0, m, -m, 2m, -2m, \ldots\}$$

is a subgroup of $(Z, +)$ and, since $(Z, +)$ is abelian, (m) is a normal subgroup. Then the coset $0 + (m) = (m)$ is simply denoted by $\bar{0}$, which we recognize from Chapter 2. Similarly, we see that the residue classes of Chapter 2 are $\bar{1} = 1 + (m)$, $\bar{2} = 2 + (m)$, and so on. Addition of residue classes is accomplished via addition of coset representatives as in Definition 4.10.1. Thus

$$\bar{1} + \bar{2} = (1 + (m)) + (2 + (m))$$
$$= (1 + 2 + (m))$$
$$= 3 + (m) = \bar{3},$$

and

$$\bar{2} + \overline{m-1} = (2 + (m)) + (m - 1 + (m))$$
$$= 2 + m - 1 + (m)$$
$$= 1 + m + (m) = 1 + (m) = \bar{1},$$

since $m + (m) = (m)$.

Example 4.10.5 If $G = S_3$, it is easy to verify that the subgroup

$$N = \{(1), (1\ 2\ 3), (1\ 3\ 2)\}$$

is normal in G. Hence the cosets N and $(1\ 2)N$ are the elements of G/N. Since $(1\ 2)N \cdot (1\ 2)N = N$, the identity element of G/N, it follows that G/N is isomorphic to the cyclic group of order 2, C_2.

Example 4.10.6 Let $(C, +)$ be the additive group of complex numbers—that is, $C = \{a + bi \mid a, b \text{ real}, i^2 = -1\}$. The set R of all real numbers is a subgroup of C under addition, since the sum and difference of any two real numbers is real. Further, since $(C, +)$ is an abelian group, R is a normal subgroup of C. Hence we may form C/R as in Definition 4.10.3. C/R consists of all the cosets of the form $ri + R$, for $r \in R$. For if $a + bi \in C$, then $a \in R$, and so $a + bi \in bi + R$. This shows that these cosets exhaust all elements of C. Further it is clear that

$$bi + R = ci + R \quad \text{if and only} \quad \text{if } b = c.$$

In the group $(C/R, +)$, $(bi + R) + (ci + R) = (b + c)i + R$, and moreover $(C/R, +) \cong (R, +)$. This can be established by showing that the function $f: C/R \longrightarrow R$, defined by $f(bi + R) = b$, is an isomorphism.

Example 4.10.7 Let S denote the set of all points in the plane:

$$S = \{(x, y) \mid x, y \text{ are real numbers}\}.$$

We define the operation $+$ in S as follows:

$$(x,y) + (x',y') = (x + x', y + y').$$

It is easy to show that $(S, +)$ is an abelian group. The group S should be familiar to the reader from his study of analytic geometry. The ordered pair (x,y) can be thought of as a vector in the plane with "tail" at the origin and "head" at the point (x,y). The definition of addition given above is exactly the same as the usual "parallelogram" definition of addition for vectors illustrated in Figure 4.5.

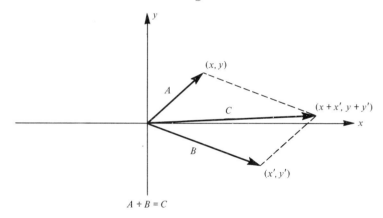

Figure 4.5

Now let

$$N = \{(rx_0, ry_0) \mid \text{for fixed real numbers } x_0, y_0 \text{ (not both 0)}$$
$$\text{and all real numbers } r\}.$$

Then N represents the set of all "multiples" of the vector (x_0, y_0), all of which lie along the straight line L determined by (x_0, y_0) (Figure 4.6). N is easily seen to be a subgroup of S, and, since S is abelian, N is a normal subgroup of S. (What is N if $x_0 = y_0 = 0$?)

Let $\alpha \in S$, $\alpha \notin N$. Then $\alpha + N$ can be described geometrically. Since each vector in $\alpha + N$ is a sum of α and a vector in N, it will occur as the diagonal of a parallelogram with one edge along L, and the opposite edge through the point α and parallel to L (Figure 4.7).

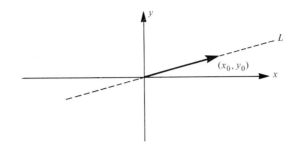

Figure 4.6

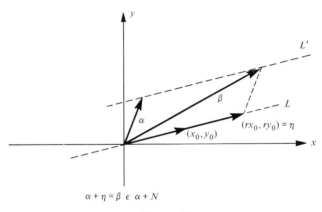

$\alpha + \eta = \beta \in \alpha + N$

Figure 4.7

If L' is the line through the point α and parallel to L, then the coset $\alpha + N$ will consist of exactly those vectors whose "heads" lie on the line L'. The set of all cosets of N can then be thought of as the set of lines in the plane parallel to L.

The addition of these cosets as given in Definition 4.10.1 can be easily described by choosing special coset representatives as follows: for each $\alpha + N$, let α' be that vector in $\alpha + N$ that is perpendicular to L. Since each coset represents a line parallel to L, all such α' will lie along a line, say P, perpendicular to all cosets. Hence, since

$$(\alpha' + N) + (\beta' + N) = (\alpha' + \beta') + N,$$

the sum of two cosets is obtained by a translation from the origin along P by a length equal to the length of $\alpha' + \beta'$ in the direction of $\alpha' + \beta'$ (Figure 4.8).

This geometrical analysis allows us now to determine the group G/N. For, if d_i represents the perpendicular distance from the coset $\alpha_i + N$ to L, with the "upward" direction considered positive and the "downward"

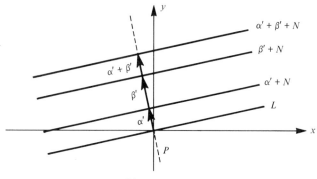

Figure 4.8

direction negative, the mapping:

$$d_i \longrightarrow \alpha_i + N$$

is an isomorphism of the additive group of real numbers onto the group G/N.

We have seen that the notion of a homomorphism leads to the concept of a normal subgroup. Normal subgroups, on the other hand, allow us to form quotient groups. We now ask if there is a direct connection between the homomorphism $f: G \longrightarrow H$, and the quotient group of G with respect to $\ker f$.

THEOREM 4.10.8 (*First Isomorphism Theorem*) Let f be a homomorphism from the group G to the group H. Then $\ker f$ is normal in G and

$$G/\ker f \cong \operatorname{Im} f.$$

Conversely, let $N \triangleleft G$. Then there exists a homomorphism

$$\sigma: G \xrightarrow{\text{onto}} G/N$$

with kernel N, defined by $\sigma(g) = gN$. We call σ the *natural homomorphism* from G to G/N.

Proof Let $f: G \longrightarrow H$ be a homomorphism, and denote $\ker f$ by K. Thus $K \triangleleft G$ and we may form $G/\ker f = G/K$. Now let

$$\nu: G/K \longrightarrow \operatorname{Im} f$$

be defined by

$$\nu(gK) = f(g), \quad \text{all } g \in G.$$

First, we must show that ν is well defined—that is, if $gK = g'K$, then $\nu(gK) = \nu(g'K)$.

4.10 FACTOR GROUPS AND THE FIRST ISOMORPHISM THEOREM

Now $gK = g'K$ implies $f(g) = f(g')$ by Theorem 4.9.14. Thus

$$\nu(g'K) = f(g') = f(g) = \nu(gK),$$

and so ν is well defined.

Next ν is a homomorphism, for

$$\nu[(gK)(g'K)] = \nu[(gg')K]$$
$$= f(gg') = f(g)f(g')$$
$$= \nu(gK)\nu(g'K).$$

Also ν is 1–1, for if

$$\nu(gK) = \nu(g'K),$$

then

$$f(g) = f(g'),$$

whence

$$gK = g'K \qquad \text{by Theorem 4.9.14.}$$

Finally, ν is indeed an isomorphism from $G/K = G/\ker f$ onto Im f.

Conversely, let $N \triangleleft G$ and form G/N. Define a mapping $\sigma\colon G \longrightarrow G/N$ by

$$\sigma(g) = gN, \qquad \text{for all } g \in G.$$

Then

$$\sigma(g_1 g_2) = (g_1 g_2)N = (g_1 N)(g_2 N) = \sigma(g_1)\sigma(g_2).$$

Thus, σ is a homomorphism, and σ is clearly onto.

$$\text{Ker } \sigma = \{g \mid \sigma(g) = N, \text{ the identity of } G/N\}.$$

Since $\sigma(g) = gN$, $\sigma(g) = N$ if and only if $gN = N$, that is, if and only if $g \in N$.

Thus ker $\sigma = N$, and the proof is completed. ∎

Example 4.10.9 Let $f\colon Z \longrightarrow \bar{Z}_m$, with $m > 0$, $f(a) = \bar{a}$. Then

$$\ker f = (m)$$

and

$$Z/(m) \cong \bar{Z}_m.$$

The reader should note that the isomorphism ν of Theorem 4.10.8 is actually the identity function in this example, since $Z/(m)$ and $\bar{Z}_m$ are the same set and $\nu[a + (m)] = \bar{a} = a + (m)$.

Example 4.10.10 Let N be the normal subgroup of S_3 described in Example 4.10.5. As we have shown, $G/N \cong C_2$, the cyclic group of order 2. Then the mapping $\sigma\colon G \longrightarrow G/N$ defined by

$$\sigma(g) = gN$$

is the homomorphism described in the converse portion of Theorem 4.10.8.

Example 4.10.11 Let G and f be as in Example 4.9.18. Then

$$\ker f = \{e,h\} \quad \text{and} \quad \operatorname{Im} f = \bar{Z}_2.$$

The isomorphism between $G/\ker f$ and $\operatorname{Im} f$ can be demonstrated explicitly by the mapping

$$\sigma: G/\ker f \longrightarrow \operatorname{Im} f, \quad \sigma(\ker f) = \bar{0}, \, \sigma(r \ker f) = \bar{1}.$$

Example 4.10.12 Let $f: D_4 \longrightarrow G$ be defined as in Example 4.9.4. Then $\ker f = \{e, r_2\}$ and

$$D_4/\ker f = \{\ker f, \ r_1 \ker f, \ v \ker f, \ d_1 \ker f\}.$$

The mapping $\sigma: D_4/\ker f \longrightarrow G$ defined by

$$\sigma(\ker f) = e, \quad \sigma(r_1 \ker f) = r, \quad \sigma(v \ker f) = v, \quad \sigma(d_1 \ker f) = h$$

is easily seen to be an isomorphism, and so $D_4/\ker f \cong \operatorname{Im} f = G$.

Example 4.10.13 Let G be the additive group of real numbers and H the set of points in the plane defined by

$$H = \{(\cos x, \sin x) \mid x \in G\}.$$

H is precisely the unit circle in the plane with center at the origin. We define a binary operation on H as follows:

$$(\cos x, \sin x) \cdot (\cos y, \sin y) = (\cos (x + y), \sin (x + y)).$$

The reader should verify that $(H, \cdot)$ is a group.

Now consider the following mapping f from G to H:

$$f(x) = (\cos x, \sin x).$$

Clearly f is onto H and is a homomorphism, since

$$\begin{aligned} f(x + y) &= (\cos (x + y), \sin (x + y)) \\ &= (\cos x, \sin x) \cdot (\cos y, \sin y) \\ &= f(x)f(y). \end{aligned}$$

By Theorem 4.10.8, G/N can be thought of as the set of real numbers where two elements are identified if they differ by an integral multiple of 2π. The mapping f has the effect of "wrapping" the real line around the unit circle.

Theorem 4.10.8 shows that any homomorphic image of a group G can be constructed from G; that is, if f is a homomorphism from G onto the group H, then H is isomorphic to a factor group of G.

EXERCISES

1. Find a subgroup H of the additive group of integers Z such that H has index 3 in Z. Describe the quotient group Z/H.
2. Find all normal subgroups of the group D_4 and describe the associated quotient groups.
3. Suppose that a is an element of a group G and that $o(a) = 3$. Let $N \triangleleft G$. What are the possibilities for the order of the coset aN as an element of the group G/N?
4. Let C_6 be the cyclic group of order 6. Find a normal subgroup N of C_6 and an element $a \in C_6$ such that $(aN)^3 = N$ in G/N, but $a^3 \neq e$ in C_6.
5. Let G be a finite group and N a normal subgroup of G. Show that for every $a \in G$, the order of aN, considered as an element of G/N, divides $o(a)$ (see Exercise 22, Section 4.9).
6. Let G be a group of order 10. Suppose G has a normal subgroup N of order 5. Show that G has an element of order 2. (*Hint:* Form the quotient group G/N. What is its order? Now use exercise 5 above.)
7. Let G be an abelian group of order pq, p and q distinct primes. Show that G contains an element of order pq (whence G is cyclic).
8. Let $H = \{(1),(1\ 2)\}$ in the group S_3. Show explicitly why the set of left cosets of H in S_3 do not form a group with respect to the operation $aH \cdot bH = abH$.
9. Let N be a subgroup of the group G. Suppose that the set of all left cosets of N form a group with respect to the operation

$$aN \cdot bN = (ab)N.$$

Show that $N \triangleleft G$.

10. Let $f\colon G \longrightarrow H$ be a homomorphism with H abelian. Let N be a subgroup of G that contains $\ker f$. Show $N \triangleleft G$.
11. Let m, n be integers such that $m \mid n$. Prove that

$$Z/(n) \Big/ (m)/(n) \cong \bar{Z}_m.$$

(*Hint:* Let $f\colon Z/(n) \longrightarrow Z/(m)$ be defined by $f(a + (n)) = a + (m)$. What is $\ker f$?)

12. Let G be a finite group such that $G = AB$ and $A \triangleleft G$. Show that if $G/A \cong B$, then $A \cap B = E$.

4.11 Permutation Groups and Cayley's Theorem

In this section we shall show that a study of groups of permutations is in fact no less than a study of all abstract groups. This idea is made precise in the famous theorem of Cayley. Several of the more elementary results in this section are simply restatements in group-theoretical terminology of theorems proved in Chapter 3.

THEOREM 4.11.1 The set of all permutations on the nonempty set X is a group with respect to the operation of composition of permutations.

Proof The proof has been given in Example 4.1.11. ∎

Definition 4.11.2 A group G is called a *permutation group* if G is a subgroup of the group of all permutations on a fixed set X.

Definition 4.11.3 The group S_n of permutations on the set $\{1,2,\ldots,n\}$ is called the *symmetric group on n letters*.[3]

In Section 3.3 we discussed the subset A_n of S_n, in which A_n consists of all even permutations on n letters.

THEOREM 4.11.4 If $n \geq 2$, A_n is a subgroup of S_n, and $|A_n| = n!/2$. (A_n is called the *alternating group on n letters*.)

Proof The proof is left as an exercise. ∎

THEOREM 4.11.5 (*Cayley*) Let $(G,\circ)$ be a group. Then the group $(G,\circ)$ is isomorphic to a permutation group on the set G.

Before giving the formal proof of Cayley's theorem, let us look briefly at the case in which G is finite, say $G = \{g_1,\ldots,g_n\}$. Then

$$gG = \{gg_1,\ldots,gg_n\} = G,$$

so that multiplication of G on the left permutes the elements of G. The elements of the permutation group in Cayley's theorem will be precisely the permutations effected by these left multiplications.

Proof of Cayley's Theorem To each element $g \in G$, we associate the function π_g on the set G defined by $\pi_g(x) = gx$, for all $x \in G$. It is easy to show that π_g is a 1–1 map from G onto G and is hence a permutation, even if G is infinite.

Now let f be a map from the group G to the group of permutations on the set G defined by $f(g) = \pi_g$. Clearly f is a function and we must show that the set $T = \{\pi_g \mid g \in G\}$ is a group and that f is an isomorphism from G onto T.

---

[3] We have defined S_n as the set of all permutation on the set of integers $X_n = \{1,\ldots,n\}$. If we consider the set of all permutations on the set $Y_n = \{\alpha_1,\ldots,\alpha_n\}$, $\alpha_i \neq \alpha_j$ for $i \neq j$, we get a group T_n isomorphic to S_n but not equal to it (see Exercise 3, Section 4.4). This distinction, however, between the elements being permuted is not important for our purposes; hence we will consider the group S_n as the set of all permutations on any set of n distinct elements.

4.11 PERMUTATION GROUPS AND CAYLEY'S THEOREM

(a) We first show that f is a homomorphism into the group of all permutations on the elements of G. Let x be an element of G. Then

$$[f(g_1g_2)](x) = \pi_{g_1g_2}(x) = (g_1g_2)x = g_1(g_2x) = \pi_{g_1}(\pi_{g_2}(x))$$
$$= (\pi_{g_1}\pi_{g_2})(x).$$

Thus
$$f(g_1g_2) = \pi_{g_1}\pi_{g_2} = f(g_1)f(g_2).$$

Hence f is a homomorphism and, since $T = \text{Im } f$, T is a group by Theorem 4.9.5.

(b) To show that f is an isomorphism from G onto T, let $g \in \ker f$. Hence $\pi_g(x) = gx = x$ for all $x \in G$. Therefore $g = e$ and f is an isomorphism by Theorem 4.9.17. ∎

If G is a finite group, then one can obtain the permutation used in the proof of Cayley's theorem by examining the multiplication table of G. The permutations are given by comparing the extreme left column of elements with each column in the table. For example, let G be the Klein group. Then $G = \{e = a_1, a_2, a_3, a_4\}$, and its multiplication table is

	a_1	a_2	a_3	a_4
a_1	a_1	a_2	a_3	a_4
a_2	a_2	a_1	a_4	a_3
a_3	a_3	a_4	a_1	a_2
a_4	a_4	a_3	a_2	a_1

Then
$$\pi_{a_1} = (a_1),$$

that is, $\pi_{a_1}(x) = x$, for all $x \in G$;

$$\pi_{a_2} = (a_1a_2)(a_3a_4),$$

that is, $\pi_{a_2}(a_1) = a_2$, $\pi_{a_2}(a_2) = a_1$, $\pi_{a_2}(a_3) = a_4$, $\pi_{a_2}(a_4) = a_3$;

$$\pi_{a_3} = (a_1a_3)(a_2a_4),$$

that is, $\pi_{a_3}(a_1) = a_3$, $\pi_{a_3}(a_3) = a_1$, $\pi_{a_3}(a_2) = a_4$, $\pi_{a_3}(a_4) = a_2$; and

$$\pi_{a_4} = (a_1a_4)(a_2a_3),$$

that is, $\pi_{a_4}(a_1) = a_4$, $\pi_{a_4}(a_4) = a_1$, $\pi_{a_4}(a_2) = a_3$, $\pi_{a_4}(a_3) = a_2$.

Replacing each element by its subscript, we have an isomorphism f from G onto a subgroup T of S_4, namely

$$T = \{(1), (12)(34), (13)(24), (14)(23)\}.$$

The isomorphism $f: G \longrightarrow T$ is defined by $f(a_1) = (1)$, $f(a_2) = (12)(34)$, $f(a_3) = (13)(24)$, $f(a_4) = (14)(23)$.

EXERCISES

1. Do Exercises 5, 7, 8, and 9, Section 3.1, using properties of groups.
2. Let C_n be a cyclic group of order n. Find a permutation group isomorphic to C_n. What is the smallest integer r such that C_n is isomorphic to a subgroup of S_r?
3. Prove Theorem 4.11.4.
4. Suppose that in the proof of Cayley's theorem we defined a permutation Γ_g on G by $\Gamma_g(x) = xg$, all $x \in G$. Does the proof of Theorem 4.11.5 still work? Explain.
5. Show that the group of symmetries of an equilateral triangle is isomorphic to a permutation group on three letters and also to a permutation group on six letters.
6. Let $f_n \colon Z \longrightarrow Z$ be defined by $f_n(x) = x + n$ for all $x \in Z$. Show that $\{f_n \mid n \in Z\}$ is a group isomorphic to Z.
7. Show that any finite group G is isomorphic to a subgroup of A_n for some n.

4.12 Direct Products

We have seen in Section 1.5 that it is possible to form a "larger" set from two given ones, say A and B, by forming the Cartesian product $A \times B$. We say that A and B are "embedded" in $A \times B$ in the sense that the set $A_1 = \{(a,b_1) \mid \text{all } a \in A \text{ and some fixed } b_1 \in B\} \subseteq A \times B$ "looks" very much like A and can, of course, be put into 1–1 correspondence with A. In a similar manner, B is "embedded" in $A \times B$.

On the other hand, many familiar sets can be considered to be the Cartesian product of "smaller" sets. For example, if K is the set of real numbers, the set of points in the Euclidean plane can be thought of as the set of all ordered pairs $\{(x,y) \mid x,y \in K\} = K \times K$. The "smaller" set in this case is the set K of real numbers. The "embedding" of K in $K \times K$ is given, for example, by the set $\{(x,0) \mid x \in K\}$.

In this section we shall show that if, instead of considering sets A and B, we consider groups G and H, then it is possible to define a new, "larger" group $G \times H$ in such a way that G and H are "embedded" as groups in $G \times H$. Also, we shall start with a group G and show when it is possible to write G in the form $H_1 \times H_2$, where H_1 and H_2 are subgroups of G.

Definition 4.12.1 Let $(G,\cdot)$ and $(H,*)$ be groups. Let $(G \times H, \circ)$ be the set $G \times H = \{(g,h) \mid g \in G, h \in H\}$ together with the binary operation $(g_1,h_1) \circ (g_2,h_2) = (g_1 \cdot g_2, h_1 * h_2)$. Then $(G \times H, \circ)$ is called the *direct product* of G and H and is denoted by $G \times H$.

Definition 4.12.2 A group G is said to be *embedded* in a group G' if there is an isomorphism from G onto a subgroup of G'.

We shall now show that $G \times H$ is a group in which G and H are embedded.

THEOREM 4.12.3 The direct product $G \times H$ of groups G and H is a group. Furthermore, $G \times H$ contains subgroups G' and H' that are isomorphic to G and H, respectively.

Proof We first show that the operation $\circ$ is a binary operation on $G \times H$. This amounts to showing that for any two elements (g_1,h_1) and (g_2,h_2) of $G \times H$, the definition of $\circ$ leads to precisely one element (g_3,h_3) of $G \times H$. The details are left to the reader.

The associativity of this binary operation follows from the associativity of $\cdot$ and $*$, for

$$\begin{aligned}((g_1,h_1) \circ (g_2,h_2)) \circ (g_3,h_3) &= (g_1 \cdot g_2, h_1 * h_2) \circ (g_3,h_3) \\ &= ((g_1 \cdot g_2) \cdot g_3, (h_1 * h_2) * h_3) \\ &= (g_1 \cdot (g_2 \cdot g_3), h_1 * (h_2 * h_3)) \\ &= (g_1,h_1) \circ (g_2 \cdot g_3, h_2 * h_3) \\ &= (g_1,h_1) \circ ((g_2,h_2) \circ (g_3,h_3)).\end{aligned}$$

Finally, if e is the identity of G and e' the identity of H, then (e,e') is the identity of $G \times H$ and (g^{-1},h^{-1}) is the inverse of the element (g,h) of $G \times H$.

To show that $G \times H$ contains subgroups $G' \cong G$ and $H' \cong H$, let $G' = \{(g,e') \mid \text{all } g \in G\}$, and $H' = \{(e,h) \mid \text{all } h \in H\}$. We leave the verification of the isomorphisms as an exercise. ∎

EXERCISES

1. Complete the proof of Theorem 4.12.3.
2. Let G' and H' be the subgroups of $G \times H$ defined in the proof of Theorem 4.12.3. Show that $g'h' = h'g'$, for all $g' \in G'$, $h' \in H'$. Show also that $G' \cap H'$ is the trivial subgroup of $G \times H$.
3. Let G and H be finite groups. What is the order of $G \times H$ in terms of the orders of G and H?
4. Show that the subgroups G' and H' defined in the proof of Theorem 4.12.3 above are normal subgroups of $G \times H$. Show also that $H \cong (G \times H)/G'$ and $G \cong (G \times H)/H'$.
5. Let G be the cyclic group of order 2 and H the cyclic group of order 3. Form $G \times H$. To what familiar group is it isomorphic?
6. Let G and H be abelian groups. Show that $G \times H$ is abelian.
7. Let $G = (\bar{Z}_m,+)$ and $H = (\bar{Z}_n,+)$. Show that if $(m,n) = 1$, then $G \times H$ is cyclic.

164   THE THEORY OF GROUPS

8. Let G and H be finite cyclic groups. Find necessary and sufficient conditions for $G \times H$ to be cyclic.
9. Let G be a cyclic group of order p^α, p a prime. Show that if

$$G \cong A \times B,$$

then either $|A| = 1$ or $|B| = 1$. That is, a cyclic group of prime power order can be represented as a direct product only in a trivial way.

10. Let $R \subseteq G \times H$. Define the projection of R to G as

$$\pi_1(R) = \{g \mid g \in G \text{ such that } (g,h) \in R \text{ for some } h \in H\}$$

and define the projection of R to H as

$$\pi_2(R) = \{h \mid h \in H \text{ such that } (g,h) \in R \text{ for some } g \in G\}.$$

Show that if R is a subgroup of $G \times H$, then $\pi_1(R)$ and $\pi_2(R)$ are subgroups of G and H, respectively. Also, show that if $R \triangleleft G \times H$, then $\pi_1(R) \triangleleft G$ and $\pi_2(R) \triangleleft H$.

11. Let $G \cong A \times B$. Prove that if $A \cong C$, then $G \cong C \times B$.

Although the direct product was defined for a pair of groups G and H, it is not difficult to see how the definition may be extended to three groups, or to any finite number of groups. That is, if G, H, and K are groups, we can form $(G \times H) \times K$ or $G \times (H \times K)$. As we shall show, these direct products are isomorphic. Thus we can write $G \times H \times K$ with no ambiguity up to isomorphism. The elements $G \times H \times K$ may be considered to be ordered triples (g,h,k) for $g \in G$, $h \in H$, $k \in K$.

THEOREM 4.12.4   Let G, H, K be groups. Then

$$(G \times H) \times K \cong G \times (H \times K).$$

Proof   The proof is left as an exercise. ∎

EXERCISES

12. Prove Theorem 4.12.4.
13. Generalize Theorem 4.12.4 to an arbitrary finite set of groups.

If G is a group, then it is very useful to know if G is isomorphic to a direct product of "smaller" groups. The next theorem provides a necessary and sufficient condition for determining when this is the case. We first prove a lemma.

LEMMA 4.12.5   Let A and B be normal subgroups of a group G such that $A \cap B = E$. Then $ab = ba$ for all $a \in A$ and all $b \in B$.

4.12 DIRECT PRODUCTS

Proof Let $a \in A$ and $b \in B$. Then $ab = bac$ for some $c \in G$, and it suffices to show that $c = e$. Since $c = a^{-1}b^{-1}ab = (a^{-1}b^{-1}a)b$ and since $B \triangleleft G$,

$$a^{-1}b^{-1}a \in B \quad \text{and therefore} \quad c = a^{-1}b^{-1}ab \in B.$$

On the other hand, $a^{-1}b^{-1}ab = a^{-1}(b^{-1}ab)$ and, since $A \triangleleft G$,

$$b^{-1}ab \in A \quad \text{and therefore} \quad c = a^{-1}b^{-1}ab \in A.$$

Therefore $c \in A \cap B$ and $c = e$. Hence $ab = ba$. ∎

THEOREM 4.12.6 Let A and B be normal subgroups of G such that

(a) $AB = G$,
(b) $A \cap B = E$.

Then

$$G \cong A \times B.$$

Proof Let f be the mapping from $A \times B$ to G defined as follows:

$$f(a,b) = ab.$$

It follows from (a) that f is a function from $A \times B$ onto G. Now

$$f((a_1,b_1) \circ (a_2,b_2)) = f(a_1 a_2, b_1 b_2) = a_1 a_2 b_1 b_2.$$

But by Lemma 4.12.5, $a_2 b_1 = b_1 a_2$; hence

$$a_1 a_2 b_1 b_2 = a_1 b_1 a_2 b_2 = f(a_1,b_1) f(a_2,b_2).$$

Thus f is a homomorphism.

To complete the proof we show that ker f is trivial and therefore f is an isomorphism.

Let $f(a,b) = ab = e$. Then $a = b^{-1}$ and $a \in A \cap B$. Hence $a = e = b$, and (a,b) is the identity of $A \times B$. ∎

Whenever it is possible to find normal subgroups A and B of a group G satisfying the hypotheses of Theorem 4.12.6, that is, $G \cong A \times B$, we shall say that G is an *internal direct product* of its subgroups A and B and we shall write $G = A \overset{\times}{} B$.

The criterion of Theorem 4.12.6 can be put in a slightly altered form that proves very useful in certain cases.

THEOREM 4.12.7 Let A and B be subgroups of a group G such that

(a) $ab = ba$, for all $a \in A$, $b \in B$, and
(b) for each $g \in G$, there are unique elements $a \in A$ and $b \in B$ such that $g = ab$.

Then $G = A \overset{\times}{} B$.

Proof It follows immediately that $G = AB$. We now show that $A \triangleleft G$ and $B \triangleleft G$. Let $a \in A$ and $g \in G$. Then $g = a_1 b_1$ for some $a_1 \in A$ and $b_1 \in B$. Hence

$$g^{-1}ag = b_1^{-1}a_1^{-1}aa_1b_1 = a_1^{-1}aa_1,$$

since $a'b' = b'a'$ for all $a' \in A$ and $b' \in B$. Since $a_1^{-1}aa_1 \in A$, $g^{-1}ag \in A$ and $A \triangleleft G$. Similarly, $B \triangleleft G$.

Finally, suppose that $g \in A \cap B$. Then $g \in A$ and $g \in B$. Since $g = a_1b_1$, for unique elements of a_1 of A and b_1 of B, $b_1 = a_1^{-1}g \in A$ and $a_1 = gb_1^{-1} \in B$. Thus, $g = a_1b_1 = a_2$ for some $a_2 \in A$ and $g = a_1b_1 = b_2$ for some $b_2 \in B$. But then

$$g = a_1b_1 = a_2e \quad \text{and} \quad g = a_1b_1 = eb_2.$$

By the unique representation for g in the form ab, $a \in A$, $b \in B$, $a_1 = a_2$, $b_1 = e$ and $a_1 = e$, $b_1 = b_2$. Thus, $g = ee = e$. That is, $A \cap B = E$. By Theorem 4.12.6, $G = A \times B$. ∎

EXERCISE

14. Show that if the condition $ab = ba$ is removed, Theorem 4.12.7 is false. (*Hint:* Consider S_3.)

If G is a group containing subgroups A and B fulfilling the conditions of Theorem 4.12.7, $G = A \times B$. If the subgroup A contains subgroups A_1, A_2 also fulfilling the conditions of the theorem with respect to A, $A = A_1 \times A_2$, and $G = (A_1 \times A_2) \times B$ (Exercise 11, Section 4.12). As we have seen, the operation of direct product is associative, and we can generalize the statement of Theorem 4.12.6 in the following manner.

THEOREM 4.12.8 Let $A_1, \ldots, A_n$ be normal subgroups of a group G, satisfying:

(a) $G = A_1 \cdots A_n$,
(b) $A_i \cap \hat{A}_i = E$, where $\hat{A}_i = A_1 \cdots A_{i-1}A_{i+1} \cdots A_n$, $i = 1, \ldots, n$.

Then $G = A_1 \times \cdots \times A_n$.

Proof The proof is left as an exercise. ∎

We conclude this section with several theorems about finite abelian groups of some special types.

THEOREM 4.12.9 Let $G \neq E$ be a finite abelian group all of whose elements different from e have order p a prime. Then, for some integer $n \geq 1$, $G \cong A_1 \times \cdots \times A_n$ where $A_i \cong C_p$, the cyclic group of order p, for $i = 1, \ldots, n$, and $|G| = p^n$.

Proof Let $S = \{a_1, a_2, \ldots, a_n\}$ be a set of generators of G with the following property: any proper subset of S does not generate G. Since G is finite, such a set must exist. We call such a set S *irredundant*. Now let $A_i = (a_i)$, and

$$\hat{A}_i = A_1 \cdots A_{i-1} A_{i+1} \cdots A_n \quad \text{for } i = 1, \ldots, n.$$

Clearly, $o(a_i) = p$, hence $A_i \cong C_p$ for all i. Also, $G = A_1 \cdots A_n$, since this product of subgroups is a subgroup and contains the set of generators $\{a_1, \ldots, a_n\}$. If $A_i \cap \hat{A}_i \neq E$, then $A_i \subseteq \hat{A}_i$, and $\{a_1, \ldots, a_n\}$ is not irredundant (see Exercise 10, Section 4.8). Therefore,

$$A_i \cap \hat{A}_i = E \quad \text{for all } i$$

and, by Theorem 4.12.8,

$$G = A_1 \dot{\times} \cdots \dot{\times} A_n.$$

Clearly, $|G| = p^n$. ∎

THEOREM 4.12.10 Let G be an abelian group of order p^2, p a prime. Then either $G \cong C_{p^2}$, the cyclic group of order p^2, or $G \cong C_p \times C_p$, the direct product of a cyclic group of order p with itself.

Proof By Lagrange's theorem each element $g \in G$, $g \neq e$, has order p or p^2. If G contains an element of order p^2, then G is cyclic and hence isomorphic to C_{p^2}. Otherwise, every element of G different from e has order p. Since G is abelian, $G \cong C_p \times C_p$ by Theorem 4.12.9. ∎

EXERCISES

15. Prove Theorem 4.12.8.
16. Let $N_1, \ldots, N_r$ be normal subgroups of a group G and assume that $G = N_1 N_2 \cdots N_r$. Show that $G = N_1 \dot{\times} N_2 \dot{\times} \cdots \dot{\times} N_r$ if and only if whenever $a_1 a_2 \cdots a_r = e$ for $a_i \in N_i$, then $a_i = e$ for $i = 1, \ldots, r$.
17. Show that there exists exactly one group of order 8 all of whose elements have order 2.
18. Let G be a direct product of its subgroups A and B. Prove that $G/A \cong B$, and $G/B \cong A$.

4.13 Automorphisms

A particularly important class of homomorphisms are the isomorphisms from a group G onto itself. Such a mapping is a permutation on the elements of G as well as a homomorphism. In this section we will study some of the properties of these mappings and show how they are used to investigate the structure of a group.

Definition 4.13.1 Let f be an isomorphism from G onto G. Then f is called an *automorphism* of G.

THEOREM 4.13.2 Let G be a group and $a \in G$. Then the function f, defined by $f(g) = a^{-1}ga$ for all $g \in G$, is an automorphism of G.

Proof To show that f is onto, let $h \in G$, and let $h' = aha^{-1}$. Then

$$f(h') = a^{-1}h'a = a^{-1}(aha^{-1})a = h,$$

and therefore f is onto. Since $a^{-1}ga = a^{-1}ha$ implies $g = h$, f is clearly 1-1.

It remains to show only that f is a homomorphism. If $u, v \in G$, then $f(u) = a^{-1}ua$, $f(v) = a^{-1}va$, and $f(uv) = a^{-1}uva$. But

$$a^{-1}uva = a^{-1}u(aa^{-1})va = (a^{-1}ua)(a^{-1}va) = f(u)f(v).$$

Hence $f(uv) = f(u)f(v)$, and f is a homomorphism. ∎

Definition 4.13.3 Let G be a group and let a be a fixed element of G. Then the automorphism f, defined by $f(g) = a^{-1}ga$ for all $g \in G$, is called the *inner automorphism* determined by a. Any automorphism that is not equal to some inner automorphism is called an *outer automorphism*. The element $a^{-1}ga$ is called a *conjugate* of g, and the function f is also called *conjugation by a*. The elements a and b are called *conjugate* if there exists $g \in G$ such that $b = g^{-1}ag$.

EXERCISES

1. Let G be a group. Show that the identity function on G is an inner automorphism.
2. Show that if the only inner automorphism of a group G is the identity function, then G is abelian, and conversely.
3. Find an outer automorphism on the
   (a) cyclic group of order 3.
   (b) cyclic group of order 6.
4. Show that the only automorphism of C_2 is the identity automorphism.
5. Let f be a function on the group G defined by $f(g) = g^{-1}$ for all $g \in G$. Show that f is an automorphism if and only if G is abelian.
6. Find the number of inner automorphisms and outer automorphisms of the following groups:
   (a) Klein group. (b) $(\bar{Z}_5, +)$.
   (c) $(\bar{Z}_m, +)$. (d) S_3.
7. Let G be a group containing an element a not of order 2. Prove that G has an automorphism different from the identity automorphism. (*Hint:* In the abelian case, consider Exercise 5.)

The notion of inner automorphism as a function defined on elements of G can be extended in a natural way to a function on sets of elements as follows:

Definition 4.13.4 Let f be the inner automorphism on G determined by $a \in G$, and let S be a subset of elements of G. Then

$$a^{-1}Sa = \{a^{-1}sa \mid s \in S\}.$$

If S is a subgroup of G, then $a^{-1}Sa$ is called a *subgroup conjugate to S*, or simply a *conjugate of S*.

THEOREM 4.13.5 Let S be a subgroup of the group G. Then $a^{-1}Sa$ is a subgroup of G isomorphic to S.

Proof Let $f: G \longrightarrow G$ be defined by $f(g) = a^{-1}ga$. Then, by Theorem 4.13.2, f is an automorphism. Thus f restricted to S is an isomorphism. ∎

It follows directly from the definition of automorphism that the identity element of a group G is invariant under any automorphism; that is, $f(e) = e$ for any automorphism f. It is also easy to describe those elements of G that are invariant under any inner automorphism of G.

Definition 4.13.6 An element $g \in G$ is called *central* in G if $a^{-1}ga = g$, or equivalently $ga = ag$, for all $a \in G$. The set of such elements is called the *center* of G, and is denoted by $C(G)$.

EXERCISES

8. Find the center of (a) C_3, (b) S_3, (c) D_4.
9. Show that the center of a group G is a normal subgroup of G.
10. Show that if $G/C(G)$ is cyclic, then G is abelian. [*Hint:* Let $G/C(G)$ be generated by the coset $aC(G)$. Then for $g_1, g_2 \in G$, there are integers n_1, n_2 such that $g_1 \in a^{n_1}C(G)$, $g_2 \in a^{n_2}C(G)$. (Why?) Show that $g_1g_2 = g_2g_1$.]
11. Let f_a and f_b be the inner automorphisms defined by $f_a(g) = a^{-1}ga$, $f_b(g) = b^{-1}gb$. Show that the inner automorphism $f_a = f_b$ if and only if $a \in bC(G)$.

If we attempt to extend the notion of a central element from elements to subgroups, we are led to investigate those subgroups N of G with the property that $a^{-1}Na = N$ for all $a \in G$. As we have already seen in Section 4.9, this is a criterion that N be a normal subgroup. It is for this reason that normal subgroups are sometimes called *invariant* subgroups.

Since an automorphism on a group G is a permutation on G, automorphisms may be multiplied as permutations.

THEOREM 4.13.7 Let $A(G)$ be the set of all automorphisms on G and $I(G)$ the set of all inner automorphisms. Then $A(G)$ is a group with respect to permutation multiplication and $I(G)$ is a normal subgroup of $A(G)$.

Proof We leave the proof of the fact that $A(G)$ and $I(G)$ are groups as an exercise, but give the proof that $I(G) \triangleleft A(G)$.

Let $f_a \in I(G)$ be defined by

$$f_a(g) = a^{-1}ga \quad \text{for all } g \in G.$$

Let τ be an arbitrary element of $A(G)$. We must show that $\tau^{-1}f_a\tau \in I(G)$. Consider

$$(\tau^{-1}f_a\tau)(g) \quad \text{for any } g \in G.$$

Then

$$(\tau^{-1}f_a\tau)(g) = [\tau^{-1}f_a](\tau(g)) = \tau^{-1}[a^{-1}(\tau(g))a].$$

Since τ is a homomorphism, so is τ^{-1}, whence

$$\tau^{-1}[a^{-1}(\tau(g))a] = [\tau^{-1}(a^{-1})][\tau^{-1}(\tau(g))][\tau^{-1}(a)] = [\tau^{-1}(a^{-1})]g[\tau^{-1}(a)]$$
$$= [\tau^{-1}(a)]^{-1}g[\tau^{-1}(a)].$$

Hence

$$\tau^{-1}f_a\tau = f_{\tau^{-1}(a)}. \qquad \blacksquare$$

Definition 4.13.8 The set of all automorphisms on a group G, $A(G)$, together with permutation multiplication, is called the *automorphism group of* G, and $I(G)$ is called the *inner automorphism group of* G.

EXERCISES

12. Compete the proof of Theorem 4.13.7.
13. Show that $G/C(G) \cong I(G)$. (*Hint:* Use the first isomorphism theorem.)
14. Express all automorphisms on S_3 as permutations. How many symbols are needed to define these permutations? To what group is $A(S_3)$ isomorphic?

The group of inner automorphisms $I(G)$ has the effect of inducing an equivalence relation and hence a partition on the group G, as is shown below.

THEOREM 4.13.9 Let G be a group. Let R be the relation of conjugacy defined on G as follows: given $a, b \in G$, aRb if there exists $g \in G$ such that $a = g^{-1}bg$. Then R is an equivalence relation on G.

Proof For any $a \in G$, $a = e^{-1}ae$, where e is the identity of G. Thus, aRa.

Now suppose aRb. Then $a = g^{-1}bg$ for some $g \in G$, whence

$$b = (g^{-1})^{-1}ag^{-1} \quad \text{and so } bRa.$$

Lastly, suppose aRb and bRc; that is, $a = g_1^{-1}bg_1$ and $b = g_2^{-1}cg_2$, for some $g_1, g_2 \in G$. Then

$$a = g_1^{-1}(g_2^{-1}cg_2)g_1 = (g_2g_1)^{-1}c(g_2g_1), \quad \text{and so } aRc. \quad \blacksquare$$

Definition 4.13.10 Let G be a group. The equivalence classes determined by the relation of conjugacy are called the *conjugate classes of G*.

These conjugate classes have many interesting and useful properties. For example, unlike the case of a partition into left cosets, the number of elements in each class is not always the same, for the identity element is always in a class by itself, as is any central element. But if an element is not central, then its conjugate class has at least two members. Finally, we note that if A is a conjugate class, and $a \in A$, then $A = \{g^{-1}ag \mid g \in G\}$.

Definition 4.13.11 Let G be a finite group of order α. Let the conjugate classes of G be $\{A_i\}$, $i = 1, \ldots, n$, and let A_i have α_i elements. Then the expression

$$\alpha = |G| = \alpha_1 + \alpha_2 + \cdots + \alpha_n$$

is called the *class equation of G*.

Definition 4.13.12 Let g be an element of the group G. Then the set

$$C_G(g) = \{x \mid x \in G, x^{-1}gx = g\}$$

is called the *centralizer* of g in G. If A is a subgroup of G, then the set

$$N_G(A) = \{x \mid x \in G, x^{-1}Ax = A\}$$

is called the *normalizer* of A in G.

We note that an element $g \in G$ is central if and only if $C_G(g) = G$ and that a subgroup $A \subseteq G$ is normal in G if and only if $N_G(A) = G$.

EXERCISES

15. Show that $C_G(g)$ is a subgroup of G for any $g \in G$.
16. Show that $N_G(A)$ is a subgroup of G for any subgroup A of G, and show that $A \triangleleft N_G(A)$.
17. Let A be a subgroup of a group G. Show that if $A \triangleleft B$, a subgroup of G, then $B \subseteq N_G(A)$.

THEOREM 4.13.13 Let G be a finite group and let α_i be the number of elements in the conjugate class A_i of G. If $a \in A_i$, then $\alpha_i = [G : C_G(a)]$.

Proof We will show there exists a 1-1 correspondence between the set A_i and the set of right cosets $\{C_G(a)g \mid g \in G\}$. For $g_1^{-1}ag_1 = g_2^{-1}ag_2$ if and only if $g_2g_1^{-1}ag_1g_2^{-1} = a$ if and only if $g_1g_2^{-1} \in C_G(a)$ if and only if $C_G(a)g_1 = C_G(a)g_2$. Hence the mapping f defined by $f(g^{-1}ag) = C_G(a)g$ is 1-1 and onto. ∎

COROLLARY 4.13.14 Let G be a finite group with class equation

$$|G| = \alpha_1 + \cdots + \alpha_k.$$

Then $\alpha_i \mid |G|$ for $i = 1, 2, \ldots, k$.

Proof It follows from the theorem that α_i is the index of the centralizer of some element of G, and hence by Lagrange's theorem, $\alpha_i \mid |G|$. ∎

We now give two important theorems about finite groups that follow readily from Corollary 4.13.14.

THEOREM 4.13.15 Let G be a finite group, $|G| = p^n$, $n > 0$, p a prime. Then G has a nontrivial center, of order at least p.

Proof Let $|G| = p^n = \alpha_1 + \cdots + \alpha_k$ be the class equation of G. By Corollary 4.13.14, $\alpha_i \mid |G|$. If $\alpha_i > 1$, then, since p is a prime, $p \mid \alpha_i$. Assume that e is the only central element of G. Then for precisely one α, say α_1, $\alpha_1 = 1$. Thus $p \mid \alpha_i$, for $i \geq 2$. But then, since $p \mid |G|$,

$$p \mid [|G| - (\alpha_2 + \cdots + \alpha_k)].$$

Since $\alpha_1 = |G| - (\alpha_2 + \cdots + \alpha_k)$, $p \mid \alpha_1$, a contradiction. Hence the center of G contains more than one element, and since $C(G)$ is a subgroup, $|C(G)| \geq p$. ∎

COROLLARY 4.13.16 Let G be a group of order p^2, p a prime. Then G is abelian.

Proof Since $|G| = p^2$, p a prime, by Theorem 4.13.15, G contains a nontrivial center $C(G)$. Since $C(G) \triangleleft G$, consider $G/C(G)$. Since $C(G)$ is nontrivial, $|G/C(G)| = p$ or 1. In either case, $G/C(G)$ is cyclic, and hence by Exercise 10, Section 4.13, G is abelian—that is, $C(G) = G$. ∎

COROLLARY 4.13.17 Let G be a group of order p^2, p a prime. Then either $G \cong C_{p^2}$, the cyclic group of order p^2, or $G \cong C_p \times C_p$, the direct product of a cyclic group of order p with itself.

Proof By Corollary 4.13.16, G is abelian. By Theorem 4.12.10, G is now of one of the two forms C_{p^2} or $C_p \times C_p$. ∎

The next theorem provides information about the subgroups of a finite group, in terms of the primes that divide its order.

THEOREM 4.13.18 Let G be a finite abelian group. If p is a prime and $p \mid |G|$, then G contains an element of order p.

We use a technique of proof in this and the following theorem that is an application of the law of well ordering and is particularly suited for finite groups. The argument goes as follows: If the theorem in question is not true, then the set of groups for which it is false is nonempty; hence the set of orders of these groups is a nonempty set of positive integers. Let n be the least such and let G be one of the groups of order n for which the theorem is false. G is therefore called a *minimal counterexample*. It must then be shown that G, in fact, does not exist; hence the set of orders is empty, and therefore the set of groups for which the theorem fails is empty. The chief tool for accomplishing this is the fact that if A is a proper subgroup of G or if G/N is a proper factor group of G, then the theorem is true for A and for G/N, since $|A| < |G|$, $|G/N| < |G|$, and G is a group of minimal order for which the theorem is false. That the theorem holds for all proper subgroups and factor groups of G is then used to bring about a contradiction.

Proof Let G be a minimal counterexample. That is, $p \mid |G|$ for some prime p, but G contains no element of order p. Further, if G' is any other group with these properties, then $|G| \leq |G'|$.

Let $g \in G$, $g \neq e$. If $o(g) = pm$, then $o(g^m) = p$, by Exercise 15, Section 4.7, a contradiction. Hence, assume that every element of G has order relatively prime to p.

Then, for any $g \neq e$, (g) is a proper subgroup of G, since otherwise G would be cyclic and $|G| = o(g)$. But $p \mid |G|$ and $p \nmid o(g)$, so G is not cyclic.

Since G is abelian, (g) is normal, and we consider $G/(g)$. Now $E \subset (g)$ and $o(g) > 1$, whence $|G/(g)| < |G|$. Since $|G/(g)| \, |(g)| = |G|$, it follows that $p \mid |G/(g)|$. Since G is a minimal counterexample, $G/(g)$ contains an element $h(g)$ of order p. But by Exercise 5, Section 4.10, the order of a coset as an element in the quotient group divides the order of any representative in the full group. Hence $p \mid o(h)$. This implies that G has an element of order p, a contradiction. Hence G is not a minimal counterexample, and the theorem is proved. ∎

THEOREM 4.13.19 (*Cauchy*) Let G be a finite group. If $p \mid |G|$, p a prime, then G contains an element of order p.

Proof Let G be a minimal counterexample. Thus $|G|$ is divisible by p, but G does not contain an element of order p. Now, if $p \mid |H|$, H a proper subgroup of G, then H contains an element of order p, and this element is also an element of G. Therefore all proper subgroups of G have orders not divisible by p.

We now consider the class equation of G:

$$|G| = \alpha_1 + \alpha_2 + \cdots + \alpha_k.$$

If $\alpha_i > 1$, then α_i is the index of a proper subgroup C_i of G. As we have seen above, $p \nmid |C_i|$; hence $p \mid [G:C_i] = \alpha_i$ for all $\alpha_i > 1$. Therefore, since $p \mid |G|$, $p \mid \alpha'$, where α' is the number of α_j's such that $\alpha_j = 1$. But α' is precisely the order of $C(G)$, the center of G, but, by assumption, p does not divide the order of any proper subgroup of G. Therefore $C(G) = G$ and G is abelian. Hence, by Theorem 4.13.18, G contains an element of order p, a contradiction. This completes the proof of the theorem.

COROLLARY 4.13.20   Let G be a finite group and p a prime. If for each $g \in G$, there is an integer $n(g) \geq 0$ such that $o(g) = p^{n(g)}$, then $|G| = p^\alpha$, for some integer $\alpha \geq 0$.

Proof   The proof is left as an exercise.

As an application of Cauchy's theorem we will give a complete classification of all groups of order $2p$, where p is a prime. We have already classified groups of order 4, and so we now assume that $p > 2$.

Let G be a group of order $2p$, p a prime. Then it follows from Cauchy's theorem that G contains a pair of elements a and b with $o(a) = 2$ and $o(b) = p$. Now the subgroup H generated by $\{a,b\}$ must be the entire group G. For H will contain elements of orders 2 and p and thus, by Lagrange's theorem, its order must be divisible by 2 and p, that is $|H| \geq 2p$. We now need to distinguish two cases.

CASE 1   If $ab = ba$, then H must be abelian by Exercise 31, Section 4.7. Thus G is an abelian group containing an element of order 2 and one of order p, an odd prime. Hence, by Exercise 14, Section 4.7, G is cyclic, which completely specifies the group G.

CASE 2   Suppose that $ab \neq ba$. The cyclic subgroup of G generated by b is a subgroup of index 2, hence is normal in G. Therefore if $N = (b)$, we have $a^{-1}Na = N$. Hence $a^{-1}ba = b^k$ for some integer k, $0 \leq k \leq p-1$, since every element of N is some power of b. Now if $a^{-1}ba = b^k$, then $a^{-1}(a^{-1}ba)a = a^{-1}b^k a$. Hence, on the one hand,

$$a^{-1}(a^{-1}ba)a = (a^{-1}a^{-1})b(aa) = b$$

since $a^2 = e$, and on the other hand, $a^{-1}b^k a = (a^{-1}ba)^k = b^{k^2}$, by Exercise 9, Section 4.5. Thus $b = b^{k^2}$, and so $k^2 \equiv 1 \pmod{p}$. This implies that either $k \equiv 1 \pmod{p}$ or $k \equiv -1 \pmod{p}$, since p is a prime. Now if $k \equiv 1 \pmod{p}$, then $b = b^k = a^{-1}ba$, which implies that $ab = ba$, which

4.13 AUTOMORPHISMS   175

we have already discussed in Case 1. Thus $k \equiv -1 \pmod{p}$, and we have the relation $a^{-1}ba = b^{-1}$.

We will now show that with $a^{-1}ba = b^{-1}$, G is isomorphic to the dihedral group D_p. For let $f\colon G \longrightarrow D_p$ be defined by (a) $f(a) = v$, a fixed reflection in D_p, and (b) $f(b) = r_1$, the rotation of $2\pi/p$ radians, and, in general, (c) $f(a^\alpha b^\beta) = v^\alpha r_1^\beta$. To show that the domain of f is the whole group G we must prove that every element of G can be written in the form

$$a^\alpha b^\beta, \quad 0 \leq \alpha \leq 1 \text{ and } 0 \leq \beta \leq p - 1.$$

Now since $a^{-1}ba = b^{-1}$, it follows that $ba = ab^{-1}$, and it is easy to conclude that $b^\beta a = ab^{-\beta}$. Thus since an arbitrary element of G can be written in the form $a^{\alpha_1}b^{\beta_1}a^{\alpha_2}b^{\beta_2}\cdots a^{\alpha_r}b^{\beta_r}$, with $0 \leq \alpha_i \leq 1$ and $0 \leq \beta_i \leq p - 1$ (see Exercise 27 Section 4.7), the a's may be gathered together to the left by the relation $b^\beta a = ab^{-\beta}$ and every element of G has the form $a^\alpha b^\beta$. Moreover if $a^\alpha b^\beta = a^{\alpha'} b^{\beta'}$, then $a^{\alpha-\alpha'} = b^{\beta'-\beta}$ and since $o(a) = 2$ and $o(b) = p$, an odd prime, it follows that $a^{\alpha-\alpha'} = b^{\beta'-\beta} = e$. Thus

$$\alpha \equiv \alpha' \pmod{2} \quad \text{and} \quad \beta \equiv \beta' \pmod{p}.$$

But $0 \leq \alpha, \alpha' \leq 1$, and $0 \leq \beta, \beta' \leq p - 1$, and so $\alpha = \alpha'$ and $\beta = \beta'$.

We also note that $b^\beta a = ab^{-\beta}$ implies that $b^\beta a^\alpha = a^\alpha b^{(-1)^\alpha \beta}$. For $b^\beta a^\alpha = ab^{-\beta}a^{\alpha-1} = ab^{-\beta}aa^{\alpha-2} = aab^\beta a^{\alpha-2} = a^\alpha b^{(-1)^\alpha \beta}$, by induction. Thus, $a^{\alpha_1}b^{\beta_1}a^{\alpha_2}b^{\beta_2} = a^{\alpha_1+\alpha_2}b^{\beta_2+(-1)^{\alpha_2}\beta_1}$.

We have already remarked in Exercise 26, Section 4.7, that every element of D_p can be written in the form

$$v^m r_1^n, \quad 0 \leq m \leq 1 \text{ and } 0 \leq n \leq p - 1.$$

Thus, it is easy to see that f is a map from G onto D_p.

It now remains to show that f is an isomorphism. Let $g_1, g_2 \in G$, and consider $f(g_1 g_2)$. Now $g_1 = a^{\alpha_1}b^{\beta_1}$ and $g_2 = a^{\alpha_2}b^{\beta_2}$ with $0 \leq \alpha_1, \alpha_2 \leq 1$, and $0 \leq \beta_1, \beta_2 \leq p - 1$. Thus,

$$f(g_1 g_2) = f(a^{\alpha_1}b^{\beta_1}a^{\alpha_2}b^{\beta_2}) = f(a^{\alpha_1}a^{\alpha_2}b^{(-1)^{\alpha_2}\beta_1}b^{\beta_2})$$
$$= f(a^{\alpha_1+\alpha_2}b^{\beta_2+(-1)^{\alpha_2}\beta_1}) = v^{\alpha_1+\alpha_2}r_1^{\beta_2+(-1)^{\alpha_2}\beta_1}.$$

But by Exercise 28, Section 4.7,

$$v^{\alpha_1+\alpha_2}r_1^{\beta_2+(-1)^{\alpha_2}\beta_1} = v^{\alpha_1}v^{\alpha_2}r_1^{(-1)^{\alpha_2}\beta_1}r_1^{\beta_2} = v^{\alpha_1}r_1^{\beta_1}v^{\alpha_2}r_1^{\beta_2}$$
$$= f(a^{\alpha_1}b^{\beta_1})f(a^{\alpha_2}b^{\beta_2}) = f(g_1)f(g_2).$$

Hence f is a homomorphism. Clearly, f is 1–1, and so we can now conclude f is an isomorphism.

We may summarize the discussion above in the following theorem.

THEOREM 4.13.21 Let G be a group of order $2p$, p a prime greater than 2. Then either

(a) G is cyclic or
(b) G is isomorphic to the dihedral group D_p. ∎

EXERCISES

18. Let G be a finite abelian group and let $G = \{a_1,\ldots,a_n\}$, where each a_j has order $\alpha_j \geq 1, j = 1, 2, \ldots, n$. Show that there exists an integer m such that any element of G is equal to precisely m expressions of the form $a_1^{\beta_1} a_2^{\beta_2} \cdots a_n^{\beta_n}$, $0 \leq \beta_i < \alpha_i$.
19. Using Exercise 18, give another proof of Theorem 4.13.18. (*Hint:* Express $m \cdot |G|$ in terms of the α_i.)
20. Let A be a subgroup of the finite group G. Show that $[G:N_G(A)]$ is the number of distinct subgroups conjugate to A.
21. Let G be a finite group with precisely two conjugate classes. Show that $|G| = 2$.
22. Let the group G contain an element a having exactly two conjugates. Show that G has a proper normal subgroup $N \neq E$.
23. Prove Corollary 4.13.20.
24. Let G be a finite group, $|G| > 2$. Show that G has a nonidentity automorphism. (See Exercise 7.)

4.14 Classification of Groups of Small Order

It has been said that the main problem in the theory of finite groups is to determine all groups of an arbitrary order n. In one sense this problem is solved, since for each order n one can, in a finite number of steps, enumerate all possible $n \times n$ multiplication tables and exhaustively check for isomorphic pairs. This solution has two serious defects. First, the effort involved in enumerating all multiplication tables and testing for isomorphic pairs becomes astronomical for even moderately large n. Second, the multiplication table of a group presents the important properties of a group in an extremely inconvenient form. The important results in the theory of groups give us information concerning specific properties for large classes of groups. Our work up to now has given us sufficient tools to completely classify groups of relatively small order without recourse to an exhaustive search of possible multiplication tables. We now give (with some gaps to be filled in by the reader) the complete classification of all finite groups of order ≤ 10.

4.14 CLASSIFICATION OF GROUPS OF SMALL ORDER

I. $|G| = 1$. Then $G \cong E$, the group with one element.
II. $|G| = 2$. Then $G \cong C_2$.
III. $|G| = 3$. Then $G \cong C_3$.
IV. $|G| = 4$. By Theorem 4.8.14 or by Corollary 4.13.17, either $G \cong C_4$ or $G \cong C_2 \times C_2$. Clearly these groups are not isomorphic.
V. $|G| = 5$. Then $G \cong C_5$.
VI. $|G| = 6$. By Exercise 22, Section 4.8, or by Theorem 4.13.21, either $G \cong C_6$ or $G \cong S_3$. Since C_6 is abelian and S_3 is not abelian, these groups are not isomorphic.
VII. $|G| = 7$. Then $G \cong C_7$.
VIII. $|G| = 8$.

A. If G is abelian, there are three possibilities:

   (1) G has an element of order 8.
   (2) G has no element of order 8, but G does have an element of order 4.
   (3) Every nonidentity element of G has order 2.

(1) Clearly $G \cong C_8$.
(2) Let $a \in G$, a of order 4. Then, since $(a) \subset G$, there is an element $b \in G$, $b \notin (a)$. If $b^2 = e$, then

$$G = (a)(b) \quad \text{and} \quad (a) \cap (b) = E.$$

Thus $G \cong C_4 \times C_2$ by Theorem 4.12.6.
If $b^2 \neq e$, then $b^2 \in (a)$, for otherwise $|(b)| = 4$ and

$$(a) \cap (b) = E$$

and $|G| \geq 16$. Hence $b^2 \in (a)$, and, since $o(b^2) = 2$, $b^2 = a^2$. Now let $c = ba^{-1}$. Clearly $c \notin (a)$, and $c^2 = b^2 a^{-2} = e$. Thus

$$G = (a)(c) \cong C_4 \times C_2.$$

(3) By Theorem 4.12.9, $G \cong C_2 \times C_2 \times C_2$.

We leave it to the reader to verify that no two of these three groups are isomorphic.

B. G is nonabelian.

We first note that G can have no element of order 8, for if it did, this would imply $G \cong C_8$, an abelian group. Also, not every element of G, $g \neq e$, can have order 2, for if this were the case, G would be abelian by Exercise 6, Section 4.2.

Thus G has an element of order 4 but none of order 8. Let $a \in G$, a of order 4. Then $[G : (a)] = 2$, and (a) is normal in G. Hence $G/(a)$ has order 2, and so there exists an element $b \in G$, $b \notin (a)$ and $b^2 \in (a)$. If $b^2 = a$, then, since a has order 4, b would have order 8, a contradiction. Similarly, $b^2 \neq a^3$. Hence (1) $b^2 = e$ or (2) $b^2 = a^2$.

(1) $b^2 = e$. In this case, we may express all the elements of G in terms of a and b, as follows:

$$a, a^2, a^3, a^4 = e, ab, a^2b, a^3b, a^4b = b.$$

It is easy to see that these are all distinct. To show that the product of any two of these elements is one of the list, we need only determine the product ba.

Since $b \notin (a)$, $ba \notin (a)$. If $ba = b$, then $a = e$, a contradiction. If $ba = ab$, then G is abelian, another contradiction. If $ba = a^2b$, then $bab^{-1} = a^2$. But bab^{-1} has order 4, while a^2 has order 2, again a contradiction. Thus, the only possibility that can hold is $ba = a^3b$.

Hence G has two generators, a and b, and $a^4 = e$, $b^2 = e$, and $ba = a^3b$. The group D_4 contains the generators r_1 and h that respectively satisfy the properties of a and b above. The mapping $f: D_4 \longrightarrow G$ defined by

$$f(r_1{}^n h^m) = a^n b^m$$

(see Exercise 26, Section 4.7) can be shown to be an isomorphism. We leave this for the reader.

(2) $b^2 = a^2$. Again

$$G = \{e, a, a^2, a^3, b, ab, a^2b, a^3b\}.$$

As in (1) we must determine the product ba. Exactly as in (1), we find that we must have $ba = a^3b$. We now know that G is generated by a and b, where $b^2 = a^2$, $a^4 = e$, $ba = a^3b$. These relations are sufficient to determine a multiplication table for the group G. From the table, writing the elements in the order $e, a, a^2, a^3, b, ab, a^2b, a^3b$, we can find a permutation group that is isomorphic to G, as we did in the example in Section 4.11. Indeed, as in the proof of Cayley's theorem, a corresponds to

$$(1\ 2\ 3\ 4)(5\ 6\ 7\ 8)$$

and b to

$$(1\ 5\ 3\ 7)(2\ 8\ 4\ 6).$$

The group generated by $(1\ 2\ 3\ 4)(5\ 6\ 7\ 8)$ and $(1\ 5\ 3\ 7)(2\ 8\ 4\ 6)$ is a group of order 8 isomorphic to G. This group is called the *quaternion group*, and in the exercises the reader will show that it is not isomorphic to D_4.

To summarize, we have displayed five nonisomorphic groups of order 8 and have proved that any group of order 8 is isomorphic to one of these.

4.14 CLASSIFICATION OF GROUPS OF SMALL ORDER

They are the following:
(1) C_8.
(2) $C_4 \times C_2$.
(3) $C_2 \times C_2 \times C_2$.
(4) Group of symmetries of the square, D_4.
(5) Quaternion group—consisting of the elements e, a, a^2, a^3, b, ab, a^2b, a^3b, in which the product of elements can be computed from $a^4 = b^4 = e$, $a^2 = b^2$, $ba = a^3b$.

IX. $|G| = 9$. By Corollary 4.13.17, either $G \cong C_9$ or $G \cong C_3 \times C_3$. These groups are clearly nonisomorphic.

X. $|G| = 10$. By Theorem 4.13.21, $G \cong C_{10}$ or $G \cong D_5$.

EXERCISES

1. Show that no two of the groups C_8, $C_4 \times C_2$, and $C_2 \times C_2 \times C_2$ are isomorphic.
2. Show that D_4 is not isomorphic to the quaternion group. (*Hint:* Show that the quaternion group contains only one element of order 2.)
3. Let $G = \{1, -1, i, -i, j, -j, k, -k\}$, with $-1, 1 \in Z$; the remaining symbols are distinct and satisfy:

$$ij = k, \quad jk = i, \quad ki = j,$$
$$i^2 = j^2 = k^2 = -1,$$
$$(-a)b = a(-b) = -(ab) = ba \quad \text{for all } a, b \in G, a \neq \pm 1, b \neq \pm 1,$$
$$-a = (-1)a = a(-1).$$

Show that G is isomorphic to the group of quaternions.
4. Classify the groups of order 12.

5 The Theory of Rings

In our study of groups we have referred to properties of the integers Z, both for purposes of illustration and to aid in proving several theorems. As a source of examples we considered additive properties of the set Z and multiplicative properties of $\bar{Z}_p - \{\bar{0}\}$, for any prime p. Some of the most important properties of the integers, however, are those that simultaneously involve the operations of addition and multiplication; for example, the division algorithm.

The integers are one example of an algebraic system involving two binary operations that are "intertwined" in some way. For Z, this intertwining takes the form of the distributive law: $a(b + c) = ab + ac$. Many other familiar number systems also satisfy this general description: the rational numbers and the real numbers, for example.

In this chapter we shall study an abstract algebraic system, known as a *ring*, that involves two binary operations and that includes as special cases the integers, rational numbers, and real numbers. Like the concept of a group, that of a ring is basic in the study of mathematics.

5.1 Definition of Ring and Examples

Definition 5.1.1 Let R be a nonempty set and let $+$ and $\cdot$ denote two binary operations on R, which we refer to as "addition" and "multiplication," respectively. Then $(R, +, \cdot)$ is called a *ring* if the following conditions hold:

(a) $(R, +)$ is a commutative group.
(b) $\cdot$ is associative. That is, $(a \cdot b) \cdot c = a \cdot (b \cdot c)$ for all $a, b, c \in R$.

5.1 DEFINITION OF RING AND EXAMPLES

(c) $a \cdot (b + c) = (a \cdot b) + (a \cdot c)$ (left distributive law)
and
$(a + b) \cdot c = (a \cdot c) + (b \cdot c)$ (right distributive law)
for all $a, b, c \in R$.

When no ambiguity exists, we shall refer to the ring $(R, +, \cdot)$ simply as R. As with groups, we shall usually drop the dot notation for products and write ab for $a \cdot b$. The identity of $(R, +)$ is denoted by 0 and is called the *zero* of the ring R. If there exists an element e in R such that for all a in R, $ea = ae = a$, then e is called a *multiplicative identity* of R.

We now proceed to some examples of rings.

Example 5.1.2 $(Z, +, \cdot)$, the ring of integers, with the usual interpretation of $+$ and $\cdot$.

Example 5.1.3 $(W, +, \cdot)$, the ring of even integers with the usual interpretation of $+$ and $\cdot$. This ring has no multiplicative identity.

Example 5.1.4 $(\bar{Z}_m, +, \cdot)$, the ring of residue classes modulo m, with $+$ and $\cdot$ as defined in Definition 2.12.1.

Example 5.1.5 $(Q, +, \cdot)$, the ring of rational numbers,[1] that is, real numbers of the form p/q, $p, q \in Z$, $q \neq 0$, with the usual interpretation of $+$ and $\cdot$.

Example 5.1.6 $(K, +, \cdot)$, the ring of real numbers, with the usual interpretation of $+$ and $\cdot$.

Example 5.1.7 Let S be a set, and let $\mathcal{S}$ be the collection of all subsets of S. For any $A, B \in \mathcal{S}$ we define $+$ and $\cdot$ as follows:

$$A + B = (A \cup B) - (A \cap B),$$
$$A \cdot B = A \cap B.$$

The set $\mathcal{S}$ is a ring with respect to these operations of $+$ and $\cdot$, with $\emptyset$ serving as the identity of $(\mathcal{S}, +)$. The set $S \in \mathcal{S}$ is a multiplicative identity of the ring $\mathcal{S}$.

There are two moderately complicated steps in verifying that $(\mathcal{S}, +, \cdot)$ is a ring. One is in showing that the operation $+$ is associative; the other is in verifying the distributive laws. As an aid in constructing proofs of these facts, the reader should interpret the various set equations in terms of Venn diagrams.

---

[1] Although we often discuss the rational and real number systems for purposes of illustration, the formal development in this chapter does not depend on a complete definition of these number systems.

Example 5.1.8 Let $R = \{(a,b) \mid a, b \in Z\}$. For (a,b) and (c,d) in R we define an addition and multiplication on R as follows:

$$(a,b) + (c,d) = (a + c, b + d)$$
$$(a,b) \cdot (c,d) = (ac + bd, ad + bc).$$

The reader should note that $(R,+)$ is simply the additive group $Z \times Z$.

Example 5.1.9 Let H be the set of formal elements

$$a_0 + a_1 i + a_2 j + a_3 k,$$

where a_0, a_1, a_2, and a_3 are real numbers. Two elements,

$$a_0 + a_1 i + a_2 j + a_3 k \quad \text{and} \quad b_0 + b_1 i + b_2 j + b_3 k,$$

are said to be equal if $a_0 = b_0$, $a_1 = b_1$, $a_2 = b_2$, and $a_3 = b_3$. We define addition and multiplication on H as follows:

$(a_0 + a_1 i + a_2 j + a_3 k) + (b_0 + b_1 i + b_2 j + b_3 k)$
$\quad = (a_0 + b_0) + (a_1 + b_1)i + (a_2 + b_2)j + (a_3 + b_3)k,$
$(a_0 + a_1 i + a_2 j + a_3 k) \cdot (b_0 + b_1 i + b_2 j + b_3 k)$
$\quad = (a_0 b_0 - a_1 b_1 - a_2 b_2 - a_3 b_3) + (a_0 b_1 + a_1 b_0 + a_2 b_3 - a_3 b_2)i$
$\quad + (a_0 b_2 + a_2 b_0 + a_3 b_1 - a_1 b_3)j + (a_0 b_3 + a_3 b_0 + a_1 b_2 - a_2 b_1)k.$

Then $(H,+,\cdot)$, called the ring of *real quaternions*, is a ring with zero element $0 + 0i + 0j + 0k$, which we shall write as 0, and the element $1 + 0i + 0j + 0k$ is a multiplicative identity, which we shall write as 1.

The ring H is our first example of a noncommutative ring, that is, a ring in which $\cdot$ is not a commutative operation.

The definition of multiplication in H could also be given in another way. We compute the product of two quaternions by first multiplying according to the distributive laws and associative laws and then using the rules for i, j, k given in Exercise 3, Section 4.14, together with the condition that $ax = xa$ for $x = i, j, k$ and all real numbers a. For example,

$(0 + i + j + 0k)(2 + 0i + j + 0k)$
$= 0 \cdot 2 + 0(0i) + 0j + 0(0k) + i2 + i(0i) + ij + i(0k)$
$\quad + j2 + j(0i) + jj + j(0k) + (0k)2 + (0k)(0i) + (0k)j + (0k)(0k)$
$= i2 + ij + j2 + jj$
$= 2i + k + 2j - 1$
$= -1 + 2i + 2j + k.$

Example 5.1.10 Let K denote the set of all real numbers, and let F denote the set of all functions $f: K \longrightarrow K$ such that f is continuous on K. For $f, g \in K$ we define $+$ and $\cdot$ as follows:

$$(f + g)(x) = f(x) + g(x),$$
$$(f \cdot g)(x) = f(x) \cdot g(x).$$

(Note that $f \cdot g$ is not the composite function $f \circ g$.) Then $(F,+,\cdot)$ forms a ring whose zero element is the function θ, $\theta x = 0$ for all $x \in K$. The additive inverse of f is the function g defined by $g(x) = -(f(x))$.

Example 5.1.11 Let $(G,+)$ be any abelian group written additively, with identity element 0. We define a multiplication on G as follows:
$$g_1 \cdot g_2 = 0, \quad \text{all } g_1, g_2 \in G.$$
Then $(G,+,\cdot)$ is a ring.

Example 5.1.12 A two-by-two "array" of real numbers,
$$\begin{bmatrix} a_{11} & a_{12} \\ a_{21} & a_{22} \end{bmatrix},$$
is called a 2×2 *matrix* over the set of real numbers K. Let
$$K_2 = \left\{ \begin{bmatrix} a_{11} & a_{12} \\ a_{21} & a_{22} \end{bmatrix} \,\middle|\, a_{ij} \in K \right\}.$$
We define addition and multiplication of matrices as follows:
$$\begin{bmatrix} a_{11} & a_{12} \\ a_{21} & a_{22} \end{bmatrix} + \begin{bmatrix} b_{11} & b_{12} \\ b_{21} & b_{22} \end{bmatrix} = \begin{bmatrix} a_{11}+b_{11} & a_{12}+b_{12} \\ a_{21}+b_{21} & a_{22}+b_{22} \end{bmatrix},$$
$$\begin{bmatrix} a_{11} & a_{12} \\ a_{21} & a_{22} \end{bmatrix} \cdot \begin{bmatrix} b_{11} & b_{12} \\ b_{21} & b_{22} \end{bmatrix} = \begin{bmatrix} c_{11} & c_{12} \\ c_{21} & c_{22} \end{bmatrix},$$
where $c_{ij} = \sum_{l=1}^{2} a_{il} b_{lj}$. Then $(K_2,+,\cdot)$ is a ring.

Example 5.1.13 For any integer $m > 0$, let $(\bar{Z}_m)_2$ denote the set of 2×2 matrices with entries from $\bar{Z}_m$; that is, in Example 5.1.12, we replace the real numbers by the elements of $\bar{Z}_m$. Then, using the analogous definitions of $+$ and $\cdot$, $((\bar{Z}_m)_2,+,\cdot)$ is a ring that is both finite and noncommutative.

For example, when $m = 3$, $\bar{Z}_3$ has three elements, and so $(\bar{Z}_3)_2$ has 3^4 elements, as each of the four positions in a 2×2 matrix can be assigned any of the three elements from $\bar{Z}_3$. Two matrices that do not commute with respect to multiplication are
$$\begin{bmatrix} \bar{0} & \bar{1} \\ \bar{2} & \bar{0} \end{bmatrix} \quad \text{and} \quad \begin{bmatrix} \bar{1} & \bar{1} \\ \bar{1} & \bar{0} \end{bmatrix},$$
where the two possible products are
$$\begin{bmatrix} \bar{0} & \bar{1} \\ \bar{2} & \bar{0} \end{bmatrix} \begin{bmatrix} \bar{1} & \bar{1} \\ \bar{1} & \bar{0} \end{bmatrix} = \begin{bmatrix} \bar{1} & \bar{0} \\ \bar{2} & \bar{2} \end{bmatrix}$$
and
$$\begin{bmatrix} \bar{1} & \bar{1} \\ \bar{1} & \bar{0} \end{bmatrix} \begin{bmatrix} \bar{0} & \bar{1} \\ \bar{2} & \bar{0} \end{bmatrix} = \begin{bmatrix} \bar{2} & \bar{1} \\ \bar{0} & \bar{1} \end{bmatrix}.$$

EXERCISES

1. Show that the multiplication in Example 5.1.7 is associative.
2. Show that the multiplications in Examples 5.1.8, 5.1.9, 5.1.12, and 5.1.13 are associative.
3. Prove the distributive laws in the rings of Examples 5.1.8, 5.1.9, 5.1.12, and 5.1.13.
4. Find quaternions q_1 and q_2 such that $q_1 \cdot q_2 \neq q_2 \cdot q_1$.
5. Let q be a quaternion $\neq 0$. Show that there exists a quaternion q' such that $q \cdot q' = 1$. [*Hint:* Let
$$q = a_0 + a_1 i + a_2 j + a_3 k.$$
Let
$$q' = \frac{a_0}{D} - \frac{a_1}{D} i - \frac{a_2}{D} j - \frac{a_3}{D} k,$$
with $D = a_0^2 + a_1^2 + a_2^2 + a_3^2$.]
6. Prove that in a ring R
$$(a + b + c)d = ad + bd + cd$$
for all $a, b, c, d \in R$.
7. In Example 5.1.9 replace the real numbers by $\bar{Z}_2$ and show that the resulting ring is commutative. Find a nonzero element in this ring that does not have a multiplicative inverse.

5.2 Isomorphism of Rings

In Chapter 4 we discussed what it means for two groups to be abstractly identical—that is, isomorphic. In this section we extend this notion to rings, noting that here we must have a single one-to-one correspondence that preserves both operations.

Definition 5.2.1 Let $(R_1, +, \cdot)$ and $(R_2, \oplus, \odot)$ be two rings. Then R_1 and R_2 are *ring isomorphic* (or, simply, *isomorphic*) if there is a 1-1 function f from R_1 onto R_2 such that
$$f(a + b) = f(a) \oplus f(b)$$
and
$$f(a \cdot b) = f(a) \odot f(b) \quad \text{for all } a, b \in R.$$

The function f is called a *ring isomorphism*, and we write $R_1 \cong R_2$.

To illustrate, let $\bar{Z}_m$ be the ring of residue classes modulo m (Example 5.1.4). Let $Z_m = \{0, 1, \ldots, m - 1\}$, the set of least positive residues modulo m. We define addition $\oplus$ and multiplication $\odot$ on Z_m as follows:

5.2 ISOMORPHISM OF RINGS

$a \oplus b = c$, the least positive residue congruent to $a + b$ modulo m. Similarly $a \odot b = d$, the least positive residue congruent to ab. Then it is easy to see that $(Z_m, \oplus, \odot)$ is a ring and that this ring is isomorphic to $(\bar{Z}_m, +, \cdot)$. The function that establishes the isomorphism is $f: Z_m \longrightarrow \bar{Z}_m$ defined by $f(a) = \bar{a}$. Thus, abstractly, these two rings are indistinguishable. In the sequel we shall usually work with the ring $(Z_m, \oplus, \odot)$, replacing $\oplus$ and $\odot$ by the more usual $+$ and $\cdot$. If the rings $(R_1, +, \cdot)$ and $(R_2, \oplus, \odot)$ are isomorphic, then the groups $(R_1, +)$ and $(R_2, \oplus)$ are isomorphic as groups. The converse of this statement is not true; that is, there exist pairs of rings with isomorphic additive groups but that are not isomorphic as rings. The following example shows such a pair.

Example 5.2.2 Let $R_1 = (Z, +, \cdot)$ be the ring of integers with the usual addition and multiplication. Let $R_2 = (Z, +, \odot)$ be the ring of integers with the usual addition but with $\odot$ defined by $a \odot b = 0$ for all $a, b \in Z$. Clearly, R_1 and R_2 are isomorphic as additive groups. Now suppose there is a function $f: R_1 \longrightarrow R_2$ satisfying the requirements of Definition 5.2.1 for R_1 and R_2. Then, for any a in R_1,

$$f(a) = f(a \cdot 1) = f(a) \odot f(1) = 0.$$

Hence f is not onto, and therefore the ring R_1 is not isomorphic to the ring R_2.

Example 5.2.3 If K is the field of all real numbers, we see that the set D of those 2×2 matrices over K of the form

$$\begin{bmatrix} a & b \\ -b & a \end{bmatrix}$$

is a ring with respect to the operations of $+$ and $\cdot$ defined in Example 5.1.12. We leave the verification of this fact to the exercises. We now assume a familiarity with the system of complex numbers C,

$$C = \{a + bi \mid a, b \in K, i^2 = -1\}.$$

Then the function $f: C \longrightarrow D$ defined by

$$f(a + bi) = \begin{bmatrix} a & b \\ -b & a \end{bmatrix}$$

is an isomorphism between C and D. For the mapping is clearly 1-1 and onto D, and we need only verify that the ring operations are preserved. We have

$$f((a + bi)(c + di)) = f(ac - bd + (bc + ad)i)$$
$$= \begin{bmatrix} ac - bd & bc + ad \\ -(bc + ad) & ac - bd \end{bmatrix}.$$

186 THE THEORY OF RINGS

But
$$\begin{bmatrix} a & b \\ -b & a \end{bmatrix} \begin{bmatrix} c & d \\ -d & c \end{bmatrix} = \begin{bmatrix} ac - bd & bc + ad \\ -(bc + ad) & ac - bd \end{bmatrix}$$
and so
$$f((a + bi)(c + di)) = f(a + bi)f(c + di).$$

It is quite easy to see that
$$f((a + bi) + (c + di)) = f(a + bi) + f(c + di).$$

Thus, the system of complex numbers C could be defined as the ring D.

EXERCISES

1. Let $R = \{e, f, g, h\}$. Define $+$ and $\cdot$ on R as follows:

+	e	f	g	h		·	e	f	g	h
e	e	f	g	h		e	e	e	e	e
f	f	e	h	g		f	e	f	e	f
g	g	h	e	f		g	e	e	g	g
h	h	g	f	e		h	e	f	g	h

Show that $(R, +, \cdot)$ is a ring. Can you find any proper subsets of R that form a ring with respect to $+$ and $\cdot$?

2. Let $S = \{a, b, c, d\}$. Define $\oplus$ and $\odot$ on S, as follows:

$\oplus$	a	b	c	d		$\odot$	a	b	c	d
a	a	b	c	d		a	a	a	a	a
b	b	a	d	c		b	a	b	a	b
c	c	d	a	b		c	a	a	c	c
d	d	c	b	a		d	a	b	c	d

Show that $(S, \oplus, \odot)$ is a ring.

3. Let $T = \{q, r, s, t\}$. Define $+$ and $\cdot$ on T, as follows:

+	q	r	s	t		·	q	r	s	t
q	q	r	s	t		q	q	q	q	q
r	r	s	t	q		r	q	r	s	t
s	s	t	q	r		s	q	s	q	s
t	t	q	r	s		t	q	t	s	r

Show that $(T, +, \cdot)$ is a ring.

4. Are any pairs of the above rings R, S, T isomorphic? Can you find a ring of four elements not isomorphic to any of those above?
5. Find all possible rings (up to isomorphism) with two elements and with three elements. (Use the fact, to be proved later, that $a \cdot 0 = 0 \cdot a = 0$ for all $a \in R$, R an arbitrary ring.)
6. Let $m, n \in Z$. For $a \in Z$, $(a) = \{ak \mid k \in Z\}$. Show that $((m), +, \cdot)$ is a ring with respect to the usual $+$ and $\cdot$ of Z. Show that

$$((m), +, \cdot) \cong ((n), +, \cdot)$$

if and only if $|m| = |n|$.
7. Verify that $(D, +, \cdot)$ of Example 5.2.3 is a ring.

5.3 Elementary Properties of Rings

In this section we derive some of the properties of rings that follow readily from the ring axioms. Before we proceed, a word about notation and terminology.

For the integer 0 and for all $a \in R$, $0a$ is the additive identity of the group $(R, +)$, which we also denote by 0. Notice that we use the same symbol for the integer zero as for the additive identity of the ring. This is justified, since, whether z is the integer 0 or the additive identity 0, we shall show that $za = 0 \in R$ for all $a \in R$ (see Theorem 5.3.1). We must, however, make a careful distinction between na, an integral multiple of the ring element a, with $n \in Z$, and ra, a ring multiple of a, with $r \in R$. There are rings in which every integral multiple na of a ring element a is equal to a ring multiple of a. A ring R has this property if it contains a multiplicative identity. This will be proved in Corollary 5.3.5.

We also note that the symbol 0 will usually be used simultaneously to represent the additive identity of several different rings when no ambiguity exists.

To illustrate these ideas, let K_2 be the ring of all 2×2 matrices over the real number system K. Then for

$$\begin{bmatrix} a & b \\ c & d \end{bmatrix} \in K_2,$$

if n is a positive integer,

$$n \begin{bmatrix} a & b \\ c & d \end{bmatrix} = \underbrace{\begin{bmatrix} a & b \\ c & d \end{bmatrix} + \begin{bmatrix} a & b \\ c & d \end{bmatrix} + \cdots + \begin{bmatrix} a & b \\ c & d \end{bmatrix}}_{n \text{ summands}} = \begin{bmatrix} na & nb \\ nc & nd \end{bmatrix}.$$

Since
$$\begin{bmatrix} 1 & 0 \\ 0 & 1 \end{bmatrix} \in K_2,$$
$$n\begin{bmatrix} 1 & 0 \\ 0 & 1 \end{bmatrix} = \begin{bmatrix} n & 0 \\ 0 & n \end{bmatrix} \text{ and } n\begin{bmatrix} a & b \\ c & d \end{bmatrix} = \begin{bmatrix} n & 0 \\ 0 & n \end{bmatrix}\begin{bmatrix} a & b \\ c & d \end{bmatrix},$$
and so each integral multiple is a ring multiple.

On the other hand, in the ring E of all even integers the element $2 + 2 + 2 = 3 \cdot 2 = 6$ and is an integral multiple of 2, but 6 is not a ring multiple of 2, since $3 \notin E$.

EXERCISE

1. Let $(R,+,\cdot)$ be a ring. Let $a, b \in R$ and $m, n \in Z$. Show that
   (a) $(m + n)a = ma + na$.
   (b) $(mn)a = m(na) = n(ma)$.
   The reader should note that (a) and (b) are restatements in $(R,+)$ of laws of exponents for groups. If $(G,*)$ is a multiplicative group, (a) translates to $a^{n+m} = a^m * a^n$, and (b) translates to $a^{mn} = (a^m)^n = (a^n)^m$.

For the remainder of this section, R denotes the ring $(R,+,\cdot)$. The reader should compare Theorems 5.3.1, 5.3.2, 5.3.3 with Theorems 2.1.2 and 2.1.4 and with Corollary 2.1.5.

THEOREM 5.3.1 Let 0 be the zero element of R. Then for any $a \in R$, $a \cdot 0 = 0 \cdot a = 0$.

Proof $0 = 0 + 0$, since $(R,+)$ is a group. Then
$$a \cdot 0 = a \cdot (0 + 0) = a \cdot 0 + a \cdot 0$$
by the left distributive law. Hence, by Exercise 2, Section 4.2, $a \cdot 0 = 0$, the identity of $(R,+)$. Similarly $0 \cdot a = 0$. ∎

In view of Theorem 5.3.1, whether 0 is the identity of $(R,+)$ or the integer 0, $0a = 0$ (the identity of $(R,+)$) for all $a \in R$. When 0 is the integer 0, $0a = 0$, since this is the analogue of $a^0 = e$ in multiplicative groups.

THEOREM 5.3.2 For any $a, b \in R$,
$$a(-b) = (-a)b = -(ab).$$

Proof $a(b + (-b)) = a \cdot 0 = 0$, by Theorem 5.3.1. But by the left distributive law
$$0 = a(b + (-b)) = ab + a(-b).$$

5.3 ELEMENTARY PROPERTIES OF RINGS

Hence $a(-b)$ is a right inverse of ab. But $-(ab)$ is an additive inverse of ab, and, since $(R,+)$ is a group, ab has a unique inverse $-(ab)$. Hence, $a(-b) = -(ab)$.

Similarly $(-a)b = -(ab)$. ∎

THEOREM 5.3.3 For any $a, b \in R$,
$$(-a)(-b) = ab.$$

Proof $(-a)(-b) = -(-(a)b)$ by Theorem 5.3.2, and
$$-((-a)b) = -(-(ab)).$$
But, since $(R,+)$ is a group, $-(-(ab)) = ab$ (Corollary 4.2.4, in additive notation). ∎

Theorems 5.3.2 and 5.3.3 are often referred to as the "law of signs." In particular, with $a = b = -1$ we again see that in Z, $(-1)(-1) = 1$.

THEOREM 5.3.4 Let $a, b \in R$, and let $m \in Z$. Then
$$m(ab) = (ma)b = a(mb).$$

Proof If $m = 0$, then the theorem follows easily, using Theorem 5.3.1 and the fact that $0c = 0$, where c is any element in R. (Here the first 0 is an integer, the second the zero of the ring.)

Assume that $k(ab) = (ka)b$ for $k \geq 0$. Then
$$(k+1)(ab) = k(ab) + ab = (ka)b + ab = (ka + a)b. \text{ (Why?)}$$
But $ka + a = (k+1)a$. Hence,
$$(k+1)(ab) = [(k+1)a]b,$$
and the theorem follows by induction for all $m \geq 0$.

We leave the remainder of the proof as an exercise. ∎

COROLLARY 5.3.5 Let R contain an element e such that $ae = ea = a$ for all $a \in R$. Then, for each integer m, there exists an element $r \in R$ such that
$$ma = ra \quad \text{for all } a \in R.$$

That is, every integral multiple is equal to a ring multiple.

Proof Since $ea = a$ for all $a \in R$, $ma = m(ea) = (me)a$ by Theorem 5.3.4. But $me \in R$. Hence let $r = me$ and $ma = ra$. ∎

EXERCISES

2. Show that for $a, b, c, d \in R$, a ring,
$$(-a)[d + (-b) + (-c)] = a(b + c) - (ad).$$

3. Complete the proof of Theorem 5.3.4.
4. Prove that for all $m, n \in Z$, and for all elements $a, b \in R$, a ring, $(mn)(ab) = (ma)(nb) = (na)(mb) = (nma)b = a(nmb)$.
5. Let R be a ring such that (i) $a^2 = a$ for all $a \in R$ and (ii) whenever $2a = 0$ for some $a \in R$, then $a = 0$. Show that R has exactly one element.
6. Let R be a ring such that $a^2 = a$ for all $a \in R$. Prove that
   (a) $2a = 0$ for all $a \in R$.
   (b) $ab = ba$ for all $a, b \in R$.
   [*Hint:* Consider $(a + b)^2$.]

5.4 Some Special Types of Rings

In the definition of a ring, almost no restrictions are placed on the operation of multiplication. It is therefore possible to specialize the definition of a ring in many different ways. We could ask, for example, that · be commutative or that there exist a multiplicative identity or that the product of two nonzero elements be nonzero. In this section we shall examine certain classes of rings defined by various properties of the ring product.

Definition 5.4.1 Let $(R, +, \cdot)$ be a ring. R is called a *commutative* ring if $ab = ba$ for all $a, b \in R$. Otherwise R is called a *noncommutative* ring.

Examples 5.1.2–5.1.8, 5.1.10, and 5.1.11 are commutative rings, and Examples 5.1.9 and 5.1.12 are noncommutative.

Definition 5.4.2 Let $(R, +, \cdot)$ be a ring. R is called a *ring with identity* if there exists an element $e \in R$ such that
$$ae = ea = a \quad \text{for all } a \in R.$$

We shall frequently denote e by 1, and when R is a ring with identity, we shall often write $R \ni 1$. If $R \ni 1$, then $R = \{0\}$ if and only if $1 = 0$. Henceforth, whenever we write $R \ni 1$, we shall assume that $1 \neq 0$.

It should be clear from the examples already given that not all rings have an identity. When a ring has an identity, then the identity is unique.

Definition 5.4.3 Let R be a ring. An element $a \neq 0$ in R is called a *left zero divisor* if there exists an element $b \neq 0$ in R such that $ab = 0$. Similarly, an element $b \neq 0$ in R is called a *right zero divisor* if there exists an element $a \neq 0$ in R such that $ab = 0$. A *zero divisor* is an element that is either a right or left zero divisor.

If R is a commutative ring, then the concepts of right and left zero divisor coincide.

5.4 SOME SPECIAL TYPES OF RINGS

The rings $(Z,+,\cdot)$, $(H,+,\cdot)$, $(K,+,\cdot)$, and $(Q,+,\cdot)$ of Section 5.1 have no zero divisors with respect to multiplication. On the other hand, when m is a composite integer, then $(Z_m,+,\cdot)$ has zero divisors. For example, if $m = ab$ in Z, $0 < a < m$, $0 < b < m$, then $ab = 0$ in Z_m. In particular, 2, 3, and 4 are zero divisors in Z_6.

EXERCISES

1. Let R be a ring with identity. Prove the identity is unique.
2. Prove that a ring R has no left zero divisors if and only if R has no right zero divisors.
3. Show that the ring $\mathcal{S}$ of Example 5.1.7 has zero divisors if $S \neq \emptyset$.
4. Does the ring of Exercise 7, Section 5.1, have zero divisors?

Definition 5.4.4 A ring R is said to satisfy the *left (right) cancellation* law if whenever $ab = ac$ ($ba = ca$), $a \neq 0$, then $b = c$.

THEOREM 5.4.5 R satisfies the left and right cancellation laws if and only if R has no zero divisors.

Proof Let R satisfy the left cancellation law and suppose $ab = 0$, $a \neq 0$. Then $ab = a \cdot 0$, whence, by cancellation, $b = 0$. Thus a is not a left zero divisor. By Exercise 2, R also has no right zero divisors.

Now let $ab = ac$, $a \neq 0$, and assume a is not a left zero divisor. Then $a(b - c) = 0$, whence $b - c = 0$ and $b = c$.

A similar argument applies when "left" is replaced by "right." ∎

Definition 5.4.6 Let R be a commutative ring, $R \ni 1$, $1 \neq 0$. If R has no zero divisors (equivalently, if R satisfies either cancellation law), we call R an *integral domain*.

The ring of integers is an example of an integral domain. Further examples are the ring of real numbers and the ring $(Z_p,+,\cdot)$, for p a prime.

The next class of rings that we consider has the property that $R^* = R - \{0\}$ is a group with respect to the ring multiplication.

Definition 5.4.7 Let $(R,+,\cdot)$ be a ring and let $R^* = R - \{0\}$. Then R is called a *division ring* if $(R^*,\cdot)$ is a group. If, in addition, R is commutative—that is, $(R^*,\cdot)$ is an abelian group—then R is called a *field*. If R is not commutative, then R is called a *skew field*.

It follows from the definition that a division ring R must contain at least two elements, since $0 \in R$ and $R^* = R - \{0\} \neq \emptyset$, since R^* is a group. There is, in fact, a division ring with precisely two elements, namely, $(Z_2,+,\cdot)$.

The reader should note that some authors use the term "field" as we use "division ring," and "commutative field" as we use "field."

Example 5.4.8 The real number system, the rational number system, and the complex number system are examples of fields.

THEOREM 5.4.9 Let p be a prime. Then $(Z_p,+,\cdot)$ is a field.

Proof We have already observed in the examples that $(Z_p,+,\cdot)$ is a commutative ring. To see that $(Z_p{}^*,\cdot)$ is a group we first observe that $1 \in Z_p{}^*$, and so for each $a \in Z_p{}^*$, $a1 = 1a = a$. Next, let $a \in Z_p{}^*$. By Corollary 2.13.5, a has a multiplicative inverse b modulo p, that is, $ab \equiv 1 \pmod{p}$. Taking b as a least positive residue modulo p (see the discussion prior to Example 5.2.2), $ab = 1$ in $Z_p{}^*$ and a has an inverse in $Z_p{}^*$. This concludes the proof that $(Z_p,+,\cdot)$ is a field. ∎

THEOREM 5.4.10 Let $m \in Z$, $m > 0$. Then $(Z_m,+,\cdot)$ is a field if and only if m is a prime.

Proof If m is a prime, Theorem 5.4.9 shows $(Z_m,+,\cdot)$ is a field. Assume that m is not a prime. Then

$$m = ab, \quad \text{for } 1 < a < m, 1 < b < m.$$

Then $a, b \in Z_m - \{0\}$, but $ab = 0$ in $(Z_m,+,\cdot)$, that is, $ab \notin Z_m - \{0\}$. Thus, $(Z_m - \{0\},\cdot)$ is not a group, and so $(Z_m,+,\cdot)$ is not a field. Thus, if $(Z_m,+,\cdot)$ is a field, then m is a prime number. ∎

Example 5.4.11 Let $R = \{a + b\sqrt{2} \mid a, b \text{ rational numbers}\}$. Since R is a subset of the real numbers, we define addition and multiplication in R to be the usual $+$ and $\cdot$ of real numbers. Then it is easy to see that $(R,+,\cdot)$ is a commutative ring. In order to show that R is a field, we must show that $(R^*,\cdot)$ is a group. Since the real number 1 is a multiplicative identity, it remains only to verify that each element in R^* has a multiplicative inverse. Thus, let $x = a + b\sqrt{2}$, not both a and b equal to 0, and let $y = c + d\sqrt{2}$. Then

$$xy = (a + b\sqrt{2})(c + d\sqrt{2}) = (ac + 2bd) + (ad + bc)\sqrt{2}.$$

If y is to be the inverse of x, then $xy = 1$, that is,

$$ac + 2bd = 1 \quad \text{and} \quad ad + bc = 0.$$

Solving for c and d in terms of a and b, we get

$$c = \frac{a}{a^2 - 2b^2} \quad \text{and} \quad d = \frac{-b}{a^2 - 2b^2},$$

provided that $a^2 - 2b^2 \neq 0$. But as long as not both a and b are zero, $a^2 - 2b^2 \neq 0$, since otherwise $a^2 = 2b^2$, $2 = a^2/b^2$, and $\sqrt{2} = |a/b|$. But, since a and b are rational, $|a/b|$ is rational, contradicting the fact that $\sqrt{2}$ is not rational. (See Theorem 2.9.4.) Hence, we may assume that $a^2 - 2b^2 \neq 0$.

Thus
$$(a + b\sqrt{2})\left(\frac{a}{a^2 - 2b^2} + \frac{-b}{a^2 - 2b^2}\sqrt{2}\right) = 1,$$
and $(R, +, \cdot)$ is a field.

Example 5.4.12 Let $(H, +, \cdot)$ be the ring of quaternions (Example 5.1.9). We have already seen (Exercise 4, Section 5.1) that H is not commutative. H contains a multiplicative identity, and, by Exercise 5, Section 5.1, for every nonzero element q in H there is a multiplicative inverse q' in H. That is, if
$$q = a_0 + a_1 i + a_2 j + a_3 k,$$
then q' can be found by solving four linear equations in four unknowns, similar to our derivation of y in Example 5.4.11. Indeed,
$$q' = \frac{a_0}{D} - \frac{a_i}{D} i - \frac{a_2}{D} j - \frac{a_3}{D} k \quad \text{with } D = a_0^2 + a_1^2 + a_2^2 + a_3^2.$$
Hence $(H, +, \cdot)$ is a skew field.

THEOREM 5.4.13 A field $(F, +, \cdot)$ has an identity element.

Proof Let e be the identity of $(F^*, \cdot)$. Then $ea = ae = a$ for all $a \in F^*$. The only other element of F not in F^* is 0, and $e0 = 0e = 0$, by Theorem 5.3.1. ∎

THEOREM 5.4.14 If $(R, +, \cdot)$ is a field, then R is an integral domain.

Proof Let $a, b \in R^*$. Then $ab \in R^*$, since R^* is a group with respect to multiplication. Hence, $ab = ba \neq 0$ for all $b \neq 0$, and, therefore, a is not a zero divisor. Since $1 \in R$, $1 \neq 0$, R is an integral domain. ∎

EXERCISES

5. Give a definition of a field *directly* in terms of the operations $+$ and $\cdot$ without using the terms "group," "ring," or "division ring."
6. Let $m \in Z$. Show that $((m), +, \cdot)$ has no identity if $|m| \geq 2$.
7. Let F be a field, $a, b \in F$. Prove that if $a \neq 0$, then $ax + b = 0$ always has a solution in F.
8. Let F be a field. Let $a, b, c, d, e, f \in F$. Prove that if $ad - bc \neq 0$, then the system of equations
$$ax + by = e$$
$$cx + dy = f$$

has a unique solution in F, that is, there exist unique x_0, $y_0 \in F$ satisfying the two equations.
9. Let $(R,+,\cdot)$ be an integral domain with n elements, $n \geq 2$. Show that R is a field.
10. Let R be a ring with identity $1 \neq 0$, and with no zero divisors. Prove that $ab = 1$ if and only if $ba = 1$ for $a, b \in R$.
11. Let $(R,+,\cdot)$ be a ring with n elements and no zero divisors, $n \geq 2$. Show that R is a division ring.
12. Let $R = \{a + b\sqrt{p} \mid a, b \text{ rational numbers}, p \text{ a prime}\}$. Show that $(R,+,\cdot)$ is a field with $+$ and $\cdot$ the usual addition and multiplication.
13. Let $F = \{0,1,x,y\}$. Define operations of $+$ and $\cdot$ on F by the following tables:

+	0	1	x	y
0	0	1	x	y
1	1	0	y	x
x	x	y	0	1
y	y	x	1	0

$\cdot$	0	1	x	y
0	0	0	0	0
1	0	1	x	y
x	0	x	y	1
y	0	y	1	x

Show that $(F,+,\cdot)$ is a field.
14. Let $(F,+,\cdot)$ be a field and suppose that there is an integer $n > 0$ such that $na = 0$, for $a \in F$. Prove the following:
   (a) If n is an integer > 0 such that $n1 = 0$ (1 the identity of F), then $na = 0$, for all $a \in F$.
   (b) If n is the least positive integer such that $na = 0$, for all $a \in F$, then n is a prime.
   [*Hint:* Suppose $n = n_1 n_2$, $1 < n_1 < n$, $1 < n_2 < n$. Then
   $$n1 = (n_1 n_2)1 = (n_1 1)(n_2 1).$$
   So if $n1 = 0$, then $n_1 1$ or $n_2 1$ is 0. (Why?)]
15. Let $(F,+,\cdot)$ be a field. Prove that either (i) there is an integer $n > 0$ such that $na = 0$, for all $a \in F$, or (ii) $na = 0$, for n an integer and $a \in F$, if and only if $n = 0$ or $a = 0$ [the latter 0 is the identity of $(F,+)$].

5.5 Homomorphisms, Kernels, and Ideals

Definition 5.5.1 Let $(R,+,\cdot)$ be a ring. Let S be a nonempty subset of R. Then S is a *subring* of R if $(S,+,\cdot)$ is also a ring with respect to the opera-

5.5 HOMOMORPHISMS, KERNELS, AND IDEALS 195

tions $+$ and $\cdot$ of R. If $(F,+,\cdot)$ is a field and if the subring $(S,+,\cdot)$ is a field, we call S a *subfield* of F.

Example 5.5.2 Let $(Z,+,\cdot)$ be the ring of all integers. For each integer $m \geq 0$ in Z, let

$$L_m = \{km \mid k \in Z\}.$$

Then $(L_m,+,\cdot)$ is a subring of $(Z,+,\cdot)$. (See Exercise 6, Section 5.2.)

Example 5.5.3 The ring of all rational numbers is a subring of the ring of all real numbers.

Example 5.5.4 If K is the set of all real numbers, then the set K_D of all 2×2 matrices of the form

$$\begin{bmatrix} a & 0 \\ 0 & b \end{bmatrix}, \quad a, b \in K,$$

is a subring of the ring K_2 of all 2×2 matrices over K.

Example 5.5.5 Let F be the ring of all continuous real-valued functions defined on K. (See Example 5.1.10.) Let I be the set of all continuous functions f on K such that $f(1) = 0$. Show that $(I,+,\cdot)$ is a subring of $(F,+,\cdot)$.

Example 5.5.6 Let $(R,+,\cdot)$ be the ring of rational numbers. The set $S = \{a/2 \mid a \in Z\}$ is a subset of R and is a group with respect to addition. The product $\frac{1}{2} \cdot \frac{1}{2} = \frac{1}{4}$ is, however, not in S, and hence S is not a subring of R. The set $T = \{a/2^r \mid r, a \in Z, r \geq 0\}$ is closed with respect to $+$ and $\cdot$ and satisfies all of the properties of a ring. Hence T is a subring of R.

THEOREM 5.5.7 Let $(R,+,\cdot)$ be a ring, and let S be a nonempty subset of R. Then S is a subring of R if and only if

(a) $a - b \in S$, whenever $a \in S$, $b \in S$.
(b) $ab \in S$, whenever $a \in S$, $b \in S$.

Proof Condition (a) and the fact that $S \neq \emptyset$ is a necessary and sufficient condition for $(S,+)$ to be a subgroup of $(R,+)$ (Theorem 4.6.5). Condition (b) guarantees that S is closed under multiplication. The rest of the ring axioms follow easily, since R is a ring and $S \subseteq R$. ∎

Henceforth we shall usually use the same notations for the operations of addition and multiplication in two different rings. This is analogous to our use, in Chapter 4, of the same symbol for the group operations in different groups. Thus, except when stated to the contrary, we shall let $+$ denote addition in all rings and we shall let $\cdot$ (or juxtaposition) denote multiplication.

Definition 5.5.8 Let $(R,+,\cdot)$ and $(T,+,\cdot)$ be rings. Let $f\colon R \longrightarrow T$ satisfy

(a) $f(a + b) = f(a) + f(b)$.
(b) $f(ab) = f(a)f(b)$, for all $a, b \in R$.

Then f is called a *(ring) homomorphism* from R to T.

The concept of a ring homomorphism generalizes that of isomorphism defined in Section 5.2. We say that a homomorphism "preserves" the operations of addition and multiplication.

There is, in the theory to follow, a close parallelism between rings and groups. As in the theory of groups, in which subgroups were important, subrings play a vital role in the theory of rings. But as in the theory of groups, where normal subgroups play a very important role, there is a special type of subring, called an ideal (which occurs as the kernel of homomorphism), which is of greatest interest in the theory of rings.

The proofs of many of the subsequent theorems are almost identical to earlier proofs in Chapter 4. An occasional reference to the earlier chapter on groups will bear out these similarities.

THEOREM 5.5.9 Let $(R,+,\cdot)$ and $(T,+,\cdot)$ be rings. Let $f\colon R \longrightarrow T$ be a homomorphism from R to T. Then $\operatorname{Im} f = \{f(r) \mid r \in R\}$ forms a subring of T.

Proof $(\operatorname{Im} f, +)$ is a subgroup of $(T, +)$ by Theorem 4.9.5.
If $f(r_1)$ and $f(r_2)$ are in $\operatorname{Im} f$, then $f(r_1)f(r_2) = f(r_1 r_2)$, since f is a homomorphism. Thus $f(r_1)f(r)_2 \in \operatorname{Im} f$. By Theorem 5.5.7, $\operatorname{Im} f$ is a subring of $(T,+,\cdot)$. ∎

Definition 5.5.10 Let $f\colon R \longrightarrow T$ be a homomorphism of the ring R to the ring T. We call the set $K = \{x \mid x \in R \text{ and } f(x) = 0\}$ the *kernel* of f, denoted by $\ker f$.

THEOREM 5.5.11 Let $f\colon R \longrightarrow T$ be a homomorphism from the ring R to the ring T. Let $K = \ker f$. Then

(a) K is a subring of R.
(b) If $k \in K$ and $r \in R$, then kr and rk are both in K.

Proof (a) By Theorem 4.9.13, we see that $(K, +)$ is a subgroup of $(R, +)$.
Now let $k \in K$ and $r \in R$. Then $f(kr) = f(k)f(r) = 0 \cdot f(r) = 0$ and also $f(rk) = 0$. In particular, if $r \in K$, we see that K is closed under multiplication, whence, by Theorem 5.5.7, $(K,+,\cdot)$ is a subring of R.
(b) Since $f(rk) = f(kr) = 0$, kr and $rk \in K$. ∎

Definition 5.5.12 Let $(R,+,\cdot)$ be a ring. A nonempty subset S of R is an *ideal (two-sided)* of R if

(a) S is a subring of R, and
(b) Whenever $s \in S$, $r \in R$, then $rs \in S$ and $sr \in S$.

Thus by Theorem 5.5.11, the kernel of a ring homomorphism is an ideal.

Example 5.5.13 Let Z be the ring of integers. Then the subring

$$L_m = (m) = \{nm \mid m \text{ fixed and } n \in Z\}$$

is an ideal in Z.

Example 5.5.14 Let R be a ring. Then the set consisting of 0 alone is an ideal in R. This ideal will usually be denoted by (0), and it is called the *zero ideal*. Also, R is an ideal in R.

Example 5.5.15 Not every subring of a ring is an ideal. For example, let K be the ring of real numbers. Then Z is a subring of K, but Z is not an ideal in K, since $\frac{1}{2} \in K$, $1 \in Z$, but $\frac{1}{2} \cdot 1 \notin Z$.

EXERCISES

1. Let $f\colon Z \longrightarrow Z$ satisfy $f(n) = 6n$ for all $n \in Z$. Show that f is not a ring homomorphism. [*Hint:* Consider

$$f(2) = f(1 \cdot 2) = f(1)f(2).]$$

2. Let $f\colon Z \longrightarrow Z$ be a ring homomorphism. Show that either $f(n) = 0$, all $n \in Z$, or $f(n) = n$, all $n \in Z$.
3. Let K_D be the ring of Example 5.5.4. Define a function $f\colon K_D \longrightarrow K$ by

$$f\left(\begin{bmatrix} a & 0 \\ 0 & b \end{bmatrix}\right) = a.$$

Show that f is a ring homomorphism. What is the kernel of f?
4. Let p be a prime number. Let $f\colon Z \longrightarrow Z_p$ be defined as follows: $f(n) = r$, where $r \in Z_p$ and $n \equiv r \pmod{p}$. Show that
(a) f is a ring homomorphism from Z onto Z_p.
(b) $\ker f = L_p = \{kp \mid k \in Z\}$.
5. Show that the subring I of Example 5.5.5 is an ideal in the ring F of that example.
6. Consider the field F in Exercise 13, Section 5.4. Show that the set $T = \{0,1\}$ is a subfield of F. Show that $T \cong Z_2$.
7. Let $(S,+,\cdot)$ be the ring defined in Example 5.1.7. Let $I \subseteq S$. Let $\mathcal{I}$ be the collection of all subsets of I. Show that $(\mathcal{I},+,\cdot)$ is an ideal in S.
8. Show that any ideal I of Z has the form L_m of Example 5.5.13.

9. Let I_1 and I_2 be ideals of the ring R. Show that the following are ideals of R:
   (a) $I_1 \cap I_2$.
   (b) $I_1 + I_2 = \{a + b \mid a \in I_1, b \in I_2\}$.
   (c) $I_1 \cdot I_2 = \{a_1b_1 + a_2b_2 + \cdots + a_nb_n \mid a_i \in I_1, b_i \in I_2,$ and $n \in P\}$; that is, $I_1 \cdot I_2$ is the set of all finite sums of products a_ib_i, $a_i \in I_1$ and $b_i \in I_2$.
10. Let R be a "zero ring"—that is, a ring in which $ab = 0$, for all $a, b \in R$. Show that any subgroup of $(R, +)$ is an ideal of R.
11. Show that if I is an ideal of a ring with identity 1 such that $1 \in I$, then $I = R$.
12. Let F be a field. Show that F has only two ideals, namely F and (0).
13. Let R be a commutative ring with identity such that the only ideals of R are R and (0). Show that R is a field. [*Hint:* Consider for each $x \in R$ the set $\{rx \mid r \in R\}$.]
14. Show that the ring K_2 of 2×2 matrices over the real numbers K has only two ideals: (0) and K_2. (*Hint:* If I is an ideal and if

$$x \neq 0, x \in I,$$

then x must have at least one nonzero entry. In case x is of the form

$$\begin{bmatrix} a & b \\ c & d \end{bmatrix} \quad a \neq 0,$$

show that there are multiples of x in K_2 of the form

$$\begin{bmatrix} \alpha & 0 \\ 0 & 0 \end{bmatrix}, \begin{bmatrix} 0 & \alpha \\ 0 & 0 \end{bmatrix}, \begin{bmatrix} 0 & 0 \\ \alpha & 0 \end{bmatrix}, \begin{bmatrix} 0 & 0 \\ 0 & \alpha \end{bmatrix}, \quad \alpha \neq 0, \alpha \in K.$$

Thus the ideal containing x must be all of K_2. Consider the other cases similarly.)
15. Show that there exist commutative rings R *without identity* having as ideals only (0) and R.
16. Let R be a commutative ring with identity. Let $a \in R$. Show that $\{ra \mid r \in R\}$ is an ideal in R. (We usually denote this ideal by (a) and call it the *principal ideal generated by a*.)

5.6 Quotient Rings

We continue to parallel our study of groups, extending the concept of quotient group to that of quotient ring.

Definition 5.6.1 Let R be a ring and let I be an ideal in R. Let $(R/I, +)$ be the quotient group of $(R, +)$ modulo $(I, +)((R, +)$ is abelian, whence

5.6 QUOTIENT RINGS

$(I,+)$ is normal in $(R,+)$. On the set R/I, define the sum and product as follows:
$$(a + I) + (b + I) = (a + b) + I,$$
$$(a + I)(b + I) = (ab + I).$$

Then $(R/I,+,\cdot)$ is called the *quotient ring of R modulo the ideal I*. (We recall that $a + I = \{a + i \mid i \in I\}$ is just a group coset in additive notation, and that the addition defined here is precisely the addition in the quotient group $(R/I,+)$.)

We now show that the quotient ring is indeed a ring.

THEOREM 5.6.2 Let R be a ring and let I be an ideal in R. Then the quotient ring R/I is a ring.

Proof Since $(R,+)$ is abelian, the quotient group $(R/I,+)$ is abelian. In order to show that $(R/I,+,\cdot)$ is a ring, we must show that multiplication is well defined—that is, that multiplication does not depend on the choice of coset representative. Thus suppose
$$a_1 + I = a_2 + I \quad \text{and} \quad b_1 + I = b_2 + I.$$
Then $a_1 = a_2 + i_1$ and $b_1 = b_2 + i_2$, where i_1 and i_2 are in I. Then
$$a_1 b_1 = (a_2 + i_1)(b_2 + i_2) = a_2 b_2 + a_2 i_2 + i_1 b_2 + i_1 i_2.$$
Since I is an ideal, $a_2 i_2$, $i_1 b_2$, $i_1 i_2$ are all in I, whence
$$a_1 b_1 - a_2 b_2 \in I.$$
Thus,
$$a_1 b_1 + I = a_2 b_2 + I,$$
since $(I,+)$ is normal subgroup of $(R,+)$. Multiplication in R/I, therefore, is independent of choice of coset representatives. We leave the proof of associativity of multiplication and the proof of the distributive law as an exercise. ∎

THEOREM 5.6.3 Let R and T be rings. Let $f: R \longrightarrow T$ be a ring homomorphism. Then $R/\ker f \cong \operatorname{Im} f$.

Proof By Theorem 4.10.8, the function $\nu: R/\ker f \longrightarrow \operatorname{Im} f$, defined by $\nu(a + \ker f) = f(a)$, is an isomorphism between the groups $(R/\ker f,+)$ and $(\operatorname{Im} f,+)$. This same function is also a ring isomorphism between $(R/\ker f,+,\cdot)$ and $(\operatorname{Im} f,+,\cdot)$.

We already know ν is 1–1 and onto, so we need only verify that ν preserves multiplication. But

$$\nu[(a + \ker f)(b + \ker f)] = \nu(ab + \ker f)$$
$$= f(ab)$$
$$= f(a)f(b) \quad \text{(since } f \text{ is a homomorphism)}$$
$$= \nu(a + \ker f)\nu(b + \ker f).$$

Thus ν is a ring isomorphism. ∎

THEOREM 5.6.4 Let R be a ring, I an ideal in R. Then there exists a homomorphism $f: R \xrightarrow{\text{onto}} R/I$ defined by $f(r) = r + I$ such that

$$\ker f = I.$$

We call f the *natural homomorphism* from R to R/I.

Proof For $a \in R$, $f(a) = a + I$. (This is the map we used when we studied groups.) Then f is a homomorphism of the group $(R,+)$ onto $(R/I,+)$ with kernel I. Also,

$$f(ab) = ab + I = (a + I)(b + I) = f(a)f(b).$$

Hence f is a ring homomorphism. ∎

THEOREM 5.6.5 Let R_1 and R_2 be two rings and let f be a homomorphism from R_1 onto R_2. Then f is an isomorphism if and only if $\ker f = (0)$.

Proof The proof is left as an exercise. ∎

EXERCISES

1. Complete the proof of Theorem 5.6.2.
2. Let R be a ring. Show that $R/(0) \cong R$, as rings.
3. Let F be a field and let R be a ring. Let $f: F \longrightarrow R$ be a ring homomorphism of F onto R. Show that either $R = (0)$ or $R \cong F$.
4. Let $f: Z \longrightarrow F$ be a homomorphism from Z onto the field F. Show that F must be finite with a prime number of elements.
5. Find all possible quotient rings of Z.
6. Let R and T be rings and let $f: R \longrightarrow T$ be a ring homomorphism. Let I be an ideal in R. Show that $\{f(x) \mid x \in I\}$ is an ideal in Im f.
7. Let R be a ring, I an ideal of R, and f the natural mapping from R onto R/I.
   (a) Let $\bar{S}$ be a subring of R/I. Prove that the preimage of $\bar{S}$, $S = \{r \in R \mid f(r) \in \bar{S}\}$, is a subring of R, with $I \subseteq S$, and $\bar{S} = S/I$.
   (b) In (a), replace "subring" by "ideal."
8. Let R and S be rings, with R containing a subring F that is a field. Let $f: R \longrightarrow S$ be a homomorphism. Prove that either $F \subseteq \ker f$ or S contains a subring isomorphic to F.

5.7 Embedding Theorems and Fields of Quotients

We have, on several occasions, discussed the concept of embedding both with sets and with groups. We saw, for example, that the direct product of groups A and B contains a pair of subgroups A_1 and B_1 that are isomorphic to A and B, respectively, and hence that A and B are *embedded* in $A \times B$. In this section we study some embedding theorems concerning rings. Our principal result will be a method for embedding an integral domain in a field. A special case of our theorem will provide a method for constructing the rational number field from the integers.

Definition 5.7.1 Let R and S be rings and f an isomorphism from R onto a subring of S. Then we say that R is *embeddable* in S or R *can be embedded* in S, and we say f is an *embedding* of R in S.

Definition 5.7.2 Let R be a ring. Then the set of ordered pairs

$$\{(a,b) \mid a, b \in R, b \neq 0\}$$

is called the *set of quotients* of R.

Note that the set of quotients of R is nonempty if and only if R has at least two elements.

THEOREM 5.7.3 Let R be a commutative ring with at least two elements and with no zero divisors. Then the relation $\sim$ defined by

$$(a,b) \sim (c,d) \quad \text{if } ad = bc$$

is an equivalence relation on the set of quotients of R.

Proof (a) $(a,b) \sim (a,b)$, since $ab = ba$.
(b) If $(a,b) \sim (c,d)$, then $(c,d) \sim (a,b)$. For, if $ad = bc$, then $cb = da$.
(c) Suppose $(a,b) \sim (c,d)$ and $(c,d) \sim (e,f)$. Then $ad = bc$ and $cf = de$. If $a = 0$, then $c = 0$, since $b \neq 0$; and so $e = 0$, since $d \neq 0$. Thus $0 = af = be = 0$, and $(a,b) \sim (e,f)$.
If $a \neq 0$, then $c \neq 0$ and $e \neq 0$. Now

$$abcf = abde, \quad (bc)(af) = (ad)(be),$$

and, since $ad = bc \neq 0$, $af = be$ and $(a,b) \sim (e,f)$. ∎

The reader should note that if we let the ring R be the ring of integers and interpret the definition of quotient in the usual arithmetic sense, then the definition of $\sim$ given above is the definition of equality between fractions, that is, $a/b = c/d$ if $ad = bc$. Note how we use our familiarity with fractions in formulating the next definition.

202 THE THEORY OF RINGS

Definition 5.7.4 Let R be a commutative ring with at least two elements and no zero divisors. Let F be the set of equivalence classes of the set of quotients of R for the equivalence relation given in Theorem 5.7.3. Denote by $[a,b]$ the equivalence class containing (a,b). We define $+$ and $\cdot$ on F as follows:

$$[a,b] + [c,d] = [ad + bc, bd],$$
$$[a,b] \cdot [c,d] = [ac, bd].$$

Then $(F, +, \cdot)$ is called the *field of quotients of* R, or the *quotient field of* R.

Although we call $(F, +, \cdot)$ the *field* of quotients, we still must verify it is a field. This is done below.

THEOREM 5.7.5 Let R be a commutative ring with at least two elements and no zero divisors, and let $(F, +, \cdot)$ be the field of quotients of R. Then F is a field and R is embeddable in F.

Proof We first show that $+$ and $\cdot$ are binary operations on F. Suppose $[a,b] = [a',b']$ and $[c,d] = [c',d']$. Then

$$[a,b] + [c,d] = [ad + bc, bd]$$

and

$$[a',b'] + [c',d'] = [a'd' + b'c', b'd'].$$

But, since $[a,b] = [a',b']$, $(a,b) \sim (a',b')$ and hence $ab' = ba'$. Similarly $cd' = dc'$. Hence

$$ab'dd' = ba'dd' \quad \text{and} \quad cd'bb' = dc'bb'.$$

Thus

$$(ad)(b'd') = (a'd')(bd) \quad \text{and} \quad (bc)(b'd') = (b'c')(bd).$$

Hence,

$$(ad)(b'd') + (bc)(b'd') = (a'd')(bd) + (b'c')(bd)$$

and

$$(ad + bc)(b'd') = (a'd' + b'c')(bd).$$

Therefore,

$$[ad + bc, bd] = [a'd' + b'c', b'd']$$

and addition is a function on $F \times F$. To show that it is a function to F, that is, that F is closed with respect to $+$, we note that $b \neq 0$ and $d \neq 0$, and R has no divisors of zero. Hence, $bd \neq 0$ and $[ad + bc, bd] \in F$.

A similar argument shows that $\cdot$ is a binary operation on F.

To show that $(F, +, \cdot)$ is a field, we observe first that $(F, +)$ is an abelian group. We leave the verification that addition is associative and commutative as an exercise. The zero element of $(F, +)$ is $[0,b]$, $b \neq 0$, $b \in R$. Since

$$(0,b) \sim (0,d), \quad b \neq 0, d \neq 0,$$

5.7 EMBEDDING THEOREMS AND FIELDS OF QUOTIENTS

any nonzero element of R can be used to denote $[0,b]$. Thus, $[c,d] + [0,b] = [cb + d0, db] = [bc, bd]$, and, since $bd \neq 0$, we see that $[bc, bd] = [c,d]$. Finally, $[-a,b]$ is the additive inverse of $[a,b]$.

To show that $(F^*, \cdot) = (F - \{0\}, \cdot)$ is an abelian group, the reader should first verify that multiplication is both associative and commutative. The multiplicative identity in F is $[b,b]$, where $b \neq 0$, for $[c,d][b,b] = [cb, db] = [c,d]$. If $[c,d] \in F$ and $[c,d] \neq [0,b]$, then $c \neq 0$, whence $[d,c] \in F$. Now $[c,d][d,c] = [cd, dc] = [b,b]$, $b \neq 0$. Thus $(F^*, \cdot)$ is an abelian group.

The only other property that one must verify to show that $(F, +, \cdot)$ is a field is the distributive law. For $[a,b], [c,d], [e,f] \in F$,

$$\begin{aligned}[a,b]([c,d] + [e,f]) &= [a,b]([cf + de, df]) \\ &= [a(cf + de), bdf] \\ &= [acf, bdf] + [ade, bdf] \\ &= [ac, bd] + [ae, bf] \\ &= [a,b][c,d] + [a,b][e,f].\end{aligned}$$

To complete the proof of the theorem, it is necessary to show that R can be embedded in F. Define $f: R \longrightarrow F$ as follows: for any $a \in R$, let $f(a) = [ab, b]$, where b is any nonzero element of R. (Recall that the class $[ab, b]$ depends upon a, but not upon the choice of $b \neq 0$.) Then f is an embedding of R in F, since

(a) $f(a_1 a_2) = [a_1 a_2 b, b] = [a_1 a_2 b^2, b^2] = [a_1 b, b][a_2 b, b]$
    $= f(a_1) f(a_2)$,
(b) $f(a_1 + a_2) = [(a_1 + a_2)b, b] = [a_1 b, b] + [a_2 b, b]$
    $= f(a_1) + f(a_2)$,
and

(c) $f(a_1) = f(a_2)$ implies $[a_1 b, b] = [a_2 b, b]$, whence $a_1 b^2 = a_2 b^2$ and $a_1 = a_2$, since $b^2 \neq 0$.

This completes the proof of Theorem 5.7.5. ∎

We may, in view of the last part of Theorem 5.7.5, consider that R is a subring of F. We do this by replacing an element of the form $[ab, b]$ by a itself. For then

$$a[c,d] = [ab, b][c,d] = [abc, bd] = [ac, d].$$

Thus, although the elements of F are equivalence classes of quotients of R, we may assume that R appears in its "original" form as a subring of F.

In a very strong sense the field F can be considered the smallest field containing the ring R. Indeed, it is not difficult to prove (Exercise 9) that if R is a subring of a field H, then H must contain a subfield that is isomorphic to F and in which R is "naturally" embedded.

In Theorem 5.7.5, if $R = Z$, the ring of integers, then the field F turns out to be the familiar ring of fractions. In view of our last paragraph, whenever Z is a subring of a field H, then H contains the field of rational numbers, as defined in Definition 5.7.6.

Definition 5.7.6 The field of quotients of the integers is called the field of *rational numbers* Q. The elements of Q are called *rational numbers*.

In place of writing $[a,b]$, we shall henceforth write the more usual a/b. The role of equivalence relations in constructing rational numbers from integers should be evident. This emphasizes once again the importance of equivalence relations in mathematics.

COROLLARY 5.7.7 Let R be an integral domain. Then there exists a field F such that R can be embedded in F.

Proof The proof is left as an exercise. ∎

EXERCISES

1. Fill in the missing details in the proof of Theorem 5.7.5.
2. Prove Corollary 5.7.7.
3. What well-known field is the quotient field of the ring of even integers?
4. Let I be an ideal of Z, $I \neq (0)$. What is the field of quotients of I?
5. Let F be a field. Prove that the quotient field of F is isomorphic to F. Thus, show that if F is the field of quotients of an integral domain R, then F is isomorphic to its field of quotients.
6. Let K be the field of real numbers. Let $C = K \times K$; that is, $C = \{(a,b) \mid a, b \in K\}$. For (a,b) and (c,d), define

$$(a,b) + (c,d) = (a + c, b + d)$$

and

$$(a,b) \cdot (c,d) = (ac - bd, ad + bc).$$

Show that $(C, +, \cdot)$ is a field. [*Hint:*

$$(a,b)^{-1} = \left(\frac{a}{a^2 + b^2}, \frac{-b}{a^2 + b^2}\right), \quad \text{if } (a,b) \neq (0,0).]$$

7. In Exercise 6, write (a,b) as $a + bi$. In particular, write $(1,0)$ as $1 + 0i = 1$ and $(0,1)$ as $0 + 1i = i$. Show that $i^2 = -1$.

Exercises 6 and 7 give the usual construction of the complex number system from the real number system. The relevant ideas will be discussed more fully in Chapter 6.

8. Let A_1 and A_2 be integral domains. Let F_1 and F_2 be their corresponding field of quotients. If $F_1 \cong F_2$, is $A_1 \cong A_2$?
9. Let A be an integral domain. Let B be a field such that A is a subring of B. Let F be the quotient field of A. Show that F can be embedded in B. (Hence, we may consider F to be the "smallest" field containing A.)
10. Let R be an integral domain with quotient field F. Let f be an isomorphism from R to F. Show that F_f, the quotient field of $f(R)$, is isomorphic to F.

Although there exist rings without multiplicative identity, the next theorem shows that any such ring can be considered a subring of a ring with identity.

THEOREM 5.7.8 Let R be a ring. Then R can be embedded in a ring R' with identity.

Proof If R has identity 1, let $R' = R$.
Thus assume that R does not have an identity and let
$$R' = \{(r,n) \mid r \in R, n \in Z\}.$$
In R', define
$$(r_1, n_1) + (r_2, n_2) = (r_1 + r_2, n_1 + n_2)$$
and
$$(r_1, n_1) \cdot (r_2, n_2) = (r_1 r_2 + n_2 r_1 + n_1 r_2, n_1 n_2).$$

By direct verification, the reader can easily show that $(R', +, \cdot)$ is a ring. Also, $(0,1)$ is an identity for R', for
$$(r,n)(0,1) = (r0 + n0 + 1r, 1n) = (r,n).$$
Thus, R' is a ring with identity.

Next, we let $f: R \longrightarrow R'$ be defined as follows: $f(a) = (a,0)$, for all $a \in R$. Then f is an embedding of R in R', and the theorem is proved. ∎

EXERCISES

11. Recall Exercises 14 and 15 of Section 5.4. Let F be a field.
    (a) If there exists a least positive integer p such that $pa = 0$, all $a \in F$ [p must be a prime by Exercise 14(b) of Section 5.4], show that F has a subfield isomorphic to Z_p. Moreover, show this subfield is unique.
    (b) Under the condition of (a), show that F does not have a subfield isomorphic to Z_q, q a prime, $q \neq p$.

(c) If there is no positive integer n such that $na = 0$, all $a \in F$, show that F contains a subfield isomorphic to the field of rational numbers. Moreover, show that this subfield is unique.

(d) Under the conditions of (c), show that for each prime number q, F does not have a subfield isomorphic to Z_q.

(*Note:* Under (a) we say F is a field of *characteristic p*. Under (c) F is a field of *characteristic* 0.)

12. Complete the proof of Theorem 5.7.8.
13. Let R be a ring with no zero divisors, and assume that R has no identity. Is the ring R' of Theorem 5.7.8 a ring with no zero divisors?
14. Suppose $R \ni 1$. Is the ring $R' = \{(r,n) \mid r \in R, n \in Z\}$ of Theorem 5.7.8 isomorphic to R?

5.8 Polynomial Rings

In this section we shall develop some of the basic properties of polynomials with coefficients from a ring. For the most part we shall restrict ourselves to polynomials in a single indeterminate.

A historical note is in order here. The problem of finding the roots of any given polynomial with real coefficients has attracted the efforts of mathematicians for many centuries. By approximately A.D. 1540, formulas had been developed to compute the roots of any polynomial of degree less than or equal to 4. Moreover, each of these formulas involved only the operations of addition, subtraction, multiplication, division, and the taking of roots. The question of whether similar formulas existed for polynomials of higher degrees remained unanswered until 1824, when Abel proved that such formulas in fact did not exist. In 1832, the twenty-year-old French mathematician Evariste Galois[2] proved that for $n \geq 5$, there was no general formula for the roots of nth-degree polynomials, using techniques that gave impetus to the growth of a whole new branch of mathematics, the theory of groups. Indeed, the importance of normal subgroups first became apparent in Galois' work. The task of actually computing roots of a given polynomial in any form whatsoever now properly belongs to the branch of mathematics called approximation theory.

Definition 5.8.1 Let R be a ring and P_0 the nonnegative integers. An *infinite sequence* of elements of R is a function $f \colon P_0 \longrightarrow R$. If $r_j = f(j)$, we denote f by $(r_0, r_1, \ldots, r_n, \ldots)$, and we call r_j the *j*th *term* of the sequence.

---

[2] Galois actually wrote his results shortly before he was to die in a duel on May 31, 1832. See, for example, E. T. Bell, *Men of Mathematics*. New York: Simon and Schuster, 1937.

Definition 5.8.2 Let R be a ring. An infinite sequence

$$(a_0, a_1, a_2, \ldots, a_n, \ldots)$$

of elements of R, with at most a finite number of nonzero terms, is called a *polynomial* over R. The set of all such polynomials is denoted by $R[x]$. For example, if $R = Z$, then

$$(1,0,2,1,0,\ldots 0,\ldots) \quad \text{and} \quad (0,0,1,0,0,\ldots,0,\ldots) \in R[x],$$

but

$$(0,1,0,1,\ldots,0,1,\ldots) \notin R[x].$$

In Theorem 5.8.6, we shall precisely define the polynomial x, but until this is done, it will help the reader if he regards the polynomial

$$(a_0, a_1, \ldots, a_n, 0, 0 \ldots, 0, \ldots)$$

as the formal expression $a_0 + a_1 x + \cdots + a_n x^n$. With this in mind, the reader should easily be able to follow our more formal development of polynomials.

We have chosen to define a polynomial in the form

$$(a_0, a_1, \ldots, a_n, 0, 0, \ldots, 0, \ldots)$$

for convenience in stating and proving some of the theorems in this section. Other advantages of our definition are discussed following Theorem 5.8.13.

EXERCISE

1. Let $(a_0, a_1, \ldots, a_n, \ldots)$ be a sequence of elements of R. Show that $(a_0, a_1, \ldots, a_n, \ldots) \in R[x]$ if and only if there exists an integer N such that $a_i = 0$, for all $i \geq N$.

Since a polynomial is a sequence (that is, a function), if we let

$$(a_0, a_1, \ldots, a_n, \ldots) \quad \text{and} \quad (b_0, b_1, \ldots, b_n, \ldots) \in R[x],$$

then

$$(a_0, a_1, \ldots, a_n, \ldots) = (b_0, b_1, \ldots, b_n, \ldots)$$

if and only if $a_i = b_i$, $i = 0, 1, 2, \ldots$.

Definition 5.8.3 Let $\alpha = (a_0, a_1, \ldots, a_n, \ldots)$, $\beta = (b_0, b_1, \ldots, b_n, \ldots)$, be in $R[x]$. Define

$$\alpha + \beta = (a_0 + b_0, a_1 + b_1, \ldots, a_n + b_n, \ldots)$$

and

$$\alpha \cdot \beta = (c_0, c_1, c_2, \ldots, c_n \ldots),$$

where
$$c_k = \sum_{i=0}^{k} a_i b_{k-i}.$$
We note that
$$\sum_{i=0}^{k} a_i b_{k-i} = \sum_{\substack{i+j=k \\ i \geq 0, j \geq 0}} a_i b_j,$$
which we shall frequently write as
$$\sum_{i+j=k} a_i b_j.$$
To illustrate Definition 5.8.3, let
$$\alpha = (a_0, a_1, a_2, 0, 0, \ldots 0, \ldots)$$
and
$$\beta = (b_0, b_1, b_2, b_3, 0, 0, \ldots, 0, \ldots).$$
Then
$$\alpha\beta = (a_0 b_0, a_0 b_1 + a_1 b_0, a_0 b_2 + a_1 b_1 + a_2 b_0,$$
$$a_0 b_3 + a_1 b_2 + a_2 b_1 + 0 \cdot b_0, a_0 \cdot 0 + a_1 b_3 + a_2 b_2 + 0 \cdot b_1 + 0 \cdot b_0,$$
$$a_0 \cdot 0 + a_1 \cdot 0 + a_2 b_3 + 0 \cdot b_2 + 0 \cdot b_1 + 0 \cdot b_0, 0, 0, \ldots)$$
$$= (a_0 b_0, a_0 b_1 + a_1 b_0, a_0 b_2 + a_1 b_1 + a_2 b_0, a_0 b_3 + a_1 b_2 + a_2 b_1,$$
$$a_1 b_3 + a_2 b_2, a_2 b_3, 0, 0, \ldots).$$
That is, if we think of α as $a_0 + a_1 x + a_2 x^2$ and β as
$$b_0 + b_1 x + b_2 x^2 + b_3 x^3,$$
then
$$\alpha \cdot \beta = a_0 b_0 + (a_0 b_1 + a_1 b_0) x + (a_0 b_2 + a_1 b_1 + a_2 b_0) x^2$$
$$+ (a_0 b_3 + a_1 b_2 + a_2 b_1) x^3 + (a_1 b_3 + a_2 b_2) x^4 + a_2 b_3 x^5.$$

THEOREM 5.8.4 $(R[x], +, \cdot)$ is a ring, called the *ring of polynomials* (*in one indeterminate*) *over* R.

Proof We first show that $+$ and $\cdot$ are binary operations on $R[x]$. Thus, let
$$\alpha = (a_0, a_1, \ldots, a_n, \ldots) \quad \text{and} \quad \beta = (b_0, b_1, \ldots, b_n, \ldots).$$
Since there are integers N_1 and N_2 such that $a_k = 0$ for $k \geq N_1$ and $b_l = 0$ for $l \geq N_2$, it is clear that
$$a_n + b_n = 0 \quad \text{for } n \geq \max(N_1, N_2),$$
and
$$c_n = \sum_{i+j=n} a_i b_j = 0 \quad \text{for } n \geq N_1 + N_2 - 1.$$
Thus $\alpha + \beta$ and $\alpha \cdot \beta$ are in $R[x]$.

5.8 POLYNOMIAL RINGS

To show that $(R[x], +)$ is a commutative group, we note that

$$0 = (0, 0, \ldots, 0, \ldots)$$

is an identity and that $-\alpha = (-a_0, -a_1, \ldots, -a_n, \ldots)$ is an additive inverse of α. The associativity of $+$ follows from the associativity of addition in R.

Now we show the associativity of multiplication in $R[x]$. Thus let

$$\alpha = (a_0, a_1, \ldots, a_n, \ldots),$$
$$\beta = (b_0, b_1, \ldots, b_n, \ldots),$$

and

$$\gamma = (c_0, c_1, \ldots, c_n, \ldots).$$

Then the lth term of $\alpha \cdot \beta$ is

$$\sum_{i+j=l} a_i b_j$$

and the sth term of $(\alpha \cdot \beta) \cdot \gamma$ is

$$\sum_{l+k=s} \left(\sum_{i+j=l} a_i b_j \right) c_k = \sum_{i+j+k=s} (a_i b_j) c_k.$$

On the other hand, the lth term of $\beta \cdot \gamma$ is

$$\sum_{j+k=l} b_j c_k$$

and the sth term of $\alpha(\beta\gamma)$ is

$$\sum_{i+l=s} a_i \left(\sum_{j+k=l} b_j c_k \right) = \sum_{i+j+k=s} a_i (b_j c_k).$$

Since multiplication in R is associative, $(a_i b_j) c_k = a_i (b_j c_k)$, and

$$(\alpha\beta)\gamma = \alpha(\beta\gamma).$$

Finally, to show that $(R[x], +, \cdot)$ is a ring, we need to verify the distributive laws. This is left as an exercise. ∎

THEOREM 5.8.5 Let R be a ring. Then R can be embedded in $R[x]$.

Proof Let $f: R \longrightarrow R[x]$ be defined by

$$f(r) = (r, 0, 0, \ldots, 0, \ldots).$$

Then f is a 1–1 homomorphism (or an embedding, from R to $R[x]$), as the reader may verify. ∎

Since

$$(r, 0, \ldots, 0, \ldots)(a_0, a_1, \ldots, a_n, \ldots) = (ra_0, ra_1, \ldots, ra_n, \ldots),$$

we may replace $(r,0,\ldots,0,\ldots)$ by r and agree to write $r(a_0,a_1,\ldots,a_n,\ldots)$ in place of $(r,0,0,\ldots)(a_0,a_1,\ldots,a_n,\ldots)$. Thus, by Theorem 5.8.5, we may consider R actually to be a subring of $R[x]$.

To facilitate obtaining the usual algebraic representation of polynomials, we shall assume for the next two theorems that R is a ring with identity.

THEOREM 5.8.6 Let R be a ring with identity 1. Let x denote the polynomial $(0,1,0,\ldots,0,\ldots)$. Then

$$x^n = (0,0,\ldots,0,1,0,\ldots),$$

where 1 appears as the nth term of $(0,0,\ldots,0,1,0,\ldots)$, and any polynomial $(b_0,b_1,\ldots,b_k,0,\ldots,0,\ldots)$ can be expressed in the form

$$b_0 + b_1 x + b_2 x^2 + \cdots + b_k x^k.$$

Proof $x^1 = x = (0,1,0,\ldots,0,\ldots)$ and the theorem is true for $n = 1$. Assume now that

$$x^n = (0,0,\ldots,0,1,0,\ldots,0,\ldots),$$

where 1 is the nth term of x^n. Then

$$x^{n+1} = x^n \cdot x = (0,\ldots,0,1,0,\ldots)(0,1,0,\ldots)$$
$$= (0,\ldots,0,0,1,0,\ldots),$$

where 1 is now the $(n + 1)$st term. The first part of the theorem now holds by induction.

Since

$$(b_0,b_1,\ldots,b_k,0,\ldots,0,\ldots) = (b_0,0,\ldots,0,\ldots) + (0,b_1,0,\ldots) + \cdots$$
$$+ (0,0,\ldots,0,b_k,0,\ldots,0,\ldots),$$

and since

$$(0,\ldots,0,b_i,0,\ldots) = b_i(0,\ldots,0,1,0,\ldots) = b_i x^i,$$

the proof is now complete. ∎

Definition 5.8.7 Let S be a ring and let R be a subring of S. Let $s \in S$ and suppose that $rs = sr$, for $r \in R$. Then s is called *transcendental over* R if

$$r_0 + r_1 s + \cdots + r_{n-1} s^{n-1} + r_n s^n = 0, \qquad r_i \in R,$$

implies $r_i = 0$, $i = 0, \ldots, n$. Otherwise, s is called *algebraic over* R.

If $R = Q$, the ring of rational numbers, and $S = K$, the ring of real numbers, then π and e are examples of real numbers transcendental over Q. The number $\sqrt{2}$ is, however, algebraic over Q, since

$$1(\sqrt{2})^2 + (-2) = 0.$$

5.8 POLYNOMIAL RINGS

Our method of introducing polynomials over a ring R now enables us to construct a ring S containing an element transcendental over R.

THEOREM 5.8.8 Let R be a ring with identity 1. Then there exists a ring S such that R is a subring of S and S has an element transcendental over R.

Proof By Theorem 5.8.5, $R \subset R[x] = S$. Now consider

$$x = (0,1,0,\ldots,0,\ldots).$$

Then x is transcendental over R, for if there exist $r_0, r_1, \ldots, r_n$ in R such that

$$r_0 + r_1 x + \cdots + r_n x^n = 0,$$

then

$$r_0 + r_1(0,1,0,\ldots) + \cdots + r_n(0,\ldots,0,1,0,\ldots) = 0,$$

whence

$$(r_0, r_1, \ldots, r_n, \ldots) = 0 \quad \text{and} \quad r_i = 0, \quad i = 0, \ldots, n. \quad \blacksquare$$

THEOREM 5.8.9 Let R be a ring (not necessarily with identity). Then there exists a ring S such that R is a subring of S and such that S has an element transcendental over R.

Proof By Theorem 5.7.8, R can be embedded in a ring R^* with identity 1. In fact, $R^* = \{(r,n) \mid r \in R, n \in Z\}$. If we identify $(r,0)$ with r, R can be considered a subring of R^*. Then $R \subset R^* \subset R^*[x] = S$, and since x is transcendental over R^*, it is transcendental over R. $\quad \blacksquare$

The reader should note that if R is a ring without identity, then $R[x]$, as defined in Definitions 5.8.2 and 5.8.3, is embedded in the ring $R^*[x]$. Thus, any element of $R[x]$ has the form

$$a_0 + a_1 x + \cdots + a_n x^n, \quad a_i \in R,$$

even though x need not be an element of $R[x]$. We should note that x is in $R[x]$ if and only if R has an identity. The element x, whether it is in $R[x]$ or not, is called an "indeterminate" over R.

Henceforth we shall adopt the convention of denoting the polynomial $\alpha = (a_0, a_1, \ldots, a_n, \ldots)$ in $R[x]$ by

$$\alpha(x) = a_0 + a_1 x + \cdots + a_n x^n$$
$$= a_n x^n + \cdots + a_1 x + a_0,$$

and we will call a_i the *coefficient* of x^i. Then, given two polynomials

$$r(x) = r_n x^n + r_{n-1} x^{n-1} + \cdots + r_0$$

and

$$s(x) = s_m x^m + s_{m-1} x^{m-1} + \cdots + s_0,$$

it follows readily from our earlier definitions of addition and multiplication that we add and multiply these polynomials exactly as in high-school algebra.

THEOREM 5.8.10   Let R be an integral domain. Then $R[x]$ is an integral domain.

Proof   It suffices to verify that
(a) $R[x]$ has an identity different from 0.
(b) $R[x]$ is commutative.
(c) $R[x]$ has no zero divisors.

(a) Since $R \ni 1$, it can easily be shown that
$$(1,0,0,0,\ldots,0,\ldots) \neq (0,0,\ldots,0,\ldots)$$
is the identity of $R[x]$, that is,
$$1 = 1 + 0x + \cdots + 0x^n$$
is the identity.

(b) Since R is commutative, observe that if
$$\alpha(x) = a_0 + a_1 x + \cdots + a_n x^n$$
and
$\beta(x) = b_0 + b_1 x + \cdots + b_m x^m$, then the kth term of $\alpha(x) \cdot \beta(x)$ is
$$\sum_{i+j=k} a_i b_j = \sum_{i+j=k} b_j a_i = \sum_{j+i=k} b_j a_i = \sum_{i+j=k} b_i a_j,$$
which is the kth term of $\beta(x) \cdot \alpha(x)$.

(c) Finally, suppose $\alpha(x) \neq 0$ and $\beta(x) \neq 0$ are in $R[x]$. Then, there are integers N_1 and N_2 such that $a_{N_1} \neq 0$, $b_{N_2} \neq 0$, but $a_k = 0$ for $k > N_1$ and $b_k = 0$ for $k > N_2$. Then, in $\alpha(x) \cdot \beta(x)$, the coefficient of $x^{N_1+N_2}$ is
$$\sum_{i+j=N_1+N_2} a_i b_j = a_{N_1} b_{N_2} \neq 0,$$
and so $R[x]$ has no zero divisors. ∎

Definition 5.8.11   If $\alpha(x) \in R[x]$, we define *degree of* $\alpha(x)$ [denoted deg α or deg $\alpha(x)$] as follows:

If $\alpha(x) \neq 0$, then deg $\alpha = N$, provided $a_N \neq 0$, but $a_k = 0$, for $k > N$.
If $\alpha(x) = 0$, then deg $\alpha = -\infty$.

If deg $\alpha = 0$ or $-\infty$, we say that $\alpha(x)$ is a *constant* (or *constant polynomial*).

We agree to make these conventions concerning $-\infty$:
$$n + (-\infty) = -\infty,$$

for all integers n, $(-\infty) + (-\infty) = -\infty$ and $-\infty < n$, for all integers n.

THEOREM 5.8.12 If R has no zero divisors, and if $\alpha(x)$ and $\beta(x)$ are in $R[x]$, then deg $(\alpha(x) \cdot \beta(x)) = \deg \alpha + \deg \beta$.

Proof The proof is left as an exercise. ∎

THEOREM 5.8.13 Let $\alpha(x), \beta(x) \in R[x]$. Then

$$\deg (\alpha(x) + \beta(x)) \leq \max (\deg \alpha, \deg \beta).$$

Proof The proof is left as an exercise. ∎

We have had three main reasons for introducing polynomials as infinite sequences. One was for convenience in stating and proving theorems. A second was to show in a simple way the existence of an element transcendental over R. The third was to emphasize that a polynomial is a sequence formally constructed from a ring and not a function on the ring. Later we shall use a polynomial to define a polynomial function. That is the context in which the reader has utilized polynomials in college algebra and calculus.

EXERCISES

2. Complete the proof of Theorem 5.8.4.
3. Prove Theorem 5.8.12.
4. Prove Theorem 5.8.13.
5. Let $R = Z_4$ and let $\alpha(x) \in R[x]$, $\beta(x) \in R[x]$. Is it true that deg $(\alpha(x) \cdot \beta(x)) = \deg \alpha + \deg \beta$?
6. Let S be a set consisting of the following objects: Γ, ⊤, L, ⌐; that is, $S = \{Γ,⊤,L,⌐\}$. Let R be the ring of all subsets of S, defined according to Example 5.1.7. List all polynomials in $R[x]$ of degree 1 or less. For example, $\{Γ,⌐\} + \{Γ,⊤\}x$ is such a polynomial.
7. Show that if R and S are isomorphic rings, then $R[x]$ and $S[x]$ are isomorphic.
8. Let R be a commutative ring with identity, let $S = R[x]$ and let $T = S[y]$. We denote T by $R[x,y]$, the ring of polynomials in the two indeterminates x, y over R. Let $R[y] = U$ and let $U[x] = V$. Show that $T \cong V$.
9. Let S be a ring with subring R, and let a be an element of S transcendental with respect to R. Show that S contains a subring isomorphic to $R[x]$.
10. Let $(Q^*,\cdot)$ be the multiplicative group of positive rational numbers. Show that $(Q^*,\cdot) \cong (Z[x],+)$. (*Hint:* Show that every element of Q^*

can be written in the form $p_1^{\alpha_1}\cdots p_k^{\alpha_k}$ for distinct primes p_i and unique integers α_i.)

11. Let p be a prime. Show that the quotient field of $Z_p[x]$ has characteristic p, yet is infinite. [This field, denoted $Z_p(x)$, shows that an infinite field may have characteristic p—see Exercise 11, Section 5.7.]

5.9 Division Algorithm for Polynomials

Although arbitrary polynomial rings are of great interest, some of the more important applications arise in the case of a polynomial ring over a field. Hence, for the remainder of this chapter, we shall consider polynomial rings of the form $F[x]$, where F is a field.

Definition 5.9.1 Let $\alpha(x) \in F[x]$, say

$$\alpha(x) = a_0 + a_1 x + a_2 x^2 + \cdots + a_n x^n,$$

with $a_n \neq 0$. We call a_n the *leading coefficient* of $\alpha(x)$ and we call a_0 the *constant term* of $\alpha(x)$. If $a_n = 1$ (the identity of F), $\alpha(x)$ is called a *monic* polynomial.

It turns out that the theory of factorization of integers (see Chapter 2) has a strong analogue in the theory of polynomials in an indeterminate x over a field F. Our point of departure is to prove a division algorithm for polynomials in $F[x]$. This division algorithm, and its proof, is very similar to that for integers, only now the degree of the polynomial replaces the absolute value of the integer.

THEOREM 5.9.2 (*Division Algorithm for $F[x]$*) Let F be a field. Let $\alpha(x) = a_0 + a_1 x + \cdots + a_n x^n \neq 0$, and $\beta(x)$ be elements of $F[x]$. Then there exist unique polynomials $q(x)$ and $r(x)$ in $F[x]$ such that

$$\beta(x) = \alpha(x)q(x) + r(x), \quad \deg r(x) < \deg \alpha(x).$$

Proof I. *Existence.* Let $S = \{\beta(x) - \alpha(x)\gamma(x) \mid \gamma(x) \in F[x]\}$. Then $S \neq \emptyset$, since $\beta(x) \in S$. If the 0 polynomial is in S, then there is a polynomial $\gamma(x)$ such that

$$\beta(x) - \alpha(x)\gamma(x) = 0 \quad \text{and} \quad \beta(x) = \alpha(x)\gamma(x).$$

Then $q(x) = \gamma(x)$ and $r(x) = 0$ satisfy the conditions of the theorem.

Thus assume $0 \notin S$. Then every polynomial in S has nonnegative degree. Choose a polynomial in S of least degree and call it $r(x)$. Thus there exists a $q(x) \in F(x)$ such that

$$\beta(x) - \alpha(x)q(x) = r(x), \quad \beta(x) = \alpha(x)q(x) + r(x).$$

5.9 DIVISION ALGORITHM FOR POLYNOMIALS

Now $\alpha(x)$ has degree n, since $a_n \neq 0$. Thus, assume that

$$\deg r(x) = m \geq n,$$

say $r(x) = r_m x^m + r_{m-1} x^{m-1} + \cdots + r_0$. Since

$$r_m a_n^{-1} x^{m-n} \alpha(x) = r_m x^m + r_m a_n^{-1} a_{n-1} x^{m-1} + \cdots + r_m a_n^{-1} a_0 x^{m-n},$$

we see that $r(x) = r_m a_n^{-1} x^{m-n} \alpha(x) + h(x)$, where

$$h(x) = -r_m a_n^{-1} a_{n-1} x^{m-1} - \cdots - r_m a_n^{-1} a_0 x^{m-n} + r_{m-1} x^{m-1} + \cdots + r_0,$$

and $\deg h(x) \leq \deg r(x) - 1$. Thus,

$$\beta(x) - \alpha(x)[q(x) + r_m a_n^{-1} x^{m-n}] = h(x) \quad \text{and} \quad h(x) \in S.$$

But $\deg h(x) < \deg r(x)$, which contradicts our choice of $r(x)$. Thus, $\deg r(x) < \deg \alpha(x)$, and we have found polynomials $q(x)$ and $r(x)$ such that

$$\beta(x) = \alpha(x)q(x) + r(x), \qquad \deg r(x) < \deg \alpha(x).$$

II. *Uniqueness.* Suppose

$$\beta(x) = q(x)\alpha(x) + r(x) \quad \text{and} \quad \beta(x) = q_1(x)\alpha(x) + r_1(x),$$

where $\deg r(x) < \deg \alpha(x)$ and $\deg r_1(x) < \deg \alpha(x)$. Then

$$[q(x) - q_1(x)]\alpha(x) = r_1(x) - r(x).$$

Now, $\deg (r_1(x) - r(x)) < \deg \alpha(x)$. But

$$\deg [(q(x) - q_1(x))\alpha(x)] \geq \deg \alpha(x),$$

unless $q(x) - q_1(x) = 0$. Thus, we must have

$$q(x) - q_1(x) = 0 \quad \text{and also} \quad r_1(x) - r(x) = 0.$$

Hence $q(x) = q_1(x)$ and $r(x) = r_1(x)$. ∎

The reader should review the proof of the division algorithm in Chapter 2 and observe its similarity with the proof just concluded.

Definition 5.9.3 Let $f(x), g(x) \in F[x]$, F a field. We say $f(x)$ *divides* $g(x)$ if there is a polynomial $h(x) \in F[x]$ such that $g(x) = f(x)h(x)$. We write $f(x) \mid g(x)$. If $f(x)$ does not divide $g(x)$, we write $f(x) \nmid g(x)$.

Definition 5.9.4 Let $f(x) \in F[x]$, F a field, $\deg f(x) \geq 1$. Then $f(x)$ is called *irreducible* if, whenever $h(x) \mid f(x)$, either $h(x) = c$ or $h(x) = cf(x)$, where c is a constant.

The development of the polynomial analogue of a prime number and the formulation of a unique factorization theorem for polynomials is left to Exercises 5 and 6. Irreducible polynomials will take the place of prime numbers.

EXERCISES

1. Where does the proof of the division algorithm fail in the case of (a) $Z_4[x]$, (b) $Z[x]$?
2. Let $\alpha(x) = 4x^3 + x^2 - 2$ and $\beta(x) = 3x + 1$. Find $q(x), r(x) \in Q[x]$, Q the field of rational numbers, such that
$$\alpha(x) = \beta(x)q(x) + r(x).$$
3. Let F be a field. Let $f(x), g(x) \in F[x]$. Define gcd $(f(x), g(x))$ in analogy with Definition 2.7.2. Show that there exist polynomials $s(x)$ and $t(x) \in F[x]$ such that gcd $(f(x),g(x)) = f(x)s(x) + g(x)t(x)$.
4. In Exercise 3 let $F = Q$, the rational numbers. Find a gcd of
   (a) $x^3 + x - 2$ and $x^4 - x^3 - 2x + 2$.
   (b) $x^2 + x + 2$ and $x^5 - x^4 + x^3 - x^2 + x - 1$.
5. Prove Euclid's lemma: if $f(x) \mid g(x)h(x)$ and if $f(x)$ is irreducible, then either $f(x) \mid g(x)$ or $f(x) \mid h(x)$.
6. Prove the unique factorization theorem for $F[x]$: a monic polynomial $f(x), \deg f(x) \geq 1$, has a unique factorization (up to order of the factors) into irreducible monic polynomials.

5.10 Consequences of the Division Algorithm

Definition 5.10.1 Let F be a field, $f(x) \in F[x]$, with
$$f(x) = a_0 + a_1x + \cdots + a_nx^n.$$
We define a function $f: F \longrightarrow F$ as follows: Let $s \in F$, then
$$f(s) = a_0 + a_1s + \cdots + a_ns^n.$$
The function f is called the *polynomial function* corresponding to the polynomial $f(x)$.

In this section we shall use properties of the polynomial function f to investigate factorization properties of the polynomial $f(x)$.

Definition 5.10.2 Let $f(x), g(x) \in F[x]$, F a field. Denote the polynomial function corresponding to $f(x) + g(x)$ by $f + g$ and the polynomial function corresponding to $f(x) \cdot g(x)$ by $f \cdot g$.

The reader should notice that in general neither $f + g$ nor $f \cdot g$ is the composite function $f \circ g$.

5.10 CONSEQUENCES OF THE DIVISION ALGORITHM

THEOREM 5.10.3 Let $f(x), g(x) \in F[x]$, where F is a field, and let $s \in F$. Then

(a) $[f + g](s) = f(s) + g(s)$.
(b) $[f \cdot g](s) = f(s) \cdot g(s)$.

Proof (a) Let
$$f(x) = a_0 + a_1 x + \cdots + a_n x^n,$$
$$g(x) = b_0 + b_1 x + \cdots + b_m x^m.$$

Without loss of generality, we may assume that $m = n$. (Why?) Then
$$f(x) + g(x) = (a_0 + b_0) + (a_1 + b_1)x + \cdots + (a_n + b_n)x^n.$$
But
$$f(s) = a_0 + a_1 s + \cdots + a_n s^n,$$
$$g(s) = b_0 + b_1 s + \cdots + b_n s^n,$$
and hence
$$f(s) + g(s) = (a_0 + b_0) + \cdots + (a_n + b_n)s^n = (f + g)(s),$$
by definition of the polynomial function $f + g$.

(b) This is left as an exercise. ∎

THEOREM 5.10.4 (*The Remainder Theorem*) Let $f(x) \in F[x]$, F a field. Let $s \in F$, and let $q(x)$ and $r(x)$ be the unique elements of $F[x]$ such that
$$f(x) = q(x) \cdot (x - s) + r(x), \quad \deg r(x) < \deg (x - s).$$
Then $r(x) = d$, d an element of F, and moreover, $d = f(s)$.

Proof Since $\deg (x - s) = 1$, $\deg r(x)$ is either 0 or $-\infty$. If
$$\deg r(x) = 0,$$
then $r(x) = c \neq 0$, for some $c \in F$. If $\deg r(x) = -\infty$, then $r(x) = 0$, $0 \in F$. In either case $r(x) = d$ for some $d \in F$.
Since $f(x) = q(x)(x - s) + r(x)$, we have, by Theorem 5.10.3,
$$f(s) = [q(s)](s - s) + r(s).$$
Thus $f(s) = r(s)$. But, since $r(x) = d$, an element of F, $r(s) = d$. ∎

COROLLARY 5.10.5 Let $f(x) \in F[x]$, F a field. Then the polynomial $x - s$ is a factor of $f(x)$ if and only if $f(s) = 0$.

Proof By Theorem 5.10.4, if $f(s) = 0$, then, writing
$$f(x) = q(x)(x - s) + r(x),$$
we see that $r(x) = 0$, hence $f(x) = q(x)(x - s)$, that is, $(x - s) \mid f(x)$.

On the other hand, if $(x - s) \mid f(x)$, then there exists $q(x)$ such that
$$f(x) = q(x)(x - s),$$
and then
$$f(s) = [q(s)](s - s) = q(s) \cdot 0 = 0.$$

Definition 5.10.6 Let $f(x) \in F[x]$. An element $s \in F$ is called a *root* of the polynomial $f(x)$ or of f, if $f(s) = 0$. We say s *satisfies* the polynomial equation $f(x) = 0$.

THEOREM 5.10.7 Let $f(x) \in F[x]$, F a field, $\deg f(x) = n \geq 0$. Then f has at most n distinct roots in F.

Proof We prove the theorem by induction on n. If $\deg f(x) = 0$, then $f(x) = c \neq 0$, so f has no roots in F. Also, trivially, if $\deg f(x) = 1$, then $f(x) = ax + b$, $a \neq 0$, and clearly $-a^{-1}b$ is the only root of f.

Thus we assume that the theorem is true for $n - 1 \geq 0$. Then for $\deg f(x) = n, f(x) = a_0 + a_1 x + \cdots + a_n x^n$. Suppose f has a root, say s_1. Then by Corollary 5.10.5, $(x - s_1) \mid f(x)$, and so there exists a polynomial $f_1(x)$ of degree $n - 1$ such that $f(x) = f_1(x)(x - s_1)$. Now, any root of $f(x)$ is a root of either $f_1(x)$ or $(x - s_1)$. If s_2 is any other root of f such that $s_1 \neq s_2$, then $f_1(s_2) = 0$. (Why?) Thus, any root of f different from s_1 must satisfy the equation $f_1(x) = 0$. By induction, f_1 has at most $n - 1$ distinct roots, and so f has at most n distinct roots.

EXERCISE

1. Define s to be a *root of multiplicity m* of f if
$$f(x) = f_1(x) \cdot (x - s)^m \quad \text{but} \quad f_1(s) \neq 0.$$
By modifying the proof of Theorem 5.10.7, prove the following:
If $f(x) \in F[x]$, F a field, $\deg f(x) = n$, and if $s_1, s_2, \ldots, s_k$ are all the distinct roots of $f(x)$, with multiplicities $m_1, m_2, \ldots, m_k$, respectively, then $m_1 + m_2 + \cdots + m_k \leq n$.

A fundamental problem in the theory of polynomials is to determine when a given polynomial over a ring can be factored. The fundamental theorem of algebra, which we state here without proof, asserts that every polynomial $f(x) \in C[x]$, C the field of complex numbers, can be *completely* factored, that is, $f(x) = b(x - a_1) \cdots (x - a_n)$, where $b, a_1, \ldots, a_n \in C$ and $n = \deg f(x)$. Equivalently, we have

THEOREM 5.10.8 (*Fundamental Theorem of Algebra*) Let $f(x) \in C[x]$, where C denotes the field of complex numbers, with $\deg f(x) \geq 1$. Then f has a root in C.

No proof of this theorem exists that does not use techniques of analysis or topology. One can, however, prove by elementary calculus that any odd-degree polynomial over the real numbers has a linear factor with real coefficients. From this point it is possible to prove Theorem 5.10.8 using only algebraic techniques.[3]

EXERCISES

2. Prove (b) of Theorem 5.10.3.
3. Show that in $Z_p[x]$, p a prime, $(x - a) \mid (x^{p-1} - 1)$, for all $a \in Z_p$, $a \neq 0$. Verify that
$$(x - 1)(x - 2) \cdots (x - p + 1) = x^{p-1} - 1.$$
4. Prove Wilson's theorem: Let $n \in Z, n > 1$. Then $(n - 1)! \equiv -1 \pmod{n}$ if and only if n is a prime number. (*Hint:* Consider the roots of $x^{n-1} - 1$.)
5. Show that Theorem 5.10.7 does not hold in $Z_m[x]$, if m is not a prime number and $m \neq 4$.

Although a polynomial function over the ring of real numbers is different from the associated polynomial, there does exist a natural 1-1 correspondence between the set of polynomial functions and the set of polynomials. In fact, the polynomial functions over the real numbers form a ring, with respect to the addition and multiplication in Definition 5.10.2, that is isomorphic to $K[x]$, where K denotes the real numbers. However, this relation between a ring of polynomials and the ring of associated polynomial functions does not hold for all rings R. For example, if $R = Z_p$, p a prime, then the polynomial $\alpha(x) = x^p - x$ is certainly different from the polynomial $\beta(x) = 0$, but the associated polynomial function of each of these is the zero function θ, where $\theta(s) = 0$, all $s \in Z_p$. It is true, however, that the ring of polynomial functions associated with $R[x]$, R commutative, is always a homomorphic image of $R[x]$ (Exercise 8).

EXERCISES

6. Let $f(x), g(x) \in Z_5[x]$. Show that if
$$f(x) = x^2 - 2 \quad \text{and} \quad g(x) = x^5 + x^2 - x - 2,$$
then $\bar{f} = \bar{g}$, where $\bar{f}$ and $\bar{g}$ are the polynomial functions corresponding to $f(x)$ and $g(x)$, respectively.

---
[3] See B. L. van der Waerden, *Modern Algebra* (rev. English ed.). New York: Ungar, 1953, sec. 70.

220   THE THEORY OF RINGS

7. Find all positive integers m such that $x^2 + 2$ is a divisor of $x^5 - 10x + 12$ in $Z_m[x]$.
8. Let R be a commutative ring and let

$$f(x) = a_0 + a_1 x + \cdots + a_n x^n$$

be a polynomial in $R[x]$. Let the function $f\colon R \longrightarrow R$, defined by

$$f(s) = a_0 + a_1 s + \cdots + a_n s^n, \quad \text{all } s \in R,$$

be the *polynomial function associated with the polynomial* $f(x)$. Let T be the set of all polynomial functions associated with $R[x]$. Show that T is a ring and that T is the homomorphic image of $R[x]$. See Definitions 5.10.1 and 5.10.2 and Theorem 5.10.3.

5.11 Prime and Maximal Ideals

It is easily seen that in Z, the ring of integers, the ideal

$$(p) = \{kp \mid k \in Z\},$$

where p is a prime number, satisfies these two properties: (a) if the product $ab \in (p)$, then either $a \in (p)$ or $b \in (p)$; and (b) if $a \notin (p)$, then the set

$$\{ma + np \mid m, n \in Z\} = Z.$$

In this section we shall study commutative rings R that contain ideals satisfying either property (a) or (b), and we shall investigate the relationship between (a) and (b).

Definition 5.11.1   Let R be a commutative ring. An ideal P in R is called a *prime ideal* if, whenever $ab \in P$, $a \in R$, $b \in R$, then either $a \in P$ or $b \in P$.

An ideal M in R is called a *maximal ideal* in R if (a) $M \subset R$ and (b) whenever N is an ideal in R such that $M \subseteq N \subseteq R$, then either $N = M$ or $N = R$.

It is now clear that the ideal (p), where p is a prime, is both a prime ideal and a maximal ideal in the ring Z.

In the ring $F[x]$, F a field, the ideal

$$I = \{f(x)(x - s) \mid f(x) \in F[x]\},$$

generated by the polynomial $x - s$, $s \in F$, is both a prime ideal and a maximal ideal.

The reader should not conclude from these two examples, however, that there is no distinction between prime and maximal ideals. We consider the ring $Z[x]$. Let A be the set of all polynomials in $Z[x]$ that have constant term zero. Then A is an ideal in $Z[x]$. Moreover, A is a prime ideal, for if $f(x)g(x) \in A$, then either $f(x) \in A$ or $g(x) \in A$. (We leave the proof of this fact for the exercises.) A is not, however, a maximal ideal, for the ideal of all polynomials with even constant term properly contains A yet is not equal to $Z[x]$.

On the other hand, it is also possible for a ring R to have an ideal M that is a maximal ideal but that is not a prime ideal. For example, let p be a prime number, and on the set $Z_p = \{0, 1, \ldots, p-1\}$ define the usual addition modulo p and define multiplication by setting $ab = 0$, for all $a, b \in Z_p$. Then (0) is a maximal ideal in this ring (why?), but it is clearly not a prime ideal.

Definition 5.11.2 Let R be a commutative ring with 1. Let $A \subseteq R$. Then the *ideal generated by* A is the intersection of all ideals I_α of R such that $A \subseteq I_\alpha$. We denote this ideal by (A). If $A \subseteq R$ and $a \in R$, we shall denote $(A \cup \{a\})$ by (A,a). If $A = \varnothing$ and $a \in R$, the ideal $(\varnothing, a)$ will be denoted simply by (a) and is called the *principal ideal in R generated by a*.

EXERCISES

1. Let R be a ring. Let $\{I_\alpha\}$ be a collection of ideals in R. Show that $\cap\, I_\alpha$ forms an ideal in R.
2. Let R be a commutative ring with 1. Let $A \subseteq R$.
   (a) Show that
   $$(A) = \{\Sigma\, r_i a_{\alpha_i} \mid r_i \in R,\, a_{\alpha_i} \in A\},$$
   that is, (A) is the set of all finite sums of the form
   $$r_1 a_{\alpha_1} + \cdots + r_k a_{\alpha_k}.$$
   (b) Show that if $A = \{a\}$, then $(A) = (a) = \{ra \mid r \in R\}$.

THEOREM 5.11.3 Let R be a commutative ring with identity 1. Let I be an ideal in R. Let $a \in R$. Then for $x \in (I,a)$, there exist $i \in I$ and $r \in R$ such that $x = i + ra$.

Proof (I,a) is the intersection of all ideals in R containing the set $I \cup \{a\}$. Since I is an ideal, the set $\{i + ra \mid i \in I,\, r \in R\}$ also is an ideal in R. Clearly, this set is (I,a), and so any $x \in (I,a)$ has the form $i + ra$, for some $i \in I,\, r \in R$. ∎

THEOREM 5.11.4 Let R be a commutative ring with 1. Let M be a maximal ideal in R. Then M is a prime ideal.

Proof Let $a, b \in R$, and suppose $ab \in M$. To show that M is a prime ideal, we need to show that either $a \in M$ or $b \in M$. Thus, assume $a \notin M$. Then the ideal $N = (M,a)$ properly contains M. Since M is a maximal ideal in R, $(M,a) = R$, whence $1 \in (M,a)$. Thus, there exists an element $m \in M$ and an element $r \in R$ such that $1 = m + ar$. But then $b = bm + bar$, that is, $b = bm + r(ab)$. Now, $ab \in M$, whence $bm + r(ab) \in M$, that is, $b \in M$ and M is a prime ideal. ∎

The reader should observe how closely this proof resembles the proof of Euclid's lemma in Chapter 2.

THEOREM 5.11.5 Let R be a commutative ring, $R \ni 1$. Let M be a maximal ideal in R. Then the quotient ring R/M is a field.

Proof Since R is a commutative ring, so is R/M. Thus, the only properties that need verification are (a) R/M has an identity, and (b) every nonzero element of R/M has an inverse in R/M.

To show (a), we merely observe that $1 + M$ is an identity for R/M.

To show (b), suppose $b + M \neq M$, that is, $b + M$ is a nonzero element in R/M. Since M is a maximal ideal, $(b,M) = R$; that is, there is an $r \in R$ and an $m \in M$ such that $1 = br + m$. Then $r + M$ is the inverse of $b + M$, and R/M is a field. ∎

THEOREM 5.11.6 Let R be a commutative ring, $R \ni 1$. Let M be an ideal in R. If R/M is a field, then M is a maximal ideal in R.

Proof Since R/M is a field, it has at least two elements. Hence there exist elements in R that are not in M. Therefore, let $a \in R$, $a \notin M$. We must show $(M,a) = R$. Now, since $a \notin M$, $a + M$ is a nonzero element in the field R/M. Hence, there exists an element $b + M$ in R/M such that $(a + M)(b + M) = 1 + M$, the identity in the field R/M. That is, there are elements $m_1, m_2, m_3 \in M$ such that

$$(a + m_1)(b + m_2) = 1 + m_3.$$

Thus, $ab + m' = 1$, where $m' = am_2 + bm_1 + m_1m_2 - m_3$ is in M. Finally, for any $r \in R$,

$$r(ab) + rm' = r \cdot 1$$
$$(rb)a + rm' = r, \text{ that is, } r \in (a,M).$$

Thus $(a,M) = R$, and M is a maximal ideal. ∎

Theorems 5.11.5 and 5.11.6 are summarized in the following.

THEOREM 5.11.7 Let R be a commutative ring with 1. Let M be an ideal in R. Then R/M is a field if and only if M is a maximal ideal in R.

5.11 PRIME AND MAXIMAL IDEALS

Example 5.11.8 To illustrate Theorem 5.11.7 in a familiar setting, let $R = Z$, the ring of integers. Let (m) be the ideal of all multiples of the integer m, $m > 0$. We know that the rings $Z/(m)$ and Z_m are isomorphic. Now (m) is a maximal ideal in Z if and only if $m = p$, a prime. Thus, $Z/(m)$ is a field if and only if $m = p$, a prime, by Theorem 5.11.7. Hence Z_m is a field if and only if $m = p$, a prime. This result, of course, is already known to us.

To further illustrate the usefulness of these theorems, we show a method of constructing the field of complex numbers from the real numbers. (See Exercises 6 and 7 in Section 5.7. The real and complex numbers will be discussed more fully in Chapter 6.)

Let K denote the field of real numbers. Then, clearly, the polynomial function corresponding to the polynomial $x^2 + 1$ has no roots in K, since $a^2 \geq 0$ for all $a \in K$. Hence, $x^2 + 1$ has no factorization into first-degree polynomials in $K[x]$.

Now let $f(x) \in K[x]$, $f(x) \notin (x^2 + 1)$, the principal ideal generated by $x^2 + 1$. Since the only factors of $x^2 + 1$ have the form either of $c \in F$, $c \neq 0$, or of $d(x^2 + 1)$, $d \in F$, $d \neq 0$, it follows that $f(x)$ and $x^2 + 1$ are relatively prime in $K[x]$, that is, gcd $(f(x), x^2 + 1) = 1$ (see Exercise 3, Section 5.9), and so there are polynomials $s(x)$ and $t(x) \in K[x]$ such that $1 = f(x)s(x) + (x^2 + 1)t(x)$, whence $1 \in ((x^2 + 1), f(x))$ and $K[x]$ is equal to the ideal generated by $x^2 + 1$ and $f(x)$. Thus, $(x^2 + 1)$ is a maximal ideal in $K[x]$.

To avoid confusion between the polynomial $x^2 + 1$ and the principal ideal $(x^2 + 1)$, we shall denote the latter by I.

By Theorem 5.11.5, $K[x]/I$ is a field that we now show is isomorphic to the field of complex numbers (see Exercises 6 and 7, Section 5.7). Its elements are of the form $bx + a + I$, where a, b take on all values in K. We denote this element by $\bar{a} + \overline{bx}$. Now let $\bar{a} + \overline{bx}$ and $\bar{c} + \overline{dx} \in K[x]/I$. Then

$$(\bar{a} + \overline{bx}) + (\bar{c} + \overline{dx}) = (a + bx + I) + (c + dx + I)$$
$$= \overline{(a + c) + (b + d)x} + I$$
$$= \overline{a + c} + \overline{(b + d)x}.$$

Also,

$$(\bar{a} + \overline{bx})(\bar{c} + \overline{dx}) = (a + bx + I)(c + dx + I)$$
$$= ac + bdx^2 + (ad + bc)x + I.$$

Now,

$$bdx^2 = bd(x^2 + 1) - bd,$$

whence

$$bdx^2 + I = -bd + I,$$

so that
$$ac + bdx^2 + (ad + bc)x + I = ac - bd + (ad + bc)x + I.$$
Thus
$$(\bar{a} + \overline{bx})(\bar{c} + \overline{dx}) = \overline{ac - bd} + \overline{(ad + bc)x}.$$
By considering the cases $(\bar{a} + \overline{0x}) + (\bar{c} + \overline{0x})$ and $(\bar{a} + \overline{0x})(\bar{c} + \overline{0x})$, we see that $\bar{a} + \bar{c} = \overline{a + c}$ and $\bar{a}\bar{c} = \overline{ac}$, for all $a, c \in K$. If we now let $\bar{a} + \overline{bx} \longrightarrow a + bi$, $i^2 = -1$, we see that this is an isomorphism. Thus the field $K[x]/I$ is isomorphic to the field of complex numbers.

EXERCISES

3. Let p be a prime number. Show that the ideal (p) is both a maximal ideal and a prime ideal in Z.
4. Let F be a field. Let $f(x)$ be irreducible over F. Prove that $(f(x))$ is a maximal ideal in $F[x]$.
5. Let Z_2 be the field with two elements. Show that the polynomial $x^2 + x + 1$ is irreducible over Z_2. Form the field
$$F = Z_2[x]/(x^2 + x + 1),$$
where $(x^2 + x + 1)$ denotes the principal ideal in $Z_2[x]$ generated by $x^2 + x + 1$. Denote the elements of F by
$$0 = 0 + (x^2 + x + 1),$$
$$1 = 1 + (x^2 + x + 1),$$
$$\bar{x} = x + (x^2 + x + 1),$$
and
$$\bar{y} = x + 1 + (x^2 + x + 1).$$
Write the multiplication table for F and verify directly that F is a field. (See Exercise 13, Section 5.4.)
6. Prove that the set of all polynomials with constant term zero is a prime ideal in $Z[x]$.
7. Prove that the set of all polynomials with even coefficients is a prime ideal in $Z[x]$. [*Hint:* Let
$$f(x) = a_n x^n + a_{n-1} x^{n-1} + \cdots + a_1 x + a_0$$
and
$$g(x) = b_m x^m + b_{m-1} x^{m-1} + \cdots + b_1 x + b_0.$$
Let a_i be chosen so that $2 \nmid a_i$, but $2 \mid a_k$, for all $k < i$. Let b_j be chosen so that $2 \nmid b_j$, but $2 \mid b_k$ for all $k < j$. Show that $2 \nmid c_{i+j}$, where c_{i+j} is the coefficient of x^{i+j} in the product $f(x) \cdot g(x)$.]

8. (Eisenstein) Let $f(x) \in Z[x]$,

$$f(x) = a_n x^n + a_{n-1} x^{n-1} + \cdots + a_0.$$

Assume there exists a prime p such that $p \mid a_i$, for $0 \le i < n$, $p \nmid a_n$, $p^2 \nmid a_0$. Prove that there do not exist polynomials $g(x)$ and $h(x)$ of degree ≥ 1 in $Z[x]$ such that $f(x) = g(x)h(x)$.

5.12 Euclidean Domains and Unique Factorization

In Section 2.9 we proved that if $n \in Z$, $n > 1$, then there exists a factorization of n into a product of primes, and that this factorization is essentially unique. In addition, in the Exercises of Section 5.9 we indicate that a similar theorem holds in $F[x]$, F a field. Indeed, the division algorithm for polynomials is the tool that we can use to prove that any polynomial $f(x) \in F[x]$, $\deg f(x) \ge 1$, can be factored into irreducible polynomials. Such a proof would mimic the proof of the fundamental theorem of arithmetic. This leads to the following definitions, which are in the same spirit as those in Chapter 2.

Definition 5.12.1 Let R be an integral domain. Let $a, b \in R$. We say that a *divides* b, denoted $a \mid b$, if there exists $c \in R$ such that $b = ac$. We say that $a \in R$ is a *unit* in R if there exists $b \in R$ such that $ab = 1$. The elements a and b are called *associates* if there exists a unit u such that $b = au$.

A nonunit element a is called *prime*[4] if, whenever a is written $a = bc$, then either b is a unit or c is a unit.

In this terminology, we see that the only units in Z are $+1$ and -1, and that both p and $-p$ are now called prime, if p is a prime according to Definition 2.8.1. We see also that the irreducible elements in $F[x]$ are also called primes in the present terminology.

Definition 5.12.2 Let R be an integral domain. R is called a *unique factorization domain* (UFD) if

(a) for each nonunit a in R, $a \ne 0$, there exists a factorization $a = b_1 b_2 \cdots b_k$, where the b_i are primes, and
(b) if $a = b_1 b_2 \cdots b_k = c_1 c_2 \cdots c_l$ are two factorizations of a into primes, then $k = l$, and for some permutation π of the subscripts j of the c_j's we have that b_i and $c_{\pi i}$ are associates, $i = 1, \ldots, k$.

---

[4] Frequently, a nonunit element a is called *prime* if, whenever $a \mid bc$, either $a \mid b$ or $a \mid c$, and a nonunit a is called *irreducible* if, whenever $a = bc$, either b or c is a unit. In the context of Euclidean domains, however, the two concepts coincide.

Clearly, both Z and $F[x]$ are UFD's. The similarity in the proofs that Z and $F[x]$ are UFD's encourages us to look for those properties common to Z and $F[x]$ that are used in the proof of the unique factorization theorems.

Definition 5.12.3 Let E be an integral domain and let $\delta\colon E \longrightarrow P_0$, P_0 the set of nonnegative integers. We call E a *Euclidean domain* if

(a) $\delta(a) = 0$ if and only if $a = 0$,
(b) $\delta(ab) = \delta(a)\delta(b)$,
(c) for any $a \in E$, $b \in E$, $b \neq 0$, there exist $q, r \in E$ such that

$$a = bq + r, \qquad \delta(r) < \delta(b).$$

Examples of Euclidean domains are (a) Z, with $\delta(m) = |m|$, for all $m \in Z$; (b) $F[x]$, F a field, where $\delta[f(x)] = 2^{\deg f(x)}$ (we adopt the convention that $2^{\deg 0} = 2^{-\infty} = 0$); (c) a field F, with $\delta(a) = 1$, $a \neq 0$ and $\delta(0) = 0$.

That a Euclidean domain E is a UFD can be proved by imitating the proof that Z is a UFD.

THEOREM 5.12.4 Let E be a Euclidean domain. Then if I is a nonzero ideal in E, there exists an element $a \in I$ such that $a \mid b$, for all $b \in I$.

Proof Since $I \neq 0$, there exists an element $a \in I$ such that (a) $\delta(a) \neq 0$ and (b) for all $b \neq 0$ in I, $\delta(a) \leq \delta(b)$, since $\delta(x)$ is a nonnegative integer, for each x in I. Now for any $b \in I$, by (c) of Definition 5.12.3, there exist q and $r \in I$ such that $b = aq + r$, $\delta(r) < \delta(a)$. By choice of a, $\delta(r) = 0$, whence $r = 0$, and so $a \mid b$. ∎

Definition 5.12.5 Let $a, b \in E$, E a Euclidean domain. A greatest common divisor (gcd) of a and b is an element g in E such that

(a) $g \mid a$ and $g \mid b$,
(b) if $e \mid a$ and $e \mid b$, then $e \mid g$.

THEOREM 5.12.6 Let E be a Euclidean domain. Let $a, b \in E$. Then (a) there exists a gcd g of a and b in E, and $g = as + bt$, for some $s, t \in E$, and (b) if g_1 and g_2 are both gcd's of a and b, there is a unit u such that $g_2 = ug_1$.

Proof (a) Let $I = \{ax + by \mid x, y \in E\}$. It is easy to see that I is an ideal in E. Then by Theorem 5.12.4, there is an element $g \in I$ such that $g \mid c$, for all $c \in I$. Clearly, $a = a \cdot 1 + b \cdot 0 \in I$, so $g \mid a$, and similarly, $g \mid b$.

Since $g \in I$, there are s and $t \in E$ such that $g = as + bt$. So if $e \mid a$ and $e \mid b$, then $e \mid (as + bt)$, that is, $e \mid g$. Thus, g is a gcd of a and b.

(b) Let g_1, g_2 both be gcd's of a and b. Then by (b) of Definition 5.12.5, $g_1 \mid g_2$ and $g_2 \mid g_1$. Since E is an integral domain, this implies $g_2 = ug_1$, for some unit u in E. ∎

THEOREM 5.12.7 (*Euclid's Lemma for Euclidean Domains*) Let E be a Euclidean domain. Let p be a prime element in E and let $p \mid ab$. Then $p \mid a$ or $p \mid b$.

Proof Suppose $p \nmid a$, and let g be the gcd of a and p. Since $g \mid p$, $p = gh$, for some $h \in E$. Since p is a prime, either g or h is a unit. Suppose that h is a unit. Then $g = ph^{-1}$, so $ph^{-1} \mid a$, whence $p \mid a$, a contradiction. Thus, g must be a unit, say $g = u$. By (a) of Theorem 5.12.6, $u = as + pt$, some s and t. Then $1 = uu^{-1} = a(su^{-1}) + p(tu^{-1})$. Multiplying through by b, $b = ba(su^{-1}) + bp(tu^{-1})$. Since $p \mid ba$ and $p \mid p$, clearly, we now have $p \mid b$. ∎

THEOREM 5.12.8 A Euclidean domain E is a UFD.

Proof Let $a \in E$, $\delta(a) > 1$. Either there exist $b, c \in E$ such that $a = bc$, $1 < \delta(b) < \delta(a)$, $1 < \delta(c) < \delta(a)$, or b or c must be a unit. In the latter case, a is already a prime. In the former case, a can be factored into a product bc, with both $\delta(b) < \delta(a)$, and $\delta(c) < \delta(a)$. By an induction argument on $\delta(a)$, we can argue that b and c can be factored into primes, and so a can be factored into primes.

Now let $a = p_1 p_2 \cdots p_r = q_1 q_2 \cdots q_s$ be two factorizations of a into primes. By Theorem 5.12.7 (generalized), $p_1 \mid q_i$, some i. Without loss of generality, $i = 1$, so $p_1 \mid q_1$. Since p_1 and q_1 are both primes, $q_1 = p_1 u$, u a unit. Dividing by p_1, we get

$$p_2 \cdots p_r = uq_2 \cdots q_s = q_2' q_3 \cdots q_s,$$

where

$$q_2' = uq_2$$

is a prime also. By induction, $r = s$, and we can conclude $p_2, \ldots, p_r$ and $q_2', \ldots, q_r$ can be arranged in pairs of associates. Thus, we can conclude $p_1, \ldots, p_r$ are associates, under proper rearrangement, of $q_1, \ldots, q_r$, respectively, and so E is a UFD. ∎

Given the theory derived in this section, one may still ask whether there are UFD's that are not Euclidean domains. Such is indeed the case. For example, if R is the field of real numbers, then $R[x,y]$, the ring of all polynomials in the indeterminates x and y over R, is a UFD. $R[x,y]$ is not a Euclidean domain. To see this, one need only show that there is not an element $h(x,y)$ in the ideal (x,y) generated by x and y that divides both x and y, as would be required by Theorem 5.12.4.

EXERCISES

1. Let E be a Euclidean domain. Prove each of the following.
   (a) a is a unit in E if and only if $\delta(a) = 1$. [*Hint:* Consider $\delta(aa^{-1})$.]
   (b) If a and b are associates, then $\delta(a) = \delta(b)$.
2. Let E be a Euclidean domain. Let $F = \{a \in E \mid \delta(a) \leq 1\}$. Suppose that for $a, b \in F$, $a + b \in F$. Prove that F is a field.
3. Let E be a Euclidean domain. Let p be a nonzero, nonunit element such that when $p \mid ab$, then also $p \mid a$ or $p \mid b$. Prove that p is a prime.
4. Let p be a prime in an integral domain. Let q be an associate of p. Show that q is a prime.
5. Show that $R[x,y]$ is not a Euclidean domain.

5.13 Direct Sums of Rings

At this point we again parallel a useful construction of Chapter 4 (see Section 4.12) to enable us to create "larger" rings from "smaller" ones. That is, given rings R_1 and R_2, we define the direct sum of R_1 and R_2.

Definition 5.13.1 Let R_1, R_2 be rings. Then the set $R_1 \times R_2$ with operations $+$ and $\cdot$ defined by

$$(a_1,a_2) + (b_1,b_2) = (a_1 + b_1, a_2 + b_2) \quad \text{and} \quad (a_1,a_2) \cdot (b_1,b_2) = (a_1b_1, a_2b_2)$$

is called the *direct sum* of R_1 and R_2. It is denoted by $R_1 \oplus R_2$.

THEOREM 5.13.2 Let R_1 and R_2 be rings. Then the direct sum $R_1 \oplus R_2$ is a ring.

Proof The zero of $R_1 \oplus R_2$ is simply $(0,0)$. The additive inverse of (a_1,a_2) is $(-a_1,-a_2)$. We leave the remainder of the proof to the reader. ∎

Clearly, the map $f: R_1 \longrightarrow R_1 \oplus R_2$ defined by $f(a) = (a,0)$, for all $a \in R_1$, is an embedding of R_1 in $R_1 \oplus R_2$. Also, R_2 is embedded in $R_1 \oplus R_2$. Moreover, R_1 and R_2 can be considered ideals in $R_1 \oplus R_2$ in the sense that $\{(a,0) \mid a \in R_1\}$ is an ideal in $R_1 \oplus R_2$ and $\{(0,b) \mid b \in R_2\}$ is an ideal in $R_1 \oplus R_2$.

We also have the following theorem analogous to Theorem 4.12.6.

THEOREM 5.13.3 Let R be a ring. Let I_1 and I_2 be ideals of R such that

(a) if $r \in R$, there exist elements $a_1 \in I_1$ and $a_2 \in I_2$ such that $r = a_1 + a_2$, and
(b) $I_1 \cap I_2 = \{0\}$.

Then $R \cong I_1 \oplus I_2$.

Proof Suppose $r = a_1 + a_2 = b_1 + b_2$, where $a_i, b_i \in I_i$, $i = 1, 2$. Then
$$a_1 - b_1 = b_2 - a_2 \in I_1 \cap I_2,$$
whence $a_1 - b_1 = b_2 - a_2 = 0$, that is, $a_1 = b_1$ and $a_2 = b_2$. Thus, any element in R has a unique representation as a sum of elements from I_1 and I_2.

Thus if we define a map $\varphi: I_1 \oplus I_2 \longrightarrow R$ by $\varphi(a_1, a_2) = a_1 + a_2$, we see that this map is well defined, onto by (a), and 1–1 by the paragraph above.

It only remains to show that φ is a homomorphism. Thus
$$\varphi[(a_1,b_1) + (a_2,b_2)] = \varphi[(a_1 + a_2, b_1 + b_2)]$$
$$= a_1 + a_2 + b_1 + b_2$$
$$= a_1 + b_1 + a_2 + b_2$$
$$= \varphi(a_1,b_1) + \varphi(a_2,b_2).$$

Now, on the one hand,
$$\varphi[(a_1,b_1)(a_2,b_2)] = \varphi[(a_1 a_2, b_1 b_2)] = a_1 a_2 + b_1 b_2.$$

On the other hand,
$$\varphi(a_1,b_1) \cdot \varphi(a_2,b_2) = (a_1 + b_1)(a_2 + b_2)$$
$$= a_1 a_2 + b_1 a_2 + a_1 b_2 + b_1 b_2.$$

Since $a_1 b_2$ and $b_1 a_2$ are in $I_1 \cap I_2$, they are zero and so
$$\varphi[(a_1,b_1)(a_2,b_2)] = \varphi(a_1,b_2) \cdot \varphi(a_2,b_2).$$

This concludes the proof that φ is an isomorphism. ∎

Frequently, under the conditions of this theorem, we say that R is the *internal direct sum* of the ideals I_1 and I_2 and we write:
$$R = I_1 \dotplus I_2.[5]$$

EXERCISES

1. In Theorem 5.13.3 replace conditions (a) and (b) by the single condition: if $r \in R$, then there exist unique element $a_1 \in I_1$ and $a_2 \in I_2$ such that
$$r = a_1 + a_2.$$

---

[5] Our use of $\dotplus$ is not universal.

2. Let R_1 and R_2 be rings with identity. Prove that $R_1 \oplus R_2$ is a ring with identity.
3. Let R be a ring. Let $I = \{a \in R \mid ar = 0, \text{ for all } r \in R\}$. Prove that I is an ideal in R.
4. Let R be a commutative ring with identity 1. Let R have an element e such that $e^2 = e$, $e \neq 0$, $e \neq 1$. Prove that
    (a) The set $Re = \{re \mid r \in R\}$ is an ideal of R.
    (b) There exists an ideal R' in R such that $R = R' \dotplus Re$.
5. Let R_1 and R_2 be rings. If R_1 and R_2 are both _____, then $R_1 \oplus R_2$ is _____. Replace the blanks by each of the properties listed below and either prove or disprove the four resulting assertions:
    (a) commutative   (b) integral domain(s)
    (c) field(s)   (d) finite
6. Let A, B, C, D be rings such that $A \cong B$ and $C \cong D$. Prove that $A \oplus C \cong B \oplus D$.
7. Let m and n be relatively prime positive integers. Prove that the ring Z_{mn} is isomorphic to $Z_m \oplus Z_n$. (*Hint:* Show that Z_{mn} contains ideals I_1 and I_2 satisfying Theorem 5.13.3, and then use Exercise 6. Or establish the isomorphism directly.)
8. Let m_1 and m_2 be positive integers $(m_1, m_2) = 1$. Let (m_i) be the principal ideal generated by m_i, $i = 1, 2$. Prove that

$$Z/[(m_1) \cap (m_2)] \cong Z/(m_1) \oplus Z/(m_2).$$

9. Let R be a ring and let I_1 and I_2 be distinct maximal two-sided ideals in R. Prove that

$$R/(I_1 \cap I_2) \cong R/I_1 \oplus R/I_2 \quad \text{(Chinese remainder theorem)}.$$

(*Hint:* See Exercise 8.)

6 The Real and Complex Number Fields

6.1 The Need for the Real Number System

So far in this book our description of the integers and our construction of the rational number system have centered on the notions of sets and binary operations. The reader, in his study of analytic geometry and calculus, has also viewed systems of numbers from a geometric point of view—that is, as points on a coordinate line. The usual procedure for constructing such a number line is to designate some point on a horizontal straight line as the origin, or zero, and then to choose a fixed length for a "unit length." Next, both to the left and right of the origin, points are marked off that are integral multiples of the unit length from the origin. These points represent the negative and positive integers, respectively, arranged in ascending order of the integers to the right. (See Figure 6.1.)

The rational number $1/n$, for the integer $n > 0$, is then represented by subdividing the interval from 0 to 1 into n equal parts, with the first mark to the right of 0 denoted by $1/n$. (By a well-known construction of Euclidean geometry using straight edge and compass, it is possible to divide a straight line segment into n equal parts.) The rational number a/b, with $b > 0$, is represented by laying off $|a|$ repetitions of the length $1/b$ starting from the origin and proceeding either to the right or the left, according as a is positive or negative (see Figure 6.1).

It was noted by the ancient Greeks that although the rational numbers are dense on the line (that is, between any two rational numbers, there is a third rational number), there are points on the line that do not correspond to any rational number. For example, if we construct an isosceles right triangle whose legs have a length of one unit on our number line, then the hypotenuse has, by the Pythagorean theorem, a length $\sqrt{2}$. However, when this length is laid off on our number line, with one end at the origin, then the other end point does not correspond to any rational number a/b (see Theorem 2.9.4). Thus there is a "gap" in the rational number system.

(This same difficulty with the system of rational numbers can be expressed algebraically by saying that the polynomial equation $x^2 - 2 = 0$ does not have a solution within the system of rational numbers.)

Figure 6.1

This discussion thus raises a question: Does there exist a "number system" that contains the rational number system and that faithfully "represents" the set of all points on the number line. More precisely, does there exist a field **R** together with an order relation such that (a) **R** contains a subfield isomorphic to the field of rational numbers, (b) **R** can be put into a one-to-one correspondence with the points on the number line, and (c) this correspondence preserves the order and arithmetic properties of the rational number system?

The first part of this chapter will give an affirmative answer to this question by constructing the field of real numbers and indicating how it has the properties stated above.

EXERCISES

1. Let p be a prime number. Show that there is no rational number x such that $x^2 = p$. That is, prove $\sqrt{p}$ is irrational.
2. Consider the field Z_p, p a prime. Prove that exactly $(p - 1)/2$ elements of Z_p are perfect squares. Thus, conclude that the equation $x^2 - a = 0$, $a \in Z_p$, does not always have a solution in Z_p. [*Hint:* Let $Z_p{}^* = Z_p - \{0\}$. $Z_p{}^*$ is a group with respect to multiplication. Let $f: Z_p{}^* \longrightarrow Z_p{}^*$ be defined by $f(x) = x^2$, all $x \in Z_p{}^*$. Consider ker f.]
3. Let p be a prime number. Use Wilson's theorem to prove that there is an element a in Z_p such that $a^2 = -1$ if and only if p is of the form $4k + 1$, for some k.
4. Show that there is a right triangle that has a hypotenuse with length $\sqrt{5}$; with length $\sqrt{10}$; with length $\sqrt{13}$.

6.2 Ordered Fields

Definition 6.2.1 A field F is called an *ordered field* if there exists a subset P of F such that

(a) for each a in F, exactly one of the following holds: $a \in P$, $a = 0$, $-a \in P$, and
(b) for a and b in P, $a + b$ and ab are both in P.

P is called the *set of positive elements* of F.

THEOREM 6.2.2 The field of rational numbers is an ordered field.

Proof Let a/b be a rational number. Without loss of generality, we may assume that the integer $b > 0$ in the ordering defined for the integers. We define P to be the set of all rational numbers a/b, $b > 0$, such that $a > 0$. Clearly, with this definition, we see that property (a) of the definition of ordered field is satisfied. Next, let a/b and c/d both be positive. Then assuming that all of a, b, c, d are > 0, as integers, we see that $bd > 0$, and $ac > 0$, whence $(a/b)(c/d) \in P$, and also

$$\frac{a}{b} + \frac{c}{d} = \frac{ad + bc}{bd} \quad \text{is in } P.$$

Thus, the set of rational numbers is an ordered field. ∎

We now define the relation $<$ for rational numbers. This definition is analogous to the definition of $<$ for integers. We let P be the set defined in the proof of Theorem 6.2.2.

Definition 6.2.3 Let a/b and c/d be rational numbers. Then we say that $a/b < c/d$ if $c/d - a/b \in P$, that is, if $c/d - a/b$ is positive.

We note that if $b, d > 0$, then $a/b < c/d$ if and only if $ad < bc$ as integers.

We use $a/b > c/d$ interchangeably with $c/d < a/b$. The term "negative rational number" is defined in the obvious way.

The next lemma was given as Exercise 3 in Section 2.4, but we give it here for completeness.

LEMMA 6.2.4 Let a and b be positive integers. Then there exists an integer n such that $na > b$.

Proof If $a > 1$, then clearly $n = b$ will work. If $a = 1$, then $n = b + 1$ will work. Thus, the lemma is easily proved. ∎

Definition 6.2.5 Let F be an ordered field with P the set of positive elements. F is called an *Archimedean ordered field* if for any a, b in P, there exists a positive integer n such that $na > b$.

THEOREM 6.2.6 The field of rational numbers is an Archimedean ordered field.

Proof Let a/b and c/d be two positive rational numbers. We must show that there is a positive integer n such that $n(a/b) > c/d$. This is equivalent to showing that there is an integer n such that $n(ad) > bc$. But by Lemma 6.2.4, such an n exists. ∎

As we remarked in Section 6.1, the rational number system is dense in the sense that between any two rational numbers there is a third rational number. For example, the rational number

$$\tfrac{1}{2}(a/b + c/d) = (ad + bc)/2bd$$

is between a/b and c/d. This "betweenness" can be interpreted in terms of the order relation defined in Definition 6.2.3 as well as in geometrical terms. It follows that between any two distinct rational numbers there are infinitely many rational numbers. Thus, it may appear that the points on the number line corresponding to rational numbers "fill up" the number line in some intuitive sense.

Now we have already seen that there are numbers, such as $\sqrt{2}$, that are not rational. Yet our initial thought may still be that such numbers are rather exceptional and that the rational numbers fill up the bulk of the number line. This bit of intuition is wildly incorrect, and in fact the rational numbers occupy an "insignificant" portion of the number line. We will show this below. First, we need a theorem that gives us a measure of the "size" of the set of rational numbers.

THEOREM 6.2.7 The set Q of rational numbers can be put into 1–1 correspondence with the set of positive integers; that is, Q is countably infinite.

Proof Let a/b be a rational number written in lowest terms. Let $h(a/b)$ be the integer $|a| + |b|$. For example, $h(0/1) = |0| + |1| = 1$, $h(1/1) = |1| + |1| = 2$, and $h(-1/1) = |-1| + |1| = 2$, and so forth. For each positive integer n, there are just finitely many rational numbers a/b such that $h(a/b) = n$. For example, with $n = 4$, only $1/3$, $3/1$, $-1/3$, and $-3/1$ satisfy this condition. Thus we can enumerate the rational numbers as follows: $0/1$, $1/1$, $-1/1$, $1/2$, $2/1$, $-1/2$, $-2/1$, $1/3$, $3/1$, $-1/3$, $-3/1$, $1/4$, $2/3$, $3/2$, $4/1$, $-1/4$, $-2/3$, $-3/2$, $-4/1$, ..., where we list all those rational numbers with a given value $h(a/b) = n$, and give some arbitrary ordering to the finite set of rational numbers associated with each integer n. Clearly, this sets up the desired one-to-one correspondence with the positive integers, since we just have to "count" the elements as listed; that is, 0/1 corresponds to 1, 1/1 to 2, $-1/1$ to 3, 1/2 to 4, 2/1 to 5, $-1/2$ to 6, $-2/1$ to 7, and so forth. ∎

The next theorem formalizes what is meant by saying that the rational numbers take up very little of the coordinate number line.

THEOREM 6.2.8 For each integer $n > 0$ the entire set of rational numbers can be contained within a set of intervals having lengths whose sum is less than $1/n$.

Proof We shall use the listing of the rational numbers that was given in the proof of the last theorem. We first fix the positive integer n, and define I_1 to be the interval of length $1/(2n)$ and center point 0/1. I_2 is the interval of length $1/(4n)$ and center point 1/1, I_3 is the interval of length $1/(2^3 n)$ and center point $-1/1$, and in general I_j has length $1/(2^j n)$ and center point the jth rational number in our list of all rational numbers. This gives us an infinite set of intervals whose union contains all of the rational numbers. Now consider the sum of the lengths of these intervals as

$$\frac{1}{2n} + \frac{1}{2^2 n} + \cdots + \frac{1}{2^j n} + \cdots = \left(\frac{1}{n}\right)\left(\frac{1}{2} + \frac{1}{4} + \frac{1}{8} + \cdots + \frac{1}{2^j} + \cdots\right)$$

$$= \left(\frac{1}{n}\right)(1) = \frac{1}{n},$$

since the infinite series within parentheses sums to 1.

(Thus if $n = 100$, for example, then the total length of a set of intervals that contain all rational numbers is 1/100 of the length of the interval [0,1].) ∎

EXERCISES

1. Show that the field Z_p, p is prime, cannot be made into an ordered field.
2. Show that $Q(x)$, the quotient field of $Q[x]$, can be made into an ordered field.
3. Show that $Q(x)$ cannot be made into an Archimedean ordered field.
4. (a) Show that the field

$$Q[\sqrt{p}] = \{a + b\sqrt{p} \mid a, b \in Q, p \text{ a fixed prime}\}$$

can be made into an ordered field.
  (b) Show that $Q[\sqrt{p}]$ is Archimedean ordered.

5. Show that any subfield of an Archimedean ordered field is also an Archimedean ordered field.

6.3 Dedekind Cuts

We are now ready to begin the formal construction of the field of real numbers from the field of rational numbers. Among several different constructions that are available, we will use the one due to R. Dedekind that was developed in 1872. Although the method is essentially algebraic (and set-theoretic), it is best discussed in terms of the geometry of the real line. The essential feature is that since the rational numbers are dense on the line, any point can be "approximated" by rational points.

To motivate our definition, let us take a close look at the number $\sqrt{2}$.

Let S be the set of all rational numbers r such that (a) $r > 0$ and (b) $r^2 < 2$. Certainly, $S \neq \emptyset$ for $1 \in S$. We claim that S has no largest element. For let $a \in S$. Then $a^2 < 2$, and so $2 - a^2 > 0$. Also, $2a + 1$ is positive, and so by Theorem 6.2.6 there is a positive integer n such that $n(2 - a^2) > 2a + 1$. Thus

(*) $$\frac{2a+1}{n} < 2 - a^2.$$

We also note that since n is positive,

(**) $$\frac{1}{n^2} \leq \frac{1}{n}.$$

With the use of (*) and (**), we shall show that $a + 1/n$ is in S, and, of course, $a < a + 1/n$.

Now

$$\left(a + \frac{1}{n}\right)^2 = a^2 + \frac{2a}{n} + \frac{1}{n^2} \leq a^2 + \frac{2a+1}{n} \quad \text{[by (**)]}$$
$$< a^2 + 2 - a^2 \quad \text{[by (*)]}$$
$$= 2.$$

What this discussion shows is that there is not a greatest rational number less than $\sqrt{2}$. A similar argument would also show that there is not a smallest positive rational number greater than $\sqrt{2}$. (We leave this to the reader.)

Given the density of the rational number system and that there is neither a largest rational number $< \sqrt{2}$ nor a smallest rational number $> \sqrt{2}$, it is easy to convince oneself that the set S has rational numbers that are as close to $\sqrt{2}$ as we desire. Hence, the set S "determines" the point that represents $\sqrt{2}$, and indeed we can use S to define $\sqrt{2}$. As the reader will notice, the set S is defined purely in terms of rational numbers. With this motivation, we now define the notion of a Dedekind cut, which is the basic idea we use to define the real number system.

Definition 6.3.1 Let Q be the field of rational numbers. A subset D of Q is called a *Dedekind cut* if it satisfies the following properties:

(a) $D \neq \emptyset$.
(b) If $a \in D$ and if $b \in Q$, $b < a$, then $b \in D$.
(c) If $a \in D$, then there exists an element b in D such that $a < b$.
(d) There exists an element b in Q such that $a < b$, for every a in D.

We note that (b) states that D is infinite and that all numbers to the "left" of any number in D are also in D. Property (c) tells us that D does not have a largest element, and (d) tells us that D is not the set of all rational numbers. From (d), we see also that there is an infinite number of rational numbers not in D, for once we have some b greater than every element in D, every rational number greater than b is also not in D.

The idea contained in our definition is that we "cut" the coordinate line in such a way that all those rational numbers to the left of our "cut" form a Dedekind cut. We note that even though D does not have a largest element, there may nonetheless be a smallest rational number not in D. For example, if D is the set of all rational numbers less than 1, then D satisfies the definition of a Dedekind cut and 1 is the smallest rational number not in D. Thus, we can think of a Dedekind cut as "cutting" the coordinate line at a point that lies between two rational numbers, or at a point that corresponds to a rational number. It will turn out that the first type of cut corresponds to irrational numbers and the second kind to rational numbers. If A is a cut of the second type, say

$$A = \{r \in Q \mid r < s, \text{ for a fixed } s \text{ in } Q\},$$

we shall denote A by s^*. In Figure 6.2, the set of all rational numbers to the left of the point indicating $\sqrt{2}$ make up the Dedekind cut that defines $\sqrt{2}$, and the set 1^* of all rational numbers to the left of 1 is the Dedekind cut that defines the real number 1.

Definition 6.3.2 A *real number* is a Dedekind cut. The *set of all real numbers* is the set of all Dedekind cuts. We denote this set by **R**.

We will show in the next several sections that it is possible to define operations of $+$, $\cdot$, $-$, $\div$ on the set of real numbers, and that under

238   THE REAL AND COMPLEX NUMBER FIELDS

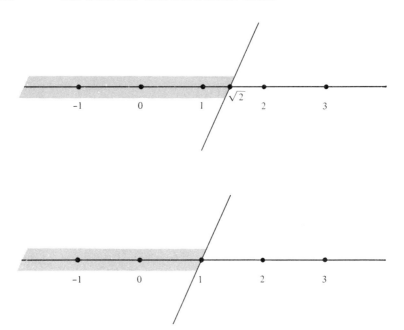

Figure 6.2

these operations the subset of the real numbers consisting of all a^*, $a \in Q$, behaves like the rational numbers and that the set of all real numbers forms an Archimedean ordered field.

One of the many important consequences of our construction will be that the equation $x^2 = a$ always has a solution in the field of real numbers if $a \geq 0$.

EXERCISES

1. Show that there is no least positive rational number a such that $a^2 > 2$.
2. Show that there is neither a smallest positive rational number a such that $a^2 > 3$ nor a largest rational number b such that $b^2 < 3$.
3. Let D be a Dedekind cut. Let $\bar{D} = Q - D$. Show that for each $a \in D$ and $b \in \bar{D}$, $a < b$.
4. Let $a \in Q$, $a > 0$. Let $D = a^* = \{s \in Q \mid s < a\}$. Show that the set $\{ns \mid s \in Q, s < a\}$ is the same as the set $(na)^*$, for $n \in Z$, $n > 0$.

6.4 The Arithmetic of the Real Numbers: (R, +) Is an Abelian Group

In this section, we define the various arithmetic operations on the set **R** of real numbers. We will omit some of the details, since many of them are easily done. But we will do enough to give the reader the flavor of the complete story. The reader can refer to W. Rudin, *Principles of Mathematical Analysis* (New York: McGraw-Hill, Inc., 1953), for an unedited version.

Elements of **R**, that is, Dedekind cuts of rational numbers, will be denoted by capital letters, generally from the front of the alphabet.

Definition 6.4.1 Let C, D be real numbers. Then

$$C + D = \{c + d \mid c \in C, d \in D\}.$$

THEOREM 6.4.2 For $C, D \in \mathbf{R}$, $C + D$ is also in **R**. Thus, + is a binary operation on **R**.

Proof We must show that $C + D$ is again a Dedekind cut. Since each c in C and d in D is a rational number, $c + d$ is a rational number, so $C + D$ is a set of rational numbers. Also, it is clear that $C + D \neq \emptyset$. Let $x \in C + D$ and let $y < x$, y in Q. We need to show that $y \in C + D$. Now $x = c + d$, for some c in C and d in D. We write $y = c + (y - c)$. Clearly, $y - c < d$, for if $y - c \geq d$, then

$$y = c + (y - c) \geq c + d = x,$$

contradicting that $y < x$. Since $y - c < d$, we have that $y - c \in D$, by the definition of D, and so y is the sum of elements from C and D, namely c and $y - c$, respectively.

Next, let x be in $C + D$. We must show that there is an element y in $C + D$ such that $x < y$. Let $x = c + d$, for some c in C and d in D. Then there are elements c' and d' in C and D, respectively, such that $c < c'$ and $d < d'$. Then with $y = c' + d'$, we get $x < y$ and $y \in C + D$.

Finally, we must show that there is an element y in Q such that $y > c + d$, for all c in C and d in D. There are elements c' and d' such that $c < c'$, for all c in C and $d < d'$, for all d in D. Put $y = c' + d'$. We must show that $c + d < y$, for all c in C and all d in D. Since $c < c'$ and $d < d'$, it follows that $c + d < c' + d' = y$. Thus, we have completed the proof that $C + D$ is a Dedekind cut. Since the sum of C and D is defined directly in terms of C and D, the cut $C + D$ is well defined and + is a binary operation on **R**. ∎

It is easy to see that addition of Dedekind cuts is associative and commutative. We leave the verification of these facts to the reader.

To see that **R** is an abelian group with respect to our definition of addition, then, it is only necessary to show the existence of a zero element and the existence of additive inverses.

THEOREM 6.4.3 The Dedekind cut $0^* = \{r \mid r \in Q \text{ and } r < 0\}$ is a zero element with respect to addition.

Proof Let C be any Dedekind cut. Then

$$C + 0^* = \{r + s \mid r \in C, s \in Q, s < 0\}.$$

We need to show that this set is exactly C. Let $x \in C$. There exists an a in C such that $x < a$. Then $(x + a)/2 < a$, and so $(x + a)/2$ is also in C. Now $(x + a)/2 - x$ is positive; that is, $(a - x)/2$ is positive. Thus $(x - a)/2$ is negative; that is, $(x - a)/2 \in 0^*$. Finally,

$$x = \frac{x+a}{2} + \frac{x-a}{2}$$

is the sum of elements from C and 0^*. Thus, $C \subseteq C + 0^*$. For c in C and d in 0^*, $d < 0$, so $c + d < c$, whence $c + d$ is in C, by property (b) of the definition of Dedekind cut. Thus

$$C + 0^* \subseteq C \quad \text{and so} \quad C = C + 0^*. \qquad \blacksquare$$

Definition 6.4.4 Let C be a Dedekind cut. Let

$$\bar{C} = \{r \in Q \mid r \notin C\} = Q - C.$$

Define $-C = \{s \in Q \mid s < -r, \text{ for some } r \text{ in } \bar{C}\}$.

We shall show that $-C$ is a Dedekind cut and that $C + -C = 0^*$. Before doing this, let us examine two specific examples.

First, let $A = 2^*$, that is, $A = \{r \in Q \mid r < 2\}$. Then

$$\bar{A} = \{r \in Q \mid r \geq 2\}.$$

Then, for any r in $\bar{A}$, $-r \leq -2$. Since $-A = \{s \in Q \mid s < -r\}$, for some r in $\bar{A}$, we see that $-A = \{s \in Q \mid s < -2\}$, as we would expect. Thus if $A = 2^*$, $-A = (-2)^*$.

Next, let $A = \{r \in Q \mid r \leq 0 \text{ or } r^2 < 2\}$. This, of course, is the Dedekind cut that corresponds to $\sqrt{2}$. Then

$$-A = \{s \in Q \mid s < -r \text{ for some } r \text{ in } \bar{A}\}.$$

Since intuitively $\bar{A}$ is the set of all rational numbers $r > \sqrt{2}$, $-r < -\sqrt{2}$. And since there is no smallest rational number greater than $\sqrt{2}$,

6.4 (R, +) IS AN ABELIAN GROUP 241

there is no largest rational number $-r$ less than $-\sqrt{2}$. Thus $-A$ is the set of all rational numbers $r < -\sqrt{2}$. Thus $-A$ will correspond to $-\sqrt{2}$. We have, of course, relied heavily on intuition in this example, and we will not verify all of the details explicitly.

THEOREM 6.4.5 Let C be a Dedekind cut. Then $-C$ is a Dedekind cut.

Proof It is clear that $-C \neq \emptyset$. Next, let $a \in -C$ and let $b < a$. We need to show that $b \in -C$. Since $a \in -C$, there is an r in $\bar{C}$ such that $a < -r$. But then $b < -r$, and so by definition $b \in -C$.

Suppose now $b \in -C$. We must show that there exists $c \in -C$, with $b < c$. There is an r in $\bar{C}$ such that $b < -r$. But $c = (b + (-r))/2$ satisfies two properties: $c < -r$, so $c \in -C$, and $b < c$.

To conclude the proof that $-C$ is a Dedekind cut, it is necessary to show that there exists a rational number b such that $a < b$, for all a in $-C$. Let $d \in C$. If $-d \in -C$, then $-d < -c$, for some element c in $\bar{C}$; that is, $c < d$, where $d \in C$. This implies that $c \in C$, but we know that $c \in \bar{C}$. Thus $-d \notin -C$. Now let $b = -d$, and suppose there is an $a \in -C$ such that $b \leq a$. Since $a \in -C$, there is an $r \in \bar{C}$ such that $a < -r$. Thus, $-d = b \leq a < -r$, whence $r < d$. Since $d \in C$ and C is a Dedekind cut, we have $r \in C$, contradicting that $r \in \bar{C}$. Thus, $a < b = -d$ for all $a \in -C$, and so $-C$ is a Dedekind cut. ∎

Before proving the next theorem, namely, that $-C$ is the additive inverse of C, we need a lemma stating that there are elements in the Dedekind cut C and elements not in C that are arbitrarily close. That is, the lemma will show that if l is a positive rational number representing a distance, then there is an element x in C and an element y in $\bar{C}$ such that $y - x$ is less than l.

LEMMA 6.4.6 Let C be a Dedekind cut and let r be a positive rational number. Then there exists a rational number c in C such that $c + r \in \bar{C}$.

Proof

CASE 1 C contains positive rational numbers. If $r \in \bar{C}$, then, since C has some positive rational numbers, $0 \in C$, and so $0 + r = r \in \bar{C}$, that is, we can take $c = 0$. If $r \notin \bar{C}$, then $r \in C$. Let b be some element of $\bar{C}$. Then b is positive, and by the Archimedean property, there is a positive integer n such that $nr > b$. Thus $nr \in \bar{C}$. Hence there is a least positive integer m such that $mr \in \bar{C}$, and so $(m-1)r \in C$. But then take $c = (m-1)r$. Clearly,

$$c + r = (m-1)r + r = mr \in \bar{C},$$

and Case 1 is completed.

CASE 2 C has no positive rational numbers. Suppose $-q \in C$, with $q > 0$. Then the set

$$C' = \{c + (q + 1) \mid c \in C\}$$

is easily seen to be a Dedekind cut. And certainly C' has positive elements, since if $c > -q$, then $c + q + 1 > 1$. Then there is an element c' in C' such that $c' + r \in \overline{C'}$; that is, $c' + r > c''$, for all c'' in C'. Now

$$c' = c + (q + 1), \quad \text{for some } c \text{ in } C.$$

We claim that $c + r \in \bar{C}$. If not, $c + r \in C$, whence

$$c + r + (q + 1) = c' + r \in C',$$

a contradiction. Thus the theorem is proved for Case 2. ∎

THEOREM 6.4.7 If C is a Dedekind cut, then

$$C + (-C) = (-C) + C = 0^*.$$

Proof We have already mentioned that the commutativity of addition of Dedekind cuts is easily proven. Thus, it suffices to show

$$C + (-C) = 0^*.$$

Let $a \in C$ and $b \in -C$. Then $b < -c$, for some $c \in \bar{C}$, and so $a + b < a + (-c)$. Since $c \in \bar{C}$, $d < c$, for all d in C. In particular, $a < c$. Thus, $a - c < 0$, so

$$a + b < a + (-c) < 0, \quad \text{and so} \quad a + b \in 0^*.$$

We have shown $C + (-C) \subseteq 0^*$.

Now suppose that $d \in 0^*$. Then $d < 0$, and $-d/2 > 0$. By Lemma 6.4.6, there is a c in C such that $c + (-d/2) \in \bar{C}$, say

$$c + \frac{-d}{2} = c' \in \bar{C}.$$

Then $c + (d/2) = c' + d$, whence

$$d = c + \left(\frac{d}{2} - c'\right).$$

But $d/2 < 0$, so $d/2 - c' < -c'$, and since $c' \in \bar{C}$, $d/2 - c' \in -C$. Thus,

$$d = c + \left(\frac{d}{2} - c'\right) \in C + (-C),$$

and so
$$0^* \subseteq C + (-C).$$
Thus we now conclude that $0^* = C + (-C)$. ∎

We can summarize Theorems 6.4.2 through 6.4.7 as follows.

THEOREM 6.4.8 The set of Dedekind cuts is an abelian group with respect to the addition of Dedekind cuts. The zero element of the group is 0^*, and the additive inverse of C is $-C$.

From this theorem, we can conclude such things as $-(0^*) = 0^*$, and $A + B = A + C$ implies $B = C$. That is, anything we know about abelian groups holds with respect to addition of Dedekind cuts.

EXERCISES

1. Show explicitly without using Theorem 6.4.7 that $2^* + (-2)^* = 0^*$.
2. Prove that addition of Dedekind cuts is associative.
3. Prove that addition of Dedekind cuts is commutative.
4. Prove without use of group properties that
$$-(A + B) = (-A) + (-B),$$
for Dedekind cuts A and B.
5. Let S be a Dedekind cut. Suppose that $\bar{S} = Q - S$ has no least element. Show that
$$-S = \{-x \in Q \mid x \in \bar{S}\}.$$
How can you similarly describe $-S$ in case $\bar{S}$ has a least element?

6.5 The Arithmetic of Real Numbers: Multiplication and Order

It will be convenient to define an order for the set of all Dedekind cuts and derive some of its properties before we define multiplication.

Definition 6.5.1 A Dedekind cut C will be called *positive* if C contains at least one positive rational number.

Notice that a positive cut must therefore contain infinitely many positive rational numbers.

THEOREM 6.5.2 (*Law of Trichotomy*) Let $\mathcal{P}$ be the set of all positive Dedekind cuts. Then for the Dedekind cut C, exactly one of the following is true: $C \in \mathcal{P}$, $C = 0^*$, $-C \in \mathcal{P}$.

Proof If C has positive elements, clearly it is in $\mathcal{P}$. Suppose C has no positive elements. If C consists of exactly all the rational numbers less than 0, then $C = 0^*$. Clearly, the first two possibilities are exclusive. Finally, suppose C has no positive rational numbers and C does not contain all the negative rational numbers. Then there is a rational number q not in C such that $q < 0$. Then $-q > 0$, and by definition of $-C$, each rational number $r < -q$ is in $-C$. Since $-q/2 < -q$, $-q/2$ is in $-C$, and $-q/2$ is positive. Thus, $-C$ is positive; that is $-C \in \mathcal{P}$. We must still show that no C can satisfy two of the above properties. Clearly, if $C \in \mathcal{P}$, then $C \neq 0^*$, since C has positive elements and 0^* does not. Also, if $C = 0^*$, then $-C = -0^* = 0^*$, so the only possibility left that needs examination is $C \in \mathcal{P}$ and $-C \in \mathcal{P}$. Now if $C \in \mathcal{P}$, then C has positive elements. Let $x \in -C$. Then $x < -r$, for some r not in C. But r not in C implies r is positive, whence $-r$ is negative. Thus, x is also negative, and so $-C$ does not have any positive elements. That is, $-C \notin \mathcal{P}$. ∎

Definition 6.5.3 Let A and B be Dedekind cuts. We define $A < B$ if $B + (-A) \in \mathcal{P}$.

In terms of this definition, we can state an expected theorem.

THEOREM 6.5.4 Let A be a Dedekind cut. Then exactly one of the following is true: $0^* < A$, $0^* = A$, $0^* > A$ (where $A < B$ and $B > A$ are equivalent).

Proof It is easy to see that this is just a restatement of the law of trichotomy. ∎

For $A < 0^*$, we shall call A a *negative* Dedekind cut (or a *negative* real number).

A convenient characterization of $A < B$ is given in the next result. We shall make use of it later.

THEOREM 6.5.5 Let A and B be Dedekind cuts. Then $A < B$ if and only if $A \subset B$.

Proof Suppose that $A < B$. Then $B + (-A) \in \mathcal{P}$. Thus there is a rational number r in $B + (-A)$, $r > 0$. But r is of the form $b + a$, where $b \in B$ and $a \in -A$; that is, $a < -a'$, for some $a' \in \bar{A}$. Since $b + a = r > 0$, we have $b > -a$, whence $b > -a > a'$, whence $b \notin A$—for otherwise $a' \in A$, contradicting $a' \in \bar{A}$. Thus we have shown that $A \neq B$. Now for any c in A, $c < a'$ and $a' < b$ for a' and b as above. Hence $c < b$, and by the definition of B as a Dedekind cut, $c \in B$. Thus $A \subset B$.

Now suppose that $A \subset B$. Then there exists b in B such that $b \notin A$, that is, such that $a < b$, for all a in A. Now since B is a Dedekind cut, we can find rational numbers b_1 and b_2 in B such that $b < b_1 < b_2$. Then $-b_1 < -b$, so by definition of $-A$, $-b_1 \in -A$. Thus $b_2 + (-b_1) \in B + (-A)$, but clearly $b_2 - b_1$ is positive, so $B + (-A)$ is in $\mathcal{P}$, that is, $A < B$. ∎

Definition 6.5.6 Let A be a Dedekind cut. We shall denote the set of positive rational numbers contained in A by $p(A)$.

We note that if $A \in \mathcal{P}$, then A is completely determined by the subset $p(A)$. For A is the union of $p(A)$ and the set of all nonpositive rational numbers.

We are now in a position to define the product of two Dedekind cuts.

Definition 6.5.7

(a) Let A and B be Dedekind cuts such that $A \in \mathcal{P}$ and $B \in \mathcal{P}$. Then AB is the Dedekind cut whose set of positive elements is $p(AB) = \{ab \mid a \in p(A), b \in p(B)\}$.
(b) Let A be a Dedekind cut. Then $A0^* = 0^*$.
(c) Let A, B be Dedekind cuts such that $A \in \mathcal{P}$ and $-B \in \mathcal{P}$. Then $AB = -(A(-B))$.
(d) Let A, B be Dedekind cuts such that $-A \in \mathcal{P}$ and $B \in \mathcal{P}$. Then $AB = -((-A)(B))$.
(e) Let A and B be Dedekind cuts such that $-A \in \mathcal{P}$, $-B \in \mathcal{P}$. Then $AB = (-A)(-B)$.

We observe that parts (c) through (e) of Definition 6.5.7 use the fact that if A is a Dedekind cut, then so is $-A$, as defined in Definition 6.4.4. And when $-A \in \mathcal{P}$, then we can use part (a) of the definition to make the definitions of parts (c) through (e) make sense.

It is still necessary to verify that the product of part (a) is indeed a Dedekind cut. Once that is done, then all the other products must be Dedekind cuts.

THEOREM 6.5.8 Let A, B be Dedekind cuts such that $A \in \mathcal{P}$ and $B \in \mathcal{P}$. Then AB is a Dedekind cut.

Proof Clearly, AB is not empty. Now let $c \in AB$. We must show that if $d \in Q$ satisfies $d < c$, then $d \in AB$. If c or d is negative, then by the way we defined AB as the set of rational numbers such that AB is the union of the nonpositive rational numbers and the positive numbers ab, where $a \in p(A)$ and $b \in p(B)$, we see that $d \in AB$. Thus, the interesting case is when c and d are both positive. Since $c \in AB$ and is positive, $c = ab$, where a and b are positive elements in A and B, respectively.

Thus, we have $0 < d < ab$. Then d/a is positive and $d/a < b$. Thus $d/a \in B$. From this, $d = a(d/a) \in AB$.

Next, let $c \in AB$. We must show that there is an element d in AB such that $c < d$. Again, there is no loss in generality in assuming that $c > 0$. Again, $c = ab$, with $a \in A$ and $b \in B$, with a and b both positive. Then there exists a' in A with $a < a'$ and b' in B with $b < b'$. Then $d = a'b' > ab = c$ and $d \in AB$.

Finally, we must show that there exists d not in AB such that $c < d$, for all c in AB. Once more using the fact that A and B are Dedekind cuts, there exists a' not in A such that $a < a'$, for all a in A, and there exists b' not in B such that $b < b'$, for all b in B. Necessarily, a' and b' are positive. Let $d = a'b'$. Then for all positive a and b in A and B, respectively, $ab < a'b' = d$. From this, we conclude that $c < d$, for all c in AB. This concludes the proof that AB is indeed a Dedekind cut. ∎

Now that we have the definition of multiplication, it is possible to prove the following.

THEOREM 6.5.9 The set **R** of all real numbers, that is, the set of all Dedekind cuts, is a commutative ring with identity with respect to the definitions of addition and multiplication given above.

Proof Most of the details of this proof will be left to the reader. We make the following observations.

First, $(\mathbf{R}, +)$ was shown to be an abelian group earlier. Second, multiplication is a binary operation and is obviously commutative. Third, the identity of the ring is the Dedekind cut $1^* = \{r \in Q \mid r < 1\}$. Multiplication is easily shown to be associative. Finally the distributive law, although nontrivial, also holds and is left to the reader. ∎

EXERCISES

1. Verify that $A1^* = A$, for all real numbers A.
2. Verify that multiplication of real numbers (Dedekind cuts) is associative.
3. Verify that $A(B + C) = AB + AC$, for Dedekind cuts A, B, and C.

6.6 The Arithmetic of the Real Numbers: R Is a Field

Our final goal, so far as arithmetic is concerned, is to prove that division can be performed within **R** and that **R** is a field. To do this, it is sufficient to show that the Dedekind cut $A \neq 0^*$ has an inverse A^{-1}, with respect to multiplication, where A^{-1} is also a Dedekind cut. It is necessary to

6.6 THE ARITHMETIC OF THE REAL NUMBERS: R IS A FIELD

define A^{-1} only for the case that $A > 0^*$, for when $-A > 0^*$ we shall simply put $A^{-1} = -(-A)^{-1}$.

Definition 6.6.1 Let A be a Dedekind cut, $A \neq 0^*$.

(a) If $A > 0^*$, then A^{-1} is the Dedekind cut whose set of positive elements is $p(A^{-1}) = \{r \in Q, r > 0 \mid r < 1/a'$, for some a' not in $A\}$. [Note that for $r > 0$, $r \in p(A^{-1})$ if and only if there is a rational number a' that is not in A and such that $r < 1/a'$.]

(b) If $-A > 0^*$, then $A^{-1} = -(-A)^{-1}$.

To show that our definition is appropriate, we must verify two things: First, we must show that A^{-1} is in fact a Dedekind cut, and second, we must show that $AA^{-1} = 1^*$. We first consider two examples.

Suppose that the Dedekind cut is $A = 2^* = \{r \in Q \mid r < 2\}$. To determine the positive rational numbers in the Dedekind cut A^{-1}, let a' not be in A. Then $a' \geq 2$, and $1/a' \leq \frac{1}{2}$. By definition, we now have that any rational number $r < \frac{1}{2}$ must be in A^{-1}. And indeed, the set of all such r determines the set of positive elements of A^{-1}. Thus $A^{-1} = (\frac{1}{2})^*$, which is what we would expect.

Now let $A = (-2)^* = \{r \in Q \mid r < -2\}$. By definition, A^{-1} is $-(-A)^{-1}$. Now $-A$ is exactly the Dedekind cut 2^* of the above paragraph, and it is easy to see that the Dedekind cut A^{-1} in this paragraph is $(-\frac{1}{2})^*$.

THEOREM 6.6.2 If A is a Dedekind cut, $A \neq 0^*$, then A^{-1} is a Dedekind cut.

Proof We need only consider the case that $A > 0^*$, for that case and the second part of Definition 6.6.1 immediately imply the theorem for $A < 0^*$.

(a) Since A is a positive Dedekind cut, there exists a positive rational number a' such that $a' \notin A$. Then $1/a' > 0$ also, so each rational number $r < 1/a'$ is in A^{-1}. Thus $A^{-1} \neq \emptyset$.

(b) Let $r \in A^{-1}$ and let $b < r$, b rational. Then if $r < 1/a'$ for some $a' \notin A$, clearly, $b < 1/a'$, whence $b \in A^{-1}$.

(c) Suppose $r \in A^{-1}$. We must show there is an element $s \in A^{-1}$ such that $r < s$. Now again there is an $a' > 0$ such that $a' \notin A$ and such that $r < 1/a'$. Then, by the density of the rational numbers, there is a rational number s such that $r < s < 1/a'$. Thus $s \in A^{-1}$.

(d) Finally, we must show that there is an element $s \in Q$ such that $r < s$ for every $r \in A^{-1}$. Since A is a Dedekind cut, for every $a' \notin A$ and every $w \in A$, we have $w < a'$. Then $1/a' < 1/w$, so $1/w \notin A^{-1}$. We set $s = 1/w$, for some fixed $w \in A$, $w > 0$. To see that $r < s$, for all $r \in A^{-1}$, suppose that $r > 0$ and

$r \geq s = 1/w$. Since $r \in A^{-1}$, $r > 0$, there is an $a' \in A$ such that $r < 1/a'$. Thus $1/a' > r \geq 1/w$, whence $w > a'$. But this is impossible, for $w \in A$ and $a' \notin A$ implies $w < a'$. Thus $r < 1/w$, for all $r \in A^{-1}$, $r > 0$, and hence $r < 1/w = s$, for all $r \in A^{-1}$.

This completes the proof that A^{-1} is a Dedekind cut. ∎

THEOREM 6.6.3   Let A be a Dedekind cut, $A \neq 0^*$. Then $AA^{-1} = 1^*$.

Proof   It suffices to assume $A > 0^*$. Then, in this case, $A^{-1} > 0^*$ also. Let $a \in p(A)$, $b \in p(A^{-1})$. Then $b < 1/a'$ for some $a' \notin A$. Thus

$$0 < ab < a\left(\frac{1}{a'}\right) < 1,$$

since $a < a'$, also. Thus, $p(AA^{-1}) \subseteq 1^*$. And so $AA^{-1} \subseteq 1^*$.

Now let $c \in 1^*$, $c > 0$. Then $0 < c < 1$, as rational numbers. Let $a \in p(A)$. Then $1 - c > 0$ and $(1 - c)a > 0$. By Lemma 6.4.6, there exists $b \in A$ such that $b + (1 - c)a \notin A$, that is, $b + (1 - c)a = d$, for some $d \notin A$. We may assume $b > 0$, for if $b \leq 0$, we can always replace it by a larger value.

Thus since $d \notin A$, we have $b < d$ and $a < d$, so

$$0 < d - b = (1 - c)a < (1 - c)d,$$

whence

$$d - b + cd < (1 - c)d + cd = d.$$

From this, $cd - b < 0$, so $cd < b$, whence $d < b/c$. Thus, $c/b < 1/d$, and as $d \notin A$, we have $c/b \in A^{-1}$. Thus, $c/b = e$, for some $e \in A^{-1}$, so $c = be$, where $b \in A$, $e \in A^{-1}$. Thus, $1^* \subseteq AA^{-1}$ and so $AA^{-1} = 1^*$, as we set out to prove. ∎

With Theorem 6.6.3 proven, we can now summarize the previous results of this chapter.

THEOREM 6.6.4   The set **R** of all real numbers is a field.

EXERCISES

1. Show directly that $2^*(\frac{1}{2})^* = 1^*$.
2. Show directly that $(-2)^*(-\frac{1}{2})^* = 1^*$.
3. Prove that for the Dedekind cut A, $A < 0^*$, $AA^{-1} = 1^*$.

6.7   The Real Numbers as an Ordered Field Extension of the Rational Numbers

We showed in the previous section that the field of real numbers can be defined as the set of all Dedekind cuts of rational numbers.

6.7 THE REAL NUMBERS AS AN ORDERED FIELD EXTENSION 249

In this section we first show that the set of real numbers is an Archimedean ordered field. Next, we show that the rational numbers are embedded in the field of real numbers in a very natural way.

THEOREM 6.7.1 The field **R** of real numbers is an Archimedean ordered field.

Proof First, we remark that the field **R** of real numbers is ordered. The set $\mathcal{P}$ of positive real numbers is the set of positive elements in **R**. It is easy to see that for A and B in $\mathcal{P}$, $A + B$ is in $\mathcal{P}$. And Definition 6.5.7 immediately implies that for A and B in $\mathcal{P}$, the product AB is in $\mathcal{P}$. The law of trichotomy was proven in Theorem 6.5.2. Thus, **R** is an ordered field.

Now we must show the Archimedean property of **R**. Let A and $B \in$ **R**. We must find an integer n such that $nA > B$. [Here nA is interpreted as the nth multiple of A in the additive group (**R**, +).] If $B < A$, we let $n = 1$. Thus, assume that $A \leq B$, or equivalently, that $A \subseteq B$. There exists a rational number b' such that $b < b'$, for all b in B. Clearly, $b' > a$, for all a in A. Let a be a fixed element of A, $a > 0$. Then, by the Archimedean property for the rational numbers, there exists an integer n such that $na > b'$. This implies that the Dedekind cut $nA > B$. Thus **R** is an Archimedean ordered field. ∎

THEOREM 6.7.2 The field of rational numbers can be embedded in the field of real numbers so that addition, multiplication, and order are preserved.

Proof Let Q be the field of rational numbers. As we observed earlier, certain real numbers are of the form r^*, where r is a rational number. We define a function $f: Q \longrightarrow$ **R**, as follows: $f(r) = r^*$. The following are all easy to verify, and are thus left to the reader. For all a and b in Q, $f(a + b) = f(a) + f(b)$, that is,

$$(a + b)^* = a^* + b^*.$$

Also, $f(ab) = f(a)f(b)$, that is,

$$(ab)^* = a^*b^*.$$

Finally, if $a < b$, as rational numbers, then $f(a) < f(b)$ as real numbers; that is, $a < b$ implies $a^* < b^*$. The last property is especially easy, since $a < b$ implies that the Dedekind cut a^* is a subset of the Dedekind cut b^*. ∎

Given this embedding, we shall henceforth sometimes write a for the real number a^*, $a \in Q$. That is, although real numbers have been defined as sets of rational numbers, we shall use that rational number a

(which is *not* in a^*) as the symbol denoting a^*, when no confusion may arise. Also, we shall refer to the subset of all elements in **R** of the form a^* as *the* subset (of **R**) of rational numbers. For example, we shall not distinguish between the rational number $\frac{1}{3}$ and the real number

$$(\tfrac{1}{3})^* = \{a \in Q \mid a < \tfrac{1}{3}\}.$$

EXERCISE

1. Prove Theorem 6.7.2.

6.8 Completeness of the Real Numbers

As we remarked in Section 6.1, one of our main purposes in seeking to enlarge the field of rational numbers was to produce an ordered field whose elements could be put into one-to-one correspondence with the points on a number line, preserving at the same time certain properties of the rational number system. In particular we needed to determine a "number" that would correspond to the length of the isosceles right triangle whose legs had length 1.

THEOREM 6.8.1 There exists a real number S such that $S^2 = 2$.

Proof Let S be the set of rational numbers defined as follows: S is the union of the nonpositive rational numbers, and the set of all positive rational numbers s such that $s^2 < 2$.

From our discussion in Section 6.3, it is clear that this set S is a Dedekind cut. We will show that this real number S satisfies $S^2 = 2$. Thus, we must show that

$$S^2 = \{r \in Q \mid r < 2\} = 2^*.$$

S^2 is determined by its positive elements, which are of the form st, $s, t \in p(S)$. Since $s^2 < 2$ and $t^2 < 2$, $(st)^2 < 4$, and so $st < 2$. Thus,

$$p(S^2) \subseteq 2^*, \quad \text{whence } S^2 \subseteq 2^*.$$

The proof is finished if we show that the inclusion is not proper.

In case $S^2 \subset 2^*$, there is a rational number r such that $st < r < 2$, for all s, t in $p(S)$; that is, $r \notin S^2$. This implies that $s^2 < r < 2$, for all s in $p(S)$. We shall show this last statement cannot hold.

First observe that $s < r$, for all s in $p(S)$, for if $r \leq s$, then $r^2 \leq s^2 < 2$, and $r \in S$. But then $r = r \cdot 1$ would also be in S^2, contradicting that $r \notin S^2$.

6.8 COMPLETENESS OF THE REAL NUMBERS

Since $r < 2$, $2 - r > 0$. Also, $r > 0$ and so $2r + 1 > 0$. By the Archimedean property for rational numbers, there exists a positive integer n such that $(2 - r)n > 2r + 1$, whence

$$\frac{2r+1}{n} < 2 - r.$$

Since $1/n^2 \leq 1/n$, it follows that

$$\frac{2r}{n} + \frac{1}{n^2} < 2 - r.$$

Now since S is a Dedekind cut, there is an element s in S such that $s + (1/n) \notin S$, by Lemma 6.4.6. Thus

$$\left(s + \frac{1}{n}\right)^2 > 2.$$

On the other hand,

$$\left(s + \frac{1}{n}\right)^2 = s^2 + \frac{2s}{n} + \frac{1}{n^2} < s^2 + \frac{2r}{n} + \frac{1}{n^2}$$
$$< r + \left(\frac{2r}{n} + \frac{1}{n^2}\right) < r + (2 - r)$$
$$= 2,$$

which is a contradiction. Thus there is no r such that $s^2 < r$, for all r in $p(S)$, and hence $S^2 = 2^*$. ∎

In a similar manner we could now prove that for each rational $q > 0$, and for each integer $n > 0$, there exists a real number T such that $T^n = q$. We will not do this, but will leave some special cases for the exercises.

In learning calculus, the reader undoubtedly studied the least-upper-bound property of the real number system, probably taking this property as an axiom of the real numbers. We shall prove that the system we have developed satisfies this property.

Definition 6.8.2 Let X be a nonempty subset of the set of all real numbers. Then X is said to have an *upper bound* U if (a) $U \in \mathbf{R}$, and (b) $A \leq U$, for all A in X. An upper bound U is called the *least upper bound* for X if $U \leq U'$, for any upper bound U'.

It is easy to see that if a least upper bound exists, then it is unique. On the other hand, there are many unbounded sets of real numbers; for example, the set of all integers, or the set of all rational numbers, or the set of all prime numbers.

THEOREM 6.8.3 *(Completeness)* Let X be a nonempty subset of the real number system such that X has an upper bound. Then X has a least upper bound.

Proof Let X be a set of real numbers. Each real number A in the set X is a Dedekind cut. Thus, the union of all real numbers A in X is a set of rational numbers. We denote this union by $\cup\, X$, and we will show that $\cup\, X$ is both a Dedekind cut and the least upper bound for the set X. First, it is clear that $\cup\, X$ is nonempty, since X has at least one element; that is, there is at least one Dedekind cut that is a member of X. Next, let $a \in \cup\, X$, and let $b \in Q$ satisfy $b < a$. Since $a \in \cup\, X$, there is an A in X such that $a \in A$. Since $b < a$, then $b \in A$, as A is a Dedekind cut. Thus, $b \in \cup\, X$, also. Next, if $a \in \cup\, X$, then again $a \in A$, for some A in X, whence there is a b in A such that $a < b$. But then $b \in \cup\, X$, and so for $a \in \cup\, X$, there is a $b \in \cup\, X$, such that $a < b$. To finally conclude that $\cup\, X$ is a Dedekind cut, we must show that there exists a rational number b such that $a < b$, for all a in $\cup\, X$. Since X has an upper bound, let C be such an upper bound. Then, since C is a Dedekind cut, there is a rational number b such that $c < b$, for all rational numbers c in C. Since C is an upper bound for X, $A \leq C$, for all A in X, that is, $A \subseteq C$, for all A in X. Thus, $a < b$, for all a in A. But then clearly, $a < b$, for all a in $\cup\, X$. Thus we can now conclude that $\cup\, X$ is a Dedekind cut.

It is immediate that $\cup\, X$ is an upper bound for X, since each A in X is a subset of $\cup\, X$. Finally, to see that $\cup\, X$ is the least upper bound for X, observe that if D is any upper bound for X, then $A \subseteq D$, for all A in X, whence $\cup\, X \subseteq D$. Thus, the least-upper-bound property holds for the set **R**. ∎

Another way of expressing that **R** satisfies the least-upper-bound property is

Definition 6.8.4 An ordered field F is *complete* if each bounded subset X of F has a least upper bound.

Definition 6.8.5 Let A be a real number, and let n be a positive integer. Then $A^{1/n}$ is that real number B, if it exists, such that $B^n = A$.

THEOREM 6.8.6 Let A be a real number, and let n be a positive integer. Then

(a) If A is positive, $A^{1/n}$ exists.
(b) If A is negative, then $A^{1/n}$ exists if n is odd, and $A^{1/n}$ does not exist if n is even.
(c) If $A = 0^*$, then $A^{1/n} = 0^*$.

6.9 THE REAL NUMBER SYSTEM AND DECIMAL NOTATION

Proof (a) Let S be the set of all those real numbers q^*, q rational, such that $(q^*)^n \leq A$. Clearly, S has an upper bound, and so by the completeness property of the real number system, S has a least upper bound. It is this least upper bound that we denote $A^{1/n}$. We leave as an exercise the verification that $(A^{1/n})^n = A$, for the real number $A^{1/n}$.

(b) The definition of $A^{1/n}$ for the case that $A < 0$ and n is odd is similar to that used in (a), and we leave the details to the reader.

(c) The case is clear. ∎

With the construction of the field of real numbers now completed, it is natural to ask whether it is possible to further enlarge the real number system by repeating the process of forming Dedekind cuts, where now we use the real numbers in place of the rational numbers. The answer is that such a process would result in a field that is isomorphic to the real number system; hence nothing new is obtained. This can be considered another interpretation of the word "complete." The reader should recall that a similar phenomenon occurred when the quotient field of the field of rational numbers turned out to be isomorphic to the field of rational numbers.

We conclude the section with an important theorem that we state without proof. This theorem gives a complete characterization of the field of real numbers.

THEOREM 6.8.7 Let F be a complete Archimedean ordered field. Then $F \cong \mathbf{R}$, the field of real numbers.

EXERCISES

1. Show that if $\chi \subseteq \mathbf{R}$ has a least upper bound U, U is unique.
2. A *lower bound* for a set $\chi \subseteq \mathbf{R}$, $\chi \neq \emptyset$, is a real number L such that $L \leq X$, for all $X \in \chi$. A *greatest lower bound* for χ is a lower bound L such that $L \geq L'$, for any lower bound L' of χ. Show that if $\chi \neq \emptyset$, and if χ has a lower bound, it has a greatest lower bound.
3. Let A be a real number, $A > 0$. Let $n \in \mathbf{Z}$, $n > 0$. Show that $\{q \in \mathbf{Q} \mid q^n < A\} \cup 0^*$ is a Dedekind cut.
4. Complete the proof of Theorem 6.8.6.

6.9 The Real Number System and Decimal Notation

To this point we have defined the set of real numbers as the set of all Dedekind cuts of rational numbers. In this section we will relate the Dedekind cut definition of real number to the more traditional description of a real number as a decimal expansion.

First, suppose that we are given an infinite decimal expansion. To simplify we shall assume that the expansion is positive and that it lies between 0 and 1. Then we can assume that the expansion looks like $.a_1a_2a_3\ldots$, where each a_i is one of the digits 0, 1, 2, 3, 4, 5, 6, 7, 8, 9. For each subscript n, we form a real number A_n as follows: A_n is the Dedekind cut consisting of all rational numbers q such that

$$q < \frac{a_1}{10} + \frac{a_2}{100} + \frac{a_3}{1000} + \cdots + \frac{a_n}{(10)^n},$$

that is, if p is this last rational number, $A_n = p^*$. Then let χ be the set of all A_n's. χ has an upper bound, so χ has a least upper bound A. It is this least upper bound A that is the real number that corresponds to this infinite decimal expansion.

Next, suppose that we are given a real number A, where A lies strictly between 0 and 1. We derive a decimal expansion from A in the following manner. Since $A < 1$, A lies between $\frac{0}{10}$ and $\frac{1}{10}$, or A lies between $\frac{1}{10}$ and $\frac{2}{10}$, and so forth, until we get $\frac{9}{10}$ and 1. Say that A satisfies $.k \leq A < .(k+1)$, where k is one of the integers 0 through 9 and $.(k+1) = 1$ if $k = 9$. If $A \neq .k$, then by the same argument we can find an integer j, $0 \leq j \leq 9$, such that $.kj \leq A < .k(j+1)$, with $.k(j+1) = .(k+1)$ if $j = 9$. Again, if $A \neq .kj$, we continue. If at some stage of this process we achieve equality, then we say that A has a finite decimal expansion. Otherwise, we say that A has an infinite decimal expansion. For example, to get the decimal expansion for $\sqrt{2}$, we see that $(1.4)^2 < 2$, whereas $(1.5)^2 > 2$, so 1.4 is the beginning of the decimal expansion for $\sqrt{2}$. Next, we can check and see that $1.41 < \sqrt{2} < 1.42$. Next, we find that $1.414 < \sqrt{2} < 1.415$, and so forth, getting explicitly as many of the digits as we have time to find.

To further illustrate, suppose we wish to use the present technique to find the decimal expansion for $\frac{1}{3}$. We observe directly that $\frac{3}{10} < \frac{1}{3} < \frac{4}{10}$, so the expansion will begin $.3$. Next, we again observe directly that $.33 < \frac{1}{3} < .34$, so now we know that the expansion begins with $.33$. And at each state, by elementary techniques only, we can observe directly that $.333\ldots333 < \frac{1}{3} < .333\ldots334$. In this manner, we see that the correct expansion for $\frac{1}{3}$ is $.333\ldots333\ldots$.

We now return to the starting point of our discussion back in Section 6.1. There we motivated our discussion by pointing out that there are gaps in the rational number system in the sense that there are points on a coordinate line that do not correspond to rational numbers. Our introduction of Dedekind cuts as real numbers was to overcome this deficiency. We can now easily indicate how the points on a coordinate line can be put into one-to-one correspondence with the set of all Dedekind cuts, that is, the set of all real numbers. We use the fact, shown above, that

6.10 THE COMPLEX NUMBERS C; POLYNOMIALS OVER R AND C

an infinite decimal expansion is nothing more than a representation of a real number.

Let P be a point on the coordinate line. For simplification, assume that the point P lies between 0 and 1. Then, in a manner similar to our deriving the decimal expansion for the rational number $\frac{1}{3}$, we see that P is between two rational numbers of the form $.a_1$ and $.(a_1 + 1)$, where a_1 is one of the digits 0, 1, ..., 9. [If $a_1 = 9$, interpret $.(a_1 + 1)$ as 1.0.] Then, similarly, P lies between the rational numbers $.a_1a_2$ and $.a_1(a_2 + 1)$, where a_2 is again one of the digits 0, 1, ..., 9. We continue in this manner, getting longer and longer rational numbers $.a_1a_2\cdots a_n$. Now the process stops if for some integer N it turns out that P corresponds (is equal) to one of the finite decimals $.a_1a_2\cdots a_N$, for some integer N. Otherwise, we obtain an infinite decimal expansion $.a_1a_2\cdots a_n\cdots$, and we say that the real number corresponding to this infinite decimal expansion is the real number that corresponds to our point P on the coordinate line.

One last word is needed in our discussion of the one-to-one correspondence between the set of points on a coordinate line and the set of real numbers. We have seen how a point gives rise to a real number. To show, conversely, how a real number gives rise to a point requires a more complete characterization of a coordinate line than we gave earlier. Such a characterization would be axiomatic in nature. Indeed, in certain approaches to geometry, the coordinate line *is defined* as the set of all real numbers.

EXERCISES

1. A decimal expansion is a *repeating decimal* if it has the form $a.u_1\ldots u_rv_1\ldots v_sv_1\ldots v_sv_1\ldots v_s\ldots$ with the sequence of digits $v_1\ldots v_s$, called the *repeating block*, repeating indefinitely. Prove that r is a rational number if and only if r has a repeating decimal expansion (possibly with 0 repeating forever).
2. Find the first six digits to the right of the decimal point in the decimal expansion for $\sqrt{2}$.
3. Find the first four digits to the right of the decimal point in the decimal expansion for $\sqrt{3}$; for $\sqrt{5}$.

6.10 The Complex Numbers C; Polynomials over R and C

Even though our motivation for extending the field of rational numbers Q to the field of real numbers R was primarily geometrical, we have been able to reap important algebraic fruits. For example, Theorem 6.8.6 shows that the equation $x^n = A$ always has a solution in the field of real numbers if A is a positive real number and n is any positive integer.

Thus, by enlarging the field of rational numbers we can now solve certain polynomial equations within the field **R** that were not solvable within the field Q. [Recall that we constructed the field Q from the integral domain of integers Z so that equations of the form $ax = b$, $a \neq 0$, $a, b \in Z$, could always be solved (within Q).]

There are, however, polynomials with coefficients from **R** that do not have any real roots. For example, $x^2 + 1$ does not have a real root. We may ask, therefore, whether there is a way to extend the real number system to another field in order that such a polynomial, or for that matter all polynomials, will have a root in the extension field.

There are two ways in which we will obtain such an extension. One has already been carried out in Section 5.11. We simply observe that the polynomial $x^2 + 1$ is irreducible over **R** and then form the quotient ring $\mathbf{R}[x]/(x^2 + 1)$. This must be a field, since the ideal $(x^2 + 1)$ is maximal. The polynomial equation $y^2 + 1 = 0$ has a solution over this field, namely, the coset $x + (x^2 + 1)$. Also, in Section 5.11 we indicate how we can relate the elements in this field with the usual way of writing complex numbers in the form $a + bi$, where $i^2 = -1$.

The other way is also quite easy, and we shall indicate it here. We shall let lower case letters represent real numbers.

Definition 6.10.1 A *complex number* is an ordered pair of real numbers (x,y). Two complex numbers are *added* as follows:

$$(x,y) + (s,t) = (x + s, y + t).$$

Two complex numbers are *multiplied* according to:

$$(x,y)(s,t) = (xs - yt, xt + ys).$$

The set of all *complex numbers* is denoted by **C**.

THEOREM 6.10.2 The set **C** of all complex numbers forms a field with respect to the operations of addition and multiplication defined in 6.10.1.

Proof Most of the argument is routine. We leave the reader the verification of all properties, noting that the zero for addition is $(0,0)$ and the multiplicative identity is $(1,0)$. For multiplicative inverses, if $(a,b) \neq (0,0)$, then

$$(a,b)^{-1} = \left(\frac{a}{\sqrt{a^2 + b^2}}, \frac{-b}{\sqrt{a^2 + b^2}} \right). \blacksquare$$

The reader is perhaps familiar with the more conventional notation $a + bi$ for complex numbers (a,b). This can be obtained from our notation as follows. Denote the complex number $(0,1)$ by i. Then $i^2 = (0,1)(0,1) = (0-1,0) = (-1,0)$. Agree to write (a,b) in the form $a(1,0) + b(0,1) =$

6.10 THE COMPLEX NUMBERS C; POLYNOMIALS OVER R AND C

$a(1,0) + bi$. Agree also to write 1 for $(1,0)$, so that we get $(a,b) = a1 + bi = a + bi$. It is now easy to see that the definition of multiplication given in Definition 6.10.1 yields the same results as the multiplication usually defined for numbers of the form $a + bi$. Note that the term "complex number" includes "real number" as a special case, namely, when $b = 0$.

The remarkable theorem known as the *fundamental theorem of algebra* states that the extension of the real number system to the complex number system achieves precisely what we had hoped. (See Theorem 5.10.8.)

THEOREM 6.10.3 (*Fundamental Theorem of Algebra*) Let $f(x)$ be a polynomial of degree ≥ 1 with complex coefficients. Then $f(x)$ can be factored into linear factors over the field of complex numbers. In particular, the polynomial equation $f(x) = 0$ always has a solution in C.

There are many known proofs of this theorem, and all involve some aspect of that branch of mathematics known as "analysis." One proof with a heavy algebraic flavor requires only the result that a polynomial of odd degree with real coefficients has a real root. This result follows readily from the intermediate value theorem for continuous real valued functions. In this book we will omit the proof of the fundamental theorem of algebra.

From the fundamental theorem of algebra and the result about odd-degree polynomials over R having a real root, we can easily prove the next theorem.

THEOREM 6.10.4 Let $f(x)$ be an irreducible polynomial over the real number system. Then $f(x)$ is of degree 1 or 2. Moreover, the second degree polynomial $ax^2 + bx + c$ is irreducible over the real number system if and only if $b^2 - 4ac < 0$.

Proof If $f(x)$ is of odd degree, it has a real root, and so is not irreducible, unless the degree is 1. Thus, the only odd degree irreducible polynomials over the reals are of degree 1.

Next, observe that for any polynomial $f(x)$ with real coefficients, $f(a + bi) = 0$ if and only if $f(a - bi) = 0$. This is easily verified by direct calculation. Now if both $a + bi$ and $a - bi$ are roots of the polynomial, then both $(x - (a + bi))$ and $(x - (a - bi))$ are factors of $f(x)$, but the product of these two factors is $x^2 - 2ax + (a^2 + b^2)$, and this has real coefficients. What we have shown is that complex roots of $f(x)$ come in *conjugate pairs*, and the corresponding factors give rise to a quadratic factor with real coefficients. Since by the fundamental theorem of algebra we assume that a polynomial with real roots can be completely factored into linear factors, we pair the linear factors as above, and so we see that the irreducible even degree polynomials must be of degree 2.

From the quadratic formula, we know that the solutions to

$$ax^2 + bx + c = 0$$

are

$$\frac{-b \pm \sqrt{b^2 - 4ac}}{2a},$$

and so we have real roots if $b^2 - 4ac \geq 0$ and complex roots if

$$b^2 - 4ac < 0.$$

This latter observation completes the proof. ∎

We mention that there are several ways to do what we have discussed here. A proof of the fundamental theorem of algebra could be given using Theorem 6.10.4. This would necessitate our proving Theorem 6.10.4 in a different manner, for otherwise we would have a circular argument. But a proof of the fundamental theorem can be given that does not depend on Theorem 6.10.4, and then our proof of Theorem 6.10.4 would be acceptable.

EXERCISES

1. Find all solutions of each of the following equations:
   (a) $x^4 - 1 = 0$.
   (b) $x^6 - 2x^5 + 4x^4 - x^2 + 2x - 4 = 0$. (*Hint:* First search for all real solutions.)
   (c) $x^6 - 2x^5 + 6x^4 - 4x^3 + 9x^2 - 2x + 4 = 0$.
2. Prove Theorem 6.10.2 completely.
3. Show that the field $\mathbf{R}[x]/(x^2 + 1)$ is isomorphic to the field $\mathbf{C}$ defined in Definition 6.10.1.
4. Prove that if $f(x)$ is a polynomial with real coefficients, then

$$f(a + bi) = 0$$

if and only if $f(a - bi) = 0$. [*Hint:* Use the facts that

$$i^2 = (-i)^2 = -1,$$

$$i^3 = -i = -(-i)^3, \qquad i^4 = (-i)^4 = 1.]$$

6.11 Algebraic and Transcendental Numbers

It is interesting to classify real and complex numbers according to whether they are solutions to polynomial equations having rational coefficients.

6.11 ALGEBRAIC AND TRANSCENDENTAL NUMBERS

Definition 6.11.1 A real or complex number a is called *algebraic* if there is a polynomial $f(x) \in Q[x]$ such that $f(a) = 0$. Otherwise, a is called *transcendental*.

Numbers such as $\sqrt{2}$ and i are algebraic, since they satisfy equations $x^2 - 2 = 0$ and $x^2 + 1 = 0$, respectively. Although we shall not prove it, the well-known real numbers e and π are transcendental, as is also the real number

$$.010010000001000000000000000000000000010\cdots010\cdots01\cdots .$$
$$\underbrace{}_{\substack{5! \\ 0\text{'s}}} \underbrace{}_{\substack{6! \\ 0\text{'s}}}$$

We can, however, easily prove that there exist (infinitely many) transcendental real numbers. To do this, note that if a satisfies a polynomial equation $f(x) = 0$ with rational coefficients, then a also satisfies a polynomial with integer coefficients, for we need only multiply $f(x) = 0$ through by the least common multiple of all the denominators appearing in the coefficients of $f(x)$. Thus we can state the following.

THEOREM 6.11.2 A real or complex number a is algebraic if and only if a satisfies a polynomial $f(x)$ with integer coefficients.

THEOREM 6.11.3 The set of algebraic numbers can be put into one-to-one correspondence with the set of positive integers; that is, the set of algebraic numbers is countably infinite.

Proof This proof is similar in concept to that of Theorem 6.2.7, where we proved that the set of rational numbers is countable. With each polynomial $f(x)$ with integer coefficients and of degree ≥ 1, associate an integer $N(f(x))$ defined as follows: If

$$f(x) = a_n x^n + a_{n-1} x^{n-1} + a_{n-2} x^{n-2} + \cdots + a_0,$$

with each a_i an integer, put

$$N(f(x)) = n + |a_n| + |a_{n-1}| + |a_{n-2}| + \cdots + |a_0|.$$

For example, $N(x - 6) = 1 + |1| + |-6| = 8$ and

$$N(x^3 - 7x^2 + 2x - 6) = 3 + 1 + |-7| + 2 + |-6| = 19.$$

Also, for a given positive integer, there are only a finite number of polynomials with that value for $N(f(x))$. For example, we see that the only polynomials $f(x)$ for which $N(f(x)) = 4$ are

$$x + 2, 2x + 1, 3x, x - 2, 2x - 1, -x + 2, -2x + 1, -3x,$$
$$-x - 2, -2x - 1, x^2 + 1, x^2 + x, x^2 - 1, x^2 - x, -x^2 + 1,$$
$$-x^2 + x, -x^2 - 1, -x^2 - x, 2x^2, -2x^2, x^3, -x^3.$$

Although the number of polynomials $f(x)$ such that $N(f(x)) = N$ gets large rapidly with increasing N, the point is that there is just a finite number of these associated with each value of N. Thus with each value of N we associate a finite number of polynomials, and then in turn with each polynomial we associate just a finite number of algebraic numbers, namely, all the solutions of the polynomial, both real and complex. Thus with each number N we associate a finite set of algebraic numbers. We now can describe how to put the set of all algebraic numbers into one-to-one correspondence with the positive integers. Write down in some order the finite list of algebraic numbers we associate with $N = 2$; next write down the finite list of algebraic numbers we associate with $N = 3$; then the list we associate with $N = 4$, and so forth. Then we can just go through this very lengthy list and assign the integers 1, 2, 3, ... to the elements in the list in order, omitting algebraic numbers already having been counted, and this will give us the desired assignment. ∎

THEOREM 6.11.4   The set of all real numbers cannot be put into one-to-one correspondence with the set of positive integers.

Proof   This is a famous argument due to Cantor, the so-called Cantor diagonal proof. First, note that if we could put all the real numbers into such a correspondence, then it would be possible to put all the real numbers strictly between 0 and 1 into a one-to-one correspondence with the positive integers. What we shall show is that the real numbers between 0 and 1 cannot be placed in one-to-one correspondence with the positive integers.

First, we observe that some real numbers have two decimal representations. For example, using our discussion of Section 6.9, we could conclude that $.5000\cdots 000\cdots$ and $.4999\cdots 999\cdots$ are two different representations for $\frac{1}{2}$. Also, the only time there are two possible representations is when one ends in all 0's, the other in all 9's. For standardization we shall choose the representation ending in 0's when there is a choice.

Now, if the real numbers strictly between 0 and 1 could be put into one-to-one correspondence with the positive integers, then in essence we could write out the list of decimal representations in the following manner:

$$A_1 = .a_{11}a_{12}a_{13}a_{14}\cdots a_{1n}\cdots$$
$$A_2 = .a_{21}a_{22}a_{23}a_{24}\cdots a_{2n}\cdots$$
$$A_3 = .a_{31}a_{32}a_{33}a_{34}\cdots a_{3n}\cdots$$
.
.
.
$$A_k = .a_{k1}a_{k2}a_{k3}a_{k4}\cdots a_{kn}\cdots$$
.
.
.

6.11 ALGEBRAIC AND TRANSCENDENTAL NUMBERS

We form another real number b as follows. We put $b_1 = 3$, if $a_{11} \neq 3$. Otherwise, put $b_1 = 4$. Put $b_2 = 3$, if $a_{22} \neq 3$, otherwise put $b_2 = 4$. Continue going down the diagonal in this manner, putting $b_k = 3$, if $a_{kk} \neq 3$, otherwise putting $b_k = 4$. In this manner we get a real number

$$b = .b_1 b_2 b_3 \cdots b_n \cdots.$$

The question is where is b in our list? If the list includes all real numbers strictly between 0 and 1, then b must equal A_m, for some m. But we note that by our choice of b_i's, $b_m \neq a_{mm}$, and so $b \neq A_m$, for any m. Thus it is impossible to put the real numbers between 0 and 1 into one-to-one correspondence with the set of positive integers, and so the set of all real numbers is not countable. ∎

COROLLARY 6.11.5 The set of all complex numbers is not countable.

Proof The set of real numbers can be considered a subset of the set of complex numbers. If the latter could be put into one-to-one correspondence with the positive integers, then so could the former. ∎

COROLLARY 6.11.6 There exist both complex and real transcendental numbers.

Proof The set of algebraic numbers (either real or complex) is countable. If no transcendental numbers existed, then all numbers would be algebraic, and so the complex and real number systems would both be countable, a contradiction. ∎

The proofs just given show the existence of transcendental numbers. That numbers such as e and π are transcendental must be proven explicitly using other techniques.

In Theorem 6.2.8 we proved that the set of all rational numbers could be contained within the union of a set of intervals on the coordinate line, where the sum of the lengths of all the intervals used could be made as small as was desired. In that proof, the only fact used about the rational numbers is that the set of rational numbers is countable. Since the set of all algebraic numbers is also countable, we could similarly prove

THEOREM 6.11.7 For each positive integer n, the set of all real algebraic numbers can be contained within a set of intervals having lengths whose sum is less than $1/n$.

Proof The details are left as an exercise. See Theorem 6.2.8. ∎

As a consequence of this theorem, we observe that among the points on the coordinate line, points corresponding to transcendental numbers abound, while algebraic numbers are "few and far between." Thus, not only do the rational numbers make up very little of the number line, but so do the algebraic numbers such as $\sqrt{2}$, $\sqrt{2} + \sqrt[3]{3}$, $\sqrt{4 + \sqrt[3]{4 + 2\sqrt{2}}}$.

EXERCISES

1. Prove the following: Consider the complex number $a + bi$ as being the point (a,b) of the coordinate plane. Show that for each positive integer n, the set of all complex algebraic numbers can be contained in a set of squares having areas whose sum is less than $1/n$.
2. Show that $e^2, e^3, \ldots, e^n, \ldots$ are all transcendental, assuming that e is transcendental.
3. Prove that if t is a transcendental number, then so is t^n, for every positive integer n. (*Hint:* Suppose t^n satisfies a polynomial equation with integer coefficients. What about t?)

7
Vector Spaces

7.1 Introduction to Vectors

The notion of a *vector* should be familiar to the reader in a number of different contexts. Forces, displacements, and velocities, for example, are physical concepts that must be described in terms of "magnitude" and "direction" and are called *vectors* or *vector quantities*. Thus we make a distinction between the concepts of "speed" and "velocity," the former being described by a single real number and the latter by a number denoting magnitude together with a set of numbers denoting direction. Such vector quantities can be represented pictorially by a line segment with an arrowhead at one end. Such a line segment is sometimes called a directed line segment. The length of the line segment represents the magnitude of the vector according to some preassigned scale, and the orientation of the line segment together with the arrow represents its direction. Thus we may interpret Figure 7.1 as depicting the velocity of the point P in the plane as 20 ft/sec in the indicated direction.

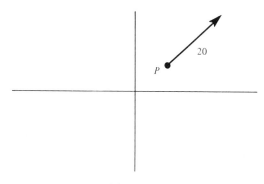

Figure 7.1

If we restrict ourselves to directed line segments whose "tails" are at the origin of a Cartesian coordinate system in the plane, then the vector can be completely described by the point in the plane at the tip of the arrowhead. Hence in Figure 7.2, several vectors are named according to their "terminal points."

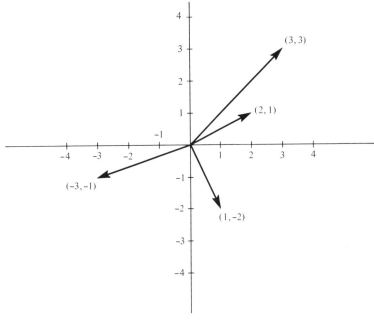

Figure 7.2

The "length" or "magnitude" of a vector is then the distance of its terminal point to the origin.

In the same way, we can describe a vector in three-dimensional space by an ordered triple of numbers representing the terminal point of a directed line segment with tail at the origin.

The distinction that we have drawn between speed and velocity, or more generally, between a real number and vector quantity, emphasizes that the arithmetic of vectors, although a generalization of the arithmetic of real numbers, is a distinctly new notion. That is, we have a clear idea of the addition of two vectors in a physical sense, and we know that this notion of addition is definitely different from the addition of real numbers. To illustrate, if an automobile moves due north with a velocity of 40 mph and we increase the velocity by 30 mph due north, then the resulting velocity will be 70 mph due north, simply by addition of 40 and 30, since both vectors have the same direction. But if we add a velocity of 30 mph due east to the velocity of 40 mph due north, then we know that the resulting velocity will have a direction that is neither due north nor due east and its magnitude will not be 70 mph. The reader is probably familiar with the addition of vectors by the parallelogram law, which we illustrate in Figure 7.3 for the above example.

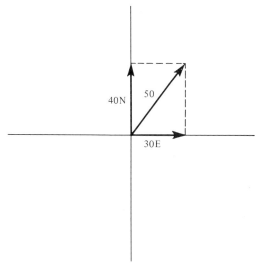

Figure 7.3

The method of addition is to produce a parallelogram (which is a rectangle in this special case) whose four sides include the two original vectors. The sum of the vectors is then the directed line segment that is the diagonal of the parallelogram with arrowhead away from the origin. (See Figure 7.4.) In this case, the magnitude of the sum is 50 mph and the direction of the sum is given by the angle it makes with the positive x axis.

The parallelogram law can be very easily described when the vectors are pairs of real numbers, or triples of real numbers, or more generally n-tuples of real numbers. Thus if $v = (a,b)$ and $u = (c,d)$ with a, b, c, d real numbers, then $v + u = (a,b) + (c,d) = (a + c, b + d)$. That is, we obtain the coordinates, or "components," of $v + u$ by adding associated components of v and u, respectively. Thus the vectors in Figure 7.3 are (0,40) and (30,0) and the diagonal is (30,40). In this way we may view the addition of vectors as a generalization of the addition of real numbers, since a given real number may be thought of as a vector specified by a single component. An advantage of this method of adding vectors over the geometrical approach, other than its quick computability, is that we can extend the notion of addition to n-tuples of real numbers. We will

266 VECTOR SPACES

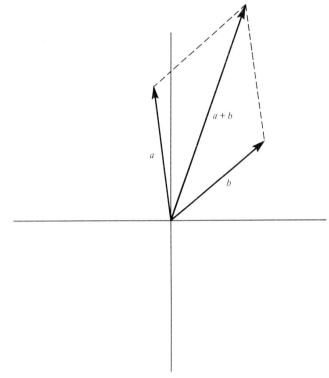

Figure 7.4

presently give some examples to indicate how these objects arise in a very natural way.

Another operation performed with vectors is "stretching" or "shrinking," usually referred to as scalar multiplication. This operation affects the magnitude of a vector, and it will also affect its direction if the scalar multiplier is negative. In the latter case, direction of the vector is changed by 180 degrees. Thus, for example, if v is a vector, then $2v$ will be a vector

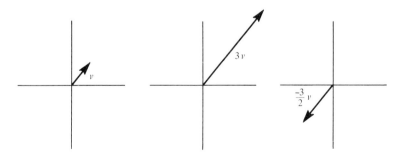

Figure 7.5

whose magnitude is twice that of v, and with the same direction. We give additional illustrations in Figure 7.5.

Scalar multiplication also has a very direct interpretation if the vectors are n-tuples. If $v = (a,b,c)$, for example, and α is a real number, then $\alpha v = (\alpha a, \alpha b, \alpha c)$. The reader will recall from analytic geometry that the vector v is a line segment with direction cosines

$$\frac{a}{r}, \frac{b}{r}, \frac{c}{r}, \quad \text{where } r = \sqrt{a^2 + b^2 + c^2}.$$

We can sum up the arithmetic operations of addition and scalar multiplication in the concept of a "linear combination." If u and v are vectors, and a and b are real numbers, then the vector $w = au + bv$ is called a *linear combination* of u and v.

EXERCISES

1. Let $u = (0,1)$ and $v = (\sqrt{3}, 0)$ be two vectors in the plane. Find $u + v$ by the parallelogram method and also by the method of adding components.
2. Find the sum of $u = (1, -1)$ and $v = (-1, -1)$.
3. Let $u = (1,2)$ and $v = (3, -2)$. Find $3u + v$.
4. If $u = (8,6)$, then show that the magnitude of u is 10.
5. Let $u = (0,1)$, $v = (1,0)$, and $w = (3, -4)$. Find real numbers a, b so that $w = au + bv$.
6. Let $u = (0,1)$, $v = (1,1)$, and $w = (1,2)$. Find real numbers a, b so that $w = au + bv$.
7. Let $u = (1,1)$, $v = (1,2)$, and $w = (1,3)$. Find three real numbers a, b, and c, not all zero, so that $au + bv + cw = (0,0)$. This vector is called the *zero vector*.
8. Let P be a parallelogram placed on a Cartesian coordinate system with vertices as shown. Prove, using plane geometry, that $d = a + b$.

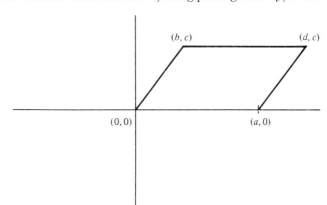

268   VECTOR SPACES

9. Let Q be a parallelogram placed on a Cartesian coordinate system similarly to P in Exercise 8, but with corresponding sides twice the length of those in P. Prove that the diagonal of Q is twice as long as the one of P.
10. Show that $(1.5)(1,-2,\sqrt{2}) = (1,-2,\sqrt{2}) + \frac{1}{2}(1,-2,\sqrt{2})$.
11. Show that

$$3[(1,-81,0) + (2,\sqrt{3},-\sqrt{5})] = 3(1,-81,0) + 3(2,\sqrt{3},-\sqrt{5}).$$

12. Show that the set of vectors $\{(a,b) \mid a, b \text{ real numbers}\}$ forms an abelian group with respect to vector addition.
13. Show that the set of vectors $\{2(a,b) \mid a, b \text{ real numbers}\}$ is the same as the set in Exercise 12.

7.2 Definition, Examples, and Elementary Properties of Vector Spaces

In Section 7.2 we saw that the essential properties of vectors are reflected in the operations of addition and multiplication by scalars. In this section we give a formal definition of vector space in terms of these arithmetic properties. As in the theories of groups and rings, we thereby include a very wide variety of mathematical systems that share these properties. In essence, we require first that we can define an addition (and subtraction) of vectors. Then we require that the elements of a field act as scalar multipliers, and that the arithmetic properties of the scalar field be compatible with those of the set of vectors. Such a collection of vectors will be called a *vector space*. This use of the word space and its obvious allusion to geometry will make good sense in the examples to follow.

Definition 7.2.1 Let $(V,+)$ be an abelian group and let $(F,\oplus,\cdot)$ be a field, such that for each $a \in F$ and $v \in V$, $av \in V$. That is, each $a \in F$ induces a function from V to V. Then V is called a *vector space over F* if the following properties are satisfied.

(a) $a(v_1 + v_2) = av_1 + av_2$ for $a \in F$, $v_1, v_2 \in V$.
(b) $(a_1 \oplus a_2)v = a_1v + a_2v$, for $a_1, a_2 \in F$, $v \in V$.
(c) $(a_1 \cdot a_2)v = a_1(a_2v)$ for $a_1, a_2 \in F$, $v \in V$.
(d) $1v = v$, where 1 is the multiplicative identity of F and $v \in V$.

The elements of V are called *vectors* and the elements of F are called *scalars*. The reader should notice that in property (b), we have carefully distinguished between addition in F, denoted by $\oplus$, and addition in V, denoted by $+$. We will soon drop this distinction, for it will always be clear from the context which addition is being considered. Notice that since multiplication is commutative in a field, property (c) can be stated $(a_1 a_2) v = a_1(a_2 v) = a_2(a_1 v) = (a_2 a_1) v$. Because of this property we will usually write $a_1 a_2 v$ without any parentheses.

We now give some examples to illustrate the diversity of mathematical systems that are vector spaces.

Example 7.2.2 Let V be the set of directed line segments in the Cartesian plane with "tail" at the origin. If addition is defined by the parallelogram law, then it is easy to prove that addition of vectors is an associative, commutative binary operation with the zero vector as an additive identity. Thus $(V, +)$ is an abelian group. If F is the field of real numbers, then av is the vector with magnitude equal to $|a|$ times the magnitude of v. The direction of av is the same as that of v, if a is positive; and the direction of av is opposite to that of v if a is negative. We leave to the reader the verification that the properties (a), (b), (c), and (d) are satisfied.

Example 7.2.3 Let $\mathbf{R}^n$ be the set of all n-tuples of real numbers,

$$\mathbf{R}^n = \{(a_1, \ldots, a_n) \mid a_i \text{ real numbers}, i = 1, \ldots, n\}.$$

If we define $(a_1, \ldots, a_n) + (b_1, \ldots, b_n) = (a_1 + b_1, \ldots, a_n + b_n)$, then it is easy to see that $(\mathbf{R}^n, +)$ is an abelian group, with $(0, \ldots, 0)$ the identity element and $(-a_1, \ldots, -a_n)$ the additive inverse of $(a_1, \ldots, a_n)$. Now if we define scalar multiplication by $c(a_1, \ldots, a_n) = (ca_1, ca_2, \ldots, ca_n)$ for any real number c, then it follows that $\mathbf{R}^n$ is a vector space over the real field $\mathbf{R}$. We leave the verification of the details to the reader.

Example 7.2.4 Let $\mathbf{R}$ be the field of real numbers and let T be the set of all functions with domain and codomain $\mathbf{R}$. Now let f and g be two elements of T and define $f + g$ by $(f + g)(x) = f(x) + g(x)$ for all $x \in \mathbf{R}$. Next, for $f \in T$ and $c \in \mathbf{R}$, define cf by $(cf)(x) = c(f(x))$. To show that $(T, +)$ is an abelian group, we remark that associativity follows directly from the associativity for $(\mathbf{R}, +)$. The function θ whose range is the single element 0 is clearly the identity of $(T, +)$, and for $f \in T$, the function $g \in T$, defined by $g(x) = -f(x)$, is the inverse of the function f. (Notice that this notion of inverse differs from that defined in relation to function composition in Section 1.7.) The function, $c(f_1 + f_2)$ with $c \in \mathbf{R}$, f_1, $f_2 \in T$, has the property

$$[c(f_1 + f_2)](x) = c[(f_1 + f_2)(x)] = c(f_1(x) + f_2(x))$$
$$= cf_1(x) + cf_2(x),$$

since $f_1(x)$ and $f_2(x)$ are real numbers for any fixed x and the real field satisfies the distributive law. Hence $c(f_1 + f_2) = cf_1 + cf_2$ and (a) is satisfied. The verifications that (b), (c), and (d) hold are quite similar and we leave them to the reader.

270   VECTOR SPACES

Example 7.2.5   We next consider the system of linear homogeneous equations in three unknowns:

$$x + y + z = 0$$
$$x - y - z = 0.$$

A solution of such a system is a triple of real numbers (a,b,c) such that if we make the substitution $x = a$, $y = b$, and $z = c$, then both equations will be satisfied. If we define addition and scalar multiplication of triples as in Example 7.2.3, then we claim that the set of all solutions to the given system of equations is a vector space.

Suppose that (a_1,a_2,a_3) and (b_1,b_2,b_3) are two solutions to the system. Then

$$a_1 + a_2 + a_3 = 0, \quad a_1 - a_2 - a_3 = 0$$

and

$$b_1 + b_2 + b_3 = 0, \quad b_1 - b_2 - b_3 = 0.$$

Thus $(a_1 + b_1, a_2 + b_2, a_3 + b_3)$ will also be a solution, since if we let $x = a_1 + b_1$, $y = a_2 + b_2$, and $z = a_3 + b_3$, then

$$a_1 + b_1 + a_2 + b_2 + a_3 + b_3 = (a_1 + a_2 + a_3) + (b_1 + b_2 + b_3)$$
$$= 0 + 0 = 0$$

and

$$(a_1 + b_1) - (a_2 + b_2) - (a_3 + b_3) = (a_1 - a_2 - a_3) + (b_1 - b_2 - b_3)$$
$$= 0 + 0 = 0.$$

Hence addition of solutions is a binary operation on the set of solutions. Now if c is any real number and (a_1,a_2,a_3) is a solution, then $c(a_1,a_2,a_3) = (ca_1,ca_2,ca_3)$ is also a solution, since

$$ca_1 + ca_2 + ca_3 = c(a_1 + a_2 + a_3) = c \cdot 0 = 0$$

and $ca_1 - ca_2 - ca_3 = c(a_1 - a_2 - a_3) = c \cdot 0 = 0$. Thus it follows easily that the set of all solutions of this system of equations is a vector space.

Of course, our choice of these two particular equations has no special significance. Our result holds for any system of equations of the form.

$$a_{11}x_1 + a_{12}x_2 + \cdots + a_{1n}x_n = 0$$
$$a_{21}x_1 + a_{22}x_2 + \cdots + a_{2n}x_n = 0$$
$$\vdots$$
$$a_{m1}x_1 + a_{m2}x_2 + \cdots + a_{mn}x_n = 0$$

with a_{ij} real numbers.

Example 7.2.6 Let

$$a(x)\frac{d^2y}{dx^2} + b(x)\frac{dy}{dx} + c(x)y = 0$$

be a differential equation with fixed real-valued functions $a(x)$, $b(x)$, and $c(x)$, and x real. We claim that the set of all functions over the field of real numbers that are solutions to this equation form a vector space with respect to the operations of addition and scalar multiplication, as defined in Example 7.2.4. We need only verify that the sum of any two solutions is a solution, and that a real number times a solution is also a solution. Thus let $y_1(x)$ and $y_2(x)$ be two functions that are solutions to the differential equation above and consider $y(x) = y_1(x) + y_2(x)$. We now recall from calculus that

$$\frac{dy}{dx} = \frac{d(y_1 + y_2)}{dx} = \frac{dy_1}{dx} + \frac{dy_2}{dx},$$

and that

$$\frac{d^2y}{dx^2} = \frac{d^2y_1}{dx^2} + \frac{d^2y_2}{dx^2}.$$

Hence

$$a(x)\frac{d^2y}{dx^2} + b(x)\frac{dy}{dx} + c(x)y$$

$$= a(x)\left(\frac{d^2y_1}{dx^2} + \frac{d^2y_2}{dx^2}\right) + b(x)\left(\frac{dy_1}{dx} + \frac{dy_2}{dx}\right) + c(x)(y_1 + y_2)$$

$$= a(x)\frac{d^2y_1}{dx^2} + b(x)\frac{dy_1}{dx} + c(x)y_1 + a(x)\frac{d^2y_2}{dx^2} + b(x)\frac{dy_2}{dx} + c(x)y_2$$

$$= 0 + 0 = 0,$$

since both $y_1(x)$ and $y_2(x)$ are solutions. It is also easy to show directly that if y is a solution, then ry is a solution, where r is a real number.

Example 7.2.7 Let $(G,+)$ be an *additive abelian p-group of exponent p*, p a prime. That is, $(G,+)$ is an abelian group such that for each $g \in G$, $pg = 0$. Let F be the field Z_p of least positive residues defined in Section 5.4. We define scalar multiplication on G in terms of integral multiples—that is, $ag = g + \cdots + g$ "a times" with $g \in G$ and $a \in F$. Now the conditions (a), (b), (c), and (d) follow easily from the laws of exponents translated to additive notation.

Notice that in the previous example we were able to produce a vector space simply by changing our point of view about an algebraic system without introducing any new operations or special conditions. This procedure of viewing a mathematical system in a new way is very common and can sometimes be an extremely powerful tool in solving problems.

Example 7.2.8 Let $(F,+,\cdot)$ be a field and let F^n be the set of all n-tuples with elements in F, where n is a fixed positive integer. Then

$$F^n = \{(a_1,\ldots,a_n) \mid a_i \in F, i = 1, \ldots, n\}.$$

We define the sum of two n-tuples and the scalar product of an element of F with an n-tuple as in Example 7.2.3, where F was the field of real numbers. Then the verification that F^n is a vector space over F is exactly the same as in the real case.

We conclude this section with some elementary properties of vector spaces, which follow directly from the definition.

THEOREM 7.2.9 Let V be a vector space over the field F.

(a) If 0 is the identity of $(V,+)$, that is, 0 is the *zero vector*, and if $a \in F$, then $a0 = 0$, the zero vector.
(b) If 0 is the additive identity of F and $v \in V$, then $0v = 0$, the zero vector. (We will use the symbol 0 to represent both the zero element of F and the zero vector in V when no ambiguity exists.)
(c) If $a \in F$ and $v \in V$, then

$$-(av) = (-a)v = a(-v).$$

(d) If $a \in F$ and $v \in V$, then $av = 0$ if and only if either $a = 0$ or $v = 0$.

Proof (a) The zero vector is by definition the additive identity of $(V,+)$. Now $a0 + av = a(0 + v) = av$ for any vector v. Thus $a0$ must be the additive identity of V, and so $a0 = 0$, the zero vector.

(b) If 0 is the zero of F and 1 is the multiplicative identity of F, then $0v + v = 0v + 1v = (0 + 1)v = 1v = v$ for any $v \in V$. Thus $0v$ is the zero vector.

(c) The element $-(av)$ is by definition the inverse in $(V,+)$ of the element av, that is, $av + (-(av)) = 0$. Since inverses in the group $(V,+)$ are unique, it will suffice to establish $av + (-a)v = 0$ and $av + a(-v) = 0$. But by (b) $av + (-a)v = (a - a)v = 0v = 0$ and by (a)

$$av + a(-v) = a(v - v) = a0 = 0.$$

(d) Suppose that $av = 0$ for $a \in F$, $v \in V$, and $a \neq 0$. Then, since F is a field, there is $b \in B$ such that $ba = 1$. Thus $0 = b0 = b(av) = (ba)v = 1v = v$. Hence if $a \neq 0$, then $v = 0$, and (d) follows. ∎

The reader should note that in the proof of (d) we have used for the first time that nonzero elements in F have multiplicative inverses. Parts

(a) and (b) of the theorem provide the justification for the use of 0 as both the zero vector and the zero of the field.

Definition 7.2.10 Let V be a vector space over the field F. A subset U of V is called a *subspace of V* if U is a vector space over F with respect to the restriction of vector addition and scalar multiplication to U.

Every vector space V has a subspace called the *trivial space* 0. This subspace consists of the zero vector alone. Note that we now use the same symbol 0 to denote (a) the scalar 0 in F, (b) the additive identity 0 of $(V,+)$, and (c) the trivial subspace 0 of V.

A subspace that is different from the whole vector space is called a *proper* subspace.

Example 7.2.11 Let $V = \mathbf{R}^2$, where $\mathbf{R}$ is the field of real numbers. Let $U = \{(a,0) \mid a \in \mathbf{R}\}$. It is easy to verify that U is a proper subspace of V.

A very simple criterion can be given to determine if a subset U of V is a subspace.

THEOREM 7.2.12 Let V be a vector space over the field F. Let U be a subset of V. Then U is a subspace of V if and only if

(a) $(U,+)$ is a subgroup of $(V,+)$, and
(b) with au defined by the scalar multiplication of V, $au \in U$ for all $a \in F$ and all $u \in U$.

Proof First, let U be a subspace of V. Then $(U,+)$ is a group where $+$ on U is the restriction of $+$ on V. Thus $(U,+)$ is a subgroup of $(V,+)$. Since U is a vector space over F, then clearly $au \in U$ for all $a \in F$ and $u \in U$.

Now suppose that U is a subset of V satisfying (a) and (b). Then $(U,+)$ is an abelian group and $au \in U$ for all $a \in F$ and $u \in U$. Thus all the properties required by the definition of a vector space are satisfied, and U is a vector space over F. ∎

COROLLARY 7.2.13 Let V be a vector space over the field F. Let U be a subset of V, $U \neq \emptyset$. Then U is a subspace of V if and only if for all u_1, $u_2 \in U$ and all α, $\beta \in F$, $\alpha u_1 + \beta u_2 \in U$.

Proof Suppose U is a subspace of V. Then U is itself a vector space, whence $\alpha u_1 + \beta u_2 \in U$, for all u_1, $u_2 \in U$, α, $\beta \in F$.

Conversely, suppose $\alpha u_1 + \beta u_2 \in U$, for all u_1, $u_2 \in U$, α, $\beta \in F$. With $\beta = 0$, we have $\alpha u_1 + 0 u_2 = \alpha u_1 \in U$, for all $\alpha \in F$, $u_1 \in U$, since $0 u_2 = 0$ by Theorem 7.2.9. With $\alpha = 1$, $\beta = -1$, $1 u_1 + (-1) u_2 = u_1 - u_2 \in U$, by Theorem 7.2.9. Thus, $(U,+)$ is a subgroup of $(V,+)$, and so by Theorem 7.2.12, U is a subspace of V. ∎

Many of the vector spaces given in the examples are subspaces of others, and we will leave to the exercises the identification of some of these.

THEOREM 7.2.14 Let A and B be subspaces of the vector space V over the field F. Then $A \cap B$ is also a subspace of V.

Proof For both A and B, $0 \in A$, $0 \in B$, where 0 is the zero vector. Thus $A \cap B \neq \varnothing$.
If $u, v \in A \cap B$, then $u \in A$, $u \in B$, $v \in A$, $v \in B$. Thus, $\alpha u + \beta v \in A$ and $\alpha u + \beta v \in B$, for all $\alpha, \beta \in F$. Thus, $\alpha u + \beta v \in A \cap B$. By Theorem 7.2.13, $A \cap B$ is a subspace of V. ∎

EXERCISES

1. Show that addition of vectors in the plane defined by the parallelogram law is associative.
2. Let $\mathbf{R}^2$ be the vector space of pairs of real numbers as defined in Example 7.2.3. Calculate the following vectors:
   (a) $(3,3) + 2(4,-2)$.
   (b) $(0,\tfrac{1}{2}) + (-1,\sqrt{2}), + \sqrt{3}(1,1)$.
   (c) $(4 + \pi)(2,3) + (1,1)$.
   (d) $2[(\tfrac{1}{2},\tfrac{1}{2}) + (3,4)] + 2(3(1,1))$.
3. Show that the set of all continuous functions with domain the closed interval [0,1] forms a vector space over the field $\mathbf{R}$ of real numbers if addition and scalar multiplication are defined as in Example 7.2.4.
4. Consider the system of equations

$$x + 2y = 0$$
$$2x + 3y = 0.$$

Show that the set of all solutions forms a vector space with addition and scalar multiplication as defined in Example 7.2.3.

5. Show why the set of all solutions to the system

$$x + y = 0$$
$$2x + 3y = 1$$

does not form a vector space with respect to the usual definitions of addition and scalar multiplication.

6. Does the set of all solutions to

$$x^2 - y = 0$$
$$x + y = 0$$

form a vector space?

7. Complete the verification that $\mathbf{R}^n$ is a vector space over the field $\mathbf{R}$.

8. Arrange the vector spaces of the examples given in this section in pairs when possible, so that the first listed is a subspace of the second.
9. Which of the following subsets of $\mathbf{R}^3$ are subspaces of $\mathbf{R}^3$?
   (a) $S = \{(0,a,b) \mid \text{all real numbers } a, b\}$.
   (b) $S = \{(a, b, c) \mid a \in Z, b, c \in \mathbf{R}\}$.
   (c) $S = \{(a,-a,b) \mid a, b \in \mathbf{R}\}$.
   (d) $S = \{(a,b,c) \mid a, b, c \in \mathbf{R} \text{ with either } a = 0 \text{ or } b = 0\}$.
   (e) $S = \{(a,b,c) \mid a, b, c \in \mathbf{R} \text{ with } a + b = 1\}$.
10. Let $\{A_\alpha\}$ be a collection of subspaces of the vector space V. Show that $\cap\, A_\alpha$ is a subspace of V.
11. Which of the following sets of real-valued functions on $\mathbf{R}$ are subspaces of the vector space of all such functions defined in Example 7.2.4?
    (a) All functions with positive range.
    (b) All polynomial functions of degree 2.
    (c) All functions satisfying $f(x) = f(1 - x)$.
12. Show that the set $\{(a,b) \mid a, b \in \mathbf{R} \text{ with } \alpha a + \beta b = 0 \text{ for fixed nonzero real numbers } \alpha, \beta\}$ is a proper subspace of $\mathbf{R}^2$. Describe these subspaces geometrically. Show that every nontrivial proper subspace is of this form.
13. Generalize Exercise 12 by replacing $\mathbf{R}$ by an arbitrary field F.
14. List all subspaces of the vector space F^2 with F the field of two elements. (See Example 7.2.8.)
15. Find a subgroup of the group $(\mathbf{R}^2, +)$ which is not a subspace of $\mathbf{R}^2$ when considered as a vector space.
16. Find a subset S of the vector space $\mathbf{R}^2$ with the property that $rs \in S$ for all $r \in \mathbf{R}$ and $s \in S$, but S is not a subspace of $\mathbf{R}^2$ (that is, S is not a subgroup of $(\mathbf{R}^2, +)$).
17. Let F be a field and E a subfield. Show that F is a vector space over E if addition and scalar multiplication are the addition and multiplication of the field F.

7.3 Linear Combinations and Spanning Sets

The concept of linear combination defined previously can be extended to arbitrary finite sets of vectors.

Definition 7.3.1 Let V be a vector space over the field F and let $v_1, v_2, \ldots, v_n \in V$. Then if $a_1, a_2, \ldots, a_n \in F$, the vector

$$v = a_1v_1 + a_2v_2 + \cdots + a_nv_n$$

is called a *linear combination over F* of $\{v_1, v_2, \ldots, v_n\}$, or simply a *linear combination* of $\{v_1, \ldots, v_n\}$.

As usual the sum $a_1v_1 + a_2v_2 + \cdots + a_nv_n$ has a unique interpretation because of the associative law in the group $(V, +)$.

If S is a set of vectors in the vector space V over the field F, then we can define the subspace of V that is generated by S similarly to the way we defined the subgroup of a group generated by a set of group elements.

Definition 7.3.2 Let S be a subset of the vector space V over the field F. Let $\{T_\alpha\}$ be the collection of all subspaces of V that contain the set S. Then $T = \cap\, T_\alpha$ is the *subspace of V spanned* (or *generated*) *by S*, and we denote this subspace by $\langle S \rangle$. S is a *spanning set* for the vector space $T = \langle S \rangle$.

We have already seen in Exercise 10, Section 7.2, that $T = \cap\, T_\alpha$ is a subspace of V. In analogy with subgroups and groups, the subspace T can be described in terms of linear combinations of elements of S. By this, we mean linear combinations of all nonempty finite subsets of S, for S may be an infinite set and we have defined linear combinations for finite subsets of V only.

THEOREM 7.3.3 Let S be a nonempty subset of the vector space V over the field F. Then the subspace $\langle S \rangle$ spanned by S is the set T of all linear combinations of all nonempty finite subsets of S.

Proof If $v_1, \ldots, v_r$ is a set of vectors in a subspace U of the vector space V, then any linear combination of these vectors is still in U, since U is closed with respect to addition and scalar multiplication. Hence any subspace of V that contains S contains all linear combinations of any nonempty finite subset of S. Thus the set T of all linear combinations of finite subsets of S is contained in every subspace containing S. Thus T is contained in the intersection of these subspaces. But this intersection is precisely the subspace $\langle S \rangle$ spanned by S. Now the proof will be complete if we can show that the set T of all linear combinations of finite subsets of S is a subspace of V. According to Theorem 7.2.12 this is equivalent to showing that $(T, +)$ is a subgroup of $(V, +)$ and that $at \in T$ for all $a \in F$ and all $t \in T$, where at is the scalar multiple of t by the scalar a.

In order to show that $(T, +)$ is a subgroup of $(V, +)$, it suffices to show that for any pair of vectors t_1, t_2 of T, the vector $t_1 - t_2 \in T$. Now since $t_1, t_2 \in T$, there is a finite subset of S, say $s_1, \ldots, s_r$, and a set of elements $a_1, \ldots, a_r$ of F such that

$$t_1 = a_1 s_1 + \cdots + a_r s_r$$

and similarly $s_1', s_2', \ldots, s_n' \in S$ and $a_1', \ldots, a_n' \in F$ such that

$$t_2 = a_1' s_1' + \cdots + a_n' s_n'.$$

7.3 LINEAR COMBINATIONS AND SPANNING SETS

Thus $t_1 - t_2 = a_1 s_1 + \cdots + a_r s_r - a_1' s_1' - a_2' s_2' - \cdots - a_n' s_n'$. But if $a_i' \in F$, then $-a_i' \in F$ and

$$t_1 - t_2 = a_1 s_1 + \cdots + a_r s_r + (-a_1') s_1' + \cdots + (-a_n') s_n',$$

which is a linear combination over F of the finite subset

$$\{s_1, \ldots, s_r, s_1', \ldots, s_n'\}$$

of S. Thus $(T, +)$ is a subgroup of $(V, +)$.

Finally, if $t \in T$ and $a \in F$, then $t = a_1 s_1 + \cdots + a_r s_r$ for $a_i \in F$, $s_i \in S$, $i = 1, \ldots, r$, and

$$at = a(a_1 s_1) + \cdots + a(a_r s_r) = (aa_1) s_1 + \cdots + (aa_r) s_r$$

by property (c) of Definition 7.2.1. Thus $at \in T$, whence T is a subspace of V. Thus, $T = \langle S \rangle$. ∎

COROLLARY 7.3.4 Let $S = \{v_1, v_2, \ldots, v_n\}$ be a finite nonempty subset of the vector space V over the field F. Then the set

$$\{a_1 v_1 + \cdots + a_n v_n \mid a_i \in F, i = 1, \ldots, n\}$$

is the subspace $\langle S \rangle$ spanned by S.

Example 7.3.5 Let $\mathbf{R}^2$ be the vector space of pairs of real numbers over the field $\mathbf{R}$, as defined in Example 7.2.3. Let S consist of the single element $(0,1)$. Then the space spanned by S is the set of all linear combinations of $(0,1)$, which is simply the set of all scalar multiples of $(0,1)$. That is, $\langle S \rangle = \{(0,a) \mid a \in \mathbf{R}\}$. This set is, of course, simply the y axis. If S_1 consists only of the vector $(1,3)$, then the subspace spanned by S_1 is $\{(a, 3a) \mid a \in \mathbf{R}\}$. This is just the set of points in the plane whose y coordinate is three times its x-coordinate, that is, the line $y = 3x$.

Example 7.3.6 Let T be the vector space of all real-valued functions with domain and codomain the field $\mathbf{R}$ (Example 7.2.4). Let S be the set $\{x^n \mid n \text{ is a nonnegative integer}\}$, where each x^n defines a function $\mathbf{R} \longrightarrow \mathbf{R}$ in the obvious manner. Then the subspace $\langle S \rangle$ spanned by S is the set of all linear combinations of elements of S; that is $\langle S \rangle$ is the set of all polynomial functions.

EXERCISES

1. Find the subspace of $\mathbf{R}^2$ spanned by $\{(1,2), (-2,-1)\}$. Can this space be spanned by a single vector?
2. Find a finite set of vectors that spans all of $\mathbf{R}^2$.
3. Described the subspaces of $\mathbf{R}^3$ spanned by
   (a) $\{(1,1,1)\}$.
   (b) $\{(0,1,1), (0,-1,1)\}$.

278   VECTOR SPACES

(c) $\{(1,0,0),(0,0,1)\}$.
(d) $\{(1,2,3),(3,2,1),(4,4,4)\}$.
(e) $\{(1,1,1),(1,1,2),(1,2,2)\}$.

4. Show that the subspace spanned by S in Example 7.3.6 cannot be spanned by a finite set of elements. Describe all subspaces of this subspace.
5. Prove Corollary 7.3.4.

7.4   Linear Dependence and Independence

Suppose that v is a vector in a given vector space V and that A is a subspace of V. It is natural to ask: is v an element of A? This is equivalent to: if A is the subspace spanned by S, then is v a linear combination of a finite subset of S? In order to develop methods to answer such a question, we need the concept of linear dependence.

Definition 7.4.1 Let $\{v_1,\ldots,v_n\}$ be a nonempty set of (distinct) vectors of the vector space V over the field F. We say that $\{v_1,\ldots,v_n\}$ is a *linearly dependent set* if there are scalars $a_1, \ldots, a_n \in F$ not all equal to the 0 of F, such that $a_1v_1 + a_2v_2 + \cdots + a_nv_n = 0$, the zero vector. If a vector

$$v = a_1v_1 + \cdots + a_nv_n \quad \text{for some } a_i \in F, i = 1, \ldots, n,$$

then we say that v is *linearly dependent* on $\{v_1,\ldots,v_n\}$.

In other words, a set $S = \{v_1,\ldots,v_n\}$ of vectors is linearly dependent if the zero vector can be written as a linear combination of $\{v_1,\ldots,v_n\}$ and not all the coefficients are zero. The reader should note that if a vector v is linearly dependent on $\{v_1,\ldots,v_r\}$, that is

$$v = a_1v_1 + \cdots + a_rv_r, \quad a_i \in F,$$

then v is also linearly dependent on the set $\{v_1,\ldots,v_r,v_{r+1},\ldots,v_s\}$, since

$$v = a_1v_1 + \cdots + a_rv_r + 0v_{r+1} + \cdots + 0v_s.$$

Definition 7.4.2 A nonempty set of vectors that is not linearly dependent is called *linearly independent*.

Note that a set of vectors $\{v_1,\ldots,v_n\}$ is linearly independent if whenever $a_1v_1 + \cdots + a_nv_n = 0$ for some $a_1, \ldots, a_n \in F$, then

$$a_1 = a_2 = \cdots = a_n = 0.$$

An *infinite set of vectors* is said to be *linearly independent* if every finite subset is a linearly independent set.

7.4 LINEAR DEPENDENCE AND INDEPENDENCE

We list some elementary properties that follow from the definition of linear dependence.

THEOREM 7.4.3 The set of (distinct) vectors $\{v_1,\ldots,v_n\}$, $n \geq 2$, from the vector space V over F is linearly dependent if and only if for some i, v_i is linearly dependent on $\{v_1,\ldots,v_{i-1},v_{i+1},\ldots,v_n\}$. If $n = 1$, then the set $\{v_1\}$ is linearly dependent if and only if $v_1 = 0$.

Proof Suppose that $v_1, \ldots, v_n$ is a linearly dependent set with $n \geq 2$. Then there are elements $a_1, \ldots, a_n \in F$ such that $a_1v_1 + \cdots + a_nv_n = 0$ and at least one of the a_j is different from 0. Assume that $a_i \neq 0$ for some i. Then $-a_iv_i = a_1v_1 + \cdots + a_{i-1}v_{i-1} + a_{i+1}v_{i+1} + \cdots + a_nv_n$, with $a_i \neq 0$. We can multiply both sides by the scalar $(-a_i)^{-1} = b$ and get

$$b(-a_i)v_i = 1v_i = v_i$$
$$= ba_1v_1 + \cdots + ba_{i-1}v_{i-1} + ba_{i+1}v_{i+1} + \cdots + ba_nv_n.$$

Hence v_i is linearly dependent on $\{v_1,\ldots,v_{i-1},v_{i+1},\ldots,v_n\}$.

If $n = 1$, $\{v_1\}$ is linearly dependent means there is an $a \in F$, $a \neq 0$, such that $av_1 = 0$. But we have seen previously that if $av_1 = 0$, then either $a = 0$ or $v_1 = 0$. Hence $v_1 = 0$.

Suppose conversely that v_i is linearly dependent on

$$\{v_1,\ldots,v_{i-1},v_{i+1},\ldots,v_n\}, \quad n \geq 2.$$

Then

$$v_i = av_1 + \cdots + a_{i-1}v_{i-1} + a_{i+1}v_{i+1} + \cdots + a_nv_n \quad \text{for } a_j \in F.$$

Hence

$$0 = a_1v_1 + \cdots + a_{i-1}v_{i-1} - v_i + a_{i+1}v_{i+1} + \cdots + a_nv_n$$

and the coefficient of v_i is -1, which is not 0, and the set $\{v_1,\ldots,v_n\}$ is linearly dependent by definition. If $n = 1$ and $v_1 = 0$, then choose any $a \in F$, $a \neq 0$, and we have $av_1 = 0$. Thus $\{v_1\}$ is linearly dependent. ∎

As we have remarked, the notion of linear dependence has an interpretation in terms of subspaces. Thus we have

COROLLARY 7.4.4 A set of (distinct) vectors $\{v_1,\ldots,v_n\}$ of the vector space V over F, $n \geq 2$, is linearly dependent if and only if the subspace spanned by $\{v_1,\ldots,v_n\}$ is spanned by a proper subset of $\{v_1,\ldots,v_n\}$.

Proof Since $n \geq 2$, if $\{v_1,\ldots,v_n\}$ is linearly dependent, then v_i, say, is a linear combination of the remaining vectors. Thus the subspace spanned by the set of $n - 1$ elements $\{v_1,\ldots,v_{i-1},v_{i+1},\ldots,v_n\}$ will contain the vector v_i and hence will span the same subspace.

Conversely, if the space spanned by $\{v_1,\ldots,v_n\}$ is spanned by a proper subset S, then at least one of the v_j's, say v_i, is not in S and is a linear combination of the vectors of S. Thus $\{v_1,\ldots,v_i,\ldots,v_n\}$ is linearly dependent. ∎

COROLLARY 7.4.5 Let S be a finite set of vectors of the vector space V over F, where S contains at least one nonzero vector. Then S contains a linearly independent subset T that spans the same space as S.

Proof The proof is by induction on the number of elements of S. If S has only one element, say v, then $v \neq 0$; hence S is itself linearly independent. Now suppose that S has n elements, $n > 1$, and the theorem is true for all integers $< n$. If S is itself a linearly independent set, then there is nothing to prove. Thus assume that S is linearly dependent. Hence by Corollary 7.4.4 the space spanned by S is spanned by a proper subset of S, say P. But P has fewer elements than S, and by the induction assumption, the theorem is true for P. Thus P contains a linearly independent subset (possibly P itself) that spans the same space as spanned by P, which is in turn the subspace spanned by S. Hence the theorem follows by induction. ∎

We have been careful to emphasize when working with linearly dependent sets of vectors that the vectors are *distinct*. Our purpose is to avoid the following type of confusion: Suppose the set $S = \{v_1,v_2\}$ and it turns out that $v_1 = v_2 \neq 0$. Then on the one hand the set S is linearly dependent because $1v_1 + (-1)v_2 = v_1 - v_2 = 0$ and both coefficients, 1, -1, are not zero. On the other hand, if $v_1 = v_2 \neq 0$, then the set $S = \{v_1,v_2\}$ is equal to the set $\{v_1\}$, and $\{v_1\}$ is a linearly independent set. For if $av_1 = 0$, then either $a = 0$ or $v_1 = 0$, and since $v_1 \neq 0$, we have $a = 0$, that is, $\{v_1\}$ is linearly independent. Thus when we speak of a set of vectors $\{v_1,\ldots,v_n\}$ being linearly dependent, it should be clear that we are assuming the vectors are distinct. Unfortunately this usage is not uniform throughout all texts. (See, for example, H. Paley and P. M. Weichsel, *A First Course in Abstract Algebra*. New York: Holt, Rinehart and Winston, Inc., 1966.)

Example 7.4.6 The subset $\{(1,-2),(-3,6)\}$ of $\mathbf{R}^2$ is a linearly dependent set over $\mathbf{R}$, since $3(1,-2) + (-3,6) = (0,0)$, while the set $\{(1,0),(1,1)\}$ is linearly independent. For if $a(1,0) + b(1,1) = (0,0)$, then

$$a(1,0) + b(1,1) = (a,0) + (b,b) = (a+b,b),$$

whence $a + b = 0$ and $b = 0$, which implies that $a = 0$.

Example 7.4.7 Let $S = \{(1,2,-3),(0,\tfrac{1}{2},-1),(3,7,-11)\}$ be a subset of $\mathbf{R}^3$. Then S is linearly dependent over $\mathbf{R}$, since

$$3(1,2,-3) + 2(0,\tfrac{1}{2},-1) - (3,7,-11) = (0,0,0).$$

7.4 LINEAR DEPENDENCE AND INDEPENDENCE

Therefore the subspace $\langle S \rangle$ spanned by S is spanned by

$$\{(1,2,-3),(0,\tfrac{1}{2},-1)\},$$

which is a linearly independent set. It is also easy to verify directly that the sets $\{(1,2,-3),(3,7,-11)\}$ and $\{(0,\tfrac{1}{2},-1),(3,7,-11)\}$ are also linearly independent sets and span the same subspace $\langle S \rangle$.

Example 7.4.8 Let $S = \{(1,0,1),(0,1,2),(-2,1,1)\}$ be a set of vectors in $\mathbf{R}^3$. To test for linear dependence, we let a, b, c be real numbers such that

$$a(1,0,1) + b(0,1,2) + c(-2,1,1) = (0,0,0).$$

Then

$$(a,0,a) + (0,b,2b) + (-2c,c,c) = (0,0,0),$$

whence

$$(a - 2c,\ b + c,\ a + 2b + c) = (0,0,0).$$

Thus

(1) $\quad a \qquad\quad - 2c = 0$
(2) $\quad\qquad\quad b + \quad c = 0$
(3) $\quad a + 2b + \quad c = 0.$

We can easily solve this system by eliminating one variable at a time. Thus subtracting (1) from (3), we get $2b + 3c = 0$, and subtracting twice (2) from this, we get $c = 0$. Substituting into (1) and (2), this gives $a = 0$ and $b = 0$. Hence since $a = b = c = 0$, S is linearly independent.

We see that questions involving linear dependence usually lead to the necessity for solving a system of linear equations. We will have more to say about this in Chapter 10.

Example 7.4.9 The set of monomials

$$S = \{x^n \mid n \text{ is a nonnegative integer}\}$$

of Example 7.3.6 is an example of an infinite linearly independent set. For if we take a finite subset of S, say $\{x^{n_1},\ldots,x^{n_r}\}$ with

$$n_1 > n_2 > \cdots > n_r \geq 0,$$

then any linear combination of these, say $a_1 x^{n_1} + a_2 x^{n_2} + \cdots + a_r x^{n_r}$, with $a_i \in \mathbf{R}$, is a polynomial function. It follows from Theorem 5.10.7 that such a function can be zero for at most n_1 values of x, unless all coefficients are zero. Hence if the function is identically zero, it follows that $a_1 = a_2 = \cdots = a_r = 0$ and the set $\{x^{n_1},\ldots,x^{n_r}\}$ is linearly independent. Therefore the set S is linearly independent.

EXERCISES

1. Show that the set of vectors $\{(1,2),(-\frac{1}{2},1),(0,1)\}$ in $\mathbf{R}^2$ is linearly dependent and find real numbers a, b, c not all zero such that $a(1,2) + b(-\frac{1}{2},1) + c(0,1) = (0,0)$.
2. Show the set of vectors $\{(1,1,1),(1,2,1),(2,1,1)\}$ is a linearly independent set. Show that $(1,0,0)$ is in the space spanned by this set.
3. Show that each of the sets

$$\{(1,2,-3),(3,7,-11)\} \quad \text{and} \quad \{(0,\tfrac{1}{2},-1),(3,7,-11)\}$$

of Example 7.4.7 is linearly independent.
4. Show that the vectors in Example 7.4.8 span the whole space $\mathbf{R}^3$.
5. Show that any set of vectors that contains the zero vector is linearly dependent.
6. Let S be a set of n distinct vectors in the vector space V over F, $n > 0$. Prove that if S is linearly independent and if T is a nonempty subset of S, then T is also linearly independent.
7. Let S be a set of n distinct vectors in the vector space V over F, $n > 0$. Prove that if there is a linearly dependent nonempty subset T of S, then S is linearly dependent.
8. Let S be a linearly independent set in the vector space V over F. Show that if T is a proper subset of S, then the subspace $\langle T \rangle$ spanned by T is a proper subspace of the subspace $\langle S \rangle$ spanned by S.
9. Give an example of a set of four vectors in $\mathbf{R}^3$ that is linearly dependent and such that any subset of three vectors is linearly independent.
10. Show that if $\{v_1,v_2,v_3\}$ is a linearly independent set in $\mathbf{R}^3$, then $\{v_1 + v_2, v_2 + v_3, v_1 + v_3\}$ is also a linearly independent set in $\mathbf{R}^3$.
11. Let v be a vector that is linearly dependent on a subset T of a vector space V over F. Show that if P is a subset of V that contains T, then v is linearly dependent on P.
12. Let S be a linearly independent subset of the vector space V. Let v be an element of V not in the subspace $\langle S \rangle$ spanned by S. Show that the set $S \cup \{v\}$ is linearly independent.

7.5 Bases

We have seen in the preceding sections that linearly independent sets of vectors often play a special role in describing vector spaces. For example, in the vector space $\mathbf{R}^n$, any vector v can be uniquely expressed in the form $a_1v_1 + \cdots + a_nv_n$, where v_i is the vector $v_i = (0,\ldots,0,1,0,\ldots,0)$, where the 1 appears in the ith position. In this section we study sets of vectors that play a role in an arbitrary vector space V similar to that of the set $\{v_1,\ldots,v_n\}$ in $\mathbf{R}^n$.

Definition 7.5.1 Let V be a vector space over the field F. A *basis for* V is a subset B of V such that

(a) B is a linearly independent set, and
(b) B spans V.

Since a basis B spans the vector space V, it follows that every element of V can be represented as a linear combination of a finite subset of B. This representation is unique.

THEOREM 7.5.2 Let B be a basis for the vector space V over F. Then if $v \in V$, $v \neq 0$, there exists a unique set of (distinct) vectors $v_1,\ldots,v_r \in B$ and a unique set of nonzero scalars $a_1, \ldots, a_r \in F$ such that

$$v = a_1 v_1 + \cdots + a_r v_r.$$

Proof Suppose there exist two sets of vectors and scalars, $v_1, \ldots, v_r$, $a_1, \ldots, a_r$ and $w_1, \ldots, w_s$, $b_1, \ldots, b_s$, with $v_i, w_j \in B$ and $a_i, b_j \in F$, $a_i \neq 0$, $b_j \neq 0$, $i = 1, \ldots, r, j = 1, \ldots, s$, such that

$$v = a_1 v_1 + \cdots + a_r v_r = b_1 w_1 + \cdots + b_s w_s.$$

Then $0 = a_1 v_1 + \cdots + a_r v_r - b_1 w_1 - \cdots - b_s w_s$. Now if $v_i \neq w_j$ for some i and all $j = 1, \ldots, s$, then $a_i = 0$, since the v's and w's come from an independent set. But by hypothesis $a_i \neq 0$. Hence, for each i, there is a j such that $v_i = w_j$. Similarly, for each j, there is an i such that $w_j = v_i$. Thus there is a one-to-one correspondence between the v_i's and the w_j's. Therefore, by renumbering, if necessary, we get $r = s$ and $v_i = w_i$, $i = 1, \ldots, r$. Hence $(a_1 - b_1)v_1 + \cdots + (a_r - b_r)v_r = 0$ and since $\{v_1,\ldots,v_r\}$ is a linearly independent set, $a_i = b_i, i = 1, \ldots, r$. This completes the proof of the theorem. ∎

The reader should observe that the set B in Theorem 7.5.2 may be either finite or infinite.

Definition 7.5.3 Let V be a vector space over F. A *minimal set of generators* is a set S of vectors such that

(a) $V = \langle S \rangle$,
(b) if $T \subset S$, $\langle T \rangle$ is a proper subspace of V.

THEOREM 7.5.4 Let V be a vector space over the field F. Let S be a minimal set of generators for V. Then S is a basis for V.

Proof Clearly, $V = \langle S \rangle$, so we need only prove that S is linearly independent.

284   VECTOR SPACES

Suppose S is linearly dependent. Then there exists a nonempty subset $T = \{v_1,\ldots,v_r\}$ of S such that T is linearly dependent. By Theorem 7.4.3, there is a $v_i \in T$ such that v_i is a linear combination of the elements of $T - \{v_i\}$. Thus, $S - \{v_i\}$ spans the same vector space as S. This contradicts that S is a minimal set of generators, and so S is linearly independent. Thus S is a basis for V. ∎

COROLLARY 7.5.5   Let S be a finite subset of the vector space V such that $V = \langle S \rangle$. Then S contains a subset B that is a basis for V.

Proof   Since S is a finite set of generators for V, S clearly has a smallest subset B that also spans V. Then B is a minimal set of generators for V, and so B is a basis for V, by Theorem 7.5.4. ∎

Example 7.5.6   Let $V = \mathbf{R}^3$. The set

$$S = \{(1,0,0),(0,1,0),(0,0,1),(1,-1,1)\}$$

is clearly a generating set for V. It is not minimal, however, since $S - \{(1,-1,1)\}$ also generates $\mathbf{R}^3$. The subset $B = \{(1,0,0),(0,1,0),(0,0,1)\}$ is clearly a minimal set of generators for $\mathbf{R}^3$ and is a basis for $\mathbf{R}^3$.

Another important concept is that of a maximal linearly independent subset of V.

Definition 7.5.7   Let V be a vector space over the field F.
A subset S of V is called a *maximal linearly independent* subset of V if

(a) S is linearly independent, and
(b)   if $S \subset T$, then T is linearly dependent.

The three notions of basis, minimal set of generators, and maximal linearly independent set are all equivalent.

THEOREM 7.5.8   Let V be a vector space over the field F. Let B be a subset of V. Then the following are equivalent:

(a) B is a basis for V.
(b) B is a maximal linearly independent subset of V.
(c) B is a minimal set of generators for V.

Proof   (a) implies (b). Let B be a basis. By definition, B is already linearly independent. We must show therefore that if $B \subset T$, then T is linearly dependent.

Suppose $B \subset T$ and T is linearly independent. Then there is a vector $v \in T$ such that $v \notin B$. Since B is a basis, $V = \langle B \rangle$, so

$$v = a_1 v_1 + \cdots + a_r v_r$$

for a unique choice of vectors $v_1, \ldots, v_r$ from B and scalars $a_1, \ldots, a_r$ from F. Then

$$0 = (-1)v + a_1v_1 + \cdots + a_rv_r,$$

whence $\{v_1,\ldots,v_r,v\}$ is linearly dependent. Thus T has a dependent subset, so T is itself linearly dependent. This implies B is a maximal linearly independent subset of V.

(b) implies (c). Let B be a maximal linearly independent subset of V. We must show B is a minimal set of generators.

First, let $v \in V$. If $v \notin \langle B \rangle$, then $B \cup \{v\}$ is a linearly independent subset of V, by Exercise 12, Section 7.4. This contradicts that B is a maximal linearly independent set. Thus $v \in \langle B \rangle$, and B spans V.

Second, suppose that B is not a minimal set of generators. Then there is an element $v \in B$ such that $V = \langle B \rangle = \langle B - \{v\} \rangle$. But $v \in V$, so $v \in \langle B - \{v\} \rangle$. This implies $v = a_1v_1 + \cdots + a_rv_r$, for scalars $a_1, \ldots, a_r \in F$ and vectors $v_1, \ldots, v_r \in B - \{v\} \subset B$. But then

$$0 = (-1)v + a_1v_1 + \cdots + a_rv_r,$$

and B is not independent. Thus, B is a minimal set of generators.

(c) implies (a). This is simply Theorem 7.5.4. ∎

We now suppose that the vector space V has a basis B. It is natural to ask whether B is unique. This is easily answered in the negative. For let $V = \mathbf{R}^2$. Then the set $B = \{(1,0),(0,1)\}$ clearly spans V, for the arbitrary vector $(a,b) = a(1,0) + b(0,1)$, and B is certainly minimal, for no subset of B spans $\mathbf{R}^2$. But the set $A = \{(1,0),(1,1)\}$ is also a basis for V, as is easily verified.

We note that the two sets A and B have the same number of elements. This leads us to a fundamental result in the theory of vector spaces.

THEOREM 7.5.9   Let V be a vector space over F with a basis B. If B has n elements, then every basis of V has n elements.

Before we can prove this theorem, we need the following lemma.

LEMMA 7.5.10   Let $A = \{v_1,\ldots,v_n\}$ be a linearly independent set of n vectors of the vector space V over F. Then any set B of linearly independent vectors in the subspace $\langle A \rangle$ spanned by A has at most n elements.

Proof   Let $B = \{u_1,\ldots,u_r\}$ be a linearly independent set of vectors in the subspace $\langle A \rangle$. Then u_1 is dependent on the set A; hence by Theorem 7.4.3 the set $\{v_1,\ldots,v_n,u_1\}$ is a linearly dependent set. Hence there are field elements, $a_1, \ldots, a_{n+1}$, not all zero, such that

$$a_1v_1 + \cdots + a_nv_n + a_{n+1}u_1 = 0.$$

Now if all of $a_1, \ldots, a_n$ are zero, then $a_{n+1} \neq 0$ and therefore $u_1 = 0$, a contradiction. Thus, at least one of $a_1, \ldots, a_n$ is not zero. Hence assume, by reindexing if necessary, that $a_1 \neq 0$. Thus v_1 is dependent on $A_1 = \{v_2,\ldots,v_n,u_1\}$, and A_1 spans the same space as A.

Next, we see that u_2 is in the space $\langle A_1 \rangle$ spanned by A_1, and so $\{v_2,\ldots,v_n,u_1,u_2\}$ is a linearly dependent set. As before there are field elements $b_1, \ldots, b_{n+1}$, not all zero, such that

$$b_1 v_2 + \cdots + b_{n-1} v_n + b_n u_1 + b_{n+1} u_2 = 0.$$

Again, if $b_1 = \cdots = b_{n-1} = 0$, then $b_n u_1 + b_{n+1} u_2 = 0$ and at least one of b_n or b_{n+1} is not zero. This contradicts that the subset $\{u_1,u_2\}$ of B is a linearly independent set. Thus, we may assume, by reindexing if necessary, that $b_1 \neq 0$, and so v_2 is dependent on $A_2 = \{v_3,\ldots,v_n,u_1,u_2\}$, whence $\langle A_2 \rangle = \langle A_1 \rangle = \langle A \rangle$.

Now suppose that $r > n$. Then, continuing in this way for a total of n steps, we get $A_n = \{u_1,\ldots,u_n\}$ spans the same subspace $\langle A \rangle$. But then u_{n+1} is in the space $\langle A_n \rangle$, and therefore the set $\{u_1,\ldots,u_n,u_{n+1}\} \subseteq B$ is linearly dependent—a contradiction, since a nonempty subset of a linearly independent set is linearly independent. Therefore, $r \leq n$. ∎

The procedure used in the proof above in obtaining the set $\{v_2,\ldots,v_n,u_1\}$ from the set $\{v_1,\ldots,v_n\}$ is usually called the *Steinitz exchange principle*. It allows us to replace one vector from a linearly independent set that spans a vector space V with another from a second linearly independent set that spans V, so that the resulting set still spans V.

Proof (of Theorem 7.5.9) Let B be a basis for V with m elements. Let A be another basis for V with n elements. From Lemma 7.5.10 $m \leq n$. But now, reversing the roles of A and B, it follows that $n \leq m$. Thus, $m = n$. ∎

COROLLARY 7.5.11 Let V be a vector space over the field F. Let V have a basis B with n elements. Let S be a subspace of V such that S has a basis A with n elements. Then $S = V$.

Proof If $S \subset V$, then there is a vector v in V such that $v \notin \langle A \rangle$. Then $A \cup \{v\}$ is a linearly independent set, and $A \cup \{v\}$ has $n + 1$ elements. But this is impossible by Lemma 7.5.10. Thus $S = V$. ∎

COROLLARY 7.5.12 Let V be a vector space over F, and suppose that V has a basis B that is infinite. Then every basis of V is infinite.

Proof The proof is left as an exercise. ∎

EXERCISES

1. Show that $\{(0,1),(1,1)\}$ is a basis for $\mathbf{R}^2$.
2. Show that $\{(1,1,1),(1,2,1),(2,1,1)\}$ is a basis for $\mathbf{R}^3$. Write $(0,0,1)$, $(0,1,0)$, and $(1,0,0)$ as linear combinations of the three basis vectors.
3. Find a basis for the vector space whose elements are real numbers of the form $a + b\sqrt{2}$, where a and b are rational numbers, over the field of rational numbers (vector addition and scalar multiplication are given by the usual addition and multiplication of real numbers).
4. Test the following sets of vectors in $\mathbf{R}^6$ for linear independence. In each case, find a basis for the subspace each set spans.
   (a) $\{(2,4,3,-1,-2,1),(1,1,2,1,3,1),(0,-1,0,3,6,2)\}$.
   (b) $\{(2,1,3,-1,4,-1),(-1,1-2,2,-3,3),(1,5,0,4,-1,7)\}$.
5. Express $(1,0,1)$ as a linear combination of the vectors
$$\{(2i,1,0),(2,-i,1),(0,1+i,1-i)\}$$
in $\mathbf{C}^3$ over $\mathbf{C}$, where $\mathbf{C}$ is the field of complex numbers.
6. In each of (a) and (b), give a basis for $\mathbf{R}^3$ that includes the given vectors.
   (a) $\{(0,0,1),(0,1,1)\}$.
   (b) $\{(1,1,2),(-2,0,1)\}$.
7. Show that the sets S and T of vectors in $\mathbf{R}^3$ span the same subspace.
$$T = \{(1,2,-3),(0,1,1)\},$$
$$S = \{(1,1,-4),(3,8,-7)\}.$$

By the Steinitz exchange principle, the vector $(1,1,-4)$ in S can be replaced by a vector in T so that the resulting set spans the same subspace. Find a vector in T that can replace $(1,1,-4)$.
8. (a) Let $E = \{(1,1,0,1),(1,1,0,0),(1,0,0,1)\}$ be a set of vectors in Q^4, the vector space of 4-tuples over the field Q of rational numbers. Show that E is a linearly independent set and find a vector v so that $E \cup \{v\}$ is a basis for Q^4. Is the choice of v unique?
   (b) Once v has been chosen in part (a), express the following vectors as linear combinations of the basis vectors: $(2,3,\frac{1}{2},5)$, $(1,2,3,4)$.
9. Prove Corollary 7.5.12.

7.6 Dimension

Theorem 7.5.9 asserts that the number of vectors in a basis of a vector space V is an invariant of that space. We assign a special name to that number.

Definition 7.6.1 Let V be a vector space over the field F. If V has a basis with n elements, then we say that V is an *n-dimensional vector space* or that V has *dimension n* over F. We denote this by $\dim_F V = n$, or more simply by $\dim V = n$ when the field F is clear from context. If V does not have a finite basis, then we say that V is infinite-dimensional over F, and we denote this by $\dim_F V = \infty$. The trivial vector space $V = 0$ is said to have dimension 0.

The use of the word dimension is in accord with ordinary usage. For we have already seen that the vector space $\mathbf{R}^2$ of points in the plane is spanned by the basis $\{(0,1),(1,0)\}$ and hence has dimension 2. Similarly it is easy to see that the space $\mathbf{R}^3$ of all points in Euclidean "3-space" is spanned by the basis $\{(1,0,0),(0,1,0),(0,0,1)\}$. Hence $\mathbf{R}^3$ has dimension 3, as we would expect.

The inclusion of the phrase "over F" in the definition of dimensionality is no mere pedantic adornment. For if we change the field of scalars to a proper subfield, then the dimension of the space may change.

Example 7.6.2 Let $\mathbf{C}^2$ be the vector space of pairs of complex numbers $\{(a,b) \mid a, b \in \mathbf{C}$, the field of complex numbers$\}$ over the field $\mathbf{C}$. Then this space has dimension 2, since it is spanned by $\{(0,1),(1,0)\}$. This would be the case for any space of the type F^2, F a field. Now let $\overline{\mathbf{C}^2}$ be the space of the same ordered pairs of complex numbers over the field $\mathbf{R}$ of real numbers. To see how these two spaces differ, consider the two vectors $(1,1)$, (i,i). They are linearly dependent over $\mathbf{C}$, since

$$i(1,1) + (-1)(i,i) = (0,0).$$

But the same vectors are linearly independent over $\mathbf{R}$. To see this, if $a(1,1) + b(i,i) = (a + bi, a + bi) = (0,0)$, then $a + bi = 0$, which implies $a = b = 0$, since a and b are real. Thus the vectors $(1,1)$ and (i,i) are linearly independent over $\mathbf{R}$. We claim that the dimension of $\overline{\mathbf{C}^2}$ over $\mathbf{R}$ is 4 and that a basis is $\{(0,1),(1,0),(0,i),(i,0)\}$. It is easy to see that this set spans $\overline{\mathbf{C}^2}$, since an arbitrary vector in $\overline{\mathbf{C}^2}$ has the form $(a + bi, c + di)$, with a, b, c, d real, and is equal to $a(1,0) + c(0,1) + b(i,0) + d(0,i)$. Moreover this set is linearly independent, since the linear combination above is the zero vector if and only if $a + bi = 0$ and $c + di = 0$, which can only occur if $a = b = c = d = 0$.

Since a basis is a linearly independent set, it is natural to ask whether a given linearly independent set of vectors can be enlarged to a basis, by the addition of sufficiently many vectors.

THEOREM 7.6.3 Let $A = \{v_1,\ldots,v_r\}$ be a linearly independent set in the n-dimensional vector space V over F. Then if $r < n$, there are vectors $v_{r+1}, \ldots, v_n$ such that $\{v_1,\ldots,v_n\}$ is a basis for V.

Proof We first remark that $r \leq n$ by Lemma 7.5.10. If $r = n$, then A is already a basis, since it is a maximal linearly independent set. Thus, assume $r < n$. If we consider all sets of linearly independent vectors that contain A, then each has at most n elements. Thus there must be one, say B, with a maximal number of elements. But then B is a maximal linearly independent set, and therefore B is a basis that contains A. ∎

We have seen that for finite-dimensional vector spaces the dimension is an invariant of the space. In order to see just how strong an invariant dimension is, we will consider the notion of isomorphism for vector spaces.

Definition 7.6.4 Let U and V be vector spaces over the same field F. Then U and V are said to be *isomorphic*, $U \cong V$, if there is a function $f\colon U \xrightarrow[\text{onto}]{1\text{-}1} V$ such that

(a) $f(u_1 + u_2) = f(u_1) + f(u_2)$ for $u_1, u_2 \in U$, and
(b) $f(au) = af(u)$ for $u \in U$ and $a \in F$.

Note that the function f establishes an isomorphism between the abelian groups $(U, +)$ and $(V, +)$ and "commutes" with the scalar multiplication.

We can now demonstrate the importance of the invariance of dimension.

THEOREM 7.6.5 Let U and V be two finite-dimensional vector spaces over the field F. Then $U \cong V$ if and only if they have the same dimension.

Proof Suppose first that U and V are isomorphic with $f\colon U \xrightarrow[\text{onto}]{1\text{-}1} V$, $f(u_1 + u_2) = f(u_1) + f(u_2)$ and $f(au) = af(u)$, for $u, u_1, u_2 \in U$ and $a \in F$. Let $\{u_1, \ldots, u_n\}$ be a basis for U and consider $A = \{f(u_1), \ldots, f(u_n)\}$. We claim that A is a basis for V. We first show that A is a linearly independent set. Thus suppose that

$$a_1 f(u_1) + \cdots + a_n f(u_n) = 0 \quad \text{for } a_1, \ldots, a_n \in F.$$

Then

$$\begin{aligned} 0 &= a_1 f(u_1) + \cdots + a_n f(u_n) \\ &= f(a_1 u_1) + \cdots + f(a_n u_n) \\ &= f(a_1 u_1 + \cdots + a_n u_n). \end{aligned}$$

Hence f maps the element $a_1 u_1 + \cdots + a_n u_n$ of U to the 0 of V. But f is an isomorphism from the abelian group $(U, +)$ to $(V, +)$; hence

$$a_1 u_1 + \cdots + a_n u_n = 0.$$

But the set $\{u_1, \ldots, u_n\}$ is linearly independent in U, and so

$$a_1 = a_2 = \cdots = a_n = 0.$$

Therefore $A = \{f(u_1),\ldots,f(u_n)\}$ is linearly independent. If A is not a basis for V, then there is $v \in V$ such that $A \cup \{v\} = \{f(u_1),\ldots,f(u_n),v\}$ is a linearly independent set in V. Since f is onto, there is an element $u \in U$ such that $v = f(u)$, and so $\{f(u_1),\ldots,f(u_n),f(u)\}$ is linearly independent. Now there exist scalars $a_1, \ldots, a_n \in F$ such that

$$u = a_1 u_1 + \cdots + a_n u_n,$$

whence $f(u) = f(a_1 u_1) + \cdots + f(a_n u_n) = a_1 f(u_1) + \cdots + a_n f(u_n)$, so v is linearly dependent on A, a contradiction. Thus A is a basis for V, and V also has dimension n.

Suppose conversely that U and V have the same dimension n. Let $A = \{u_1,\ldots,u_n\}$ be a basis for U and let $B = \{v_1,\ldots,v_n\}$ be a basis for V. Thus every element of U can be written uniquely in the form

$$u = a_1 u_1 + \cdots + a_n u_n, \quad \text{for } a_1, \ldots, a_n \in F.$$

Now define a mapping $f: U \longrightarrow V$ as follows:

$$f(u_i) = v_i, \quad i = 1, \ldots, n,$$

and in general $f(a_1 u_1 + \cdots + a_n u_n) = a_1 v_1 + a_2 v_2 + \cdots + a_n v_n$. We need to show now that f is 1-1 and onto and that f satisfies the two properties: (a) $f(u + u') = f(u) + f(u')$ and (b) $f(au) = af(u)$, for $u, u' \in U$ and $a \in F$. We leave the verification of these for the exercises. ∎

If F is a field and n is a positive integer, then the vector space F^n of all n-tuples of elements of F is often taken as a prototype of vector spaces of dimension n over F, since every vector space of dimension n over F is isomorphic to it. It should be clear that the set $\{\epsilon_1, \epsilon_2, \ldots, \epsilon_n\}$, where $\epsilon_1 = (1,0,\ldots,0)$, $\epsilon_2 = (0,1,0,\ldots,0)$, $\epsilon_3 = (0,0,1,0,\ldots,0)$, $\ldots$,

$$\epsilon_n = (0,\ldots,0,1),$$

is a basis for F^n and is sometimes called the *canonical basis*.

An important question about bases and dimension, which has been somewhat glossed over to this point, is the question of existence. That is, when does a vector space have a basis? A partial answer has been given in Corollary 7.5.5. We repeat it here.

THEOREM 7.6.6 (7.5.5) Let V be a vector space over F, $V \neq 0$, and let S be a finite subset of V that spans V. Then V has a finite basis.

The answer to the existence question for infinite dimensional vector spaces is also affirmative—that is, *every* vector space has a basis. The proof, however, requires the use of an axiom of set theory known as the axiom of choice. A discussion of this axiom is beyond the scope of this book. The interested reader is referred to H. Paley and P. M. Weichsel,

A First Course in Abstract Algebra. New York: Holt, Rinehart and Winston, Inc., 1966.

EXERCISES

1. Complete the proof of Theorem 7.6.5.
2. Let V be an infinite-dimensional vector space over the field F. Show that V contains an infinite linearly independent set.
3. Let V be an infinite-dimensional vector space over the field F. Is each infinite linearly independent set of vectors in V a basis for V? (*Hint:* See Example 7.3.6.)
4. Let **R** be the vector space of real numbers over the field Q of rational numbers.
   (a) Find a subspace of **R** over Q of dimension n, for each integer $n > 0$.
   (b) Find a proper subspace of **R** over Q of infinite dimension.
5. Let **C** be the complex numbers and **R** the real numbers. Use the fact that **C** is a vector space over **R** to prove that there does not exist a field F such that $\mathbf{R} \subset F \subset \mathbf{C}$.
6. Test for linear dependence of the set of vectors

$$\{(1,1,0),(1,0,1),(0,1,1)\}$$

   in $(Z_2)^3$ over the field Z_2.
7. Test for linear dependence of the set of vectors

$$\{(1,1,0),(1,0,1),(0,1,1)\}$$

   in $(Z_3)^3$ over the field Z_3.
8. How many distinct vectors are there in the vector space $(Z_2)^n$, for any positive integer n?
9. Find all subspaces of $(Z_3)^2$.
10. Let V be an n-dimensional vector space over the field F. Show that for each integer m, $m < n$, V has an m-dimensional subspace.
11. Let V be an n-dimensional vector space over the field F. Let U be a subspace of V. Show that U is a proper subspace of V if and only if $\dim_F U < n = \dim_F V$.

7.7 Theorems Relating Dimensions

We conclude this chapter with a useful result concerning subspaces of finite-dimensional vector spaces.

292 VECTOR SPACES

Definition 7.7.1 Let U_1 and U_2 be two subspaces of a vector space V. Then the *sum* $U_1 + U_2$ is defined to be the set

$$\{u_1 + u_2 \mid u_1 \in U_1 \text{ and } u_2 \in U_2\}.$$

THEOREM 7.7.2 Let U_1 and U_2 be subspaces of the vector space V. Then $U_1 + U_2$ is the subspace of V spanned by $U_1 \cup U_2$; that is, $U_1 + U_2$ is the intersection of all subspaces of V that contain both U_1 and U_2.

Proof If W is a subspace that contains both U_1 and U_2, then clearly W contains $U_1 + U_2$. Thus the theorem will be proved if we show that $U_1 + U_2$ is a subspace. We must show that $(U_1 + U_2, +)$ is a subgroup of $(V, +)$ and that $au \in U_1 + U_2$ for each $u \in U_1 + U_2$ and $a \in F$. Now let $u, u' \in U_1 + U_2$. Then $u = u_1 + u_2$ and $u' = u_1' + u_2'$ for $u_1, u_1' \in U_1$ and $u_2, u_2' \in U_2$. Thus

$$u - u' = u_1 + u_2 - (u_1' + u_2')$$
$$= (u_1 - u_1') + (u_2 - u_2') \in U_1 + U_2,$$

since U_1 and U_2 are subspaces. Hence $(U_1 + U_2, +)$ is a subgroup of $(V, +)$. If $u = u_1 + u_2$, then $au = a(u_1 + u_2) = au_1 + au_2$. Also, $au_1 \in U_1$ and $au_2 \in U_2$, since U_1 and U_2 are subspaces. Thus $au \in U_1 + U_2$ whenever $u \in U_1 + U_2$. This concludes the proof that $U_1 + U_2$ is a subspace. ∎

If V is an n-dimensional vector space and U_1 and U_2 are subspaces, then the dimensions of U_1 and U_2 are each at most n. The dimension of $U_1 \cap U_2$ is also bounded by the dimensions of U_1 and U_2. The following theorem relates the dimensions of U_1, U_2, $U_1 \cap U_2$, and $U_1 + U_2$. We will write dim W in place of $\dim_F W$ for a vector space W over F.

THEOREM 7.7.3 Let V be a vector space over F and let U_1 and U_2 be finite-dimensional subspaces of V. Then

$$\dim U_1 + \dim U_2 = \dim (U_1 + U_2) + \dim (U_1 \cap U_2).$$

Proof Let $\dim (U_1 \cap U_2) = r$, and let $\{v_1, \ldots, v_r\}$ be a basis for $U_1 \cap U_2$. It then follows from Theorem 7.6.3 that U_1 has a basis of the form $\{v_1, \ldots, v_r, u_1, \ldots, u_s\}$ and U_2 has a basis $\{v_1, \ldots, v_r, w_1, \ldots, w_t\}$, where $\dim U_1 = r + s$ and $\dim U_2 = r + t$. We now claim that

$$B = \{v_1, \ldots, v_r, u_1, \ldots, u_s, w_1, \ldots, w_t\}$$

is a basis for $U_1 + U_2$. Clearly, B spans $U_1 + U_2$. It only remains, therefore, to show that B is linearly independent.

Let

$$a_1 v_1 + \cdots + a_r v_r + b_1 u_1 + \cdots + b_s u_s + c_1 w_1 + \cdots + c_t w_t = 0.$$

Then
$$-\sum_{i=1}^{t} c_i w_i = \sum_{i=1}^{r} a_i v_i + \sum_{i=1}^{s} b_i u_i$$

and hence $-\sum_{i=1}^{t} c_i w_i$ is in U_1 as well as in U_2; that is

$$-\sum_{i=1}^{t} c_i w_i \text{ is in } U_1 \cap U_2.$$

Thus
$$-\sum_{i=1}^{t} c_i w_i = d_1 v_1 + \cdots + d_r v_r,$$

for scalars $d_1, \ldots, d_r \in F$, whence

$$\sum_{i=1}^{r} d_i v_i + \sum_{i=1}^{t} c_i w_i = 0,$$

and so, since the basis $\{v_1,\ldots,v_r,w_1,\ldots,w_t\}$ for U_2 is a linearly independent set, each d_i and c_i is 0. Hence

$$0 = \sum_{i=1}^{r} a_i v_i + \sum_{i=1}^{s} b_i u_i.$$

Since the basis $\{v_1,\ldots,v_r,u_1,\ldots,u_s\}$ for U_1 is also a linearly independent set, each a_i and b_i is 0. This proves that $\{v_1,\ldots,v_r,u_1,\ldots,u_s,w_1,\ldots,w_t\}$ is a basis for $U_1 + U_2$. Therefore dim $(U_1 + U_2) = r + s + t$. Since dim U_1 + dim $U_2 = (r + s) + (r + t)$ and

$$\dim (U_1 + U_2) + \dim (U_1 \cap U_2) = (r + s + t) + r,$$

the theorem is proved. ∎

Example 7.7.4 Let U_1 and U_2 be two subspaces of the vector space $\mathbf{R}^4$ spanned by

$$\{(1,-1,2,-3),(1,1,2,0),(3,-1,6,-6)\} \quad \text{and} \quad \{(0,-2,0,-3),(1,0,1,0)\},$$

respectively. We will compute the dimension of U_1, U_2, $U_1 \cap U_2$, and $U_1 + U_2$. Clearly dim $U_2 = 2$, since the two vectors listed are linearly independent.

Next we test the generators of U_1 for linear independence. Let

$$a(1,-1,2,-3) + b(1,1,2,0) + c(3,-1,6,-6) = (0,0,0,0).$$

Then

(1) $\qquad a + b + 3c = 0$
(2) $\qquad -a + b - c = 0$
(3) $\qquad 2a + 2b + 6c = 0$
(4) $\qquad -3a - 6c = 0.$

Notice that (3) may be discarded, since it is simply twice (1). From (4) we get $a = -2c$, and, substituting this relation in (1) and (2), we have

$$-2c + b + 3c = b + c = 0$$
$$2c + b - c = b + c = 0.$$

Thus our system reduces to

$$a + 2c = 0$$
$$b + c = 0$$

which is satisfied by $b = 1$, $c = -1$, $a = 2$. Thus each of the three vectors being considered is dependent on the other 2. Hence U_1 is spanned by $\{(1,-1,2,-3),(1,1,2,0)\}$, and it is easy to show that the two vectors $(1,-1,2,-3)$ and $(1,1,2,0)$ are linearly independent. Thus dim $U_1 = 2$.

Finally, we check the dimension of $U_1 \cap U_2$ by determining the elements in the vector space $U_1 \cap U_2$. We thus determine all vectors that satisfy a relation

$$a(0,-2,0,-3) + b(1,0,1,0) = c(1,-1,2,-3) + d(1,1,2,0).$$

Adding and setting components equal, we obtain

(1) $\qquad a \cdot 0 + b \cdot 1 = b = c \cdot 1 + d \cdot 1 = c + d,$
(2) $\qquad a(-2) + b \cdot 0 = -2a = c(-1) + d \cdot 1 = -c + d,$
(3) $\qquad a(0) + b \cdot 1 = b = c(2) + d(2) = 2c + 2d,$
(4) $\qquad a(-3) + b(0) = -3a = c(-3) + d(0) = -3c.$

(1) gives us that $b = c + d$ and (3) yields $b = 2(c + d)$. Hence $b = 0$, and so the only possible vectors common to both U_1 and U_2 are multiples of $(0,-2,0,-3)$. In fact we get from (4) that $a = c$ and from (1) that $c = -d$. Thus for $a = c = -d = 1$ we get

$$(0,-2,0,-3) = (1,-1,2,-3) - (1,1,2,0)$$

and $U_1 \cap U_2$ is the 1-dimensional space spanned by $(0,-2,0,-3)$. Now from Theorem 7.7.3,

$$\dim (U_1 + U_2) = \dim U_1 + \dim U_2 - \dim (U_1 \cap U_2),$$

which in this example yields dim $(U_1 + U_2) = 2 + 2 - 1 = 3$. Thus, the set of vectors $\{(1,0,1,0),(1,-1,2,-3),(1,1,2,0)\}$ is a basis for $U_1 + U_2$, since the union of bases for U_1 and U_2 spans $U_1 + U_2$, and $(0,-2,0,-3)$ is dependent on $\{(1,-1,2,-3),(1,1,2,0)\}$.

EXERCISES

1. Let U_1 and U_2 be subspaces of $\mathbf{R}^3$ spanned by $\{(1,1,1),(1,2,-1)\}$ and $\{(1,0,1),(0,-1,-1)\}$, respectively. Find the dimensions of U_1, U_2, $U_1 \cap U_2$, and $U_1 + U_2$.
2. Let S and T be subspaces of a vector space V with dim $S = s$ and dim $T = t$. What are the largest and smallest possible values of dim $(S + T)$ and dim $(S \cap T)$, in terms of s and t?
3. Let S and T be subspaces of a vector space V, such that

    (a) $S + T = V$ and (b) $S \cap T = 0$.

    (a) Show that every vector of V can be written uniquely as a sum $x + y$, where $x \in S$ and $y \in T$.
    (b) Show that the set union of a basis of S and a basis of T is a basis of V.
4. Let V be a finite-dimensional vector space. Show that if U is a proper subspace of V, then V contains another subspace T such that

    (a) $U + T = V$ and (b) $U \cap T = 0$.

5. Let V be a vector space over a field F, with dim $V = n$. Let U_1 and U_2 be subspaces of V. Prove that $U_1 \cap U_2 = 0$ if and only if $V = U_1 + U_2$ and dim $V = $ dim $U_1 + $ dim U_2.
6. In Exercise 4, show by example that the subspace T need not be unique.
7. Let V be an n-dimensional vector space over the field F. Suppose $V = U_1 + T_1$ and $V = U_2 + T_2$, with $U_1 \cap T_1 = 0$, $U_2 \cap T_2 = 0$. Show that dim $U_1 \leq$ dim U_2 if and only if dim $T_2 \leq$ dim T_1.

8 Linear Transformations and Matrices

8.1 Examples of Linear Transformations and Elementary Results

It is not an exaggeration to say that much of mathematics is the study of functions. The reader will find ample evidence to support this claim by recalling his study of calculus and the abstract algebra already presented in this book. The binary operations used to define groups and rings are examples of functions and other important functions are homomorphisms and isomorphisms.

In this section we will define a special class of functions that map vector spaces to vector spaces and that have the important property that they are "compatible" with the vector-space operation of forming linear combinations.

Definition 8.1.1 Let U and V be vector spaces over a field K. The function $T\colon U \longrightarrow V$ is called a *linear transformation* if

(a) $T(au) = aT(u)$ for all $a \in K$ and all $u \in U$, and
(b) $T(u_1 + u_2) = T(u_1) + T(u_2)$.

Notice that since $T(u) \in V$, $aT(u)$ is simply a scalar multiple of $T(u)$ in V. In (b), we have, as usual, used the same symbol $+$ to represent vector addition in both U and V.

Properties (a) and (b) are often combined together in the equivalent form

$$T(a_1u_1 + a_2u_2) = a_1T(u_1) + a_2T(u_2)$$

for all $a_1, a_2 \in K$ and $u_1, u_2 \in U$.

THEOREM 8.1.2 Let $T\colon U \longrightarrow V$ be a function from the vector space U over the field K to the vector space V over the field K. Then T is a linear

8.1 LINEAR TRANSFORMATIONS AND ELEMENTARY RESULTS

transformation from U to V if and only if

$$T(a_1 u_1 + a_2 u_2) = a_1 T(u_1) + a_2 T(u_2),$$

for all $a_1, a_2 \in K$ and all $u_1, u_2 \in U$.

Proof If T is a linear transformation, then by (b) of Definition 8.1.1, $T(a_1 u_1 + a_2 u_2) = T(a_1 u_1) + T(a_2 u_2)$, since $a_1 u_1$ and $a_2 u_2$ are elements of U. Now by (a), $T(a_1 u_1) = a_1 T(u_1)$ and $T(a_2 u_2) = a_2 T(u_2)$. Hence

$$T(a_1 u_1 + a_2 u_2) = a_1 T(u_1) + a_2 T(u_2).$$

Conversely let T satisfy $T(a_1 u_1 + a_2 u_2) = a_1 T(u_1) + a_2 T(u_2)$. Then if we let $a_1 = a_2 = 1$, where 1 is the multiplicative identity of K, then $T(u_1 + u_2) = T(u_1) + T(u_2)$, which is (b) of the definition. Now choose $a_2 = 0$. Then $T(a_1 u_1 + 0 u_2) = T(a_1 u_1 + 0) = T(a_1 u_1)$. On the other hand,

$$T(a_1 u_1 + 0 u_2) = a_1 T(u_1) + 0 T(u_2) = a_1 T(u_1) + 0 = a_1 T(u_1).$$

Therefore, $T(a_1 u_1) = a_1 T(u_1)$, and hence T is a linear transformation. ∎

It is easy to see that the linearity property extends to arbitrarily long finite sums. Thus if T is a linear transformation,

$$T\left(\sum_{i=1}^{n} a_i u_i\right) = T(a_1 u_1 + \cdots + a_n u_n) = a_1 T(u_1) + \cdots + a_n T(u_n)$$
$$= \sum_{i=1}^{n} a_i T(u_i), \quad \text{for } u_i \in U \text{ and } a_i \in K, \quad i = 1, \ldots, n.$$

See Exercise 4.

The list of examples that follows will not only illustrate the wide variety of functions that are linear transformations but will also give some geometrical justification for the use of the term linear.

Example 8.1.3 Let U and V be the same vector space $\mathbf{R}^2$ over the field of real numbers $\mathbf{R}$ ($\mathbf{R}^2$ is the usual Cartesian plane), and let L be a straight line in $\mathbf{R}^2$. We may think of L as a set of vectors with tails at the origin and heads on the line L as illustrated in Figure 8.1.

If u and v are two of these vectors ($u \neq v$), then the set of vectors with heads on the line L is precisely the set $\mathcal{L} = \{au + (1 - a)v \mid a \in \mathbf{R}\}$. We leave the verification of this fact to the exercises. Now let T be a linear transformation from $\mathbf{R}^2$ to $\mathbf{R}^2$. Then

$$T(au + (1 - a)v) = aTu + (1 - a)Tv.$$

Thus the image of the set under T is just the set

$$\mathfrak{I} = \{aTu + (1 - a)Tv \mid a \in \mathbf{R}\}.$$

298 LINEAR TRANSFORMATIONS AND MATRICES

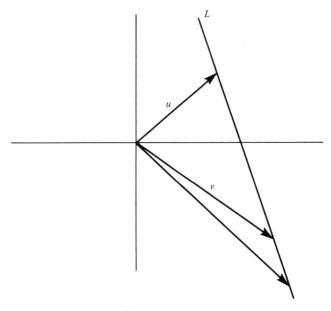

Figure 8.1

Now if either Tu or Tv is not the zero vector, then $\mathfrak{I}$ is a straight line. Hence a linear transformation either maps straight lines into straight lines or, if $Tu = Tv = 0$, maps a straight line into the origin (the zero vector).

Example 8.1.4 Again let $U = V = \mathbf{R}^2$ and consider the mapping $T: \mathbf{R}^2 \longrightarrow \mathbf{R}^2$ defined by $T(a,b) = (b,a)$, with (a,b) a vector in $\mathbf{R}^2$ (Figure 8.2). T is a linear transformation, since

$$\begin{aligned} T(r(a,b) + s(c,d)) &= T((ra,rb) + (sc,sd)) \\ &= T(ra + sc, rb + sd) \\ &= (rb + sd, ra + sc) \\ &= (rb,ra) + (sd,sc) \\ &= r(b,a) + s(d,c) \\ &= rT(a,b) + sT(c,d). \end{aligned}$$

The reader should notice that this mapping has the effect of reflecting the plane through the line $y = x$.

Example 8.1.5 Let T_θ be the mapping that rotates the plane through an angle of θ in the counterclockwise direction about the origin (Figure 8.3). If (x,y) is a point (vector) in $\mathbf{R}^2$, then $T_\theta(x,y)$ is given by $(x \cos \theta - y \sin \theta, x \sin \theta + y \cos \theta)$. (The reader should consult a text in analytic geometry, where this rotation is discussed in detail.) We will

8.1 LINEAR TRANSFORMATIONS AND ELEMENTARY RESULTS

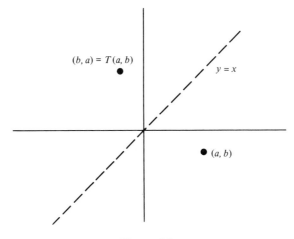

Figure 8.2

show, in two steps, that T_θ is a linear transformation. First,

$$T_\theta(a(x,y)) = T_\theta(ax,ay) = (ax \cos \theta - ay \sin \theta, ax \sin \theta + ay \cos \theta)$$
$$= a(x \cos \theta - y \sin \theta, x \sin \theta + y \cos \theta) = aT_\theta(x,y).$$

Second,

$$T_\theta((x,y) + (u,v)) = T_\theta(x + u, y + v)$$
$$= ((x + u) \cos \theta - (y + v) \sin \theta,$$
$$(x + u) \sin \theta + (y + v) \cos \theta)$$
$$= (x \cos \theta - y \sin \theta + u \cos \theta - v \sin \theta,$$
$$x \sin \theta + y \cos \theta + u \sin \theta + v \cos \theta)$$
$$= (x \cos \theta - y \sin \theta, x \sin \theta + y \cos \theta)$$
$$+ (u \cos \theta - v \sin \theta, u \sin \theta + v \cos \theta)$$
$$= T_\theta(x,y) + T_\theta(u,v).$$

Example 8.1.6 Let U be the set of all real-valued differentiable functions with domain the set $\mathbf{R}$ of all real numbers and let V be the set of all real-valued functions with the same domain R. We have already shown that these two sets are vector spaces over the field of real numbers. (See Section 7.2.)

Now define the function $T: U \longrightarrow V$ by $T(f) = f'$ for $f \in U$, where f' is the derivative of f. Then the statement that T is a linear transformation is equivalent to two theorems that are usually studied in a beginning calculus course. The first theorem says that the derivative of a function that has been multiplied by a constant is that constant times the derivative of the function, that is, $(af)' = af'$, or equivalently $T(af) = aT(f)$. The second theorem says that the derivative of a sum is the sum of the

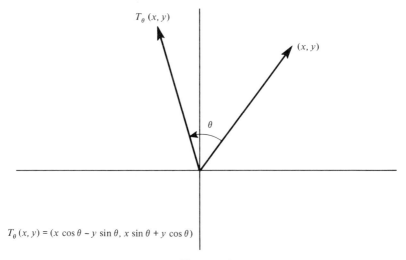

Figure 8.3

derivatives, that is,

$$(f + g)' = f' + g' \quad \text{or} \quad T(f + g) = T(f) + T(g).$$

Thus T is a linear transformation.

Example 8.1.7 Let U be the vector space of continuous real-valued functions with domain the set A of nonnegative real numbers, and let V be the set of all real-valued functions with the same domain. Define $T: U \longrightarrow V$ as follows: for $f \in U$,

$$[T(f)](x) = \int_0^x f(t)\, dt, \quad \text{for all } x \text{ in } A.$$

That is, if $f \in U$, then $T(f)$ is the function whose value at $a \geq 0$ is the number

$$\int_0^a f(t)\, dt.$$

T is well defined, since the continuity of f guarantees that the integral exists. T is a linear transformation, since

$$\int_0^x af(t)\, dt = a \int_0^x f(t)\, dt,$$

that is, $T(af) = aT(f)$, and since

$$\int_0^x (f(t) + g(t))\, dt = \int_0^x f(t)\, dt + \int_0^x g(t)\, dt,$$

8.1 LINEAR TRANSFORMATIONS AND ELEMENTARY RESULTS

that is, $T(f + g) = T(f) + T(g)$. Thus, for example,

$$T(x^2) = \frac{x^3}{3} \quad \text{and} \quad T(\sin x) = -\cos x + 1.$$

The remainder of this section gives several important properties of linear transformations that we will need shortly.

First, we prove an easy result, that will be used from time to time.

THEOREM 8.1.8 Let V and U be vector spaces over the field K and let $T\colon V \longrightarrow U$ be a linear transformation. Then

(a) $T(0) = 0$.
(b) $T(-v) = -T(v)$, for every $v \in V$.

Proof (a) $0 = 0v$, for any vector $v \in V$. Then

$$T(0) = T(0v) = 0T(v) = 0.$$

(b) In V, $-v = (-1)v$, for every $v \in V$. Then

$$T(-v) = T((-1)v) = (-1)T(v) = -T(v). \quad \blacksquare$$

THEOREM 8.1.9 Let U and V be vector spaces over the field K. Let U have a basis $\{u_1,\ldots,u_n\}$. Let $\{v_1,\ldots,v_n\}$ be a set of vectors in V. Then there is a unique linear transformation $T\colon U \longrightarrow V$ such that $T(u_i) = v_i$, for all i, $i = 1, \ldots, n$.

Proof To show that such a linear transformation exists, define T as follows: if

$$u = \sum_{i=1}^{n} a_i u_i,$$

with $a_i \in K$, put

$$T(u) = \sum_{i=1}^{n} a_i v_i.$$

We will use Theorem 8.1.2 to show that T is a linear transformation. Let

$$w_1 = \sum_{i=1}^{n} c_i u_i, \qquad c_i \in K,$$

and let

$$w_2 = \sum_{i=1}^{n} d_i u_i, \qquad d_i \in K.$$

Let $a, b \in K$, also. Then

$$T(aw_1 + bw_2) = T\left(a\sum_{i=1}^{n} c_i u_i + b\sum_{i=1}^{n} d_i u_i\right)$$

$$= T\left(\sum_{i=1}^{n}(ac_i)u_i + \sum_{i=1}^{n}(bd_i)u_i\right)$$

$$= T\left(\sum_{i=1}^{n}(ac_i + bd_i)u_i\right)$$

$$= \sum_{i=1}^{n}(ac_i + bd_i)v_i$$

$$= \sum_{i=1}^{n}(ac_i)v_i + \sum_{i=1}^{n}(bd_i)v_i$$

$$= a\sum_{i=1}^{n} c_i v_i + b\sum_{i=1}^{n} d_i v_i$$

$$= aT(w_1) + bT(w_2),$$

so T is a linear transformation.

Next, observe that the basis element u_j has a unique representation as a sum of basis elements, namely,

$$u_j = 0u_1 + \cdots + 0u_{j-1} + 1u_j + 0u_{j+1} + \cdots + 0u_n,$$

so $T(u_j) = 0v_1 + \cdots + 1v_j + \cdots + 0v_n = v_j$.

We now need only verify that T is unique. Thus if S is a linear transformation that satisfies $S(u_i) = v_i$, $i = 1, \ldots, n$, and if

$$u = \sum_{i=1}^{n} a_i u_i$$

is an arbitrary vector in U, then

$$S(u) = S\left(\sum_{i=1}^{n} a_i u_i\right) = \sum_{i=1}^{n} S(a_i u_i)$$

$$= \sum_{i=1}^{n} a_i S(u_i) = \sum_{i=1}^{n} a_i v_i.$$

Hence $S(u) = T(u)$, and so T is unique. ∎

Theorem 8.1.9 may be paraphrased as follows: a linear transformation T on a vector space U is completely determined by its action on a basis of U. The procedure used in the proof of Theorem 8.1.9 is often described

8.1 LINEAR TRANSFORMATIONS AND ELEMENTARY RESULTS 303

be saying: first, specify the action of T on a basis of U and then extend T to all of U by linearity.

An easy, but important, theorem is the next result.

THEOREM 8.1.10 Let V and U be vector spaces over the field K. Let $T: V \longrightarrow U$ be a linear transformation. Then the set

$$T(V) = \{T(v) \mid v \in V\}$$

is a subspace of U.

Proof The proof is left as an exercise. ∎

THEOREM 8.1.11 Let V and U be vector spaces over the field K, and let U be finite dimensional. Let $T: V \longrightarrow U$ be a linear transformation. Let $\mathcal{B}$ be a basis for V. Then the set $T(\mathcal{B}) = \{T(v) \mid v \in \mathcal{B}\}$ contains a basis for the vector space $T(V) = \{T(v) \mid v \in V\}$.

Proof $T(V)$ is a vector space, by Theorem 8.1.10.

Now let $u \in T(V)$. Then $u = T(v)$, for some $v \in V$. Since $\mathcal{B}$ is a basis for V, there are vectors $v_1, \ldots, v_r \in \mathcal{B}$ and scalars $a_1, \ldots, a_r \in K$ such that

$$v = \sum_{i=1}^{r} a_i v_i.$$

Thus

$$u = T(v) = T\left(\sum_{i=1}^{r} a_i v_i\right) = \sum_{i=1}^{r} a_i T(v_i).$$

Now each $T(v_i) \in T(\mathcal{B})$, and therefore we have shown that $T(\mathcal{B})$ spans $T(V)$. Since U is finite dimensional, $T(V) \subseteq U$ is also finite dimensional. By Corollary 7.5.5, the spanning set $T(\mathcal{B})$ of $T(V)$ contains a basis for $T(V)$. ∎

The condition that U be finite dimensional is in fact unnecessary. We needed to insert it, however, since we have only proved the existence of bases for finite dimensional vector spaces.

COROLLARY 8.1.12 Let $T: V \longrightarrow U$ be a linear transformation from the finite-dimensional vector space V to the finite-dimensional vector space U. Then $\dim V \geq \dim T(V)$.

Proof Let $\dim V = n$. Then V has a basis $\mathcal{B}$ containing n elements. It follows from Theorem 8.1.11 that $T(\mathcal{B})$ contains a basis for $T(V)$. But the number of distinct elements in $T(\mathcal{B})$ is not greater than the number of elements in $\mathcal{B}$. Hence $\dim T(V) \leq n = \dim V$. ∎

We remarked earlier in this section that a linear transformation is a mapping from one vector space to another that is "compatible" with the vector space operations. The notion of compatibility of operations is, of course, precisely the idea behind the general notion of homomorphism. In fact, the definition of linear transformation is the exact analogue of the definition of a homomorphism given for other algebraic systems. Hence we can think of linear transformations as vector space homomorphisms, and so develop all of the usual concepts, such as kernel, image, and quotient space. We will discuss all of this at the appropriate places in the sequel.

EXERCISES

1. Let u and v be nonzero vectors in the plane. Show that the heads of the vectors in the set
$$\mathcal{L} = \{au + (1-a)v \mid a \in \mathbf{R}\}$$
all lie on the same straight line.
2. Let L be a line in the plane. If u and v are two different points on the line, then show that any point z on L can be expressed in the form $z = au + (1-a)v$, for some real number a.
3. Show that the mapping $T: \mathbf{R}^2 \longrightarrow \mathbf{R}^2$ defined by $Tu = 3u$ is a linear transformation. It is usually called a "stretching." Why?
4. Prove by induction that if $T: U \longrightarrow V$ is a linear transformation and if $u_i \in U$, for $i = 1, \ldots, n$, then
$$T\left(\sum_{i=1}^{n} a_i u_i\right) = \sum_{i=1}^{n} a_i T(u_i), \quad a_i \in K, i = 1, \ldots, n.$$
5. Describe a linear transformation from $\mathbf{R}^2$ to $\mathbf{R}^2$ that maps the line $y = x$ to the line $y = 2x$.
6. Describe a linear transformation from $\mathbf{R}^2$ to $\mathbf{R}^2$ that maps the line $y = 1 - x$ to the line $y = x$.
7. Let $T: U \longrightarrow V$ be a linear transformation. Show that $T(U) = \{T(u) \mid u \in U\}$ (the image of U under T) is a subspace of V.
8. Let $U = V = \mathbf{R}^2$, with basis $e_1 = (1,0)$ and $e_2 = (0,1)$. Let
$$T: U \longrightarrow V$$
be a linear transformation defined by
$$T(e_1) = 2e_1 + e_2,$$
$$T(e_2) = e_1 - e_2.$$
   (a) Find $T(2,3)$; $T(1,1)$; $T(e_1 - 3e_2)$.
   (b) Find $T(a,b)$, for the arbitrary vector (a,b) of U.
   (c) Show that $T(U)$ has dimension 2.

8.1 LINEAR TRANSFORMATIONS AND ELEMENTARY RESULTS

9. Let $U = V = \mathbf{R}^2$ with basis $e_1 = (1,0)$ and $e_2 = (0,1)$. Let
$$T: U \longrightarrow V$$
be a linear transformation defined by
$$T(e_1) = e_1,$$
$$T(e_2) = 2e_2.$$

   (a) On the same set of coordinate axes, draw the graphs of the circle $x^2 + y^2 = 4$ and the ellipse $x^2/4 + y^2/1 = 1$.
   (b) Show that if $(x_0, y_0) = x_0 e_1 + y_0 e_2$ is a point on the ellipse, then $T(x_0, y_0)$ is a point on the circle.

10. Let $U = V = \mathbf{R}^2$, with basis $e_1 = (1,0)$ and $e_2 = (0,1)$. Let
$$x^2 + y^2 = r^2 \quad \text{and} \quad \frac{x^2}{a^2} + \frac{y^2}{b^2} = 1$$
be the equations of a circle C and ellipse E, respectively, in $\mathbf{R}^2$. Show that there is a linear transformation T such that $T(C) = E$. What are the values $T(e_1)$, $T(e_2)$ that you are using here?

11. Let $U = V = \mathbf{R}^3$, with basis $e_1 = (1,0,0)$, $e_2 = (0,1,0)$, $e_3 = (0,0,1)$. Let $T: U \longrightarrow V$ be a linear transformation defined by
$$T(e_1) = 2e_2 + e_3, \qquad T(e_2) = e_1 - e_3,$$
$$T(e_3) = -e_1 + 4e_2 + 3e_3.$$

   (a) Find $T(1,2,3)$.
   (b) Find $T(a,b,c)$, for the arbitrary vector (a,b,c) of U.
   (c) Show that $T(U)$ has dimension 2.

12. Let $U = V = \mathbf{R}^3$, with basis $e_1 = (1,0,0)$, $e_2 = (0,1,0)$, and $e_3 = (0,0,1)$. Let $T: U \longrightarrow V$ be a linear transformation defined by
$$T(e_1) = 3e_1 - e_2,$$
$$T(e_2) = 6e_1 - 2e_2,$$
$$T(e_3) = -12e_1 + 4e_2.$$

   (a) Find $T(1,1,1)$; $T(2e_1 + e_2 + e_3)$.
   (b) Find $T(a,b,c)$ for any vector (a,b,c) in $\mathbf{R}^3$.
   (c) What is the dimension of $T(U)$?

13. In $\mathbf{R}^3$, let $x^2 + y^2 + z^2 = 4$ be the equation of a sphere S and let $(x^2/4) + (y^2/4) + z^2 = 1$ be the equation of an ellipsoid E. Show that there is a linear transformation $T: \mathbf{R}^3 \longrightarrow \mathbf{R}^3$ such that $T(S) = E$.

14. In $\mathbf{R}^2$, let $x^2 - y^2 = 1$ be the equation of a hyperbola H and let $x^2 + y^2 = 1$ be the equation of a circle C. Prove that there is no linear transformation T such that $T(C) = H$. [*Hint*: For $ae_1 + be_2$

on C, note that $a^2 + b^2 = 1$; hence show that the coordinates of $T(ae_1 + be_2)$ are bounded.]

15. Let T and S be linear transformations from the vector space U over the field K to the vector space V over K. Define a function $f: U \longrightarrow V$ as follows:

$$f(u) = T(u) - S(u), \quad \text{for all } u \in U.$$

Show that f is a linear transformation from U to V.

16. Let $T: \mathbf{R}^2 \longrightarrow \mathbf{R}^2$ be a linear transformation defined by

$$T(e_1) = e_1 + e_2,$$
$$T(e_2) = e_1 - e_2,$$

where $e_1 = (1,0)$, $e_2 = (0,1)$.
 (a) Show that T is a 1–1 and onto function.
 (b) Determine $T^{-1}(e_1)$, $T^{-1}(e_2)$.
 (c) Show that $T^{-1}: \mathbf{R}^2 \longrightarrow \mathbf{R}^2$ is also a linear transformation.

17. Let V and U be vector spaces over the field K, and let $T: V \longrightarrow U$ be a linear transformation. Suppose dim $V = n$, dim $U = m$. Prove each of the following:
 (a) If $T(V) = U$, then $n \geq m$.
 (b) If $n < m$, then $T(V) \subset U$.
 (c) If $n \leq m$, and if $T(V) = U$, then $n = m$.

8.2 Range Space, Null Space, and Quotient Space of a Linear Transformation

Let U and V be vector spaces over a field K, and let $T: U \longrightarrow V$ be a linear transformation. As we mentioned in the previous section, the notions of kernel, image, and quotient group, associated with homomorphisms of groups, all carry over to linear transformations.

THEOREM 8.2.1 Let $T: U \longrightarrow V$, where U and V are vector spaces over a field K, and T is a linear transformation. Then

 (a) Im $T = \{T(u) \mid u \in U\}$ is a subspace of V.
 (b) Ker $T = \{u \in U \mid T(u) = 0\}$ is a subspace of U.

Proof (a) This is Theorem 8.1.10.
 (b) Let $u_1, u_2 \in \ker T$, and let $\alpha, \beta \in K$. Then

$$T(\alpha u_1 + \beta u_2) = \alpha T(u_1) + \beta T(u_2) = \alpha 0 + \beta 0 = 0.$$

Thus $\alpha u_1 + \beta u_2 \in K$, and so ker T is a subspace of U. ∎

We generally use special terminology for Im T and ker T.

8.2 RANGE SPACE, NULL SPACE, AND QUOTIENT SPACE

Definition 8.2.2 Let $T: U \longrightarrow V$, U, V vector spaces over a field K, and T a linear transformation. We call Im T the *range of T* and we call ker T the *null space of T*. We will denote the range of T by $R(T)$, and the null space of T by $N(T)$.

In analogy to the concept of a quotient group, we could define the notion of quotient space. We will not need this idea in the sequel, so we leave the formulation of the definition to the reader. It turns out, as one would expect, that $U/N(T) \cong R(T)$ as vector spaces.

Definition 8.2.3 Let U and V be vector spaces over a field K and let $T: U \longrightarrow V$ be a linear transformation. The *nullity of T*, denoted $n(T)$, is defined as dim $N(T)$, and the *rank of T*, denoted $r(T)$, is defined as dim $R(T)$.

In the case that U is finite dimensional (we need no restriction on V), an important relation holds between $n(T)$ and $r(T)$.

THEOREM 8.2.4 Let U and V be vector spaces over the field K. Let $T: U \longrightarrow V$ be a linear transformation, and let U be finite dimensional. Then $r(T) + n(T) = \dim U$.

Proof Since $N(T)$ is a subspace of U, $N(T)$ has a basis, say

$$\{u_1, \ldots, u_{n(T)}\}.$$

Then if dim $U = n$, this basis for $N(T)$ can be supplemented with $n - n(T)$ vectors $v_1, \ldots, v_{n-n(T)}$ to get a basis for all of U. (See Theorem 7.6.3.) It will turn out that $\{T(v_1), \ldots, T(v_{n-n(T)})\}$ is a basis for $R(T)$.

To see this, first let $x \in U$. Then there are scalars $\alpha_1, \ldots, \alpha_{n(T)}, \beta_1, \ldots, \beta_{n-n(T)}$, such that

$$x = \alpha_1 u_1 + \cdots + \alpha_{n(T)} u_{n(T)} + \beta_1 v_1 + \cdots + \beta_{n-n(T)} v_{n-n(T)}.$$

Thus,

$$T(x) = \alpha_1 T(u_1) + \cdots + \alpha_{n(T)} T(u_{n(T)}) + \beta_1 T(v_1) + \cdots \\ + \beta_{n-n(T)} T(v_{n-n(T)}).$$

But $T(u_i) = 0$, $i = 1, \ldots, n(T)$, so

$$T(x) = \beta_1 T(v_1) + \cdots + \beta_{n-n(T)} T(v_{n-n(T)}).$$

This implies, since $T(x)$ is arbitrary in $R(T)$, that $\{T(v_1), \ldots, T(v_{n-n(T)})\}$ spans $R(T)$. Thus, to complete the proof, we need only show that

$$\{T(v_1), \ldots, T(v_{n-n(T)})\}$$

is an independent set.

Thus suppose there are scalars $\gamma_1, \ldots, \gamma_{n-n(T)}$ such that

$$\gamma_1 T(v_1) + \cdots + \gamma_{n-n(T)} T(v_{n-n(T)}) = 0.$$

Then

$$T(\gamma_1 v_1 + \cdots + \gamma_{n-n(T)} v_{n-n(T)}) = 0,$$

also. This implies

$$\gamma_1 v_1 + \cdots + \gamma_{n-n(T)} v_{n-n(T)} \in N(T),$$

the null space of T. Hence there are scalars $\delta_1, \ldots, \delta_{n(T)}$ such that

$$\gamma_1 v_1 + \cdots + \gamma_{n-n(T)} v_{n-n(T)} = \delta_1 u_1 + \cdots + \delta_{n(T)} u_{n(T)},$$

whence

$$(-\delta_1) u_1 + \cdots + (-\delta_{n(T)}) u_{n(T)} + \gamma_1 v_1 + \cdots + \gamma_{n-n(T)} v_{n-n(T)} = 0.$$

But $\{u_1, \ldots, u_{n(T)}, v_1, \ldots, v_{n-n(T)}\}$ is a basis for U, so each coefficient is 0. Thus $\gamma_1 = \cdots = \gamma_{n-n(T)} = 0$, and $T(v_1), \ldots, T(v_{n-n(T)})$ are independent. Thus

$$r(T) = n - n(T) \quad \text{and so} \quad r(T) + n(T) = n = \dim U. \blacksquare$$

EXERCISES

1. Let W_1, W_2 be vector spaces over a field K, with W_2 a subspace of W_1. Define the quotient space W_1/W_2. [*Hint:* Use the fact that $(W_1, +)$ is a group, and that $(W_2, +)$ is a subgroup of $(W_1, +)$].
2. Let $T: U \longrightarrow V$, U, V vector spaces over the field K, T a linear transformation. Show that $U/N(T) \cong R(T)$.
3. Let $T: \mathbf{R}^3 \longrightarrow \mathbf{R}^2$ be defined by

$$T(1,0,0) = (1,1),$$
$$T(0,1,0) = (0,1),$$
$$T(0,0,1) = (1,-1).$$

Find $R(T)$ and $N(T)$. Verify that $r(T) + n(T) = \dim \mathbf{R}^3 = 3$.

4. Let $T: \mathbf{R}^2 \longrightarrow \mathbf{R}^3$ be defined by

$$T(1,0) = (1,1,1),$$
$$T(0,1) = (0,1,-1).$$

Find $R(T)$ and $N(T)$ and verify that $r(T) + n(T) = \dim \mathbf{R}^2 = 2$.

5. Define $T: \mathbf{R}^3 \longrightarrow \mathbf{R}^2$ by

$$T(x_1, x_2, x_3) = (2x_1 + x_2 - x_3, 3x_1 - x_2 + 2x_3).$$

Find $N(T)$ and verify that $n(T) + r(T) = \dim \mathbf{R}^3 = 3$.

6. Let V and U be vector spaces over the field K. Let $T\colon V \longrightarrow U$ be a linear transformation.
   (a) Show that T is 1-1 if and only if $N(T) = 0$.
   (b) Let dim V = dim $U = n$, n an integer.
      (1) Show that T is 1-1 if and only if $R(T) = U$.
      (2) Show T is onto if and only if $N(T) = 0$.

8.3 The Arithmetic of Linear Transformations: Linear Combinations

If T and S are linear transformations from the vector space V to the vector space U, both over the field K, then it is possible to define new linear transformations from these by forming sums and scalar multiples as follows:

(1) $T + S$ is defined as the function from V to U defined by
$$(T + S)(v) = T(v) + S(v),$$
for each $v \in V$.

Notice that the symbol "+" in $T + S$ is the new symbol being defined, whereas the "+" in $T(v) + S(v)$ is just the symbol for addition in the vector space U.

(2) If a is a scalar, $a \in K$, then aT is defined as the mapping from V to U defined by $(aT)(v) = a(T(v))$, for all $v \in V$.

It is necessary to verify that these newly defined mappings are indeed linear transformations.

THEOREM 8.3.1 Let T and S be linear transformations from the vector space V to the vector space U. Then the mappings $T + S$ and aT are also linear transformations from V to U.

Proof Let $v_1, v_2 \in V$ and $a_1, a_2 \in K$. Then
$$(T + S)(a_1v_1 + a_2v_2) = T(a_1v_1 + a_2v_2) + S(a_1v_1 + a_2v_2)$$
by definition of $T + S$. Since T and S are both linear,
$$T(a_1v_1 + a_2v_2) = a_1T(v_1) + a_2T(v_2),$$
and
$$S(a_1v_1 + a_2v_2) = a_1S(v_1) + a_2S(v_2).$$
Hence
$$\begin{aligned}(T + S)(a_1v_1 + a_2v_2) &= a_1T(v_1) + a_2T(v_2) + a_1S(v_1) + a_2S(v_2)\\ &= a_1T(v_1) + a_1S(v_1) + a_2T(v_2) + a_2S(v_2)\\ &= a_1(T(v_1) + S(v_1)) + a_2(T(v_2) + S(v_2))\\ &= a_1((T + S)(v_1)) + a_2((T + S)(v_2)).\end{aligned}$$

Thus $T + S$ is a linear transformation by Theorem 8.1.2.

Next,
$$(aT)(a_1v_1 + a_2v_2) = a(T(a_1v_1 + a_2v_2))$$
$$= a(a_1T(v_1) + a_2T(v_2))$$
$$= (aa_1)T(v_1) + (aa_2)T(v_2)$$
$$= (a_1a)T(v_1) + (a_2a)T(v_2)$$
$$= a_1(aT(v_1)) + a_2(aT(v_2))$$
$$= a_1((aT)(v_1)) + a_2((aT)(v_2)),$$

since by definition
$$a(T(v_1)) = (aT)(v_1).$$

Thus aT is a linear transformation. ∎

If S and T are linear transformations from V to U over K and if a and b are elements of the field K, it now follows that $aS + bT$ is a linear transformation from V to U. Given the definition of a linear combination of vectors, it is not difficult to see that the "arithmetic" of linear transformations is very similar to that of vectors.

Now let $\mathrm{Hom}_K(V,U)$ represent the set of all linear transformations from the vector space V to the vector space U over the field K.

THEOREM 8.3.2 Let V and U be vector spaces over the field K. Then the set $\mathrm{Hom}_K(V,U)$ is a vector space over K with sum and scalar multiples defined as follows: for S and T in $\mathrm{Hom}_K(V,U)$, and all v in V, $(S + T)(v) = S(v) + T(v)$, and for T in $\mathrm{Hom}_K(V,U)$, $a \in K$, and all v in V, $(aT)(v) = a(T(v))$.

Further, if V and U are finite-dimensional vector spaces over K with $\dim V = n$ and $\dim U = m$, then $\dim \mathrm{Hom}_K(V,U) = mn$.

Proof We have already shown that sums and scalar multiples of linear transformations are linear transformations. It is also very easy to show that if a and b are scalars, and if T and S are linear transformations, then

$$(a + b)T = aT + bT \quad \text{and} \quad a(T + S) = aT + aS.$$

To complete the proof that $\mathrm{Hom}_K(V,U)$ is a vector space, we need only show that it forms an abelian group with respect to addition. Thus the linear transformation $\theta: V \longrightarrow U$ defined by $\theta(v) = 0$, the zero vector of U, is the identity element of $(\mathrm{Hom}_K(V,U), +)$, since

$$(T + \theta)(v) = T(v) + \theta(v) = T(v) + 0 = T(v).$$

The additive inverse of T is defined as $-T$, where $(-T)(v) = -T(v)$. Then $T + (-T) = \theta$, for

$$(T + (-T))(v) = T(v) + (-T)(v) = T(v) + (-T(v)) = 0 = \theta(v),$$

for all $v \in V$. Next, for T and S in $\mathrm{Hom}_K(V,U)$,

$$(T + S)(v) = T(v) + S(v) = S(v) + T(v) = (S + T)(v),$$

whence $T + S = S + T$, that is, addition is commutative. Finally, $(T + S) + R = T + (S + R)$, for all T, S, and R in $\text{Hom}_K (V,U)$, as is easily verified. This completes the proof that $\text{Hom}_K (V,U)$ is a vector space.

Now suppose that dim $V = n$ and dim $U = m$. Let $\{v_1,\ldots,v_n\}$ be a basis for V and let $\{u_1,\ldots,u_m\}$ be a basis for U. We will define a set of mn linear transformations from V to U that will be a basis for $\text{Hom}_K (V,U)$.

Fix j and i. Define a mapping $T_{ji}: V \longrightarrow U$ as follows:

$$T_{ji}(v_i) = u_j,$$
$$T_{ji}(v_k) = 0, \quad \text{for } k \neq i.$$

By Theorem 8.1.9, extend T_{ji} by linearity to all of V, so that if

$$v = \sum_{r=1}^{n} a_r v_r,$$

then

$$T_{ji}(v) = T_{ji}\left(\sum_{r=1}^{n} a_r v_r\right) = \sum_{r=1}^{n} a_r T_{ji}(v_r)$$
$$= a_i T_{ji}(v_i) = a_i u_j.$$

Then T_{ji} is a linear transformation from V to U.

The set $\mathcal{B} = \{T_{ji} \mid 1 \leq i \leq n, 1 \leq j \leq m\}$ is thus a set of linear transformations from V to U. Moreover, $\mathcal{B}$ contains exactly mn elements, since $T_{ji} \neq T_{kl}$ unless both $j = k$ and $i = l$.

We must now show that $\mathcal{B}$ is a basis for the vector space $\text{Hom}_K (V,U)$. First, let S be any element of $\text{Hom}_K (V,U)$. Then for each v_i in the basis for V,

$$S(v_i) = \sum_{j=1}^{m} b_{ji} u_j, \quad \text{where } b_{ji} \in K, j = 1, \ldots, m,$$

since $\{u_1,\ldots,u_m\}$ is a basis for U. Now we show that the linear transformation S can be written as a linear combination of elements of $\mathcal{B}$. To do this, let

$$\bar{S} = \sum_{i=1}^{n} \sum_{j=1}^{m} b_{ji} T_{ji}$$

and consider $\bar{S}(v_k)$. Since $T_{ji}(v_k) = 0$ if $i \neq k$,

$$\bar{S}(v_k) = \left(\sum_{j=1}^{m} b_{jk} T_{jk}\right)(v_k) = \sum_{j=1}^{m} b_{jk} T_{jk}(v_k)$$
$$= \sum_{j=1}^{m} b_{jk} u_j = S(v_k).$$

Hence $\bar{S}(v_k) = S(v_k)$ for $k = 1, \ldots, n$, and S agrees with $\bar{S}$ on the basis $\{v_1,\ldots,v_n\}$ of V. Thus, by Theorem 8.1.9, $S = \bar{S}$.

Finally we must show that the set $\mathfrak{B}$ is linearly independent. Thus let

$$T = \sum_{i=1}^{n} \sum_{j=1}^{m} a_{ji} T_{ji} = \theta.$$

(Notice that the order of summation is immaterial, since vector addition is commutative.) Since $T_{ji}(v_k) = 0$, for $i \neq k$,

$$0 = T(v_k) = \sum_{j=1}^{m} a_{jk} T_{jk}(v_k) = \sum_{j=1}^{m} a_{jk} u_j.$$

But $\{u_1,\ldots,u_m\}$ is a basis for U and so $a_{1k} = a_{2k} = \cdots = a_{mk} = 0$. Since this is true for all k, it follows that $a_{ji} = 0$ for all $i = 1, \ldots, n$, and $j = 1, \ldots, m$. Thus $\mathfrak{B}$ is a linearly independent set and so is a basis for $\text{Hom}_K (V,U)$. ∎

EXERCISES

1. Let V and U be vector spaces over the field K. Let

    $$R, S, T \in \text{Hom}_K (V,U).$$

    Prove that $(R + S) + T = R + (S + T)$.

2. Let V and U be vector spaces over the field K. Let

    $$S, T \in \text{Hom}_K (V,U).$$

    Let $a, b \in K$. Verify directly, without recourse to theorems in this section, that $aS + bT$ is a linear transformation from V to U.

3. Let $V = \mathbf{R}^2$ and $U = \mathbf{R}^3$. Find a basis for $\text{Hom}_\mathbf{R} (V,U)$. Find a basis for $\text{Hom}_\mathbf{R} (U,V)$.

4. Let $T: V \longrightarrow U$ be a linear transformation from the vector space V over K to the vector space U over K. Suppose that T, considered as a function from the set V to the set U, has an inverse function $S: U \longrightarrow V$. Show that $S: U \longrightarrow V$ is also a linear transformation.

5. Let T be the linear transformation from $\mathbf{R}^2$ to $\mathbf{R}^2$ defined by $T(1,0) = (0,1)$ and $T(0,1) = (1,0)$.
    (a) Show that T is 1-1 and onto.
    (b) Find the inverse of T.

6. Let T and S be linear transformations in $\text{Hom}_K (V,U)$. Let $a, b \in K$.
    (a) Show that $(a + b)T = aT + bT$.
    (b) Show that $a(T + S) = aT + aS$.

8.4 The Arithmetic of Linear Transformations: Composition

Since linear transformations are functions, they may be combined by function composition, provided, of course, that the appropriate conditions on ranges and domains are met. Thus if U, V, and W are vector spaces over the field K, and if $S\colon U \longrightarrow V$ and $T\colon V \longrightarrow W$ are linear transformations, then we can define $T \circ S\colon U \longrightarrow W$ by

$$(T \circ S)(u) = T(S(u)), \quad \text{for all } u \in U.$$

Clearly, $T \circ S$ means: first apply S, and then apply T.

THEOREM 8.4.1 Let U, V, W be vector spaces over the field K, and let $S\colon U \longrightarrow V$ and $T\colon V \longrightarrow W$ be linear transformations. Then the composite $T \circ S\colon U \longrightarrow W$ is a linear transformation.

Proof Let $u_1, u_2 \in U$ and let $a_1, a_2 \in K$. Then

$$\begin{aligned}
(T \circ S)(a_1 u_1 + a_2 u_2) &= T[S(a_1 u_1 + a_2 u_2)] \\
&= T[S(a_1 u_1) + S(a_2 u_2)] \\
&= T[a_1 S(u_1) + a_2 S(u_2)] \\
&= T[a_1 S(u_1)] + T[a_2 S(u_2)] \\
&= a_1 T(S(u_1)) + a_2 T(S(u_2)) \\
&= a_1[(T \circ S)(u_1)] + a_2[(T \circ S)(u_2)]. \quad \blacksquare
\end{aligned}$$

Suppose that $S\colon U \longrightarrow V$ and $T\colon V \longrightarrow U$ are linear transformations with the property $(T \circ S)(u) = u$, for all $u \in U$. Then $T \circ S$ is the identity function I_U on U. Under these conditions, we call S a *right inverse* of T, and write $S = T_R^{-1}$. Similarly, we call T a *left inverse* of S, and we write $T = S_L^{-1}$.

Example 8.4.2 Let $U = \mathbf{R}^2$, $V = \mathbf{R}^3$. Let $\{e_1, e_2\}$ be a basis for R^2, where $e_1 = (1,0)$, $e_2 = (0,1)$, and let $\{u_1, u_2, u_3\}$ be a basis for R^3, where $u_1 = (1,0,0)$, $u_2 = (0,1,0)$, and $u_3 = (0,0,1)$. Let $S\colon U \longrightarrow V$ be a linear transformation such that $S(e_1) = u_1$, $S(e_2) = u_2$. Let $T\colon V \longrightarrow U$ be a linear transformation such that $T(u_1) = e_1$, $T(u_2) = e_2$, and $T(u_3) = e_1 - e_2$. By Theorem 8.1.9, the linear transformations S and T exist.

Now $T \circ S\colon \mathbf{R}^2 \longrightarrow \mathbf{R}^2$, and we have

$$(T \circ S)(e_1) = T(S(e_1)) = T(u_1) = e_1,$$

and

$$(T \circ S)(e_2) = T(S(e_2)) = T(u_2) = e_2.$$

Thus, for $u \in \mathbf{R}^2$, $u = ae_1 + be_2$, we get immediately that $(T \circ S)(u) = u$. Hence, $T \circ S = I_U$, and $T = S_L^{-1}$, $S = T_R^{-1}$.

On the other hand, $S \circ T = \mathbf{R}^3 \longrightarrow \mathbf{R}^3$, but

$$\begin{aligned}
(S \circ T)(u_3) &= S[T(u_3)] = S(e_1 - e_2) = S(e_1) - S(e_2) \\
&= u_1 - u_2 \neq u_3.
\end{aligned}$$

Thus, $S \circ T \neq I_V$. Hence, we see that although $S = T_R^{-1}$ and $T = S_L^{-1}$, it need not follow that $S = T_L^{-1}$ and $T = S_R^{-1}$. Indeed, in this example, functions T_L^{-1} and S_R^{-1} do not even exist.

The situation is much simpler in the case where $U = V$, so that $T: V \longrightarrow V$ and $S: V \longrightarrow V$. In this situation, T (also S) is called a *linear operator on the vector space* V. The composites $T \circ S$ and $S \circ T$ are always defined. Furthermore, if V is finite dimensional, then T_R^{-1} exists if and only if T_L^{-1} exists, and $T_R^{-1} = T_L^{-1}$.

THEOREM 8.4.3 Let S and T be linear operators on the vector space V, where V is finite dimensional, with dim $V = n$. Then

(a) $S \circ T = I_V$ if and only if $T \circ S = I_V$.
(b) If $T': V \longrightarrow V$ satisfies $S \circ T' = S \circ T = I_V$, then $T' = T$.
(c) If $T'': V \longrightarrow V$ satisfies $T'' \circ S = T \circ S = I_V$, then $T'' = T$.

Proof (a) If $S \circ T = I_V$, then $(S \circ T)(v) = v$, for all $v \in V$. Therefore, $S(T(V)) = V$. By Theorem 8.1.11 (or by Exercise 17, Section 8.1), dim $T(V) =$ dim $V = n$. Thus, since $T(V) \subseteq V$ and dim $T(V) =$ dim V, $T(V) = V$, so T maps V onto V.

Now consider the linear transformation $T \circ S$. For each $v \in V$, $(T \circ S)(v) = T(S(v))$, and since T is onto, there is a $w \in V$ such that $v = T(w)$. Hence,

$$(T \circ S)(v) = T(S(v)) = T(S(T(w))) = T((S \circ T)(w))$$
$$= T(w) = v.$$

Hence, $T \circ S = I_V$.

Conversely, if $T \circ S = I_V$, then by interchanging the roles of T and S in the above proof, we have $S \circ T = I_V$.

(b) If $S \circ T = S \circ T' = I_V$, then

$$T = T \circ I_V = T \circ (S \circ T) = T \circ (S \circ T') = (T \circ S) \circ T'$$
$$= I_V \circ T' = T'.$$

Note we have used here $T \circ I_V = T$, $I_V \circ T' = T'$. See Exercise 8 in this section.

(c) This is similar to (b). ∎

The hypothesis that V is finite dimensional cannot be eliminated in Theorem 8.4.3, as is seen in Example 8.4.4.

Example 8.4.4 Let V be the vector space of all polynomial functions with real coefficients. We have seen (Section 7.3) that $\{1 = x^0, x, x^2, \ldots\}$ is a basis for V, where $x^k(\alpha) = \alpha^k$, for each $\alpha \in \mathbf{R}$. Clearly, V is not finite dimensional.

Let $T: V \longrightarrow V$ be defined by

$$T(x^i) = \frac{x^{i+1}}{i+1}, \quad \text{for } i = 0, 1, 2, \ldots.$$

Thus, for $f(x) \in V$, $T(f(x))$ replaces $f(x)$ by that antiderivative of $f(x)$ with constant term equal to 0. [Note that the definition of T implies $T(0) = 0$.]

Now let $D: V \longrightarrow V$ be defined by

$$T(f(x)) = f'(x), \quad \text{for all } f(x) \in V.$$

For $f(x) = \sum_{i=0}^{r} a_i x^i$,

$$(D \circ T)(f(x)) = D[T(f(x))] = D\left(\sum_{i=0}^{r} \frac{a_i x^{i+1}}{i+1}\right)$$

$$= \sum_{i=0}^{r} a_i x^i = f(x),$$

whence $D \circ T = I_V$. On the other hand,

$$(T \circ D)(x + 1) = T(D(x + 1)) = T(1) = x,$$

so $T \circ D \neq I_V$.

Note that even though $T(V)$ is infinite dimensional, $T(V) \neq V$. For example, there is no $g(x) \in V$ such that $T(g(x)) = x + 1$. That $T(V) \neq V$ accounts for the fact that $T \circ D \neq I_V$.

Definition 8.4.5 Let T be a linear operator on the finite-dimensional vector space V. The linear operator S on V is called the *inverse of T* if $S \circ T = I_V$ (whence also $T \circ S = I_V$, by Theorem 8.4.3). We denote S by T^{-1}. If the operator T does not have an inverse, T is called a *singular* operator. If T has an inverse, T is called a *nonsingular* operator.

We note that, by Theorem 8.4.3, if T is a linear operator on the finite-dimensional vector space V, then all of T^{-1}, T_R^{-1}, T_L^{-1} exist, or none of them exists. In case one of T^{-1}, T_R^{-1}, T_L^{-1} exists, we have

$$T^{-1} = T_R^{-1} = T_L^{-1}.$$

Example 8.4.6 Let V be the vector space $\mathbf{R}^2$. Then the operator

$$T: V \longrightarrow V$$

defined by $T(1,0) = (1,2)$ and $T(0,1) = (1,1)$ is nonsingular, with T^{-1} defined by $T^{-1}(1,0) = (-1,2)$ and $T^{-1}(0,1) = (1,-1)$. To see that T^{-1}

is indeed the inverse of T, note that

$$\begin{aligned}(T^{-1} \circ T)(a,b) &= T^{-1}(T(a,b)) = T^{-1}(T(a(1,0) + b(0,1))) \\ &= T^{-1}(a(1,2) + b(1,1)) = T^{-1}(a+b, 2a+b) \\ &= T^{-1}((a+b)(1,0) + (2a+b)(0,1)) \\ &= (a+b)(-1,2) + (2a+b)(1,-1) \\ &= (-a - b + 2a + b, 2a + 2b - 2a - b) \\ &= (a,b).\end{aligned}$$

On the other hand, the operator S defined by $S(1,0) = (1,1)$ and $S(0,1) = (2,2)$ is singular. For

$$S(-2,1) = S(-2(1,0) + (0,1)) = -2(1,1) + (2,2) = (0,0),$$

and so if R is any operator, then

$$(R \circ S)(-2,1) = R(S(-2,1)) = R(0,0) = (0,0) \neq (-2,1).$$

We will devote considerable effort in the sequel to determining when a linear operator has an inverse and to seeing how to find it.

THEOREM 8.4.7 Let S be a linear operator on the finite-dimensional vector space V over K. Then the following are equivalent:

(a) S is nonsingular.
(b) If $S(v) = 0$, then $v = 0$.
(c) S is 1–1.
(d) S is onto.

Proof (a) $\Rightarrow$ (b). Let S be nonsingular with $S^{-1} = T$. Thus $T \circ S = I_V$. If $S(v) = 0$ for some $v \in V$, $v \neq 0$, then

$$v = (T \circ S)(v) = T(S(v)) = T(0) = 0 \neq v,$$

a contradiction. Hence $S(v) = 0$ only if $v = 0$.

(b) $\Rightarrow$ (c). If S is not 1–1, then there are vectors $v_1, v_2, v_1 \neq v_2$, such that $S(v_1) = S(v_2)$. Thus, $S(v_1 - v_2) = 0$, with $v_1 - v_2 \neq 0$, contradicting (b).

(c) $\Rightarrow$ (d). If S is 1–1 and not onto, then $S(V)$ is a proper subspace of V and hence has smaller dimension. Thus if $\{v_1, \ldots, v_n\}$ is a basis for V, then $\{S(v_1), \ldots, S(v_n)\}$ is a linearly dependent set of vectors. Hence there are elements $a_1, \ldots, a_n$ in K not all zero, such that

$$\sum_{i=1}^{n} a_i S(v_i) = 0.$$

Therefore

$$S\left(\sum_{i=1}^{n} a_i v_i\right) = 0, \quad \text{but} \quad \sum_{i=1}^{n} a_i v_i \neq 0,$$

since $\{v_1,\ldots,v_n\}$ is a basis for V. But S is 1-1, a contradiction. Hence S is onto.

(d) $\Rightarrow$ (a). Let $\{v_1,\ldots,v_n\}$ be a basis for V. Since S is onto, there exist vectors $w_1, \ldots, w_n$ such that $S(w_i) = v_i$ for $i = 1, \ldots, n$. Now define the linear operator T by $T(v_i) = w_i$, $i = 1, \ldots, n$. Then

$$S(T(v_i)) = S(w_i) = v_i,$$

and hence $S(T(v)) = v$ for any $v \in V$, that is, $S \circ T = I_V$. But then $T \circ S = I_V$, by Theorem 8.4.3. Of course, we have again used the fact that a linear operator on V is completely specified by its action on a basis of V. This completes the proof of the theorem. ∎

THEOREM 8.4.8 Let V be a finite-dimensional vector space and let $T: V \longrightarrow V$ be a linear operator. Then the following are equivalent:

(a) T is nonsingular.
(b) If $\{v_1,\ldots,v_n\}$ is a basis for V, then $\{T(v_1),\ldots,T(v_n)\}$ is also a basis for V.
(c) There exists a basis $\{e_1,\ldots,e_n\}$ of V such that $\{T(e_1),\ldots,T(e_n)\}$ is a basis of V.
(d) For each independent set $\{w_1,\ldots,w_k\}$ of vectors of V, the set $\{T(w_1),\ldots,T(w_k)\}$ is also independent.

Proof The proof is left as an exercise. ∎

EXERCISES

1. Let Hom (V,V) be the set of all linear operators on V. Show that (Hom $(V,V), +, \circ)$ is a ring with $+$ and $\circ$ as defined in Sections 8.3 and 8.4, respectively.
2. Let T be the linear operator on the vector space $\mathbf{R}^2$ defined by $T(1,0) = (1,1)$, $T(0,1) = (0,0)$. Show that T does not have an inverse.
3. Let $T: \mathbf{R}^2 \longrightarrow \mathbf{R}^2$ be a linear transformation, where $T(1,0) = (a,b)$ and $T(0,1) = (c,d)$.
   (a) Prove that T has an inverse T^{-1} if and only if $ad - bc \neq 0$.
   (b) If T^{-1} exists, what are $T^{-1}(1,0)$ and $T^{-1}(0,1)$?
4. Let $T: \mathbf{R}^3 \longrightarrow \mathbf{R}^3$ be a linear transformation defined by $T(1,0,0) = (1,2,0)$; $T(0,1,0) = (1,0,1)$; and $T(0,0,1) = (0,-1,1)$. Show that T has an inverse T^{-1} and find $T^{-1}(1,0,0)$, $T^{-1}(0,1,0)$, and $T^{-1}(0,0,1)$.
5. Let $T: \mathbf{R}^3 \longrightarrow \mathbf{R}^3$ be a linear transformation defined by $T(1,0,0) = (a_1,b_1,c_1)$; $T(0,1,0) = (a_2,b_2,c_2)$; and $T(0,0,1) = (a_3,b_3,c_3)$. Show that T^{-1} exists, if

$$a_1 b_2 c_3 + a_3 b_1 c_2 + a_2 b_3 c_1 - a_3 b_2 c_1 - a_2 b_1 c_3 - a_1 b_3 c_2 \neq 0.$$

6. Let T_θ be the linear operator on $\mathbf{R}^2$ discussed in Example 8.1.5. Show that T_θ has an inverse T_θ^{-1}. What is the simplest way of describing T_θ^{-1}?
7. Prove Theorem 8.4.8.
8. Let $S: V \longrightarrow V$ be a linear operator on the vector space V, and let I_V be the identity map on V. Show that I_V is a linear operator on V and that $S \circ I_V = I_V \circ S = S$.
9. Let V be n dimensional and U m dimensional, $n < m$. Let

$$T: V \longrightarrow U$$

be a linear transformation. Show that there does not exist a linear transformation $S: U \longrightarrow V$ such that $S \circ T: V \longrightarrow V$ is the identity on V. (See Exercise 17, Section 8.1.)
10. Let V be a vector space over the field K, and let $T: V \longrightarrow V$ be a linear operator on V. Define $T^2: V \longrightarrow V$ by $T^2 = T \circ T$, $T^3: V \longrightarrow V$ by $T^3 = T \circ T^2$, and in general, $T^n = T \circ (T^{n-1})$.
    (a) Show that $T^3 = T^2 \circ T$.
    (b) Show that $T^4 = T^3 \circ T = T^2 \circ T^2$.
    (c) Let $V = \mathbf{R}^3$, with basis

$$\{e_1 = (1,0,0),\ e_2 = (0,1,0),\ e_3 = (0,0,1)\}.$$

Let $T: V \longrightarrow V$ be defined by

$$T(e_1) = e_1 + e_2 + e_3,$$
$$T(e_2) = e_1 - e_2 + e_3,$$
$$T(e_3) = e_1 + e_2 - e_3.$$

Find $T^2(e_i)$, for $i = 1, 2, 3$, and find $T^4(e_i)$, for $i = 1, 2, 3$.
11. Let $T: V \longrightarrow V$, where V is finite dimensional. Let $N(T^i)$ be the null space of T^i, and let $R(T^i)$ be the range of T^i.
    (a) Show that $N(T) \subseteq N(T^2) \subseteq N(T^3) \subseteq \cdots$.
    (b) Show that $R(T) \supseteq R(T^2) \supseteq R(T^3) \supseteq \cdots$.
    (c) Prove there exists an integer k such that $N(T^k) = N(T^j)$, for all $j > k$, and $R(T^k) = R(T^j)$, for all $j > k$.

8.5 Matrices: Introduction

Let V and U be vector spaces over the field K. A linear transformation $T: V \longrightarrow U$ is completely determined by its values on a basis of V (see Theorem 8.1.9). This basic fact has been used several times in previous sections.

Now let us assume that V and U are finite-dimensional of dimensions

n and m, respectively, and let $\mathcal{B} = \{v_1, \ldots, v_n\}$ be a basis for V, and let $\mathcal{C} = \{u_1, \ldots, u_m\}$ be a basis for U. Then for each v_j,

$$T(v_j) = \sum_{i=1}^{m} a_{ij} u_i,$$

where the a_{ij}'s are uniquely determined. By Theorem 8.1.9, the linear transformation T is thus completely determined by the choice of the basis $\mathcal{B}$ for V, the basis $\mathcal{C}$ for U, and the set of mn elements

$$\{a_{ij} \mid i = 1, \ldots, m; j = 1, \ldots, n\}.$$

It is convenient, and customary, to write the set $\{a_{ij}\}$ in a rectangular array. We note that if $V = U$, that is, if $T: V \longrightarrow V$ is a linear operator, we will usually assume that $\mathcal{B} = \mathcal{C}$.

Definition 8.5.1 A rectangular array of the mn elements a_{ij}, $i = 1, \ldots, m; j = 1, \ldots, n; a_{ij} \in K$, written as

$$\begin{bmatrix} a_{11} & a_{12} & \cdots & a_{1n} \\ a_{21} & a_{22} & \cdots & a_{2n} \\ \vdots & & & \\ a_{m1} & a_{m2} & \cdots & a_{mn} \end{bmatrix},$$

is called an $m \times n$ (m by n) *matrix*. The matrix is often denoted by $[a_{ij}]_{m \times n}$ or by $[a_{ij}]$. The integer i is called the *row index* and the integer j is called the *column index*. The element a_{ij} is called the (i,j)th *entry*.

In the matrix $[a_{ij}]$, the list of elements $a_{i1}, a_{i2}, \ldots, a_{in}$, listed in that order, is called the *ith row of the matrix*. We usually will write the ith row as

$$[a_{i1} \quad a_{i2} \quad \cdots \quad a_{in}].$$

The list of elements $a_{1j}, a_{2j}, \ldots, a_{mj}$, listed in that order, is called the *jth column* of $[a_{ij}]$. We usually will write the jth column as

$$\begin{bmatrix} a_{1j} \\ a_{2j} \\ \vdots \\ a_{mj} \end{bmatrix}.$$

If a matrix has only one row, that is, if it is of the form

$$[a_{ij}]_{1 \times n} = [a_{11} \quad a_{12} \quad \cdots \quad a_{1n}]$$

then it is called a *row matrix*. If it has only one column,

$$[a_{ij}]_{m \times 1} = \begin{bmatrix} a_{11} \\ a_{21} \\ \cdot \\ \cdot \\ \cdot \\ a_{m1} \end{bmatrix},$$

then it is called a *column matrix*.

A matrix in which the number of rows is the same as the number of columns, say n, is called a *square matrix of order n*, or simply a *matrix of order n*.

Strictly speaking, a matrix can be defined as a function from the set of ordered pairs $\{(i,j) \mid i = 1, \ldots, m; j = 1, \ldots, n\}$ to the field K. Thus two matrices are equal if and only if they are both $m \times n$ and corresponding entries are equal.

Definition 8.5.2 Let $T: V \longrightarrow U$ be a linear transformation from the vector space V over K to the vector space U over K. Suppose $\mathcal{B} = \{v_1, \ldots, v_n\}$ and $\mathcal{C} = \{u_1, \ldots, u_m\}$ are bases for V and U, respectively. Suppose that

$$T(v_j) = \sum_{i=1}^{m} a_{ij} u_i, \qquad a_{ij} \in K; i = 1, \ldots, m; j = 1, \ldots, n.$$

We call the $m \times n$ matrix $[a_{ij}]$ the *matrix of the linear transformation T relative to the bases $\mathcal{B}$ and $\mathcal{C}$* and we denote this by $[T]_{\mathcal{C},\mathcal{B}} = [a_{ij}]_{m \times n}$. We also say T has matrix $[T]_{\mathcal{C},\mathcal{B}}$ relative to the bases $\mathcal{B}$ and $\mathcal{C}$.

If $U = V$ and $\mathcal{B} = \mathcal{C}$, we call $[a_{ij}]_{m \times n}$ the *matrix of the linear operator T with respect to the basis $\mathcal{B}$*, and we denote this by

$$[T]_{\mathcal{B}} = [a_{ij}]_{n \times n}.$$

One of the reasons for writing matrices as rectangular arrays will become evident when we investigate their arithmetic properties. Primarily, we use rectangular arrays for convenience in performing certain arithmetic operations.

Example 8.5.3 The matrix

$$A = \begin{bmatrix} 0 & 1 & .3 & -4 \\ 2 & 6 & 1 & .25 \\ 21 & -10 & 1 & 5 \end{bmatrix}$$

is a 3×4 matrix, with entries from the field of real numbers, since it has 3 rows and 4 columns, and all entries are real. The matrix

$$B = \begin{bmatrix} 0 & 2 & 21 \\ 1 & 6 & -10 \\ .3 & 1 & 1 \\ -4 & .25 & 5 \end{bmatrix}$$

is a 4 × 3 matrix. If we denote A by $[a_{ij}]_{3\times 4}$ and B by $[b_{ij}]_{4\times 3}$, then, for example, $a_{11} = 0$, $a_{21} = 2$, $a_{34} = 5$ and $b_{11} = 0$, $b_{21} = 1$, $b_{33} = 1$.

Example 8.5.4 Suppose that

$$A = \begin{bmatrix} 1 & 0 & -1 \\ 2 & -3 & 0 \end{bmatrix},$$

the matrix associated with the linear transformation $T: \mathbf{R}^3 \longrightarrow \mathbf{R}^2$ with respect to the bases

$$\mathcal{B} = \{v_1 = (1,0,0),\ v_2 = (0,1,0),\ v_3 = (0,0,1)\} \quad \text{of } \mathbf{R}^3$$

and

$$\mathcal{C} = \{u_1 = (1,0),\ u_2 = (0,1)\} \quad \text{of } \mathbf{R}^2.$$

Then, for example, if $v = (2,3,4)$, $T(v)$ can be computed as follows. First, represent $v = (2,3,4)$ as a linear combination of the basis vectors in $\mathcal{B}$, namely,

$$\begin{aligned} v = (2,3,4) &= 2(1,0,0) + 3(0,1,0) + 4(0,0,1) \\ &= 2v_1 + 3v_2 + 4v_3. \end{aligned}$$

Then, according to the matrix A, $T(v_1) = u_1 + 2u_2$; $T(v_2) = -3u_2$; $T(v_3) = -u_1$. Thus

$$\begin{aligned} T(v) &= T(2v_1 + 3v_2 + 4v_3) = 2T(v_1) + 3T(v_2) + 4T(v_3) \\ &= 2(u_1 + 2u_2) + 3(-3u_2) + 4(-u_1) \\ &= 2u_1 + 4u_2 - 9u_2 - 4u_1 = -2u_1 - 5u_2. \end{aligned}$$

Notice that if we change the basis of either $\mathbf{R}^3$ or $\mathbf{R}^2$, then the matrix $[T]_{\mathcal{C},\mathcal{B}}$ associated with the linear transformation will also change. Thus, if we let $\mathcal{C}' = \{u_1' = (1,1),\ u_2' = (1,2)\}$ be a second basis for $\mathbf{R}^2$, then

$$\begin{aligned} T(v_1) &= (1,2) = u_2' = 0u_1' + u_2', \\ T(v_2) &= (0,-3) = 3u_1' - 3u_2', \\ T(v_3) &= (-1,0) = -2u_1' + u_2'. \end{aligned}$$

Thus T has matrix

$$[T]_{\mathcal{C}',\mathcal{B}} = A' = \begin{bmatrix} 0 & 3 & -2 \\ 1 & -3 & 1 \end{bmatrix}.$$

We again emphasize a crucial point. The matrix $A = [a_{ij}]$ associated with the linear transformation $T: V \longrightarrow U$ depends on the bases $\mathcal{B}$ and $\mathcal{C}$ chosen for V and U, respectively. A different choice of bases ("change in bases") for V and/or U will generally result in a change in the matrix for T.

In Section 8.9 we shall study the relationship among different matrices associated with the same linear transformation T.

EXERCISES

1. Let $T: \mathbf{R}^2 \longrightarrow \mathbf{R}^3$ be defined as follows:
$$T(1,0) = (0,1,1),$$
$$T(0,1) = (1,0,-1).$$
Let $\mathcal{B}_2 = \{u_1 = (1,0),\ u_2 = (0,1)\}$ be a basis for $\mathbf{R}^2$, and
$$\mathcal{B}_3 = \{v_1 = (1,0,0),\ v_2 = (0,1,0),\ v_3 = (0,0,1)\}$$
be a basis for $\mathbf{R}^3$.
   (a) Write the matrix for T with respect to the bases $\mathcal{B}_2$ and $\mathcal{B}_3$ for $\mathbf{R}^2$ and $\mathbf{R}^3$, respectively; that is, write $[T]_{\mathcal{B}_3, \mathcal{B}_2}$.
   (b) Let $\{u_1' = (1,-1),\ u_2' = (1,1)\}$ be a second basis $\mathcal{B}_2'$ for $\mathbf{R}^2$. Write the matrix for T with respect to the bases $\mathcal{B}_2'$ and $\mathcal{B}_3$; that is, write $[T]_{\mathcal{B}_3, \mathcal{B}_2'}$.

2. Let T be a linear transformation from $\mathbf{R}^3$ to $\mathbf{R}^4$ whose matrix is
$$\begin{bmatrix} 1 & 0 & 1 \\ 2 & 1 & -1 \\ 3 & -1 & 2 \\ 1 & 0 & 1 \end{bmatrix}$$
with respect to the "canonical" bases
$$\mathcal{B}_3 = \{u_1 = (1,0,0),\ u_2 = (0,1,0),\ u_3 = (0,0,1)\}$$
and
$$\mathcal{B}_4 = \{w_1 = (1,0,0,0),\ w_2 = (0,1,0,0),\ w_3 = (0,0,1,0),\ w_4 = (0,0,0,1)\}$$
of $\mathbf{R}^3$ and $\mathbf{R}^4$, respectively.
   (a) Determine $T(1,1,1)$; $T(0,1,2)$; $T(a,b,c)$.
   (b) Show that $\mathcal{B}_3' = \{z_1 = (1,1,1),\ z_2 = (1,1,0),\ z_3 = (1,0,0)\}$ is a basis for $\mathbf{R}^3$.
   (c) What is the matrix of T with respect to the bases $\mathcal{B}_3'$ and $\mathcal{B}_4$; that is, what is $[T]_{\mathcal{B}_4, \mathcal{B}_3'}$?

3. Let V be the vector space of all polynomials of degree $\leq n$, with real coefficients, over the field $\mathbf{R}$.
   (a) Show that $\mathcal{B} = \{1, x, \ldots, x^n\}$ is a basis for V.
   (b) Show that $\mathcal{C} = \{1,\ x+1,\ (x+1)^2,\ \ldots,\ (x+1)^n\}$ is also a basis for V.

(c) Let $D: V \longrightarrow V$ be the linear operator on V defined by $D(f(x)) = f'(x)$ [the derivative of $f(x)$] for all $f(x)$ in V.
  (1) For $n = 3$, find $[D]_\mathcal{B}$, $[D]_\mathcal{C}$, and $[D]_{\mathcal{C},\mathcal{B}}$.
  (2) For $n = 4$, find $[D]_\mathcal{B}$.
  (3) What is $[D]_\mathcal{B}$ for arbitrary n?

4. Let T be the linear operator on $\mathbf{R}^4$ defined by $T(e_1) = e_2$, $T(e_2) = e_3$, $T(e_3) = e_4$, $T(e_4) = 0$, where

$$\{e_1 = (1,0,0,0),\ e_2 = (0,1,0,0),\ e_3 = (0,0,1,0),\ e_4 = (0,0,0,1)\} = \mathcal{B}$$

is a basis for $\mathbf{R}^4$.
  (a) Find $[T]_\mathcal{B}$.
  (b) Let $T^2 = T \circ T$. What is $T^2(e_i)$, for $i = 1, 2, 3, 4$?
  (c) Find $[T^2]_\mathcal{B}$.
  (d) Let $T^3 = T^2 \circ T = T \circ T^2$. Find $T^3(e_i)$, for $i = 1, 2, 3, 4$.
  (e) Find $(T^3)_\mathcal{B}$.
  (f) Define T^4 and determine $[T^4]_\mathcal{B}$.

5. Let $V = \mathbf{R}^2$, $U = \mathbf{R}^4$, and $W = \mathbf{R}^3$. Let $T: V \longrightarrow U$ be defined by $T(e_1) = f_1 + f_2$, $T(e_2) = f_3 + f_4$, where $e_1 = (1,0)$, $e_2 = (0,1)$, $f_1 = (1,0,0,0)$, $f_2 = (0,1,0,0)$, $f_3 = (0,0,1,0)$ and $f_4 = (0,0,0,1)$, Let $S: U \longrightarrow W$ be defined by $S(f_1) = g_1 - g_2$, $S(f_2) = g_1 + g_3$, $S(f_3) = g_1 + g_2 - g_3$, $S(f_4) = g_2 + 2g_3$, where $g_1 = (1,0,0)$, $g_2 = (0,1,0)$, $g_3 = (0,0,1)$. Let $\mathcal{B} = \{e_1, e_2\}$, $\mathcal{C} = \{f_1, f_2, f_3, f_4\}$, and $\mathcal{D} = \{g_1, g_2, g_3\}$.
  (a) Find $[T]_{\mathcal{C},\mathcal{B}}$, $[S]_{\mathcal{D},\mathcal{C}}$.
  (b) Find $(S \circ T)(e_i)$, for $i = 1, 2$.
  (c) What is $[S \circ T]_{\mathcal{D},\mathcal{B}}$.
  (d) If $[a_{i1},\ a_{i2},\ a_{i3},\ a_{i4}]$ is the ith row of $[S]_{\mathcal{D},\mathcal{C}}$ and if

$$\begin{bmatrix} b_{1j} \\ b_{2j} \\ b_{3j} \\ b_{4j} \end{bmatrix}$$

is the jth column of $[T]_{\mathcal{C},\mathcal{B}}$, show that the (i,j)th entry of $[S \circ T]_{\mathcal{D},\mathcal{B}}$ is $a_{i1}b_{1j} + a_{i2}b_{2j} + a_{i3}b_{3j} + a_{i4}b_{4j}$, all pairs i, j; $i = 1, 2, 3$; $j = 1, 2$.

8.6 Matrices: Linear Combinations

In this section we investigate how the arithmetic properties of linear transformations carry over to matrices. To begin, we will let V and U denote vector spaces over the field K and we will let T and S be linear transformations, $T: V \longrightarrow U$ and $S: V \longrightarrow U$. Also, we will, for the time being, let $\mathcal{B} = \{v_1, \ldots, v_n\}$ be a basis for V, and $\mathcal{C} = \{u_1, \ldots, u_m\}$ be a basis for U.

Thus suppose that T is defined by

$$T(v_j) = \sum_{i=1}^{m} a_{ij}u_i, \quad \text{for each } j = 1, \ldots, n.$$

Then T has the matrix $[T]_{\mathcal{C},\mathcal{B}} = [a_{ij}]_{m \times n}$. Similarly, if S is defined by

$$S(v_j) = \sum_{i=1}^{m} b_{ij}u_i, \quad \text{for each } j = 1, \ldots, n,$$

then S is specified by the matrix $[S]_{\mathcal{C},\mathcal{B}} = [b_{ij}]_{m \times n}$.

The sum $T + S$, of T and S, was defined as the linear transformation $T + S: V \longrightarrow U$, where $(T + S)(v) = T(v) + S(v)$, for each v in V. Thus, for each v_j in the basis $\mathcal{B}$ of V, we have

$$(T + S)(v_j) = T(v_j) + S(v_j) = \sum_{i=1}^{m} a_{ij}u_i + \sum_{i=1}^{m} b_{ij}u_i$$

$$= \sum_{i=1}^{m} (a_{ij} + b_{ij})u_i.$$

Thus the matrix associated with the sum $T + S$ has as its (i,j)th entry the element $a_{ij} + b_{ij}$. We are led to the following definition.

Definition 8.6.1 Let $A = [a_{ij}]_{m \times n}$ and $B = [b_{ij}]_{m \times n}$ be two $m \times n$ matrices with entries from the field K. We define the *sum* of the matrices as follows: $A + B = [a_{ij} + b_{ij}]_{m \times n}$. That is, $A + B$ is the $m \times n$ matrix whose (i,j)th entry is $a_{ij} + b_{ij}$.

To illustrate, if

$$A = \begin{bmatrix} 1 & 0 & 3 \\ 4 & -2 & 1 \end{bmatrix} \quad \text{and} \quad B = \begin{bmatrix} 8 & -10 & 4 \\ 2 & 1 & 1 \end{bmatrix},$$

then

$$A + B = \begin{bmatrix} 9 & -10 & 7 \\ 6 & -1 & 2 \end{bmatrix}.$$

Next, we recall that for each $a \in K$, we can define the linear transformation $aT: V \longrightarrow U$, by putting $(aT)(v) = aT(v)$, for each $v \in V$. Thus, if $[a_{ij}]_{m \times n}$ is again the matrix $[T]_{\mathcal{C},\mathcal{B}}$, then the matrix $[aT]_{\mathcal{C},\mathcal{B}}$ is $[aa_{ij}]_{m \times n}$, that is, the matrix whose (i,j)th entry is aa_{ij}. To see this, we merely observe that

$$(aT)(v_j) = aT(v_j) = a\sum_{i=1}^{m} a_{ij}u_i = \sum_{i=1}^{m} aa_{ij}u_i,$$

for each v_j in the basis.

This discussion leads to the following.

Definition 8.6.2 Let $A = [a_{ij}]_{m \times n}$ be an $m \times n$ matrix with entries in the field K. Let $a \in K$. Then the *scalar multiple of A by a* is defined as $aA = [aa_{ij}]_{m \times n}$. That is, aA is the matrix whose (i,j)th entry is aa_{ij}.

To illustrate this definition, let

$$A = \begin{bmatrix} 1 & -2 & -1 & 8 \\ 2 & 0 & 2 & 16 \\ 3 & \tfrac{1}{2} & \tfrac{3}{8} & 2 \end{bmatrix}$$

and let $a = \tfrac{3}{4}$. Then

$$\tfrac{3}{4}A = \begin{bmatrix} \tfrac{3}{4} & -\tfrac{3}{2} & -\tfrac{3}{4} & 6 \\ \tfrac{3}{2} & 0 & \tfrac{3}{2} & 12 \\ \tfrac{9}{4} & \tfrac{3}{8} & \tfrac{9}{32} & \tfrac{3}{2} \end{bmatrix}.$$

Now that we have defined the sum of two matrices A and B and we have defined the scalar multiple aA, where $a \in K$, it is easy to see that linear combinations of matrices have also been defined. For if we let A and B be $m \times n$ matrices over the field K, and if we let a and $b \in K$, then both aA and bB are defined by Definition 8.6.2, and the linear combination $aA + bB$ is now defined by Definition 8.6.1.

The next example will exemplify the connection between linear combinations of linear transformations and linear combinations of matrices.

Example 8.6.3 Let T and S be linear transformations from $\mathbf{R}^2$ to $\mathbf{R}^3$, and let the associated matrices be

$$A = \begin{bmatrix} 0 & -1 \\ 1 & 1 \\ 2 & 0 \end{bmatrix} \text{ and } B = \begin{bmatrix} 1 & 4 \\ -1 & -3 \\ 1 & -2 \end{bmatrix},$$

respectively, with respect to some fixed bases of $\mathbf{R}^2$ and $\mathbf{R}^3$. Then the linear transformation $3T - 2S$ has matrix, relative to the same bases,

$$3A - 2B = \begin{bmatrix} 0 & -3 \\ 3 & 3 \\ 6 & 0 \end{bmatrix} + \begin{bmatrix} -2 & -8 \\ 2 & 6 \\ -2 & 4 \end{bmatrix} = \begin{bmatrix} -2 & -11 \\ 5 & 9 \\ 4 & 4 \end{bmatrix}.$$

Given Definitions 8.6.1 and 8.6.2, we can now state the matrix analogue of Theorem 8.3.2. The reader will recall that in Theorem 8.3.2 we proved that $\text{Hom}_K(V,U)$ was a vector space over K of dimension nm.

THEOREM 8.6.4 Let $\mathfrak{M}$ be the set of all $m \times n$ matrices over the field K (that is, with entries from K). For matrices A and B in $\mathfrak{M}$ and scalars $a \in K$, define $A + B$ and aA, according to Definitions 8.6.1 and 8.6.2. Then $\mathfrak{M}$ is a vector space over K, with respect to these definitions of addition and scalar multiple.

Furthermore, the set of matrices $E_{ij} = [a_{st}]_{m \times n}$, where $a_{ij} = 1$ and $a_{st} = 0$ if either $s \neq i$ or $t \neq j$, is a basis for $\mathfrak{M}$, and therefore $\mathfrak{M}$ has dimension nm.

Proof We leave the proof as an exercise. We simply point out that the matrix E_{ij} is a matrix with precisely one entry equal to 1—namely, the (i,j)th entry is 1—and all the other entries are equal to 0. Also, the matrix E_{ij} is the matrix that corresponds to the linear transformation T_{ij} defined in the proof of Theorem 8.3.2. Note that T_{ij} maps the jth basis element v_j in the basis $\mathfrak{B}$ of V to the ith basis element u_i in the basis $\mathfrak{C}$ for U, and T_{ij} maps all other basis elements v_k, $k \neq j$, to 0. ∎

Example 8.6.5 The set of all 2×3 matrices with real entries is a vector space of dimension 6, with the following set as a basis.

$$E_{11} = \begin{bmatrix} 1 & 0 & 0 \\ 0 & 0 & 0 \end{bmatrix} \quad E_{12} = \begin{bmatrix} 0 & 1 & 0 \\ 0 & 0 & 0 \end{bmatrix} \quad E_{13} = \begin{bmatrix} 0 & 0 & 1 \\ 0 & 0 & 0 \end{bmatrix}$$

$$E_{21} = \begin{bmatrix} 0 & 0 & 0 \\ 1 & 0 & 0 \end{bmatrix} \quad E_{22} = \begin{bmatrix} 0 & 0 & 0 \\ 0 & 1 & 0 \end{bmatrix} \quad E_{23} = \begin{bmatrix} 0 & 0 & 0 \\ 0 & 0 & 1 \end{bmatrix}$$

The arbitrary 2×3 matrix $\begin{bmatrix} a & b & c \\ d & e & f \end{bmatrix}$ has representation

$$\begin{bmatrix} a & b & c \\ d & e & f \end{bmatrix} = aE_{11} + bE_{12} + cE_{13} + dE_{21} + eE_{22} + fE_{23},$$

and this representation is clearly unique.

We conclude this section with a theorem that summarizes a number of important facts.

THEOREM 8.6.6 Let V and U be vector spaces over K of dimensions n and m, respectively. Let $\mathfrak{B} = \{v_1, \ldots, v_n\}$ and $\mathfrak{C} = \{u_1, \ldots, u_m\}$ be fixed bases for V and U, respectively. Let $\mathfrak{M}$ be the set of all $m \times n$ matrices over K. Then:

(a) Each linear transformation $T: V \longrightarrow U$ has a unique $m \times n$ matrix $[T]_{\mathfrak{C},\mathfrak{B}}$, and $[T]_{\mathfrak{C},\mathfrak{B}} \in \mathfrak{M}$.

(b) For each matrix $A \in \mathfrak{M}$, there exists a unique linear transformation $T \in \text{Hom}_K(V,U)$ such that $[T]_{\mathfrak{C},\mathfrak{B}} = A$.

(c) If T and $S \in \text{Hom}_K(V,U)$, and if a and $b \in K$, then

$$[aT + bS]_{\mathfrak{C},\mathfrak{B}} = a[T]_{\mathfrak{C},\mathfrak{B}} + b[S]_{\mathfrak{C},\mathfrak{B}}.$$

(d) The function $f: \text{Hom}_K(V,U) \longrightarrow \mathfrak{M}$, where $f(T) = [T]_{e,\mathfrak{B}}$, is an isomorphism between the vector spaces $\text{Hom}_K(V,U)$ and $\mathfrak{M}$.

Proof (a) This is true since $[T]_{e,\mathfrak{B}}$ is an $m \times n$ matrix.
(b) Given a matrix $A = [a_{ij}]_{m \times n}$, define $T: V \longrightarrow U$ by

$$T(v_j) = \sum_{i=1}^{m} a_{ij} u_i.$$

By Theorem 8.1.9, T is uniquely defined in this manner.

(c) $[aT + bS]_{e,\mathfrak{B}} = [aT]_{e,\mathfrak{B}} + [bS]_{e,\mathfrak{B}} = a[T]_{e,\mathfrak{B}} + b[S]_{e,\mathfrak{B}}$.

(d) This is left to the reader. Note that the function f is the correspondence used in (a). ∎

We emphasize that the function f in Theorem 8.6.6 depends on the choice of fixed bases $\mathfrak{B}$ and e for V and U, respectively. With $\mathfrak{B}$ replaced by $\mathfrak{B}'$ and e replaced by e', conclusion (d) would still hold; only the correspondence f would have to be changed.

Example 8.6.7 Let $V = \mathbf{R}^2$ with canonical basis $\mathfrak{B} = \{e_1, e_2\}$, $e_1 = (1,0)$, $e_2 = (0,1)$, and let $U = \mathbf{R}^3$ with canonical basis $e = \{f_1, f_2, f_3\}$, $f_1 = (1,0,0)$, $f_2 = (0,1,0)$, $f_3 = (0,0,1)$. Let $T: V \longrightarrow U$ be defined by

$$T(e_1) = f_1 - f_2 + f_3, \qquad T(e_2) = f_2 - f_3.$$

Then

$$[T]_{e,\mathfrak{B}} = \begin{bmatrix} 1 & 0 \\ -1 & 1 \\ 1 & -1 \end{bmatrix}.$$

Now let $v_1 = e_1 + e_2$, $v_2 = e_1 - e_2$. Then $\mathfrak{B}' = \{v_1, v_2\}$ is also a basis for V. And let $u_1 = f_1 + f_2 + f_3$, $u_2 = f_2 + f_3$, $u_3 = f_3$, so that $e' = \{u_1, u_2, u_3\}$ is a basis for U.

Then

$$\begin{aligned} T(v_1) &= T(e_1 + e_2) = T(e_1) + T(e_2) \\ &= (f_1 - f_2 + f_3) + (f_2 - f_3) \\ &= f_1 = u_1 - u_2 \end{aligned}$$

and

$$\begin{aligned} T(v_2) &= T(e_1 - e_2) = T(e_1) - T(e_2) \\ &= (f_1 - f_2 + f_3) - (f_2 - f_3) \\ &= f_1 - 2f_2 + 2f_3 \\ &= u_1 - 3u_2 + 4u_3. \end{aligned}$$

Thus

$$[T]_{e',\mathfrak{B}'} = \begin{bmatrix} 1 & 1 \\ -1 & -3 \\ 0 & 4 \end{bmatrix}.$$

Referring now to Theorem 8.6.6, T corresponds to

$$\begin{bmatrix} 1 & 0 \\ -1 & 1 \\ 1 & -1 \end{bmatrix}$$

when the bases $\mathcal{B}$ and $\mathcal{C}$ are chosen, whereas T corresponds to

$$\begin{bmatrix} 1 & 1 \\ -1 & -3 \\ 0 & 4 \end{bmatrix}$$

when the bases $\mathcal{B}'$ and $\mathcal{C}'$ are selected. Thus in applying Theorem 8.6.6 it is important to remember that fixed bases for V and U must be used.

EXERCISES

1. Find the sums $A + B$ and $B + A$ for each of the following pairs of matrices:

   (a) $A = \begin{bmatrix} 1 & 2 & 1 \\ 0 & 1 & 1 \\ 1 & 2 & 3 \\ 0 & 0 & -1 \end{bmatrix}$ $B = \begin{bmatrix} -1 & -2 & 1 \\ 1 & -1 & 1 \\ -1 & -2 & -3 \\ 1 & 1 & 7 \end{bmatrix}$

   (b) $A = \begin{bmatrix} 1 & 2 \\ 1 & 3 \end{bmatrix}$ $B = \begin{bmatrix} 1 & -2 \\ -1 & 3 \end{bmatrix}$

   (c) $A = \begin{bmatrix} 1 & 1 & 1 \\ 1 & 1 & 0 \\ 0 & 1 & 1 \end{bmatrix}$ $B = \begin{bmatrix} 1 & -1 & 1 \\ 1 & -1 & 1 \\ 1 & -1 & 1 \end{bmatrix}$

   (d) $A = \begin{bmatrix} 1 & 2 & 3 & 4 \\ -3 & -1 & 11 & -2 \end{bmatrix}$ $B = \begin{bmatrix} 1 & -5 & 2 & 6 \\ 1 & 2 & 3 & -8 \end{bmatrix}$

2. Why does it not make sense to add

$$A = \begin{bmatrix} 1 & 2 & 3 \\ 0 & 1 & -1 \end{bmatrix} \quad \text{and} \quad B = \begin{bmatrix} 1 & 0 \\ -1 & 1 \\ 2 & -1 \end{bmatrix}?$$

3. Let $V = \mathbf{R}^2$, $U = \mathbf{R}^3$, and let $\mathcal{B} = \{e_1, e_2\}$ and $\mathcal{C} = \{f_1, f_2, f_3\}$ be the canonical bases of V and U, as given in Example 8.6.7. Let $T_{ij}: V \longrightarrow U$ be defined by

$$T_{ij}(e_k) = \begin{cases} f_i, & \text{if } k = j, \\ 0, & \text{if } k \neq j, \end{cases}$$

(a) In Theorem 8.6.6, the matrices corresponding to the T_{ij}, relative to $\mathfrak{B}$ and $\mathfrak{C}$, are the E_{ij}, defined in Theorem 8.6.4. Find the matrices corresponding to the same T_{ij} if the bases are changed to $\mathfrak{B}'$ and $\mathfrak{C}'$, as given in Example 8.6.7. For example,

$$T_{11}(v_1) = T_{11}(e_1 + e_2) = T_{11}(e_1) + T_{11}(e_2)$$
$$= f_1 + 0 = u_1 - u_2,$$
$$T_{11}(v_2) = T_{11}(e_1 - e_2) = T_{11}(e_1) - T_{11}(e_2)$$
$$= f_1 + 0 = u_1 - u_2.$$

Thus

$$[T_{11}]_{\mathfrak{C}',\mathfrak{B}'} = \begin{bmatrix} 1 & 1 \\ -1 & -1 \\ 0 & 0 \end{bmatrix}$$

although

$$[T_{11}]_{\mathfrak{C},\mathfrak{B}} = \begin{bmatrix} 1 & 0 \\ 0 & 0 \\ 0 & 0 \end{bmatrix}.$$

(b) Show that the six matrices, $[T_{ij}]_{\mathfrak{C}',\mathfrak{B}'}$, $i = 1, 2, 3$, $j = 1, 2$, are linearly independent over $\mathbf{R}$.

4. Prove that if A, B, and C are $m \times n$ matrices over the field K, then $(A + B) + C = A + (B + C)$. (*Hint:* Use Theorem 8.6.6 and the fact that associativity holds for addition of linear transformations, or prove this directly from Definition 8.6.1.)

5. Let A be an $m \times n$ matrix, and let $a, b \in K$. Prove that $(ab)A = a(bA) = b(aA)$.

6. Let A, B be $m \times n$ matrices, and let $a, b \in K$. Prove
   (a) $A + B = B + A$.
   (b) $a(A + B) = aA + aB$.
   (c) $(a + b)A = aA + bA$.

7. Let $T: \mathbf{R}^3 \longrightarrow \mathbf{R}^3$ have matrix

$$[T]_{\mathfrak{B}} = \begin{bmatrix} 1 & 0 & 0 \\ 0 & 1 & 0 \\ 0 & 0 & 0 \end{bmatrix}$$

with respect to the canonical basis $\mathfrak{B} = \{e_1, e_2, e_3\}$ of $\mathbf{R}^3$, where $e_1 = (1,0,0)$, $e_2 = (0,1,0)$, and $e_3 = (0,0,1)$.
(a) Find the null space $N(T)$ and the range $R(T)$ of T.
(b) Suppose $\mathfrak{B}'$ is a second basis for $\mathbf{R}^3$. Prove that

$$[T]_{\mathfrak{B}'} \neq \begin{bmatrix} 1 & 1 & 1 \\ 1 & 0 & 1 \\ 1 & 0 & 0 \end{bmatrix}.$$

[*Hint:* Let $\mathcal{B}' = \{u_1, u_2, u_3\}$. Determine $N(S)$, where S is the linear transformation such that

$$[S]_{\mathcal{B}'} = \begin{bmatrix} 1 & 1 & 1 \\ 1 & 0 & 1 \\ 1 & 0 & 1 \end{bmatrix}.$$

What is the dimension of $N(S)$?]

8.7 Matrices: Products

We have defined the composition $S \circ T$ of two linear transformations S and T, where $T\colon V \longrightarrow U$ and $S\colon U \longrightarrow W$. The composite $S \circ T$ can be considered the product of the two linear transformations. As such, it will motivate the definition of the product of two matrices A and B, provided A and B satisfy certain conditions.

Thus let V, U, and W be vector spaces with bases $\mathcal{B} = \{v_1, \ldots, v_n\}$, $\mathcal{C} = \{u_1, \ldots, u_m\}$, and $\mathcal{D} = \{w_1, \ldots, w_t\}$, respectively. Then if $T\colon V \longrightarrow U$ and $S\colon U \longrightarrow W$ are linear transformations, we know that

$$[T]_{\mathcal{C},\mathcal{B}} = [a_{ij}]_{m \times n} \quad \text{and} \quad [S]_{\mathcal{D},\mathcal{C}} = [b_{ij}]_{t \times m},$$

for uniquely determined elements a_{ij}, $i = 1, \ldots, m$; $j = 1, \ldots, n$; and b_{ij}, $i = 1, \ldots, t$; $j = 1, \ldots, m$.

Now $S \circ T\colon V \longrightarrow W$, so $[S \circ T]_{\mathcal{D},\mathcal{B}}$ will be a $t \times n$ matrix. We determine $[S \circ T]_{\mathcal{D},\mathcal{B}}$ as follows: for each $v_j \in \mathcal{B}$,

$$(S \circ T)(v_j) = S(T(v_j)) = S\left(\sum_{i=1}^{m} a_{ij} u_i\right)$$

$$= \sum_{i=1}^{m} a_{ij} S(u_i) = \sum_{i=1}^{m} a_{ij} \left(\sum_{k=1}^{t} b_{ki} w_k\right)$$

$$= \sum_{k=1}^{t} \sum_{i=1}^{m} (b_{ki} a_{ij}) w_k.$$

This implies that the coefficient of w_k is

$$\sum_{i=1}^{m} b_{ki} a_{ij},$$

whence the (k,j)th entry of $[S \circ T]_{\mathcal{D},\mathcal{B}}$ is

$$\sum_{i=1}^{m} b_{ki} a_{ij}.$$

By interchanging the i and k we see that the (i,j)th entry of $[S \circ T]_{\mathfrak{D},\mathfrak{B}}$ is

$$\sum_{k=1}^{m} b_{ik}a_{kj}.$$

This leads us to the following.

Definition 8.7.1 Let $B = [b_{ij}]_{t \times m}$ and $A = [a_{ij}]_{m \times n}$ be matrices with entries in the field K. Then the *product* BA is the matrix

$$C = [c_{ij}]_{t \times n}, \quad \text{where } c_{ij} = \sum_{k=1}^{m} b_{ik}a_{kj}.$$

The reader should note that the product BA has only been defined for pairs of matrices B and A for which the number of columns of B equals the number of rows A.

The technique of writing matrices as rectangular arrays makes possible rapid computation of the product of two matrices. Thus if $B = [b_{ij}]_{t \times m}$ and $A = [a_{ij}]_{m \times n}$, then in order to compute the (i,j)th entry of the product BA, take the ith *row* of B and the jth *column* of A and form the "scalar product" as though the ith row of B and the jth column of A were a pair of vectors (m-tuples) with coefficients in K, that is,

$$BA = [c_{ij}]_{t \times n}, \quad \text{where } c_{ij} = b_{i1}a_{1j} + b_{i2}a_{2j} + \cdots + b_{im}a_{mj}.$$

Example 8.7.2 Let

$$B = \begin{bmatrix} 1 & -1 & 3 & 0 \\ 0 & 6 & 4 & 1 \\ 3 & 4 & -2 & 1 \end{bmatrix} \quad \text{and} \quad A = \begin{bmatrix} 1 & 0 \\ 0 & -3 \\ 2 & 1 \\ -1 & 4 \end{bmatrix}.$$

Since B is 3×4 and A is 4×2, the number of columns in B is equal to the number of rows in A, so BA is defined. Then if $C = BA$, we can calculate the entries of C according to the following scheme: to find c_{11}, move along the first row of B and down the first column of A, multiplying corresponding elements and placing the sum of these products in position $(1,1)$ of C as indicated here:

$$\begin{bmatrix} 1 & -1 & 3 & 0 \\ 0 & 6 & 4 & 1 \\ 3 & 4 & -2 & 1 \end{bmatrix} \begin{bmatrix} 1 & 0 \\ 0 & -3 \\ 2 & 1 \\ -1 & 4 \end{bmatrix} = \begin{bmatrix} 7 & \\ & \\ & \end{bmatrix}_{3 \times 2},$$

since $7 = 1 \cdot 1 + (-1) \cdot 0 + 3 \cdot 2 + 0 \cdot (-1)$. To find c_{12}, move along the first row of B and down the second column of A, thus:

$$\begin{bmatrix} 1 & -1 & 3 & 0 \\ 0 & 6 & 4 & 1 \\ 3 & 4 & -2 & 1 \end{bmatrix} \begin{bmatrix} 1 & 0 \\ 0 & -3 \\ 2 & 1 \\ -1 & 4 \end{bmatrix},$$

getting

$$c_{12} = 1 \cdot 0 + (-1)(-3) + 3 \cdot 1 + 0 \cdot 4 = 3 + 3 = 6.$$

Continuing in this manner, we get

$$\begin{bmatrix} 1 & -1 & 3 & 0 \\ 0 & 6 & 4 & 1 \\ 3 & 4 & -2 & 1 \end{bmatrix} \begin{bmatrix} 1 & 0 \\ 0 & -3 \\ 2 & 1 \\ -1 & 4 \end{bmatrix} = \begin{bmatrix} 7 & 6 \\ 7 & -10 \\ -2 & -10 \end{bmatrix}.$$

Before proceeding, we note that Definition 8.7.1 and our discussion preceding it imply that if $T: V \longrightarrow U$ and $S: U \longrightarrow W$, and if V, U, and W have bases $\mathfrak{B}$, $\mathfrak{C}$, and $\mathfrak{D}$, as described before, then the following matrix equation holds:

$$[S \circ T]_{\mathfrak{D},\mathfrak{B}} = [S]_{\mathfrak{D},\mathfrak{C}}[T]_{\mathfrak{C},\mathfrak{B}}.$$

This relationship is important, and will be used repeatedly in the sequel.

Note that if T and S are linear operators on V, that is, $U = W = V$ and $\mathfrak{C} = \mathfrak{D} = \mathfrak{B}$, then this equation becomes:

$$[S \circ T]_{\mathfrak{B}} = [S]_{\mathfrak{B}}[T]_{\mathfrak{B}}.$$

Now let C, B, and A be three matrices, where C is $s \times t$, B is $t \times m$, and A is $m \times n$, all with entries in a field K. We can form the products $C(BA)$ and $(CB)A$, given the numbers of rows and columns of C, B, and A. Using Definition 8.7.1, we could prove directly that $C(BA) = (CB)A$. We prefer, however, to give a proof that relies on our knowledge of linear transformations. To facilitate this, we let K^r denote the vector space of all r-tuples, $(x_1, \ldots, x_r)$, $x_i \in K$. We note that K^r has a canonical basis $\mathfrak{B}_r$ consisting of the r-tuples of the form $\epsilon_i = (0, \ldots, 0, 1, 0, \ldots, 0)$, where the 1 is in the ith position, $i = 1, \ldots, r$.

THEOREM 8.7.3 Let $C_{s \times t}$, $B_{t \times m}$, and $A_{m \times n}$ be matrices with entries in K. Then $(CB)A = C(BA)$.

Proof We let $V = K^n$, $U = K^m$, $W = K^t$, and $Y = K^s$. We let $T: V \longrightarrow U$, $S: U \longrightarrow W$, and $R: W \longrightarrow Y$ be linear transformations whose matrices, with respect to the canonical bases of K^n, K^m, K^t, and K^s, are

$$[T]_{\mathfrak{B}_m, \mathfrak{B}_n} = A, \quad [S]_{\mathfrak{B}_t, \mathfrak{B}_m} = B, \quad \text{and} \quad [R]_{\mathfrak{B}_s, \mathfrak{B}_t} = C.$$

Such linear transformations exist by our discussion in earlier sections.

Since T, S, and R are functions, we know that $(R \circ S) \circ T = R \circ (S \circ T)$. Now on the one hand

$$[(R \circ S) \circ T]_{\mathcal{B}_s, \mathcal{B}_n} = [R \circ S]_{\mathcal{B}_s, \mathcal{B}_m} [T]_{\mathcal{B}_m, \mathcal{B}_n}$$
$$= ([R]_{\mathcal{B}_s, \mathcal{B}_t} [S]_{\mathcal{B}_t, \mathcal{B}_m}) [T]_{\mathcal{B}_m, \mathcal{B}_n}.$$

On the other hand

$$[R \circ (S \circ T)]_{\mathcal{B}_s, \mathcal{B}_n} = [R]_{\mathcal{B}_s, \mathcal{B}_t} [S \circ T]_{\mathcal{B}_t, \mathcal{B}_n}$$
$$= [R]_{\mathcal{B}_s, \mathcal{B}_t} ([S]_{\mathcal{B}_t, \mathcal{B}_m} [T]_{\mathcal{B}_m, \mathcal{B}_n}).$$

Since $(R \circ S) \circ T = R \circ (S \circ T)$,

$$[(R \circ S) \circ T]_{\mathcal{B}_s, \mathcal{B}_n} = [R \circ (S \circ T)]_{\mathcal{B}_s, \mathcal{B}_n},$$

so

$$([R]_{\mathcal{B}_s, \mathcal{B}_t} [S]_{\mathcal{B}_t, \mathcal{B}_m}) [T]_{\mathcal{B}_m, \mathcal{B}_n} = [R]_{\mathcal{B}_s, \mathcal{B}_t} ([S]_{\mathcal{B}_t, \mathcal{B}_m} [T]_{\mathcal{B}_m, \mathcal{B}_n}),$$

that is, $(CB)A = C(BA)$, as we asserted. ∎

The two distributive laws also hold for matrices, provided the orders of the given matrices permit the necessary additions and multiplications.

THEOREM 8.7.4 (a) Let $A_{m \times n}$, $B_{m \times n}$, and $C_{t \times m}$ be matrices over the field K. Then $C(A + B) = CA + CB$.

(b) Let $A_{m \times n}$, $B_{m \times n}$, and $D_{n \times s}$ be matrices over the field K. Then $(A + B)D = AD + BD$.

Proof These follow easily from the definitions, or from properties of linear transformations. We leave the verifications to the reader. ∎

We will now turn our attention to a useful application of matrix multiplication.

If V has basis $\mathcal{B} = \{v_1, \ldots, v_n\}$, then a vector $v \in V$ is completely described by the coefficients $c_1, \ldots, c_n$, where $v = c_1 v_1 + \cdots + c_n v_n$. For convenience, we will represent v as the $n \times 1$ column matrix

$$\begin{bmatrix} c_1 \\ c_2 \\ \cdot \\ \cdot \\ \cdot \\ c_n \end{bmatrix},$$

which we call the *coefficient matrix of v with respect to* $\mathcal{B}$. We will denote this matrix by $[v]_{\mathcal{B}}$. A change from the basis $\mathcal{B}$ to, say, a basis $\mathcal{B}' = \{v_1', \ldots, v_n'\}$, will result in a different column matrix. For if $v = c_1' v_1' + \cdots + c_n' v_n'$, then

$$[v]_{\mathcal{B}'} = \begin{bmatrix} c_1' \\ \cdot \\ \cdot \\ \cdot \\ c_n' \end{bmatrix}.$$

Thus it is important to know the bases that are being used in the discussion.

Now let U have basis $\mathcal{C} = \{u_1, \ldots, u_m\}$ and let $T\colon V \longrightarrow U$ be a linear transformation. If $[T]_{\mathcal{C},\mathcal{B}} = [a_{ij}]_{m \times n}$, then $[v]_{\mathcal{B}}$ and $[T(v)]_{\mathcal{C}}$ have a simple relationship.

THEOREM 8.7.5 Let V and U have bases $\mathcal{B} = \{v_1, \ldots, v_n\}$ and $\mathcal{C} = \{u_1, \ldots, u_m\}$, respectively. Let $T\colon V \longrightarrow U$ have matrix $[T]_{\mathcal{C},\mathcal{B}} = [a_{ij}]_{m \times n}$. Let $v \in V$. Then $[T(v)]_{\mathcal{C}} = [T]_{\mathcal{C},\mathcal{B}}[v]_{\mathcal{B}}$.

Proof Let $v = \sum_{j=1}^{n} c_j v_j$. Then

$$T(v) = T\left(\sum_{j=1}^{n} c_j v_j\right) = \sum_{j=1}^{n} c_j T(v_j) = \sum_{j=1}^{n} c_j \left(\sum_{i=1}^{m} a_{ij} u_i\right)$$
$$= \sum_{i=1}^{m} \sum_{j=1}^{n} (a_{ij} c_j) u_i.$$

Thus, the coefficient of u_i is

$$\sum_{j=1}^{n} (a_{ij} c_j),$$

which is just the "inner product" of row i of $[a_{ij}]_{m \times n}$ and the column matrix

$$\begin{bmatrix} c_1 \\ \cdot \\ \cdot \\ \cdot \\ c_n \end{bmatrix}.$$

Thus, $[T(v)]_{\mathcal{C}} = [T]_{\mathcal{C},\mathcal{B}}[v]_{\mathcal{B}}$, which is the product of an $m \times n$ and an $n \times 1$ matrix, that is, the product is $m \times 1$, the desired dimensions. Thus, if we have $A = [a_{ij}]_{m \times n}$ and the coefficient matrix $[v]_{\mathcal{B}}$, then the coefficient matrix of $T(v)$ with respect to $\mathcal{C}$ is $A[v]_{\mathcal{B}}$. ∎

Example 8.7.6 Let $T: \mathbf{R}^3 \longrightarrow \mathbf{R}^2$, and let $\mathfrak{B}_3$ and $\mathfrak{B}_2$ be the canonical bases for $\mathbf{R}^3$ and $\mathbf{R}^2$, respectively. Say that $\mathfrak{B}_3 = \{f_1, f_2, f_3\}$ and $\mathfrak{B}_2 = \{e_1, e_2\}$, so that in $\mathbf{R}^3$, $(a,b,c) = af_1 + bf_2 + cf_3$ and in $\mathbf{R}^2$, $(x,y) = xe_1 + ye_2$. Let T be defined by

$$T(f_1) = (2,3) = 2e_1 + 3e_2,$$
$$T(f_2) = (2,4) = 2e_1 + 4e_2,$$
$$T(f_3) = (1,-1) = e_1 - e_2.$$

Then

$$[T]_{\mathfrak{B}_2, \mathfrak{B}_3} = \begin{bmatrix} 2 & 2 & 1 \\ 3 & 4 & -1 \end{bmatrix} = A.$$

Now consider the vector $v = (1,2,3) = f_1 + 2f_2 + 3f_3$. Then

$$[v]_{\mathfrak{B}_3} = \begin{bmatrix} 1 \\ 2 \\ 3 \end{bmatrix},$$

and

$$[T(v)]_{\mathfrak{B}_2} = A[v]_{\mathfrak{B}_3} = \begin{bmatrix} 2 & 2 & 1 \\ 3 & 4 & -1 \end{bmatrix} \begin{bmatrix} 1 \\ 2 \\ 3 \end{bmatrix} = \begin{bmatrix} 9 \\ 8 \end{bmatrix}.$$

Thus, $T(v) = 9e_1 + 8e_2$.

This could also have been obtained as follows:

$$T(v) = T(f_1 + 2f_2 + 3f_3) = T(f_1) + 2T(f_2) + 3T(f_3)$$
$$= (2e_1 + 3e_2) + 2(2e_1 + 4e_2) + 3(e_1 - e_2)$$
$$= 9e_1 + 8e_2.$$

To emphasize that the coefficient matrices depend on the choice of bases, let $\mathfrak{B}_3' = \{f_1', f_2', f_3'\}$ and $\mathfrak{B}_2' = \{e_1', e_2'\}$ be different bases for $\mathbf{R}^3$ and $\mathbf{R}^2$, where $f_1' = f_1 + f_3$, $f_2' = f_2$, and $f_3' = f_2 + f_3$, and $e_1' = e_1 + e_2$, $e_2' = e_1 - e_2$. (It is easy to verify that $\mathfrak{B}_3'$ and $\mathfrak{B}_2'$ are bases for $\mathbf{R}^3$ and $\mathbf{R}^2$.)

Then

$$T(f_1') = T(f_1 + f_3) = T(f_1) + T(f_3)$$
$$= (2e_1 + 3e_2) + (e_1 - e_2)$$
$$= 3e_1 + 2e_2 = me_1' + ne_2'$$
$$= m(e_1 + e_2) + n(e_1 - e_2)$$
$$= (m+n)e_1 + (m-n)e_2,$$

whence $m + n = 3$, $m - n = 2$, and so $m = \frac{5}{2}$, $n = \frac{1}{2}$. Thus

$$T(f_1') = \tfrac{5}{2}e_1' + \tfrac{1}{2}e_2'.$$

Similarly,

$$T(f_2') = 3e_1' + e_2', \qquad T(f_3') = 3e_1' + 0e_2' = 3e_1'.$$

From this we have
$$[T]_{\mathcal{B}_2', \mathcal{B}_3'} = \begin{bmatrix} \frac{5}{2} & 3 & 3 \\ \frac{1}{2} & 1 & 0 \end{bmatrix}.$$

Now the vector $v = (1,2,3)$, already considered, can be written
$$(1,2,3) = af_1' + bf_2' + cf_3' = a(1,0,1) + b(0,1,0) + c(0,1,1),$$
and so $a = 1$, $b = 0$, and $c = 2$, as is easily verified. Thus, we now have
$$[v]_{\mathcal{B}_3'} = \begin{bmatrix} 1 \\ 0 \\ 2 \end{bmatrix},$$

whence
$$[T(v)]_{\mathcal{B}_2'} = \begin{bmatrix} \frac{5}{2} & 3 & 3 \\ \frac{1}{2} & 1 & 0 \end{bmatrix} \begin{bmatrix} 1 \\ 0 \\ 2 \end{bmatrix} = \begin{bmatrix} \frac{17}{2} \\ \frac{1}{2} \end{bmatrix},$$

so that
$$\begin{aligned} T(v) &= \tfrac{17}{2} e_1' + \tfrac{1}{2} e_2' \\ &= \tfrac{17}{2}(e_1 + e_2) + \tfrac{1}{2}(e_1 - e_2) \\ &= \tfrac{18}{2} e_1 + \tfrac{16}{2} e_2 = 9e_1 + 8e_2, \end{aligned}$$
as before.

We conclude this section by making several observations. First, although we introduced matrices via linear transformations, the actual definition of a matrix and the actual definitions of sums and products are independent of linear transformations. Thus, we may view matrices over K and matrix arithmetic as generalizations of the field K itself (although some properties are lost). For example, the set of 1×1 matrices over K have the same arithmetic behavior as K itself:
$$[a_{ij}]_{1 \times 1} + [b_{ij}]_{1 \times 1} = [a_{ij} + b_{ij}]_{1 \times 1}$$
and
$$[a_{ij}]_{1 \times 1} [b_{ij}]_{1 \times 1} = [a_{ij} b_{ij}]_{1 \times 1}.$$

Second, we may multiply $A_{m \times n}$ and $B_{t \times s}$ to form the product BA only when $m = s$, that is, only when B has the same number of columns as A has rows. Thus, we can form the two products
$$A_{m \times n} B_{t \times s} \quad \text{and} \quad B_{t \times s} A_{m \times n}$$
only when $n = t$, in the first case, and $s = m$, in the second case. Thus, $A_{m \times n}$ and $B_{n \times m}$ can be multiplied to obtain
$$A_{m \times n} B_{n \times m} = C_{m \times m} \quad \text{and} \quad B_{n \times m} A_{m \times n} = D_{n \times n}.$$

Third, the distributive laws are meaningful only when the dimensions of the various matrices match up, as in Theorem 8.7.4.

Finally, matrix multiplication need not be commutative, and usually is not. If A is 2×3 and B is 3×2, then AB is a 2×2 matrix and BA is a 3×3 matrix. Hence, matrix multiplication, just as composition of linear transformations, is not commutative. Even if A and B are square matrices of the same size, AB may be different from BA. For example, if

$$A = \begin{bmatrix} 0 & 1 \\ 0 & 0 \end{bmatrix} \quad \text{and} \quad B = \begin{bmatrix} 0 & 0 \\ 1 & 0 \end{bmatrix},$$

then

$$AB = \begin{bmatrix} 1 & 0 \\ 0 & 0 \end{bmatrix}, \quad \text{whereas} \quad BA = \begin{bmatrix} 0 & 0 \\ 0 & 1 \end{bmatrix}.$$

EXERCISES

1. Find the product AB for each of the following pairs of matrices:

   (a) $A = \begin{bmatrix} 1 & -1 \\ 0 & 0 \end{bmatrix}, B = \begin{bmatrix} 1 & 2 & 1 \\ 0 & 1 & 1 \end{bmatrix}.$

   (b) $A = \begin{bmatrix} 1 & 1 & -1 \\ 0 & 1 & \frac{1}{2} \\ \frac{1}{3} & \frac{1}{4} & 2 \end{bmatrix}, B = \begin{bmatrix} 1 & -\frac{1}{2} & \frac{1}{2} & 0 \\ 2 & -\frac{1}{2} & \frac{1}{5} & -1 \\ -1 & 2 & 1 & -\frac{1}{8} \end{bmatrix}.$

   (c) $A = \begin{bmatrix} a & 0 & 0 \\ 0 & a & 0 \\ 0 & 0 & a \\ 1 & 0 & 0 \end{bmatrix}, B = \begin{bmatrix} a_{11} & a_{12} & a_{13} & a_{14} \\ a_{21} & a_{22} & a_{23} & a_{24} \\ a_{31} & a_{32} & a_{33} & a_{34} \end{bmatrix}.$

   (d) $A = \begin{bmatrix} 0 & 1 & 0 & 0 \\ 0 & 0 & 1 & 0 \\ 0 & 0 & 0 & 1 \end{bmatrix}, B = \begin{bmatrix} 1 & 1 & 0 \\ 2 & -1 & 1 \\ 3 & 1 & -1 \\ 4 & 2 & 1 \end{bmatrix}.$

   (e) $A = \begin{bmatrix} 1 & 0 & 0 \\ 0 & 0 & 1 \\ 0 & 1 & 0 \end{bmatrix}, B = \begin{bmatrix} 1 & 1 & 1 \\ 2 & 2 & 2 \\ 3 & 3 & 3 \end{bmatrix}.$

2. Let $V = \mathbf{R}^2$ and let

$$\mathcal{B}_2 = \{e_1 = (1,0),\ e_2 = (0,1)\}$$

and

$$\mathcal{B}_2' = \{u_1 = (1,1),\ u_2 = (1,-1)\}$$

be two bases for $\mathbf{R}^2$. Let $T: \mathbf{R}^2 \longrightarrow \mathbf{R}^2$ be a linear transformation such that its matrix $[T]_{\mathcal{B}_2}\ (= [T]_{\mathcal{B}_2, \mathcal{B}_2})$ with respect to the basis $\mathcal{B}_2$ is $\begin{bmatrix} 1 & -1 \\ 1 & 1 \end{bmatrix}.$

338   LINEAR TRANSFORMATIONS AND MATRICES

(a) If the vector v has coordinates $(2,3)$ in the $\mathcal{B}_2$ basis, that is, $[v]_{\mathcal{B}_2} = \begin{bmatrix} 2 \\ 3 \end{bmatrix}$, what are the coordinates of $T(v)$ in the $\mathcal{B}_2$ basis; that is, what is $[T(v)]_{\mathcal{B}_2}$?

(b) What is the matrix of T with respect to the basis $\mathcal{B}_2'$; that is, what is $[T]_{\mathcal{B}_2'}$?

(c) If $[v]_{\mathcal{B}_2} = \begin{bmatrix} 2 \\ 3 \end{bmatrix}$, what is $[v]_{\mathcal{B}_2'}$?

(d) If $[v]_{\mathcal{B}_2} = \begin{bmatrix} 2 \\ 3 \end{bmatrix}$, what is $[T(v)]_{\mathcal{B}_2'}$?

3. Let $V = \mathbf{R}^3$ and let $U = \mathbf{R}^2$. Let $\mathcal{B}_3$ and $\mathcal{B}_2$ be the canonical bases for V and U, respectively. Let $T: V \longrightarrow U$ have matrix

$$[T]_{\mathcal{B}_2, \mathcal{B}_3} = \begin{bmatrix} 1 & -1 & 1 \\ 2 & 1 & 0 \end{bmatrix}.$$

(a) Find $T(1,-1,1)$; $T(0,0,2)$; $T(1,-3,0)$; $T(a,b,c)$.

(b) Let $(-1,5) \in U$. Is there a vector (a,b,c) in V such that $T(a,b,c) = (-1,5)$? If so, find it.

4. Let V, U, $\mathcal{B}_3$, $\mathcal{B}_2$ and T be as in Exercise 3. Let

$$\mathcal{B}_3' = \{(0,1,1),(1,0,1),(1,1,0)\} \quad \text{and} \quad \mathcal{B}_2' = \{(0,1),(-1,1)\}.$$

(a) Show that $\mathcal{B}_3'$ and $\mathcal{B}_2'$ are bases for V and U, respectively.

(b) Find the matrix of T with respect to bases $\mathcal{B}_3'$ and $\mathcal{B}_2'$; that is, find $[T]_{\mathcal{B}_2', \mathcal{B}_3'}$.

(c) If $[v]_{\mathcal{B}_3} = \begin{bmatrix} 1 \\ 2 \\ 3 \end{bmatrix}$, find $[T(v)]_{\mathcal{B}_2}$, $[v]_{\mathcal{B}_3'}$, and $[T(v)]_{\mathcal{B}_2'}$.

5. For an integer $t > 0$, let

$$I_t = \begin{bmatrix} 1 & 0 & 0 & \cdots & 0 \\ 0 & 1 & 0 & \cdots & 0 \\ 0 & 0 & 1 & \cdots & 0 \\ \vdots & & & & \vdots \\ 0 & 0 & 0 & \cdots & 1 \end{bmatrix}_{t \times t}.$$

Let A be an $n \times m$ matrix. Show that $I_n A = A I_m = A$.

6. Let $A = \begin{bmatrix} 1 & -1 \\ 0 & 1 \end{bmatrix}$ and $B = \begin{bmatrix} 1 & 1 \\ 0 & 1 \end{bmatrix}$. Show that

$$AB = BA = I_2 = \begin{bmatrix} 1 & 0 \\ 0 & 1 \end{bmatrix}.$$

8.8 INVERSE MATRICES; TRANSFORMATION OF COORDINATES

7. Give an example of 3×3 matrices A and B such that $AB \neq BA$.
8. Suppose A and B are $n \times n$ matrices such that $AB = BA$. Show that for all integers r and s, $r > 0$, $s > 0$, $A^r B^s = B^s A^r$.
9. Let $A = \begin{bmatrix} d_1 & 0 & 0 & \cdots & 0 \\ 0 & d_2 & 0 & \cdots & 0 \\ 0 & 0 & d_3 & \cdots & 0 \\ \vdots & & & & \vdots \\ 0 & 0 & 0 & \cdots & d_n \end{bmatrix}_{n \times n}$

   Let B be an $n \times m$ matrix, C an $m \times n$ matrix. Describe the products AB and CA in terms of the rows of B, in the first case, and the columns of C in the second case.
10. Prove Theorem 8.7.4 (the distributive laws).
11. Let $\alpha \in K$, and let

    $$I_n = \begin{bmatrix} 1 & 0 & 0 & \cdots & 0 \\ 0 & 1 & 0 & \cdots & 0 \\ 0 & 0 & 1 & \cdots & 0 \\ \vdots & & & & \vdots \\ 0 & 0 & 0 & \cdots & 1 \end{bmatrix}_{n \times n}$$

    (a) Show that

    $$\alpha I_n = \begin{bmatrix} \alpha & 0 & 0 & \cdots & 0 \\ 0 & \alpha & 0 & \cdots & 0 \\ \vdots & & & & \vdots \\ 0 & 0 & 0 & \cdots & \alpha \end{bmatrix}_{n \times n}$$

    (*Note:* We call αI_n a *scalar matrix*.)
    (b) Show that for $A_{n \times m}$, $\alpha A = (\alpha I_n) A$.
12. Prove Theorem 8.7.3 using Definition 8.7.1 directly.

8.8 Identity Matrices and Inverse Matrices; Transformation of Coordinates

If $T: V \longrightarrow V$ is the identity function on V, then it has a particularly simple matrix representation.

THEOREM 8.8.1 Let V be an n-dimensional vector space over K, and let $T: V \longrightarrow V$ be the linear transformation $T = I_V$, that is, $T(v) = v$, for all $v \in V$. Then for any basis $\mathcal{B}$ of V,

$$[T]_\mathcal{B} = I_n = \begin{bmatrix} 1 & 0 & 0 & \cdots & 0 \\ 0 & 1 & 0 & \cdots & 0 \\ \vdots & & & & \vdots \\ 0 & 0 & 0 & \cdots & 1 \end{bmatrix}_{n \times n}.$$

That is, $[T]_\mathcal{B}$ is the matrix $[a_{ij}]_{n \times n}$, where $a_{ii} = 1$, for all i, and $a_{ij} = 0$, for $i \neq j$.

Proof Let $\mathcal{B} = \{v_1, \ldots, v_n\}$ be any basis for V. Then $T(v_j) = v_j$, for all j. Thus, the matrix $[T]_\mathcal{B} = I_n$, as described above. ∎

Definition 8.8.2 The matrix

$$I_n = \begin{bmatrix} 1 & 0 & \cdots & 0 \\ 0 & 1 & \cdots & 0 \\ \vdots & & & \vdots \\ 0 & 0 & \cdots & 1 \end{bmatrix}_{n \times n}$$

described in Theorem 8.8.1 is called the $n \times n$ *identity matrix* or the *identity matrix of order n*.

THEOREM 8.8.3 Let I_n be the $n \times n$ identity matrix. Then

(a) If A is an $n \times m$ matrix, $I_n A = A$.
(b) If B is an $m \times n$ matrix, $B I_n = B$.

Proof These can be verified directly from the definition of the product of matrices. However, we give a proof using linear transformations.

(a) Let W be an m-dimensional space over K with basis

$$\mathcal{C} = \{u_1, \ldots, u_m\}.$$

Let V be n-dimensional, with basis $\mathcal{B} = \{v_1, \ldots, v_n\}$. Then there is a linear transformation $S: U \longrightarrow V$ with matrix $[S]_{\mathcal{B}, \mathcal{C}} = A$. If $T: V \longrightarrow V$ is the identity, $[T]_{\mathcal{B}, \mathcal{B}} = I_n$. Then $T \circ S = S$, since $T = I_V$;

$$[T \circ S]_{\mathcal{B}, \mathcal{C}} = [T]_{\mathcal{B}, \mathcal{B}} [S]_{\mathcal{B}, \mathcal{C}} = I_n A;$$

and $[T \circ S]_{\mathcal{B}, \mathcal{C}} = [S]_{\mathcal{B}, \mathcal{C}} = A$. Thus, $I_n A = A$.

(b) This is similar to (a). ∎

8.8 INVERSE MATRICES; TRANSFORMATION OF COORDINATES

Theorem 8.8.3 justifies our calling I_n an identity matrix. This now puts us in the position to define the inverse of a matrix. We mention now that not every matrix has an inverse. An important condition for the existence of the inverse of a square matrix will be discussed in Chapter 9.

Definition 8.8.4 Let $A = [a_{ij}]_{m \times n}$ and $B = [b_{ij}]_{n \times m}$ be matrices with entries from the field K. If $AB = I_m$, then B is called a *right inverse of A* and A is called a *left inverse of B*.

Example 8.8.5 Let

$$A = \begin{bmatrix} 1 & 0 & -1 \\ 2 & 1 & 1 \end{bmatrix}, \quad B = \begin{bmatrix} 2 & 1 \\ -5 & -2 \\ 1 & 1 \end{bmatrix}.$$

Then

$$AB = \begin{bmatrix} 1 & 0 & -1 \\ 2 & 1 & 1 \end{bmatrix} \begin{bmatrix} 2 & 1 \\ -5 & -2 \\ 1 & 1 \end{bmatrix} = \begin{bmatrix} 1 & 0 \\ 0 & 1 \end{bmatrix} = I_2.$$

Note that on the other hand,

$$BA = \begin{bmatrix} 2 & 1 \\ -5 & -2 \\ 1 & 1 \end{bmatrix} \begin{bmatrix} 1 & 0 & -1 \\ 2 & 1 & 1 \end{bmatrix} = \begin{bmatrix} 4 & 1 & -1 \\ -9 & -2 & 3 \\ 3 & 1 & 0 \end{bmatrix}.$$

It follows from this example that even if B is a right inverse of A, A need not be a right inverse of B. If, however, A and B are both square matrices, then the situation is analogous to that discovered for linear operators in Theorem 8.4.3. For a linear operator T on a vector space V of dimension n has as a matrix representation an $n \times n$ matrix.

THEOREM 8.8.6 Let A and B be $n \times n$ matrices. Then $AB = I_n$ if and only if $BA = I_n$.

Proof Let V be an n-dimensional vector space over the field K, with basis $\mathfrak{B} = \{v_1, \ldots, v_n\}$. Then there exist linear operators T and S on V such that $[T]_\mathfrak{B} = A$ and $[S]_\mathfrak{B} = B$.

By Theorem 8.4.3, $T \circ S = I_V$ if and only if $S \circ T = I_V$. Since $[I_V]_\mathfrak{B} = I_n$, this becomes, in matrix terminology, $[T \circ S]_\mathfrak{B} = I_n$ if and only if $[S \circ T]_\mathfrak{B} = I_n$. But we have already observed that $[T \circ S]_\mathfrak{B} = [T]_\mathfrak{B}[S]_\mathfrak{B}$ and $[S \circ T]_\mathfrak{B} = [S]_\mathfrak{B}[T]_\mathfrak{B}$, so that finally $AB = I_n$ if and only if $BA = I_n$. ∎

Definition 8.8.7 Let A be an $n \times n$ matrix with entries from a field K. Then the $n \times n$ matrix B is called the *inverse of A* if $AB = I_n$, and B is denoted by A^{-1}. If A has an inverse, then A is called a *nonsingular* (or *invertible*) matrix. If A does not have an inverse, A is called *singular*.

Example 8.8.8 Let

$$A = \begin{bmatrix} 1 & 0 & -1 \\ 2 & 1 & 1 \\ -3 & -1 & 1 \end{bmatrix} \text{ and } B = \begin{bmatrix} 2 & 1 & 1 \\ -5 & -2 & -3 \\ 1 & 1 & 1 \end{bmatrix}.$$

Then

$$AB = \begin{bmatrix} 1 & 0 & -1 \\ 2 & 1 & 1 \\ -3 & -1 & 1 \end{bmatrix} \begin{bmatrix} 2 & 1 & 1 \\ -5 & -2 & -3 \\ 1 & 1 & 1 \end{bmatrix} = \begin{bmatrix} 1 & 0 & 0 \\ 0 & 1 & 0 \\ 0 & 0 & 1 \end{bmatrix} = I_3$$

and

$$BA = \begin{bmatrix} 2 & 1 & 1 \\ -5 & -2 & -3 \\ 1 & 1 & 1 \end{bmatrix} \begin{bmatrix} 1 & 0 & -1 \\ 2 & 1 & 1 \\ -3 & -1 & 1 \end{bmatrix} = \begin{bmatrix} 1 & 0 & 0 \\ 0 & 1 & 0 \\ 0 & 0 & 1 \end{bmatrix} = I_3.$$

Hence $B = A^{-1}$ and $A = B^{-1}$.

Theorem 8.8.6 and Definition 8.8.7 allows us to speak of inverses of square matrices without reference to right or left. Also, combining Theorem 8.8.6 with parts (b) and (c) of Theorem 8.4.3, we can also use the term *the inverse*, since Theorem 8.4.3(b) and (c) show that inverses are unique.

THEOREM 8.8.9 Let $A = [a_{ij}]_{n \times n}$ be a square matrix of order n over the field K. Let V be an n-dimensional vector space over K with basis $\mathcal{B} = \{v_1, \ldots, v_n\}$. If $T: V \longrightarrow V$ is the linear operator such that $[T]_\mathcal{B} = A$, then T is a nonsingular linear transformation if and only if A is a nonsingular matrix.

Proof By Definition 8.4.5, T is nonsingular if and only if there is an operator $S: V \longrightarrow V$ such that $T \circ S = S \circ T = I_V$. Clearly, $[S]_\mathcal{B}$ is the inverse matrix of $[T]_\mathcal{B} = A$; that is, $[S]_\mathcal{B} = A^{-1}$.

Conversely, if A is a nonsingular matrix, say with $A^{-1} = B$, then the linear operator $S: V \longrightarrow V$ defined so that $[S]_\mathcal{B} = B$ satisfies $T \circ S = S \circ T = I_V$, for $AB = I_n$ implies $[T]_\mathcal{B}[S]_\mathcal{B} = I_n$, whence $[T \circ S]_\mathcal{B} = I_n$, so that $T \circ S = I_V$. ∎

THEOREM 8.8.10 Let V be a vector space over the field K, with basis $\mathcal{B} = \{v_1, \ldots, v_n\}$. Let

$$v_j' = \sum_{i=1}^n c_{ij} v_i, \quad \text{for } j = 1, \ldots, n,$$

for elements $c_{ij} \in K$. Then $\mathcal{B}' = \{v_1', \ldots, v_n'\}$ is a basis for V if and only if $C = [c_{ij}]_{n \times n}$ is nonsingular. Further, if C is nonsingular, then for each $v \in V$, $[v]_{\mathcal{B}'} = C^{-1}[v]_\mathcal{B}$.

8.8 INVERSE MATRICES; TRANSFORMATION OF COORDINATES

Proof Suppose $\{v_1', \ldots, v_n'\}$ is a basis for V. Let $S: V \longrightarrow V$ be defined by

$$S(v_j) = v_j' = \sum_{i=1}^{n} c_{ij} v_i, \quad \text{for all } v_j \in \mathcal{B}.$$

By Theorem 8.4.8, statements (a) and (c), S is nonsingular. Finally, by Theorem 8.8.9, $C = [S]_\mathcal{B}$ is nonsingular, also.

Conversely, define a linear transformation $T: V \longrightarrow V$ by

$$T(v_j) = \sum_{i=1}^{n} c_{ij} v_i.$$

Since C is nonsingular, the linear transformation T is nonsingular by Theorem 8.8.9. Now using statements (a) and (b) of Theorem 8.4.8, $\{T(v_1), \ldots, T(v_n)\}$ is a basis for V. Thus $\{v_1', \ldots, v_n'\}$ is a basis for V.

Finally, suppose C is nonsingular. Let $v \in V$.

$$v = \sum_{j=1}^{n} d_j v_j', \quad \text{so} \quad [v]_{\mathcal{B}'} = \begin{bmatrix} d_1 \\ \cdot \\ \cdot \\ \cdot \\ d_n \end{bmatrix}.$$

But

$$v = \sum_{j=1}^{n} d_j v_j' = \sum_{j=1}^{n} d_j \left(\sum_{i=1}^{n} c_{ij} v_i \right)$$

$$= \sum_{i=1}^{n} \left(\sum_{j=1}^{n} c_{ij} d_j \right) v_i,$$

whence the ith coefficient of v when written with respect to $\mathcal{B}$ is

$$\sum_{j=1}^{n} c_{ij} d_j.$$

Thus $[v]_\mathcal{B} = C[v]_{\mathcal{B}'}$, whence $[v]_{\mathcal{B}'} = C^{-1}[v]_\mathcal{B}$. ∎

The results of Theorem 8.8.10 are often referred to as a "change in coordinates," or as a "transformation of coordinates." Also, Theorem 8.8.10 gives us another way of interpreting a nonsingular matrix C.

In Section 8.5 we introduced matrices in terms of linear transformations, and this has been our main interpretation of matrices to this point.

However, in Theorem 8.8.10 we have seen how to use a nonsingular matrix

$$C = \begin{bmatrix} c_{11} & c_{12} & \cdots & c_{1n} \\ c_{21} & c_{22} & \cdots & c_{2n} \\ \cdot & & & \\ \cdot & & & \\ \cdot & & & \\ c_{n1} & c_{n2} & \cdots & c_{nn} \end{bmatrix}$$

to define a second basis $\mathcal{B}'$ in terms of a given basis $\mathcal{B}$, where if $\mathcal{B} = \{v_1, \ldots, v_n\}$, then $\mathcal{B}' = \{v_1', \ldots, v_n'\}$, with

$$v_j' = \sum_{i=1}^{n} c_{ij} v_i.$$

And then we saw how we can go from the coordinate matrix $[v]_\mathcal{B}$ of a vector v with respect to the given basis $\mathcal{B}$ to a new coordinate matrix $[v]_{\mathcal{B}'} = C^{-1}[v]_\mathcal{B}$ with respect to the new basis. Thus, in this manner, we have a transformation of coordinates.

Example 8.8.11 Let $V = \mathbf{R}^2$ and let $\mathcal{B}$ be the canonical basis $\mathcal{B} = \{e_1 = (1,0),\ e_2 = (0,1)\}$. Let

$$C = \begin{bmatrix} 0 & 1 \\ -1 & 0 \end{bmatrix}.$$

Then we can use C to define a second basis $\mathcal{B}'$ for V, $\mathcal{B}' = \{f_1, f_2\}$, where $f_1 = -e_2$ and $f_2 = e_1$.

The vector $v = (a,b) = ae_1 + be_2$ has coordinate matrix $\begin{bmatrix} a \\ b \end{bmatrix}$ with respect to the basis $\mathcal{B}$, but with respect to $\mathcal{B}'$, the same vector v has coordinate matrix

$$[v]_{\mathcal{B}'} = C^{-1}[v]_\mathcal{B} = \begin{bmatrix} 0 & -1 \\ 1 & 0 \end{bmatrix} \begin{bmatrix} a \\ b \end{bmatrix} = \begin{bmatrix} -b \\ a \end{bmatrix}$$

that is, $v = -bf_1 + af_2$.

Later, in Chapter 9, when we have a convenient method for finding the inverse of a matrix, we will give a number of exercises on this concept.

We make one final remark concerning the interpretation of a given nonsingular matrix C on the one hand as a matrix of a given linear operator with respect to the fixed basis $\mathcal{B}$ and on the other hand as a matrix defining a change of coordinates. In the first case, the basis $\mathcal{B}$ is held fixed and C tells us how to define an operator T that takes a vector v to another vector $T(v)$. In the second case, C tells us how to go from the basis $\mathcal{B}$ to the

8.8 INVERSE MATRICES; TRANSFORMATION OF COORDINATES

basis $\mathcal{B}'$, and then it tells us how to transform the coordinate matrix $[v]_\mathcal{B}$ of a fixed vector to a new coordinate matrix $[v]_{\mathcal{B}'} = C^{-1}[v]_\mathcal{B}$ of the same vector v. In other words, in the first case, the basis is fixed and the vectors are moved. In the second case, the vectors are kept fixed, but the basis is changed, so that the coordinate matrix is changed.

EXERCISES

1. Let A be an $n \times n$ matrix over the field K such that $AB = B$, for all $n \times n$ matrices B over K. Show that $A = I_n$.
2. Let A be an $r \times n$ matrix and B an $n \times r$ matrix, both over the field K, with $r < n$. Show that it is possible for AB to be equal to I_r, but that it is never possible for BA to be equal to I_n. [*Hint:* Consider A and B as matrix representatives of linear transformations S and T from K^n to K^r, and K^r to K^n, respectively. With $r < n$, show that $T \circ S$ can never be the identity on K^n.]
3. Without reference to linear transformations, suppose A, B and C are $r \times r$ matrices over the field K such that $BA = AB = I_r$, and $CA = AC = I_r$. Show that $B = C$.
4. (a) Show that $\begin{bmatrix} 1 & 1 \\ 0 & 1 \end{bmatrix}$ is the inverse of $\begin{bmatrix} 1 & -1 \\ 0 & 1 \end{bmatrix}$.

    (b) Show that $\begin{bmatrix} 1 & 2 \\ 0 & 1 \end{bmatrix}$ is the inverse of $\begin{bmatrix} 1 & -2 \\ 0 & 1 \end{bmatrix}$.

    (c) What is the inverse of $\begin{bmatrix} 1 & k \\ 0 & 1 \end{bmatrix}$, for k any element of the field K?

5. (a) Show that
$$\begin{bmatrix} 1 & 0 & a \\ 0 & 1 & b \\ 0 & 0 & 1 \end{bmatrix} \text{ is the inverse of } \begin{bmatrix} 1 & 0 & -a \\ 0 & 1 & -b \\ 0 & 0 & 1 \end{bmatrix}.$$

    (b) Show that
$$\begin{bmatrix} 1 & 0 & 0 \\ 0 & 1 & 0 \\ a & b & 1 \end{bmatrix} \text{ is the inverse of } \begin{bmatrix} 1 & 0 & 0 \\ 0 & 1 & 0 \\ -a & -b & 1 \end{bmatrix}.$$

6. (a) Show that
$$\begin{bmatrix} d_1 & 0 & \cdots & 0 \\ 0 & d_2 & \cdots & 0 \\ \cdot & \cdot & & \\ \cdot & \cdot & & \\ \cdot & \cdot & & \\ 0 & 0 & \cdots & d_n \end{bmatrix}$$

has an inverse

$$\begin{bmatrix} d_1^{-1} & 0 & \cdots & 0 \\ 0 & d_2^{-1} & \cdots & 0 \\ \cdot & & & \\ \cdot & & & \\ \cdot & & & \\ 0 & 0 & \cdots & d_n^{-1} \end{bmatrix}$$

provided $d_i \neq 0$, for all i.

(b) Show that

$$\begin{bmatrix} d_1 & 0 & \cdots & 0 \\ 0 & d_2 & \cdots & 0 \\ \cdot & & & \\ \cdot & & & \\ \cdot & & & \\ 0 & 0 & \cdots & d_n \end{bmatrix}$$

has an inverse if and only if $d_1 d_2 \cdots d_n \neq 0$.

7. An $n \times n$ matrix with exactly one 1 in each row and column and 0 in all other positions is called a *permutation* matrix.
   (a) By inspection, find the inverse of each of the following permutation matrices:

   (1) $\begin{bmatrix} 1 & 0 & 0 \\ 0 & 0 & 1 \\ 0 & 1 & 0 \end{bmatrix}$.
   (2) $\begin{bmatrix} 0 & 1 & 0 \\ 1 & 0 & 0 \\ 0 & 0 & 1 \end{bmatrix}$.

   (3) $\begin{bmatrix} 0 & 1 & 0 & 0 \\ 1 & 0 & 0 & 0 \\ 0 & 0 & 0 & 1 \\ 0 & 0 & 1 & 0 \end{bmatrix}$.
   (4) $\begin{bmatrix} 0 & 0 & 0 & 1 \\ 0 & 0 & 1 & 0 \\ 0 & 1 & 0 & 0 \\ 1 & 0 & 0 & 0 \end{bmatrix}$.

   (b) Prove that the inverse of a 3×3 permutation matrix always exists and is also a permutation matrix.
   (c) Show that there are $n!$ permutation matrices of order n.
   (d) What is the relation between permutations on the set

   $$\{1, 2, \ldots, n\}$$

   and $n \times n$ permutation matrices?

8. Let $C = \begin{bmatrix} 1 & 2 & -1 \\ 1 & -1 & 0 \\ 1 & 0 & 0 \end{bmatrix}$ be a matrix over **R**. Let $\mathcal{B}_3 = \{f_1, f_2, f_3\}$ be the canonical basis for $\mathbf{R}^3$.

8.8 INVERSE MATRICES; TRANSFORMATION OF COORDINATES

(a) Show that $\mathcal{B}_3' = \{v_1', v_2', v_3'\}$ is also a basis for $\mathbf{R}^3$, where

$$v_j' = \sum_{i=1}^{3} c_{ij} f_i.$$

(b) Write each f_j as a linear combination of the v_j's.

(c) Suppose $f_j = \sum_{i=1}^{3} d_{ij} v_i'$, for each j. Let

$$D = [d_{ij}]_{3\times 3}.$$

Show that $CD = DC = I_3$, so that $D = C^{-1}$.

(d) Let $v \in \mathbf{R}^3$ have coordinate matrix

$$[v]_{\mathcal{B}_3} = \begin{bmatrix} a \\ b \\ c \end{bmatrix}$$

with respect to the canonical basis $\mathcal{B}_3$. Find $[v]_{\mathcal{B}_3'}$, the coordinate matrix of v with respect to the basis $\mathcal{B}_3'$.

9. A matrix $[a_{ij}]_{n\times n}$ is called *symmetric* if it satisfies the condition: $a_{ij} = a_{ji}$, for all i, j.

(a) Which of the following are symmetric?

(1) $\begin{bmatrix} 1 & 2 & 1 \\ 2 & 0 & 3 \\ 1 & 3 & 1 \end{bmatrix}$. (2) $\begin{bmatrix} 1 & -1 & 0 & 1 \\ -1 & 2 & 3 & -1 \\ 0 & 3 & 5 & 0 \\ 1 & -1 & 0 & \frac{1}{2} \end{bmatrix}$.

(3) $\begin{bmatrix} 1 & 2 & 1 \\ 2 & 0 & 3 \\ 1 & 2 & 1 \end{bmatrix}$. (4) $\begin{bmatrix} 1 & 0 & -1 \\ 0 & 1 & -1 \\ 0 & 1 & -1 \end{bmatrix}$.

(b) Show that if A and B are symmetric matrices and if $AB = BA$, then AB is also symmetric.

(c) Show that if A is symmetric and has an inverse matrix A^{-1}, then A^{-1} is also symmetric.

10. Let $A_{m\times n}$ be a matrix over the field K, $A_{m\times n} = [a_{ij}]_{m\times n}$. The *transpose of A* is the matrix $A^T = [b_{ij}]_{n\times m}$, where $b_{rs} = a_{sr}$, for $r = 1, \ldots, n$; $s = 1, \ldots, m$. Find the transpose of each of the following:

(a) $A = \begin{bmatrix} 1 & 2 & 3 \\ -1 & 2 & -3 \\ 0 & -1 & 2 \end{bmatrix}$. (b) $A = \begin{bmatrix} 1 & -1 \\ 2 & -7 \\ 3 & 8 \end{bmatrix}$.

(c) $A = \begin{bmatrix} 1 & -3 & 8 \\ 2 & -7 & 6 \end{bmatrix}$. (d) $A = \begin{bmatrix} 1 & 0 \\ 1 & 1 \end{bmatrix}$.

11. Let A and B be $m \times n$ matrices over K. Show that

$$(A + B)^T = A^T + B^T.$$

12. Let A be $m \times n$ and B be $n \times r$. Show that $(AB)^T = B^T A^T$. [*Hint:* Let $A = [a_{ij}]_{m \times n}$, $B = [b_{ij}]_{n \times r}$. Compare the (i,j)th element of $(AB)^T$ with the (i,j)th element of $B^T A^T$.]
13. Let A be an $n \times n$ matrix over K.
    (a) Show that $(A^T)^T = A$.
    (b) Show that $A + A^T$ is symmetric.

8.9 Change of Basis

Let V and U be finite-dimensional vector spaces over the field K, and let $T: V \longrightarrow U$ be a linear transformation. Then whenever we associate a matrix A with T, the matrix A depends on the choice of bases $\mathcal{B}$ and $\mathcal{C}$ chosen for V and U, respectively. Thus, we can associate with T a whole class of matrices, each representing T with respect to a different pair of bases for V and U. Since each of these matrices represents the same transformation T, we should expect that they will be strongly related to each other. In this section we will investigate this relationship.

LEMMA 8.9.1 Let U be a vector space over K, with bases $\mathcal{C} = \{u_1, \ldots, u_m\}$ and $\mathcal{C}' = \{u_1', \ldots, u_m'\}$. Let

$$u_j' = \sum_{i=1}^{m} c_{ij} u_i \quad \text{and} \quad u_j = \sum_{i=1}^{m} b_{ij} u_i'.$$

Let $C = [c_{ij}]_{m \times m}$ and $B = [b_{ij}]_{m \times m}$. Then $CB = BC = I_m$, so that $B = C^{-1}$.

Proof B and C are invertible by Theorem 8.8.10. Now

$$u_j = \sum_{i=1}^{m} b_{ij} u_i' = \sum_{i=1}^{m} b_{ij} \left(\sum_{k=1}^{m} c_{ki} u_k \right)$$
$$= \sum_{k=1}^{m} \left(\sum_{i=1}^{m} c_{ki} b_{ij} \right) u_k.$$

Since $\mathcal{C}$ is a basis, the coefficient of u_k is 1, if $k = j$; 0, if $k \neq j$. But

$$\sum_{i=1}^{m} c_{ki} b_{ij}$$

is just the (k,j)th entry of the product CB. Thus, $CB = I_m$, whence $BC = I_m$ also. ∎

THEOREM 8.9.2 Let V and U be vector spaces over the field K, dim $V = n$ and dim $U = m$.

Let $T: V \longrightarrow U$ be a linear transformation.

Let $\mathcal{B} = \{v_1, \ldots, v_n\}$ and $\mathcal{C} = \{u_1, \ldots, u_m\}$ be bases for V and U, respectively.

(a) Let $\mathcal{B}' = \{v_1', \ldots, v_n'\}$ and $\mathcal{C}' = \{u_1', \ldots, u_m'\}$ be bases for V and U, respectively, with

$$v_j' = \sum_{i=1}^{n} d_{ij} v_i, \qquad j = 1, \ldots, n,$$

$$u_j' = \sum_{i=1}^{m} c_{ij} u_i, \qquad j = 1, \ldots, m,$$

and

$$u_j = \sum_{i=1}^{m} b_{ij} u_i', \qquad j = 1, \ldots, m.$$

Then

$$[T]_{\mathcal{C}', \mathcal{B}'} = B[T]_{\mathcal{C}, \mathcal{B}} D$$
$$= C^{-1}[T]_{\mathcal{C}, \mathcal{B}} D,$$

where $B = [b_{ij}]_{m \times m}$, $C = [c_{ij}]_{m \times m}$, $D = [d_{ij}]_{n \times n}$.

(b) Let $D_{n \times n}$ and $C_{m \times m}$ be nonsingular matrices. Let

$$v_j' = \sum_{i=1}^{n} d_{ij} v_i, \qquad j = 1, \ldots, n,$$

and let

$$u_j' = \sum_{i=1}^{m} c_{ij} u_i, \qquad j = 1, \ldots, m.$$

Then $\mathcal{B}' = \{v_1', \ldots, v_n'\}$ and $\mathcal{C}' = \{u_1', \ldots, u_m'\}$ are bases for V and U, respectively, and $[T]_{\mathcal{C}', \mathcal{B}'} = C^{-1}[T]_{\mathcal{C}, \mathcal{B}} D$.

Proof (a) We note that by definition the ith column of $[T]_{\mathcal{C}', \mathcal{B}'}$ has, as its entries, the entries of the coordinate matrix $[T(v_i')]_{\mathcal{C}'}$.

By Theorem 8.7.5, $[T(v_i')]_{\mathcal{C}} = [T]_{\mathcal{C}, \mathcal{B}}[v_i']_{\mathcal{B}}$. By Theorem 8.8.10, for any $u \in U$, $[u]_{\mathcal{C}'} = C^{-1}[u]_{\mathcal{C}}$ and for any $v \in V$, $[v]_{\mathcal{B}} = D[v]_{\mathcal{B}'}$. Thus we have

$$[T(v_i')]_{\mathcal{C}'} = C^{-1}[T(v_i')]_{\mathcal{C}} = C^{-1}[T]_{\mathcal{C}, \mathcal{B}}[v_i']_{\mathcal{B}}$$
$$= C^{-1}[T]_{\mathcal{C}, \mathcal{B}} D[v_i']_{\mathcal{B}'}.$$

Since $[v_i']_{\mathcal{B}'}$ is the coordinate matrix of v_i' in the basis

$$\mathcal{B}' = \{v_1', \ldots, v_n'\}, \qquad [v_i']_{\mathcal{B}'} = \begin{bmatrix} 0 \\ \cdot \\ \cdot \\ \cdot \\ 0 \\ 1 \\ 0 \\ \cdot \\ \cdot \\ \cdot \\ 0 \end{bmatrix}_{n \times 1},$$

where 1 is in the ith row and 0 is in all other positions.

Thus $C^{-1}[T]_{\mathcal{C},\mathcal{B}} D [v_i']_{\mathcal{B}'}$ is the column matrix where entries are the same as the entries of the ith column of $C^{-1}[T]_{\mathcal{C},\mathcal{B}} D$, since the product

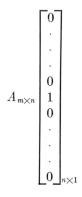

is just the ith column of A.

Hence we may now conclude that $[T]_{\mathcal{C}',\mathcal{B}'}$ and $C^{-1}[T]_{\mathcal{C},\mathcal{B}} D$ have the same ith column, $i = 1, \ldots, n$, whence

$$[T]_{\mathcal{C}',\mathcal{B}'} = C^{-1}[T]_{\mathcal{C},\mathcal{B}} D.$$

Since $B = C^{-1}$, by Lemma 8.9.1, we also have

$$[T]_{\mathcal{C}',\mathcal{B}'} = B[T]_{\mathcal{C},\mathcal{B}} D.$$

(b) That $\mathcal{B}'$ and $\mathcal{C}'$ are bases for V and U, respectively, follows from Theorem 8.8.10. Since $\mathcal{B}'$ and $\mathcal{C}'$ are bases, if we let $B = C^{-1}$, then by Lemma 8.9.1 the hypotheses of (a) are satisfied. The conclusion in (b) now follows directly from (a). ∎

A second proof of Theorem 8.9.2 can be given that only requires examining entries of products of matrices. Since, in a sense, this second proof is more self-contained than the one just given, we give it here for the interested reader. We will prove only part (a), since the proof of part (b) follows as in the proof above.

Second Proof of Theorem 8.9.2(a) Let $A = [T]_{\mathcal{C},\mathcal{B}}$. To find $[T]_{\mathcal{C}',\mathcal{B}'}$, we must express $T(v_j')$, for each j, in terms of the basis $\mathcal{C}'$. Thus

$$T(v_j') = T\left(\sum_{i=1}^{n} d_{ij}v_i\right)$$

$$= \sum_{i=1}^{n} d_{ij}T(v_i) = \sum_{i=1}^{n} d_{ij}\left(\sum_{k=1}^{m} a_{ki}u_k\right)$$

$$= \sum_{i=1}^{n}\sum_{k=1}^{m} a_{ki}d_{ij}u_k$$

$$= \sum_{i=1}^{n}\sum_{k=1}^{m} a_{ki}d_{ij}\left(\sum_{p=1}^{m} b_{pk}u_p'\right),$$

so that

(*) $$T(v_j') = \sum_{p=1}^{m}\left(\sum_{i=1}^{n}\sum_{k=1}^{m} b_{pk}a_{ki}d_{ij}\right)u_p'.$$

Now if $[T]_{\mathcal{C}',\mathcal{B}'} = A'$, we see that

$$a_{pj}' = \sum_{i=1}^{n}\sum_{k=1}^{m} b_{pk}a_{ki}d_{ij}.$$

Next, the pth row of BA has for ith entry

$$\sum_{k=1}^{m} b_{pk}a_{ki},$$

whence the (p,j)th entry of BAD is

$$\sum_{i=1}^{n}\left(\sum_{k=1}^{m} b_{pk}a_{ki}\right)d_{ij},$$

that is, $[T]_{\mathcal{C}',\mathcal{B}'} = B[T]_{\mathcal{C},\mathcal{B}}D = C^{-1}[T]_{\mathcal{C},\mathcal{B}}D$, since $C^{-1} = B$. ∎

Example 8.9.3 Let $V = \mathbf{R}^3$, $U = \mathbf{R}^2$, with $\mathcal{B} = \{f_1, f_2, f_3\}$ and $\mathcal{C} = \{e_1, e_2\}$ the canonical bases for $\mathbf{R}^3$ and $\mathbf{R}^2$, respectively. Let $T: V \longrightarrow U$ have matrix

$$[T]_{\mathcal{C},\mathcal{B}} = \begin{bmatrix} 1 & 0 & -1 \\ 2 & -3 & 0 \end{bmatrix}.$$

Let $\mathcal{B}' = \mathcal{B}$ and let $\mathcal{C}' = \{u_1', u_2'\}$, where
$$u_1' = e_1 + e_2, \qquad u_2' = e_1 + 2e_2.$$
Then $D = I_3$ and $C = \begin{bmatrix} 1 & 1 \\ 1 & 2 \end{bmatrix}$. Then
$$[T]_{\mathcal{C}',\mathcal{B}'} = C^{-1} \begin{bmatrix} 1 & 0 & -1 \\ 2 & -3 & 0 \end{bmatrix} I_3 = C^{-1} \begin{bmatrix} 1 & 0 & -1 \\ 2 & -3 & 0 \end{bmatrix}.$$

To find C^{-1}, the easiest thing for us to do at this time is to solve for e_1 and e_2 in terms of u_1' and u_2'. Thus, $u_1' = e_1 + e_2$, $u_2' = e_1 + 2e_2$, whence $e_1 = 2u_1' - u_2'$ and $e_2 = -u_1' + u_2'$, so
$$C^{-1} = B = \begin{bmatrix} 2 & -1 \\ -1 & 1 \end{bmatrix}.$$
Thus, we now obtain
$$[T]_{\mathcal{C}',\mathcal{B}'} = \begin{bmatrix} 2 & -1 \\ -1 & 1 \end{bmatrix} \begin{bmatrix} 1 & 0 & -1 \\ 2 & -3 & 0 \end{bmatrix} = \begin{bmatrix} 0 & 3 & -2 \\ 1 & -3 & 1 \end{bmatrix}.$$
To confirm that our result in this example is indeed correct, note that
$$T(f_1) = e_1 + 2e_2 = 2u_1' - u_2' + 2(-u_1' + u_2')$$
$$= 0u_1' + 1u_2',$$
$$T(f_2) = -3e_2 = -3(-u_1' + u_2') = 3u_1' - 3u_2',$$
and
$$T(f_3) = -e_1 = -(2u_1' - u_2') = -2u_1' + u_2'.$$

The relationship between the matrices $[T]_{\mathcal{C}',\mathcal{B}'}$ and $[T]_{\mathcal{C},\mathcal{B}}$ of Theorem 8.9.1 can be expressed without any reference to linear transformations.

Definition 8.9.4 Let M_1 and M_2 be two $m \times n$ matrices over the field K. M_1 and M_2 are called *equivalent* if there exist a nonsingular $m \times m$ matrix B and a nonsingular $n \times n$ matrix D such that
$$M_2 = BM_1D.$$
Equivalence of matrices will be studied in some detail in Chapter 10.

We now consider the very important special case of Theorem 8.9.2, in which $U = V$ and T is a linear operator.

THEOREM 8.9.5 Let T be a linear operator on the vector space V over the field K, where V is finite dimensional.

(a) Let $\mathcal{B} = \{v_1, \ldots, v_n\}$ and $\mathcal{B}' = \{v_1', \ldots, v_n'\}$ be two bases for V, with
$$v_j' = \sum_{i=1}^{n} d_{ij} v_i, \qquad D = [d_{ij}]_{n \times n}.$$
Then
$$[T]_{\mathcal{B}'} = D^{-1}[T]_{\mathcal{B}} D.$$

(b) Let $\mathcal{B} = \{v_1,\ldots,v_n\}$ be a basis for V. Let $D = [d_{ij}]_{n\times n}$ be nonsingular, with entries in K. Let

$$v_j' = \sum_{i=1}^n d_{ij}v_i, \quad j = 1, \ldots, n.$$

Then $\{v_1',\ldots,v_n'\} = \mathcal{B}'$ is a basis for V and

$$[T]_{\mathcal{B}'} = D^{-1}[T]_{\mathcal{B}} D.$$

Proof (a) In Theorem 8.9.2(a), let $V = U$, $\mathcal{C} = \mathcal{B}$, $\mathcal{C}' = \mathcal{B}'$. Then $C = D$, and the result follows.

(b) In Theorem 8.9.2(b), let $V = U$ and $\mathcal{C} = \mathcal{B}$. Then $\mathcal{C}' = \mathcal{B}'$, and the result follows. ∎

The relationship between the matrices $[T]_{\mathcal{B}'}$ and $[T]_{\mathcal{B}}$ in Theorem 8.9.5 can also be described without any reference to linear transformations.

Definition 8.9.6 Let A and B be $n \times n$ matrices with entries from the field K. We say that A and B are *similar* over K (or *conjugate* over K) if there is a nonsingular matrix D with entries in K such that $B = D^{-1}AD$.

Example 8.9.7 Let $A = \begin{bmatrix} 1 & 0 \\ 0 & 2 \end{bmatrix}$ and let $B = \begin{bmatrix} 1 & 0 \\ 1 & 2 \end{bmatrix}$. Then A and B are similar, for if

$$D = \begin{bmatrix} 1 & 0 \\ 1 & 1 \end{bmatrix}, \quad \text{then } D^{-1} = \begin{bmatrix} 1 & 0 \\ -1 & 1 \end{bmatrix}$$

and

$$D^{-1}AD = \begin{bmatrix} 1 & 0 \\ -1 & 1 \end{bmatrix} \begin{bmatrix} 1 & 0 \\ 0 & 2 \end{bmatrix} \begin{bmatrix} 1 & 0 \\ 1 & 1 \end{bmatrix} = \begin{bmatrix} 1 & 0 \\ 1 & 2 \end{bmatrix} = B.$$

It is easy to verify that similarity is an equivalence relation on the set of $n \times n$ matrices with entries in the field K. The reader should recall the definition of conjugacy in a group.

Theorem 8.9.5 states that (a) matrices that represent the same linear operator with respect to different bases are similar, and (b) matrices that are similar represent the same linear operator with respect to different bases. The next theorem can be considered a restatement of Theorem 8.9.5(b).

THEOREM 8.9.8 Let A and A' be similar $n \times n$ matrices over the field K. Then if V is an n-dimensional vector space over K, there are bases $\mathcal{B} = \{v_1,\ldots,v_n\}$ and $\mathcal{B}' = \{v_1',\ldots,v_n'\}$ and a linear operator $T: V \longrightarrow V$ such that $[T]_{\mathcal{B}} = A$ and $[T]_{\mathcal{B}'} = A'$.

Proof Let V be any n-dimensional vector space over K. Let $\mathcal{B}$ be a basis for V. Define $T: V \longrightarrow V$ so that $[T]_{\mathcal{B}} = A$.

Since A and A' are similar, there is an $n \times n$ matrix D such that $A' = D^{-1}AD$. Let

$$v_j' = \sum_{i=1}^{n} d_{ij}v_i.$$

Then by Theorem 8.9.5(b), $\{v_1', \ldots, v_n'\} = \mathcal{B}'$ is a basis for V, and

$$[T]_{\mathcal{B}'} = D^{-1}[T]_{\mathcal{B}}D,$$

that is,

$$[T]_{\mathcal{B}'} = A'. \qquad \blacksquare$$

Example 8.9.10 Let

$$A = \begin{bmatrix} 5 & 6 & -3 \\ 2 & 5 & -2 \\ 6 & 10 & -4 \end{bmatrix}, \qquad B = \begin{bmatrix} 1 & 0 & 0 \\ 0 & 2 & 0 \\ 0 & 0 & 3 \end{bmatrix},$$

and

$$D = \begin{bmatrix} 0 & 1 & 3 \\ 1 & 0 & 1 \\ 2 & 1 & 4 \end{bmatrix}.$$

Then it can be verified directly that D is nonsingular with

$$D^{-1} = \begin{bmatrix} -1 & -1 & 1 \\ -2 & -6 & 3 \\ 1 & 2 & -1 \end{bmatrix}.$$

Furthermore, $B = D^{-1}AD$.

Now let $\mathcal{B} = \{f_1, f_2, f_3\}$ be the canonical basis for $\mathbf{R}^3$. We define $T: \mathbf{R}^3 \longrightarrow \mathbf{R}^3$ by letting $[T]_{\mathcal{B}} = A$, that is,

$$T(f_1) = 5f_1 + 2f_2 + 6f_3,$$
$$T(f_2) = 6f_1 + 5f_2 + 10f_3,$$

and

$$T(f_3) = -3f_1 - 2f_2 - 4f_3.$$

We define a new basis $\mathcal{B}' = \{f_1', f_2', f_3'\}$ for $\mathbf{R}^3$ by using the matrix $D: f_1' = f_2 + 2f_3$, $f_2' = f_1 + f_3$, $f_3' = 3f_1 + f_2 + 4f_3$. Then, by Theorem 8.9.8,

$$[T]_{\mathcal{B}'} = D^{-1}[T]_{\mathcal{B}}D = D^{-1}AD = B = \begin{bmatrix} 1 & 0 & 0 \\ 0 & 2 & 0 \\ 0 & 0 & 3 \end{bmatrix}.$$

Thus $\mathcal{B}'$ is a basis for $\mathbf{R}^3$ such that the matrix for T has an especially simple form.

The problem of determining when $n \times n$ matrices A and A' are similar is difficult. The necessary techniques go far beyond the intended scope

EXERCISES

1. Prove that equivalency of matrices defined in Definition 8.9.4 is indeed an equivalence relation on the set of all $m \times n$ matrices over a field K.
2. Prove that similarity of matrices defined in Definition 8.9.6 is an equivalence relation on the set of all $n \times n$ matrices over the field K.
3. Let $T: \mathbf{R}^3 \longrightarrow \mathbf{R}^3$ be defined by

$$T(f_1) = f_1 - f_3, \quad T(f_2) = f_2 - f_1, \quad T(f_3) = f_3 - f_2,$$

where $\mathcal{B} = \{f_1, f_2, f_3\}$ is the canonical basis of $\mathbf{R}^3$. Let $\mathcal{B}' = \{v_1, v_2, v_3\}$ be a second basis for $\mathbf{R}^3$, where

$$v_1 = f_1 + f_2 + f_3, \quad v_2 = f_2 + f_3, \quad v_3 = f_3.$$

Find a matrix D such that $[T]_{\mathcal{B}'} = D^{-1}[T]_{\mathcal{B}} D$.

4. Let $T: \mathbf{R}^3 \longrightarrow \mathbf{R}^3$ have matrix

$$[T]_{\mathcal{B}} = \begin{bmatrix} 1 & 1 & 2 \\ 0 & -1 & -1 \\ 1 & 2 & 3 \end{bmatrix}$$

with respect to the canonical basis $\mathcal{B} = \{f_1, f_2, f_3\}$.
   (a) Find a basis $\mathcal{B}'$ for $\mathbf{R}^3$ that contains a basis for the null space of T.
   (b) Determine $[T]_{\mathcal{B}'}$.

5. Let $V = \{f(x) \in \mathbf{R}[x] \mid \deg f(x) \leq 3\}$; that is, let V be the vector space of all polynomials over $\mathbf{R}$ with degree ≤ 3. Let $D: V \longrightarrow V$ be defined by $D(f(x)) = f'(x)$, the derivative of $f(x)$.
   (a) Let $\mathcal{B} = \{1, x, x^2, x^3\}$ be a basis for V. What is $[D]_{\mathcal{B}}$?
   (b) Let $\mathcal{B}' = \{1, x - 1, x^2 - x + 1, x^3 - x^2 + x - 1\}$. Show that $\mathcal{B}'$ is also a basis for V.
   (c) Find explicitly a matrix M such that $[D]_{\mathcal{B}'} = M^{-1}[D]_{\mathcal{B}} M$.
   (d) Find explicitly a matrix N such that $[D]_{\mathcal{B}} = N^{-1}[D]_{\mathcal{B}'} N$. (Of course, $N^{-1} = M$.)

6. Let G be the set of all nonsingular $n \times n$ matrices over the field K.
   (a) Prove that G is a group with respect to matrix multiplication.
   (b) Show that the set H of all matrices of the form αI_n, $\alpha \neq 0$, $\alpha \in K$, is a subgroup of G.
   (c) Show that the group H in (b) is a normal subgroup of G.

7. Let L be a subgroup of the group G defined in Exercise 6. Show that L is normal in G if and only if for each matrix A in L, every matrix similar to A is in L.

8. Let G be the group defined in Exercise 6. Let L be the set of matrices of the form

$$\begin{bmatrix} d_1 & 0 & 0 & \cdots & 0 \\ 0 & d_2 & 0 & \cdots & 0 \\ 0 & 0 & d_3 & \cdots & 0 \\ \vdots & & & \vdots & \\ 0 & 0 & 0 & & d_n \end{bmatrix}_{n \times n}$$

with $d_i \neq 0$, all i. (These are the nonsingular diagonal matrices.)
   (a) Show that L is a subgroup of G.
   (b) Show that L is not normal for $n \geq 2$, if $1 + 1 \neq 0$ in the field K.

9. Let $A = [a_{ij}]_{n \times n}$. Define the *trace* of A as

$$\operatorname{tr} A = a_{11} + a_{22} + \cdots + a_{nn}.$$

   (a) Prove that for $A_{n \times n}$ and $B_{n \times n}$

   $$\operatorname{tr}(A + B) = \operatorname{tr} A + \operatorname{tr} B.$$

   (b) Show that $\operatorname{tr} A = \operatorname{tr}(C^{-1}AC)$, for all nonsingular $n \times n$ matrices C.

10. Let $K = Z_2$, the field with two elements. Find all the equivalence classes of 2×2 matrices over Z_2 under similarity. (*Hint:* First find all nonsingular 2×2 matrices.)

9 Determinants

9.1 Determinants

Given a matrix over a field F or a linear transformation of a finite-dimensional vector space over a field F, it is possible to associate with that matrix or linear transformation an element of F called the determinant of the matrix or linear transformation. Although there are various ways to define determinant, we shall do so in terms of a "big" sum. To motivate the definition, we first consider systems of linear equations.

Consider the two simultaneous equations

$$a_{11}x_1 + a_{12}x_2 = b_1$$
$$a_{21}x_1 + a_{22}x_2 = b_2.$$

If we solve these equations for x_1 and x_2 by the usual "elimination" procedure, we discover that

$$x_1 = \frac{b_1 a_{22} - a_{12} b_2}{a_{11} a_{22} - a_{12} a_{21}} \quad \text{and} \quad x_2 = \frac{a_{11} b_2 - b_1 a_{21}}{a_{11} a_{22} - a_{12} a_{21}}$$

provided that $a_{11}a_{22} - a_{12}a_{21} \neq 0$.

The denominator in these solutions is the element of the field F that we call the determinant of the matrix

$$\begin{bmatrix} a_{11} & a_{12} \\ a_{21} & a_{22} \end{bmatrix}.$$

Next we consider the more complicated case of three simultaneous linear equations in three unknowns:

$$a_{11}x_1 + a_{12}x_2 + a_{13}x_3 = b_1$$
$$a_{21}x_1 + a_{22}x_2 + a_{23}x_3 = b_2$$
$$a_{31}x_1 + a_{32}x_2 + a_{33}x_3 = b_3.$$

In solving for these three unknowns, we discover that $x_1 = c/d$, where

$$c = b_1(a_{22}a_{33} - a_{23}a_{32}) - b_2(a_{12}a_{33} - a_{13}a_{32}) + b_3(a_{12}a_{23} - a_{13}a_{22}),$$

and

(*) $\quad d = a_{11}a_{22}a_{33} - a_{11}a_{23}a_{32} + a_{12}a_{23}a_{31} - a_{12}a_{21}a_{33} + a_{13}a_{21}a_{32}$
$$- a_{13}a_{22}a_{31}.$$

Similar formulas hold for x_2 and x_3, with the same value for d occurring in all three denominators. Of course, the formulas here are valid only if $d \neq 0$.

As with the case of two equations in two unknowns, the value d that appears in solving three equations in three unknowns is called the determinant of the 3×3 matrix

$$\begin{bmatrix} a_{11} & a_{12} & a_{13} \\ a_{21} & a_{22} & a_{23} \\ a_{31} & a_{32} & a_{33} \end{bmatrix}.$$

We now define the determinant of a matrix of order n. Later, we will see how the determinant enters into the solution of systems of n linear equations in n unknowns.

Definition 9.1.1 Let A be the $n \times n$ matrix

$$A = \begin{bmatrix} a_{11} & a_{12} & a_{13} & \cdots & a_{1n} \\ a_{21} & a_{22} & a_{23} & \cdots & a_{2n} \\ \vdots & & & & \\ a_{n1} & a_{n2} & a_{n3} & \cdots & a_{nn} \end{bmatrix}.$$

Then the *determinant of A* is defined to be the sum

$$\sum_{\sigma \in S_n} \text{sgn}(\sigma) a_{1\sigma(1)} a_{2\sigma(2)} a_{3\sigma(3)} \cdots a_{n\sigma(n)},$$

where the symbol $\sum_{\sigma \in S_n}$ means that we let σ vary over all of the $n!$ permutations of the integers $1, 2, \ldots, n$, and where $\text{sgn}(\sigma) = 1$, if σ is even, and $\text{sgn}(\sigma) = -1$, if σ is odd. (See Definition 3.3.5.) We abbreviate determinant of A by *det A*.

To illustrate, we look at the case $n = 2$. Then

$$A = \begin{bmatrix} a_{11} & a_{12} \\ a_{21} & a_{22} \end{bmatrix}, \quad \text{and} \quad \det A = a_{11}a_{22} - a_{12}a_{21},$$

since $2! = 2$ and the only permutations of the integers 1, 2 are $\sigma = (1)$

9.1 DETERMINANTS

and $\sigma = (12)$, which are even and odd, respectively. Similarly, for $n = 3$, there are $3! = 6$ permutations of the integers 1, 2, 3, and so we should get 6 summands in evaluating det A. This will turn out to be exactly the same expression we found in (*).

We leave it as an exercise for the reader to find the value of det A, in the case that A is a 4×4 matrix.

It may be of interest to compute some numerical examples. Thus let

$$A = \begin{bmatrix} 1 & 2 \\ 3 & 4 \end{bmatrix}; \quad \text{then} \quad \det A = 1 \cdot 4 - 2 \cdot 3 = -2.$$

In the 3×3 case, the matrix

$$A = \begin{bmatrix} 1 & 2 & 3 \\ 2 & 4 & 4 \\ 0 & 1 & -1 \end{bmatrix}$$

has determinant equal to

$$1 \cdot 4 \cdot (-1) - 1 \cdot 4 \cdot 1 + 2 \cdot 4 \cdot 0 - 2 \cdot 2 \cdot (-1) + 3 \cdot 2 \cdot 1 - 3 \cdot 4 \cdot 0 = 2.$$

The reader will be given an opportunity to evaluate several other determinants in the exercises.

EXERCISE

1. Evaluate the determinant of each of the following:

(a) $\begin{bmatrix} 1 & 2 & 3 \\ -1 & 0 & -1 \\ 2 & 11 & 3 \end{bmatrix}.$ (b) $\begin{bmatrix} 1 & 1 & 1 \\ 2 & 1 & 3 \\ 3 & 5 & 1 \end{bmatrix}.$

(c) $\begin{bmatrix} 1 & 0 & 0 \\ 1 & 1 & 0 \\ 1 & 1 & 1 \end{bmatrix}.$ (d) $\begin{bmatrix} 1 & 0 & 0 \\ 0 & 0 & 1 \\ 0 & 1 & 0 \end{bmatrix}.$

(e) $\begin{bmatrix} 1 & 2 & 0 \\ 0 & 1 & 2 \\ 0 & 0 & 1 \end{bmatrix}.$ (f) $\begin{bmatrix} 1 & 0 & 0 & 0 \\ 0 & 2 & 0 & 0 \\ 0 & 0 & 3 & 0 \\ 0 & 0 & 0 & 4 \end{bmatrix}.$

(g) $\begin{bmatrix} 1 & 1 & 1 & 1 \\ 0 & 1 & 1 & 1 \\ 0 & 0 & 1 & 1 \\ 0 & 0 & 0 & 1 \end{bmatrix}.$ (h) $\begin{bmatrix} -1 & 1 & 0 & 0 \\ 1 & -1 & 0 & 0 \\ 0 & 0 & -1 & 0 \\ 1 & 0 & 0 & -1 \end{bmatrix}.$

(i) $\begin{bmatrix} 1 & 2 & 3 & 4 \\ 1 & 2 & 4 & 6 \\ 1 & 2 & 7 & 12 \\ 0 & 0 & 0 & 1 \end{bmatrix}.$

The definition of det A in Definition 9.1.1 is sometimes called the *row expansion* of det A, since in each summand the row subscripts of the elements $a_{i\sigma(i)}$ proceed in the order of the rows of A. There is another expansion of det A that is sometimes called the *column expansion* of det A. If we look at one summand

$$\text{sgn } (\sigma) a_{1\sigma(1)} a_{2\sigma(2)} \cdots a_{n\sigma(n)},$$

then we observe that because σ is a permutation of the integers $1, 2, \ldots, n$, each column subscript also appears once and only once in the summand, as does each row subscript. Thus, we can rearrange the factors $a_{1\sigma(1)}$, $a_{2\sigma(2)}, \ldots, a_{n\sigma(n)}$ that appear in a given summand so that the column subscripts now appear in their natural order $1, 2, \ldots, n$. This will have the effect of permuting the row subscripts, but that permutation will just be the inverse of σ. Thus we can rewrite the summand

$$\text{sgn } (\sigma) a_{1\sigma(1)} a_{2\sigma(2)} a_{3\sigma(3)} \cdots a_{n\sigma(n)}$$

as

$$\text{sgn } (\sigma) a_{\sigma^{-1}(1)\, 1} a_{\sigma^{-1}(2)\, 2} \cdots a_{\sigma^{-1}(n)\, n}.$$

But we also observe that sgn (σ) and sgn (σ^{-1}) are equal, since a permutation and its inverse have the same parity (see Chapter 3). Thus the sum

$$\sum_{\sigma \in S_n} \text{sgn } (\sigma) a_{1\sigma(1)} a_{2\sigma(2)} \cdots a_{n\sigma(n)} = \sum_{\sigma \in S_n} \text{sgn } (\sigma^{-1}) a_{\sigma^{-1}(1)\, 1} \cdots a_{\sigma^{-1}(n)\, n}.$$

In the latter sum, as σ varies over all permutations of $1, 2, \ldots, n$, so does σ^{-1}. Thus, the last sum can be replaced by

$$\sum_{\sigma \in S_n} \text{sgn } (\sigma) a_{\sigma(1)\, 1} \cdots a_{\sigma(n)\, n},$$

which is called the *column expansion* of det A. We have thus proven the following.

THEOREM 9.1.2 Let

$$A = \begin{bmatrix} a_{11} & \cdots & a_{1n} \\ \cdot & & \\ \cdot & & \\ \cdot & & \\ a_{n1} & \cdots & a_{nn} \end{bmatrix}$$

be an $n \times n$ matrix with entries in the field F. Then

$$\det A = \sum_{\sigma \in S_n} \text{sgn } (\sigma) a_{\sigma(1)\, 1} a_{\sigma(2)\, 2} \cdots a_{\sigma(n)\, n}.$$

Proof The proof precedes the statement of the theorem. ∎

To illustrate what has been described above in a relatively simple case, consider the case for the 3 × 3 matrix

$$A = \begin{bmatrix} a_{11} & a_{12} & a_{13} \\ a_{21} & a_{22} & a_{23} \\ a_{31} & a_{32} & a_{33} \end{bmatrix}.$$

By Definition 9.1.1,

$$\det A = a_{11}a_{22}a_{33} - a_{11}a_{23}a_{32} + a_{12}a_{23}a_{31} - a_{12}a_{21}a_{33} \\ + a_{13}a_{21}a_{32} - a_{13}a_{22}a_{31},$$

but by manipulating and rearranging the terms, we get

$$\det A = a_{11}a_{22}a_{33} - a_{11}a_{32}a_{23} + a_{31}a_{12}a_{23} - a_{21}a_{12}a_{33} \\ + a_{21}a_{32}a_{13} - a_{31}a_{22}a_{13},$$

which is the column expansion of det A.

EXERCISE

2. Evaluate each determinant of Exercise 1 in terms of the column expansion, and verify that the results are the same using either column or row expansions.

In the exercises of Section 8.8 we defined the transpose of an $m \times n$ matrix A. In the case of $n \times n$ matrices we will need, in the sequel, the fact that det (A^T) = det A, where A^T is the transpose of A.

We recall the definition of a transpose.

Definition 9.1.3 Let

$$A = \begin{bmatrix} a_{11} & \cdots & a_{1n} \\ a_{21} & \cdots & a_{2n} \\ \cdot & & \\ \cdot & & \\ \cdot & & \\ a_{m1} & \cdots & a_{mn} \end{bmatrix}$$

be an $m \times n$ matrix with entries in the field F. Then A^T is the $n \times m$ matrix $[b_{ij}]_{n \times m}$, where $b_{ij} = a_{ji}$, for $i = 1, \ldots, n; j = 1, \ldots, m$. We call A^T the *transpose of A*.

To illustrate, if

$$A = \begin{bmatrix} 1 & 2 & 3 \\ -1 & 0 & 6 \end{bmatrix}, \quad \text{then } A^T = \begin{bmatrix} 1 & -1 \\ 2 & 0 \\ 3 & 6 \end{bmatrix}.$$

We note that A^T is obtained from A by letting the entries $a_{i1}, \ldots, a_{in}$ of the ith row of A be the entries of the ith column of A^T.

It is immediate, of course, that if A is an $n \times n$ matrix, then A^T is also a matrix of order n.

THEOREM 9.1.4 Let $A = [a_{ij}]_{n \times n}$, with each $a_{ij} \in F$, a field. Then
$$\det (A^T) = \det A.$$

Proof Let $A^T = [b_{ij}]_{n \times n}$, where $b_{ij} = a_{ji}$, all i and j. Then
$$\det (A^T) = \sum_{\sigma \in S_n} \text{sgn } \sigma \; b_{1\sigma(1)} \cdots b_{n\sigma(n)}$$
$$= \sum_{\sigma \in S_n} \text{sgn } \sigma \; a_{\sigma(1)\,1} \cdots a_{\sigma(n)\,n}.$$

But this last sum is simply the column expansion of $\det A$, so
$$\det (A^T) = \det A. \qquad \blacksquare$$

EXERCISES

3. Let I_n be the $n \times n$ identity matrix. Prove that $\det I_n = 1$.
4. Let $\alpha \in F$, a field. Prove that $\det (\alpha I_n) = \alpha^n$.
5. Let
$$D = \begin{bmatrix} d_1 & 0 & \cdots & 0 \\ 0 & d_2 & \cdots & 0 \\ \cdot & \cdot & & \cdot \\ \cdot & \cdot & & \cdot \\ \cdot & \cdot & & \cdot \\ 0 & 0 & \cdots & d_n \end{bmatrix}$$
be a diagonal matrix. Prove that $\det D = d_1 d_2 \cdots d_n$.

6. Let $A = \begin{bmatrix} a_{11} & a_{12} & a_{13} \\ 0 & a_{22} & a_{23} \\ 0 & 0 & a_{33} \end{bmatrix}$. Compute $\det A$.

7. Let $A = \begin{bmatrix} a_{11} & a_{12} & a_{13} & a_{14} \\ 0 & a_{22} & a_{23} & a_{24} \\ 0 & 0 & a_{33} & a_{34} \\ 0 & 0 & 0 & a_{44} \end{bmatrix}$. Compute $\det A$.

9.2 Computing Determinants

The definition of $\det A$ affords one technique for evaluating determinants. This definition, however, is awkward to apply when the size of A becomes fairly large. For example, when $n = 5$, there are $5! = 120$ summands in the definition of $\det A$. Thus, we need to have some methods that will simplify calculations.

We note, for example, that a matrix that contains a row with many 0's will have a determinant that is relatively easy to compute. To illustrate, if

$$A = \begin{bmatrix} 1 & 2 & 4 \\ 19 & -3 & 1 \\ 0 & 1 & 0 \end{bmatrix},$$

then all summands in det A have a 0 factor, except for

$$-a_{11}a_{23}a_{32} = -1 \cdot 1 \cdot 1 = -1 \quad \text{and} \quad a_{13}a_{21}a_{32} = 4 \cdot 19 \cdot 1 = 76.$$

In this example, det $A = 76 - 1 = 75$.

It is easily seen, in fact, that an $n \times n$ matrix A with a row in which all but one term is 0 will have at most $(n - 1)!$ nonzero summands in the row expansion of det A.

The following theorems will provide techniques that will yield from a matrix A a matrix A' such that A' has many zero entries and

$$\det A' = \det A.$$

THEOREM 9.2.1 Let A be an $n \times n$ matrix with entries in a field F such that A has two identical rows. Then det $A = 0$.

Proof Assume rows r and s are identical, with $r < s$; that is $a_{rj} = a_{sj}$, $j = 1, \ldots, n$.

We form the sum

$$(*) \qquad \det A = \sum_{\sigma \in S_n} (\text{sgn } \sigma) a_{1\sigma(1)} \cdots a_{n\sigma(n)}.$$

We note that the sum in (*) can be broken up in the following manner. We let σ vary over A_n, the set of all even permutations on the integers $\{1, 2, \ldots, n\}$. Then with $\mu = (rs)$, $\sigma\mu$ will vary over all the odd permutations. Thus, we can now write

$$\det A = \sum_{\sigma \in A_n} (\text{sgn } \sigma) a_{1\sigma(1)} \cdots a_{n\sigma(n)} + \sum_{\sigma \in A_n} (\text{sgn } \sigma\mu) a_{1\,\sigma\mu(1)} \cdots a_{n\,\sigma\mu(n)}.$$

Next, we will show that for a fixed σ in A_n, the summands

$$(\text{sgn } \sigma) a_{1\sigma(1)} \cdots a_{n\sigma(n)} \quad \text{and} \quad (\text{sgn } \sigma\mu) a_{1\,\sigma\mu(1)} \cdots a_{n\,\sigma\mu(n)}$$

cancel out. Thus, as we let σ vary over A_n, the $n!/2$ pairs of summands will cancel, and we will have det $A = 0$.

We note that for $i \neq r, s$, $\sigma\mu(i) = \sigma(i)$. Also, $\sigma\mu(r) = \sigma(s)$ and $\sigma\mu(s) = \sigma(r)$. Moreover, since rows r and s of A are identical,

$$a_{r\sigma(s)} = a_{s\sigma(s)} \quad \text{and} \quad a_{s\sigma(r)} = a_{r\sigma(r)}.$$

Finally, $(\text{sgn } \sigma\mu) = -\text{sgn } \sigma$, since μ is odd.

Putting the above information together, we obtain:

$$(\text{sgn } \sigma)a_{1\sigma(1)} \cdots a_{r\sigma(r)} \cdots a_{s\sigma(s)} \cdots a_{n\sigma(n)}$$
$$+ (\text{sgn } \sigma\mu)a_{1\sigma\mu(1)} \cdots a_{r\sigma\mu(r)} \cdots a_{s\sigma\mu(s)} \cdots a_{n\sigma\mu(n)}$$
$$= (\text{sgn } \sigma)a_{1\sigma(1)} \cdots a_{r\sigma(r)} \cdots a_{s\sigma(s)} \cdots a_{n\sigma(n)}$$
$$- (\text{sgn } \sigma)a_{1\sigma(1)} \cdots a_{r\sigma(s)} \cdots a_{s\sigma(r)} \cdots a_{n\sigma(n)}$$
$$= (\text{sgn } \sigma)a_{1\sigma(1)} \cdots a_{r\sigma(r)} \cdots a_{s\sigma(s)} \cdots a_{n\sigma(n)}$$
$$- (\text{sgn } \sigma)a_{1\sigma(1)} \cdots a_{s\sigma(s)} \cdots a_{r\sigma(r)} \cdots a_{n\sigma(n)}$$
$$= 0,$$

so that det $A = 0$. ∎

Example 9.2.2 Let

$$A = \begin{bmatrix} 1 & 2 & 3 \\ -1 & 1 & 2 \\ 1 & 2 & 3 \end{bmatrix}.$$

By definition,

$$\det A = 1 \cdot 1 \cdot 3 - 1 \cdot 2 \cdot 2 + 2 \cdot 2 \cdot 1 - 2 \cdot (-1) \cdot 3 + 3 \cdot (-1) \cdot 2 - 3 \cdot 1 \cdot 1$$
$$= 3 - 4 + 4 + 6 - 6 - 3 = 0.$$

THEOREM 9.2.3 Let A be an $n \times n$ matrix with entries in the field F. Let A' be obtained from A by multiplying each element of row r of A by the constant k. Then det $A' = k$ det A.

Proof We have

$$\det A' = \sum_{\sigma \in S_n} (\text{sgn } \sigma)a_{1\sigma(1)} \cdots a_{(r-1)\sigma(r-1)}(ka_{r\sigma(r)})a_{(r+1)\sigma(r+1)} \cdots a_{n\sigma(n)}$$

$$= \sum_{\sigma \in S_n} k(\text{sgn } \sigma)a_{1\sigma(1)} \cdots a_{n\sigma(n)} = k \det A. \quad \blacksquare$$

Before stating Theorem 9.2.4, we look at an easy example. Let

$$A = \begin{bmatrix} 1 & 2 & 1 \\ 2 & 1 & 0 \\ 1 & 2 & 4 \end{bmatrix}, \quad B = \begin{bmatrix} 1 & 2 & 1 \\ 3 & 8 & -1 \\ 1 & 2 & 4 \end{bmatrix}, \quad C = \begin{bmatrix} 1 & 2 & 1 \\ 5 & 9 & -1 \\ 1 & 2 & 4 \end{bmatrix}.$$

By use of the definition, we see that det $A = -9$, det $B = 6$, det $C = -3$, so that det $C = $ det $A + $ det B. Note that A, B, and C are identical, except that each entry of the second row of C is the sum of the corresponding entries of A and B.

THEOREM 9.2.4 Let $A = [a_{ij}]$ and $B = [b_{ij}]$ be two $n \times n$ matrices with entries from the field F. For all $i \neq r$, let $a_{ij} = b_{ij}$, $j = 1, \ldots, n$. That is, let A and B be identical except possibly in the rth row.

Let $C = [c_{ij}]$ be the $n \times n$ matrix identical to A and B, except for the rth row, where

$$c_{rj} = a_{rj} + b_{rj}, \quad j = 1, \ldots, n.$$

Then det $C = $ det $A + $ det B.

Proof We prove the theorem by appealing directly to Definition 9.1.1. We have

$$\det C = \sum_{\sigma \in S_n} (\text{sgn } \sigma) c_{1\sigma(1)} \cdots c_{r\sigma(r)} \cdots c_{n\sigma(n)}$$

$$= \sum_{\sigma \in S_n} (\text{sgn } \sigma) c_{1\sigma(1)} \cdots [a_{r\sigma(r)} + b_{r\sigma(r)}] \cdots c_{n\sigma(n)}$$

$$= \sum_{\sigma \in S_n} (\text{sgn } \sigma) c_{1\sigma(1)} \cdots a_{r\sigma(r)} \cdots c_{n\sigma(n)}$$

$$+ \sum_{\sigma \in S_n} (\text{sgn } \sigma) c_{1\sigma(1)} \cdots b_{r\sigma(r)} \cdots c_{n\sigma(n)}$$

$$= \sum_{\sigma \in S_n} (\text{sgn } \sigma) a_{1\sigma(1)} \cdots a_{n\sigma(n)} + \sum_{\sigma \in S_n} (\text{sgn } \sigma) b_{1\sigma(1)} \cdots b_{n\sigma(n)}$$

$$= \det A + \det B. \qquad \blacksquare$$

THEOREM 9.2.5 Let $A = [a_{ij}]_{n \times n}$. Let A' be formed from A by adding k times the sth row of A to the rth row of A, where $r \neq s$. That is, if $A' = [b_{ij}]$, then $b_{ij} = a_{ij}$, for $i \neq r$; $j = 1, \ldots, n$, and $b_{rj} = a_{rj} + ka_{sj}$, $j = 1, \ldots, n$.
Then $\det A' = \det A$.

Proof Let $B = [b_{ij}]_{n \times n}$ be the matrix identical to A except for the rth row. For the rth row of B, let $b_{rj} = ka_{sj}$, $j = 1, \ldots, n$, that is, the rth row of B is k times the sth row of A.
By Theorem 9.2.4, $\det A' = \det A + \det B$. By Theorem 9.2.3, $\det B = k \det B'$, where B' is the same as B, except the (r,j)th element of B' is simply a_{sj}, $j = 1, \ldots, n$. But then B' has identical rth and sth rows, so $\det B' = 0$, by Theorem 9.2.1. Thus,

$$\det A' = \det A + 0 = \det A. \qquad \blacksquare$$

THEOREM 9.2.6 Let A' be the $n \times n$ matrix obtained from the $n \times n$ matrix A by interchanging the rth and sth rows of A, $r \neq s$. Then $\det A' = -\det A$.

Proof Assume that rows r and s of A are interchanged to obtain A', $r \neq s$. We will see how this can be accomplished by a number of steps that would fall under the hypotheses of earlier theorems.

Let A_1 be the matrix obtained from A by adding (-1) times row s to row r. Note that by Theorem 9.2.5, $\det A_1 = \det A$.

Next, let A_2 be obtained from A_1 by adding the rth row of A_1 to the sth row of A_1. By Theorem 9.2.5, $\det A_2 = \det A_1$. Also, we observe that row i of A_2, $i \neq r, s$, is the same as row i of A; row r of A_2 is row r of A plus (-1) times row s of A; row s of A_2 is row r of A.

For the next step, obtain A_3 by adding (-1) times row s of A_2 to row r of A_2. Again by Theorem 9.2.5, $\det A_3 = \det A_2 = \det A_1 = \det A$.

Moreover, A_3 is the same as A, except the rth row of A_3 is (-1) times the sth row of A, and the sth row of A_3 is the rth row of A. Finally, multiply the rth row of A_3 by -1, obtaining A'. By Theorem 9.2.3,

$$\det A' = (-1) \det A_3 = -\det A,$$

and, of course, A' is the matrix described in the hypotheses of the theorem. ∎

It may be helpful in visualizing the steps of this theorem to use the following diagram. R_j will denote the jth row of A:

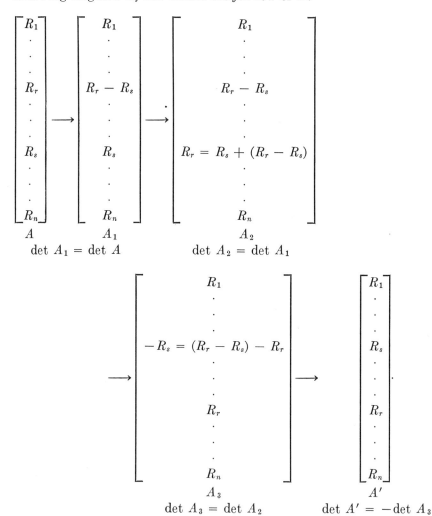

To illustrate this theorem, let

$$A = \begin{bmatrix} 1 & 2 & 3 \\ -1 & 1 & -2 \\ 1 & 0 & 1 \end{bmatrix}.$$

Then

$\det A = 1 \cdot 1 \cdot 1 - 1 \cdot (-2) \cdot 0 + 2(-2) \cdot 1 - 2 \cdot (-1) \cdot 1 + 3(-1) \cdot 0$
$ - 3 \cdot 1 \cdot 1$
$= 1 - 4 + 2 - 3 = -4.$

If we interchange rows 1 and 3, we obtain

$$A' = \begin{bmatrix} 1 & 0 & 1 \\ -1 & 1 & -2 \\ 1 & 2 & 3 \end{bmatrix},$$

and we note that

$\det A' = 1 \cdot 1 \cdot 3 - 1 \cdot (-2) \cdot 2 + 0 \cdot (-2) \cdot 1 - 0 \cdot (-1) \cdot 3$
$ + 1 \cdot (-1) \cdot 2 - 1 \cdot 1 \cdot 1$
$= 3 + 4 - 2 - 1 = 4,$

so that, indeed, $\det A' = 4 = -(-4) = -\det A$.

We will shortly apply the theorems of this section to a number of examples. First, however, we mention that whereas we have been working with rows, we could just as well have worked with columns. Thus, in any theorem in this section, replace the word "row" by the word "column" and another theorem will result. The proof would be obtained by working with the column expansion of $\det A$. For example, if the matrix A has two identical columns, then $\det A = 0$.

In the sequel, we will be free to work with the theorems as stated in the text, or with the "column" analogues.

For convenience, we introduce a standard notation that we will use frequently.

Notation 9.2.7 Let

$$A = \begin{bmatrix} a_{11} & \cdots & a_{1n} \\ \vdots & & \\ a_{n1} & \cdots & a_{nn} \end{bmatrix}$$

be an $n \times n$ matrix with entries in a field F. We will, when convenient, denote det A by

$$\begin{vmatrix} a_{11} & \cdots & a_{1n} \\ \vdots & & \vdots \\ a_{n1} & \cdots & a_{nn} \end{vmatrix}.$$

Example 9.2.8 Compute

$$\begin{vmatrix} 1 & 2 & 3 \\ 1 & 1 & -1 \\ 2 & 1 & 0 \end{vmatrix}.$$

Subtracting row 1 from row 2, and using Theorem 9.2.5, we have

$$\begin{vmatrix} 1 & 2 & 3 \\ 1 & 1 & -1 \\ 2 & 1 & 0 \end{vmatrix} = \begin{vmatrix} 1 & 2 & 3 \\ 0 & -1 & -4 \\ 2 & 1 & 0 \end{vmatrix}.$$

Subtracting two times row 1 from row 3, we get

$$\begin{vmatrix} 1 & 2 & 3 \\ 0 & -1 & -4 \\ 2 & 1 & 0 \end{vmatrix} = \begin{vmatrix} 1 & 2 & 3 \\ 0 & -1 & -4 \\ 0 & -3 & -6 \end{vmatrix}.$$

Applying Theorem 9.2.3 twice, with respect to rows 2 and 3, we see that

$$\begin{vmatrix} 1 & 2 & 3 \\ 0 & -1 & -4 \\ 0 & -3 & -6 \end{vmatrix} = (-1)(-1) \begin{vmatrix} 1 & 2 & 3 \\ 0 & 1 & 4 \\ 0 & 3 & 6 \end{vmatrix}.$$

Subtracting three times row 2 from row 3, we have

$$\begin{vmatrix} 1 & 2 & 3 \\ 0 & 1 & 4 \\ 0 & 3 & 6 \end{vmatrix} = \begin{vmatrix} 1 & 2 & 3 \\ 0 & 1 & 4 \\ 0 & 0 & -6 \end{vmatrix}.$$

Finally, using the string of equalities and Definition 9.1.1, we have

$$\det \begin{bmatrix} 1 & 2 & 3 \\ 1 & 1 & -1 \\ 2 & 1 & 0 \end{bmatrix} = \begin{vmatrix} 1 & 2 & 3 \\ 0 & 1 & 4 \\ 0 & 0 & -6 \end{vmatrix} = 1 \cdot 1 (-6) = -6.$$

Example 9.2.9 Compute

$$\det \begin{bmatrix} -1 & 2 & 4 \\ 3 & -2 & 7 \\ 1 & 2 & 15 \end{bmatrix}.$$

9.2 COMPUTING DETERMINANTS

We add three times row 1 to row 2:
$$\begin{vmatrix} -1 & 2 & 4 \\ 3 & -2 & 7 \\ 1 & 2 & 15 \end{vmatrix} = \begin{vmatrix} -1 & 2 & 4 \\ 0 & 4 & 19 \\ 1 & 2 & 15 \end{vmatrix}.$$

We add row 1 to row 3:
$$\begin{vmatrix} -1 & 2 & 4 \\ 0 & 4 & 19 \\ 1 & 2 & 15 \end{vmatrix} = \begin{vmatrix} -1 & 2 & 4 \\ 0 & 4 & 19 \\ 0 & 4 & 19 \end{vmatrix}.$$

By Theorem 9.2.6,
$$\begin{vmatrix} -1 & 2 & 4 \\ 0 & 4 & 19 \\ 0 & 4 & 19 \end{vmatrix} = 0.$$

Thus
$$\det \begin{bmatrix} -1 & 2 & 4 \\ 3 & -2 & 7 \\ 1 & 2 & 15 \end{bmatrix} = 0.$$

EXERCISES

1. In each step of each of the following examples, explain precisely which theorems in this section are being applied.

(a) $\begin{vmatrix} 1 & 2 & 3 \\ -3 & -2 & -1 \\ 0 & -1 & 0 \end{vmatrix} = \begin{vmatrix} 1 & 2 & 3 \\ 0 & 4 & 8 \\ 0 & -1 & 0 \end{vmatrix} = \begin{vmatrix} 1 & 2 & 3 \\ 0 & 0 & 8 \\ 0 & -1 & 0 \end{vmatrix}$

$= - \begin{vmatrix} 1 & 2 & 3 \\ 0 & 0 & 8 \\ 0 & 1 & 0 \end{vmatrix} = \begin{vmatrix} 1 & 2 & 3 \\ 0 & 1 & 0 \\ 0 & 0 & 8 \end{vmatrix} = 8.$

(b) $\begin{vmatrix} -1 & 0 & 2 \\ 2 & 3 & -6 \\ 4 & 1 & 2 \end{vmatrix} = \begin{vmatrix} -1 & 0 & 0 \\ 2 & 3 & -2 \\ 4 & 1 & 10 \end{vmatrix}$ (columns used here)

$= \begin{vmatrix} -1 & 0 & 0 \\ 0 & 3 & -2 \\ 4 & 1 & 10 \end{vmatrix} = \begin{vmatrix} -1 & 0 & 0 \\ 0 & 3 & -2 \\ 0 & 1 & 10 \end{vmatrix} = \begin{vmatrix} -1 & 0 & 0 \\ 0 & 0 & -32 \\ 0 & 1 & 10 \end{vmatrix} = -32.$

(c) $\begin{vmatrix} 1 & 2 & 3 & 4 \\ 5 & 6 & 7 & 8 \\ -4 & -3 & -2 & -1 \\ 7 & 8 & 9 & 0 \end{vmatrix} = \begin{vmatrix} 1 & 2 & 3 & 4 \\ 4 & 4 & 4 & 4 \\ -4 & -3 & -2 & -1 \\ 7 & 8 & 9 & 0 \end{vmatrix}$

$= 4 \begin{vmatrix} 1 & 2 & 3 & 4 \\ 1 & 1 & 1 & 1 \\ -4 & -3 & -2 & -1 \\ 7 & 8 & 9 & 0 \end{vmatrix} = 4 \begin{vmatrix} 1 & 2 & 3 & 4 \\ 1 & 1 & 1 & 1 \\ 0 & 1 & 2 & 3 \\ 7 & 8 & 9 & 0 \end{vmatrix} = 4 \begin{vmatrix} 0 & 1 & 2 & 3 \\ 1 & 1 & 1 & 1 \\ 0 & 1 & 2 & 3 \\ 7 & 8 & 9 & 0 \end{vmatrix} = 0.$

(d) $\begin{vmatrix} 1 & 2 & 3 \\ -1 & 3 & -2 \\ -3 & 2 & 1 \end{vmatrix} = \begin{vmatrix} 1 & 0 & 3 \\ -1 & 5 & -2 \\ -3 & 8 & 1 \end{vmatrix} = \begin{vmatrix} 1 & 0 & 0 \\ -1 & 5 & 1 \\ -3 & 8 & 10 \end{vmatrix}$

$= \begin{vmatrix} 1 & 0 & 0 \\ 0 & 5 & 1 \\ 0 & 8 & 10 \end{vmatrix} = -\begin{vmatrix} 1 & 0 & 0 \\ 0 & 1 & 5 \\ 0 & 10 & 8 \end{vmatrix} = -\begin{vmatrix} 1 & 0 & 0 \\ 0 & 1 & 5 \\ 0 & 0 & -42 \end{vmatrix} = 42.$

2. Compute each of the following determinants:

(a) $\begin{vmatrix} 3 & 9 & 1 \\ 2 & 5 & 7 \\ -3 & 6 & -14 \end{vmatrix}$.

(b) $\begin{vmatrix} \frac{1}{2} & 3 & \frac{4}{3} \\ \frac{5}{6} & 7 & -\frac{1}{3} \\ 2 & -1 & \frac{4}{3} \end{vmatrix}$.

(c) $\begin{vmatrix} 1 & 4 & 7 & 10 \\ 2 & 5 & 8 & 11 \\ -1 & 3 & -5 & 7 \\ -2 & 5 & -8 & 11 \end{vmatrix}$.

(d) $\begin{vmatrix} 0 & 1 & 2 & 4 & 6 \\ -1 & -7 & 8 & 9 & 1 \\ 3 & 6 & 5 & -1 & -2 \\ 4 & -3 & -6 & 8 & 9 \\ -2 & 4 & 9 & 3 & -8 \end{vmatrix}$.

3. Prove the following theorem:

   Theorem: Let A be an $n \times n$ matrix. Let A' be the matrix obtained from A in the following manner:

   Row 1 of A' is the same as row 1 of A.
   Row 2 of A' is k_1 times row 1 of A added to row 2 of A.
   Row 3 of A' is k_2 times row 1 of A added to row 3 of A.
   .
   .
   .
   Row n of A' is k_{n-1} times row 1 of A added to row n of A.

   Then $\det A' = \det A$.

4. Prove the following theorem:

   Theorem: Let A be an $n \times n$ matrix. Let A' be the matrix obtained from A in the following manner:

   For each i, $i = 1, \ldots, n$, row i of A' is k_i times row i of A.

   Then $\det A' = k_1 k_2 \cdots k_n \det A$.

5. (a) Let

$$A = \begin{bmatrix} a_{11} & a_{12} & a_{13} & \cdots & a_{1n} \\ 0 & a_{22} & a_{23} & \cdots & a_{2n} \\ 0 & 0 & a_{33} & \cdots & a_{3n} \\ \cdot \\ \cdot \\ \cdot \\ 0 & 0 & 0 & \cdots & a_{nn} \end{bmatrix}.$$

Show that $\det A = a_{11} a_{22} \cdots a_{nn}$.

(b) Show that to evaluate det B, where B is an $n \times n$ matrix, one can reduce the matrix B to a matrix of the form A in (a) using the theorems in this section.

6. Let A be an $n \times n$ matrix with entries from the field F. Let $\alpha \in F$. Show that det $(\alpha A) = \alpha^n$ det A.

7. (a) Let A be an $n \times n$ matrix such that $A^T = -A$, and let n be odd. Show that det $A = 0$.

   (b) Show by example that if n is even, it need not follow that det $A = 0$.

8. Let **C** be the field of complex numbers, and **R** the field of real numbers. For $\alpha = a + ib$, $a, b \in \mathbf{R}$, recall that $\bar{\alpha} = a - ib$. Let $A = [\alpha_{ij}]$ be an $n \times n$ matrix with entries in **C**, and let $\bar{A} = [\overline{\alpha_{ij}}]$, that is, $\bar{A}$ is obtained by replacing each entry α_{ij} of A by $\overline{\alpha_{ij}}$. Prove that det $\bar{A} = \overline{\det A}$.

9. Let F be a field, and let $x_1, \ldots, x_n \in F$. Prove by induction that

$$\begin{vmatrix} x_1^{n-1} & x_1^{n-2} & \cdots & x_1 & 1 \\ x_2^{n-1} & x_2^{n-2} & \cdots & x_2 & 1 \\ \cdot & & & & \cdot \\ \cdot & & & & \cdot \\ \cdot & & & & \cdot \\ x_n^{n-1} & x_n^{n-2} & \cdots & x_n & 1 \end{vmatrix}$$

$$= \prod_{1 \leq i < j \leq n} (x_i - x_j)$$
$$= (x_1 - x_2)(x_1 - x_3) \cdots (x_1 - x_n)(x_2 - x_3) \cdots (x_2 - x_n)$$
$$\cdots (x_{n-1} - x_n).$$

10. (a) Let (x_1, y_1) and (x_2, y_2) be points on a line L in the Euclidean plane. Prove that (x, y) is also on L if and only if

$$\begin{vmatrix} x_1 & y_1 & 1 \\ x_2 & y_2 & 1 \\ x & y & 1 \end{vmatrix} = 0.$$

   (b) Let $(3,5)$ and $(-1,7)$ be two points in the plane. Find the equation of the line through these points by using part (a).

11. (a) Let (x_1, y_1, z_1), (x_2, y_2, z_2), (x_3, y_3, z_3) be points on a plane M in 3-dimensional Euclidean space. Prove that (x, y, z) is a point on M if and only if

$$\begin{vmatrix} x_1 & y_1 & z_1 & 1 \\ x_2 & y_2 & z_2 & 1 \\ x_3 & y_3 & z_3 & 1 \\ x & y & z & 1 \end{vmatrix} = 0.$$

   (b) Let $(1,3,6)$, $(2,0,7)$, $(-1,-2,4)$ be three points in 3-space. Find the equation of the plane through these three points by using part (a).

12. Let $A = [a_{ij}]_{n \times n}$, $B = [b_{ij}]_{m \times m}$, $C = [c_{ij}]_{m \times n}$. Let D be the $(m+n) \times (m+n)$ matrix

$$D = \begin{bmatrix} A & 0 \\ C & B \end{bmatrix} = \begin{bmatrix} a_{11} & \cdots & a_{1n} & 0 & \cdots & 0 \\ \vdots & & \vdots & \vdots & & \vdots \\ a_{n1} & \cdots & a_{nn} & 0 & \cdots & 0 \\ c_{11} & \cdots & c_{1n} & b_{11} & \cdots & b_{1m} \\ \vdots & & \vdots & \vdots & & \vdots \\ c_{m1} & \cdots & c_{mn} & b_{m1} & \cdots & b_{mm} \end{bmatrix}.$$

Use the definition of determinant to show that

$$\det D = \det A \cdot \det B.$$

9.3 Cofactors

Another method used in computing $\det A$ is by cofactors. This approach will yield a computational method of computing A^{-1}, provided A^{-1} exists.

Definition 9.3.1 Let A be the $n \times n$ matrix $[a_{ij}]_{n \times n}$. For a fixed pair of indices (i,j), let θ_{ij} be the sum of those summands in

$$\det A = \sum_{\sigma \in S_n} (\text{sgn } \sigma) a_{1\sigma(1)} \cdots a_{n\sigma(n)}$$

in which a_{ij} appears as a factor. Each summand of θ_{ij} contains a_{ij} as a factor so that θ_{ij} may be written in the form $\theta_{ij} = a_{ij} A_{ij}$. We call A_{ij} the (i,j)-*cofactor* of $\det A$.

We illustrate the definition in the 3×3 case. Thus, let

$$A = \begin{bmatrix} a_{11} & a_{12} & a_{13} \\ a_{21} & a_{22} & a_{23} \\ a_{31} & a_{32} & a_{33} \end{bmatrix}.$$

Then

$\det A = a_{11}a_{22}a_{33} - a_{11}a_{23}a_{32} + a_{12}a_{23}a_{31} - a_{12}a_{21}a_{33} + a_{13}a_{21}a_{32}$
$\qquad - a_{13}a_{22}a_{31}.$

The element a_{31}, for example, appears in the terms $a_{12}a_{23}a_{31}$ and $-a_{13}a_{22}a_{31}$, so

$$\theta_{31} = a_{12}a_{23}a_{31} - a_{13}a_{22}a_{31} = a_{31}(a_{12}a_{23} - a_{13}a_{22}).$$

Hence $A_{31} = a_{12}a_{23} - a_{13}a_{22}$.

Similarly, a_{12} appears in $a_{12}a_{23}a_{31}$ and $-a_{12}a_{21}a_{33}$, and so

$$A_{12} = a_{23}a_{31} - a_{21}a_{33}.$$

It is interesting to note that

$$A_{31} = a_{12}a_{23} - a_{13}a_{22} = \det \begin{bmatrix} a_{12} & a_{13} \\ a_{22} & a_{23} \end{bmatrix}$$

and

$$A_{12} = a_{23}a_{31} - a_{21}a_{33} = \det \begin{bmatrix} a_{23} & a_{21} \\ a_{33} & a_{31} \end{bmatrix}$$

$$= -\det \begin{bmatrix} a_{21} & a_{23} \\ a_{31} & a_{33} \end{bmatrix}.$$

It is not coincidental, as we will see, that these determinants appear in the computation of the cofactors A_{31} and A_{12}.

We now illustrate how det A can be expressed in terms of cofactors. First, we fix our attention on the ith row of the matrix $A = [a_{ij}]$. For each pair $(i,1)$, $(i,2)$, ..., (i,n), we compute the cofactor A_{ij}, $j = 1, \ldots, n$. We have already seen that $a_{ij}A_{ij} = \theta_{ij}$ is the sum of all those summands in the determinant of A which contain a_{ij} as a factor.

Now notice that $a_{ir}A_{ir}$ and $a_{is}A_{is}$ will have no summands in common, $r \neq s$, for otherwise there would be a single summand containing two factors from a single row, that is, a_{ir} and a_{is}. Hence, since every summand in det A contains a factor a_{ik} from each column, it follows that every summand in det A appears among all the terms $a_{ij}A_{ij}$, as j ranges from 1 to n. Thus, it follows that

$$\det A = \sum_{j=1}^{n} a_{ij}A_{ij}, \quad \text{for any } i.$$

We state this result, and also the analogous result for columns, in the following theorem.

THEOREM 9.3.2 Let $A = [a_{ij}]_{n \times n}$. Then

$$\det A = a_{i1}A_{i1} + a_{i2}A_{i2} + \cdots + a_{in}A_{in}, \quad i = 1, \ldots, n$$

and

$$\det A = a_{1i}A_{1i} + a_{2i}A_{2i} + \cdots + a_{ni}A_{ni}, \quad i = 1, \ldots, n.$$

The first expansion is called the *expansion of the determinant of A by cofactors of the ith row*, and the second expansion is called the *expansion of the determinant of A by cofactors of the ith column*. ∎

In order to profitably utilize Theorem 9.3.2, it is important to see how the cofactor A_{ij} can be computed. If A is the $n \times n$ matrix $[a_{ij}]$, we will let $\hat{A}_{(ij)}$ be the $(n-1) \times (n-1)$ matrix obtained by striking out the ith row and jth column of A. Thus, if

$$A = \begin{bmatrix} 1 & 2 & 3 \\ 1 & 1 & 0 \\ 0 & 1 & -1 \end{bmatrix},$$

then

$$\hat{A}_{(12)} = \begin{bmatrix} 1 & 2 & 3 \\ 1 & 1 & 0 \\ 0 & 1 & -1 \end{bmatrix} = \begin{bmatrix} 1 & 0 \\ 0 & -1 \end{bmatrix}$$

and

$$\hat{A}_{(32)} = \begin{bmatrix} 1 & 2 & 3 \\ 1 & 1 & 0 \\ 0 & 1 & -1 \end{bmatrix} = \begin{bmatrix} 1 & 3 \\ 1 & 0 \end{bmatrix}.$$

Each $\hat{A}_{(ij)}$ is an $(n-1) \times (n-1)$ matrix, and so has a determinant. We will prove that the (i,j)th cofactor A_{ij} of A is equal to $(-1)^{i+j} \det \hat{A}_{(ij)}$. To do this, we first consider a special case.

THEOREM 9.3.3 Let A be an $n \times n$ matrix. Then the (n,n)th cofactor A_{nn} is equal to $\det \hat{A}_{(n,n)}$, that is, $A_{nn} = \det \hat{A}_{(nn)}$.

Proof A_{nn} is, by definition, obtained by first summing all terms in $\det A$ that contain a_{nn}, and then factoring out a_{nn}. Thus from

$$\sum_{\sigma \in S_n} (\text{sgn } \sigma) a_{1\sigma(1)} \cdots a_{n\sigma(n)},$$

we first extract the sum

$$\theta_{nn} = \sum_{\substack{\sigma \in S_n \\ \sigma(n) = n}} (\text{sgn } \sigma) a_{1\sigma(1)} \cdots a_{(n-1)\sigma(n-1)} a_{nn},$$

that is, we take the sum of all those terms in which $\sigma(n) = n$. Then

$$\theta_{nn} = a_{nn} \sum_{\substack{\sigma \in S_n \\ \sigma(n) = n}} (\text{sgn } \sigma) a_{1\sigma(1)} \cdots a_{(n-1)\sigma(n-1)},$$

and

$$\sum_{\substack{\sigma \in S_n \\ \sigma(n) = n}} (\text{sgn } \sigma) a_{1\sigma(1)} \cdots a_{(n-1)\sigma(n-1)}$$
$$= \sum_{\tau \in S_{n-1}} (\text{sgn } \tau) a_{1\tau(1)} \cdots a_{(n-1)\tau(n-1)} \quad (*)$$

since any $\sigma \in S_n$, with $\sigma(n) = n$, can be identified with $\tau \in S_{n-1}$, where $\sigma(i) = \tau(i); i = 1, 2, \ldots, n-1$.

Moreover, sgn (σ) = sgn (τ), for each σ such that $\sigma(n) = n$. This latter sum (∗), however, is simply the determinant of the $(n-1) \times (n-1)$ matrix $\hat{A}_{(nn)}$, so $A_{nn} = \det \hat{A}_{(nn)}$. ∎

THEOREM 9.3.4 Let $A = [a_{ij}]_{n \times n}$ be an $n \times n$ matrix with entries in a field F. Then

$$A_{ij} = (-1)^{i+j} \det \hat{A}_{(ij)}.$$

Proof For $i = j = n$, $i + j = 2n$, so $(-1)^{i+j} = (-1)^{2n} = 1$, and the result in this case is just Theorem 9.3.3.

We now observe that if

$$A = \begin{bmatrix} a_{11} & a_{12} & \cdots & a_{1n} \\ a_{21} & a_{22} & \cdots & a_{2n} \\ \cdot & & & \\ \cdot & & & \\ \cdot & & & \\ a_{n1} & a_{n2} & \cdots & a_{nn} \end{bmatrix},$$

then if we interchange the jth column successively with each of the columns to its right, and then interchange the ith row successively with each row beneath it, then we have the matrix

$$A' = \begin{bmatrix} a_{11} & \cdots & a_{1n} & a_{1j} \\ a_{21} & \cdots & a_{2n} & a_{2j} \\ \cdot & & & \\ \cdot & & & \\ \cdot & & & \\ a_{n1} & \cdots & a_{nn} & a_{nj} \\ a_{i1} & \cdots & a_{in} & a_{ij} \end{bmatrix}.$$

Each interchange of two rows or two columns results in one change of sign in the resulting determinant, so

$$\det A' = (-1)^{(n-i)+(n-j)} \det A = (-1)^{i+j} \det A,$$

whence $\det A = (-1)^{i+j} \det A'$.

If we now cross out the last row and last column of A', we get the same $(n-1) \times (n-1)$ submatrix obtained by crossing out the ith row and jth column of A, that is, $\widehat{A'}_{(nn)} = \hat{A}_{(ij)}$.

In A', the (n,n)th cofactor is $\det \widehat{A'}_{(nn)} = \det \hat{A}_{(ij)}$, by Theorem 9.3.3. Since $\det A = (-1)^{i+j} \det A'$, we may thus conclude that

$$A_{ij} = (-1)^{i+j} \det \hat{A}_{(ij)}. \quad \blacksquare$$

376 DETERMINANTS

Example 9.3.5 Let
$$A = \begin{bmatrix} 1 & 2 & 3 \\ 0 & 1 & -2 \\ -1 & 1 & 2 \end{bmatrix}.$$

We will compute A_{21}, A_{22}, A_{23}, and det A.

To compute A_{21}, we find that
$$\hat{A}_{(21)} = \begin{bmatrix} 2 & 3 \\ 1 & 2 \end{bmatrix},$$
whence
$$A_{21} = (-1)^{2+1} \det \begin{bmatrix} 2 & 3 \\ 1 & 2 \end{bmatrix} = -(4 - 3) = -1.$$

Similarly,
$$A_{22} = (-1)^{2+2} \det \hat{A}_{(22)} = \begin{vmatrix} 1 & 3 \\ -1 & 2 \end{vmatrix} = 2 - (-3) = 5,$$
and
$$A_{23} = (-1)^{2+3} \det \hat{A}_{(23)} = -\begin{vmatrix} 1 & 2 \\ -1 & 1 \end{vmatrix} = -(1 + 2) = -3.$$

By Theorem 9.3.2,
$$\det A = 0 \cdot (-1) + 1 \cdot (5) + (-2)(-3) = 5 + 6 = 11.$$

One last result involving cofactors will prove highly useful in computing inverses.

THEOREM 9.3.6 Let $A = [a_{ij}]_{n \times n}$ be a matrix with entries in a field F. Then
$$a_{1i}A_{1j} + a_{2i}A_{2j} + \cdots + a_{ni}A_{nj} = 0 \quad \text{if } i \neq j$$
and
$$a_{i1}A_{j1} + a_{i2}A_{j2} + \cdots + a_{in}A_{jn} = 0 \quad \text{if } i \neq j.$$

Proof Obtain a matrix A' from A by replacing the jth row of A by the ith row of A, and leaving all other rows as they were. Then in the new matrix A', rows i and j are identical, whence det $A' = 0$.

By Theorem 9.3.2, if we expand det A' by the cofactors of the jth row of A', we have
$$\det A' = a_{i1}A_{j1} + a_{i2}A_{j2} + \cdots + a_{in}A_{jn},$$
since the cofactors of the jth row of A' are the same as the cofactors of the jth row of A. Thus,
$$0 = a_{i1}A_{j1} + \cdots + a_{in}A_{jn}, \quad \text{for } i \neq j.$$

A similar argument on columns shows that
$$a_{1i}A_{1j} + a_{2i}A_{2j} + \cdots + a_{ni}A_{nj} = 0, \quad \text{for } i \neq j. \quad \blacksquare$$

EXERCISES

1. Compute each of the following determinants in two ways: by expanding by cofactors of the second column and by expanding by cofactors of the third row.

   (a) $\begin{vmatrix} 1 & 2 & -1 & 2 \\ -3 & 7 & -4 & 5 \\ 2 & 1 & 0 & 3 \\ -1 & 2 & 0 & 6 \end{vmatrix}.$

   (b) $\begin{vmatrix} 3 & 4 & 1 & 5 \\ -1 & -2 & 3 & -6 \\ 0 & 0 & 1 & 4 \\ 1 & 4 & 0 & 2 \end{vmatrix}.$

   (c) $\begin{vmatrix} 2 & 1 & 1 & 1 \\ 1 & 1 & 1 & 2 \\ 1 & 1 & 2 & 1 \\ 1 & 2 & 1 & 1 \end{vmatrix}.$

2. Why is the following diagram useful in computing the determinant of an $n \times n$ matrix in terms of cofactors?

$$\begin{bmatrix} + & - & + & - & \cdots & & & \\ - & + & - & + & \cdots & & & \\ + & - & + & - & \cdots & & & \\ - & + & - & + & \cdots & & & \\ \vdots & & & & & \vdots & & \\ & & & & \cdots & + & - & + \\ & & & & \cdots & - & + & - \\ & & & & \cdots & + & - & + \end{bmatrix}$$

3. Compute the determinant of the $n \times n$ matrix

$$\begin{bmatrix} 0 & 0 & \cdots & 0 & 0 & 1 \\ 0 & 0 & \cdots & 0 & 1 & 1 \\ \vdots & & & & & \vdots \\ 0 & 1 & \cdots & 1 & 1 & 1 \\ 1 & 1 & \cdots & 1 & 1 & 1 \end{bmatrix}$$

using any method.

9.4 Determinant of a Product of Two Matrices

Let A and B be $n \times n$ matrices. In this section, we will show that

$$\det(AB) = \det A \cdot \det B.$$

Although many proofs of this theorem are known, we give one that involves computing the determinant of a $2n \times 2n$ matrix, specially chosen to yield the result.

THEOREM 9.4.1 Let A and B be $n \times n$ matrices. Then

$$\det (AB) = \det A \cdot \det B.$$

Proof Let C be the $2n \times 2n$ matrix

$$C = \begin{bmatrix} a_{11} & a_{12} & \cdots & a_{1n} & 0 & 0 & \cdots & 0 \\ a_{21} & a_{22} & \cdots & a_{2n} & 0 & 0 & \cdots & 0 \\ \cdot & & & \cdot & & & & \\ \cdot & & & \cdot & & & & \\ \cdot & & & \cdot & & & & \\ a_{n1} & a_{n2} & \cdots & a_{nn} & 0 & 0 & \cdots & 0 \\ -1 & 0 & \cdots & 0 & b_{11} & b_{12} & \cdots & b_{1n} \\ 0 & -1 & \cdots & 0 & b_{21} & b_{22} & \cdots & b_{2n} \\ \cdot & & & \cdot & & & & \\ \cdot & & & \cdot & & & & \\ \cdot & & & \cdot & & & & \\ 0 & 0 & \cdots & -1 & b_{n1} & b_{n2} & \cdots & b_{nn} \end{bmatrix}.$$

The matrix C is often abbreviated in the form

$$C = \begin{bmatrix} A & 0 \\ -I_n & B \end{bmatrix}.$$

By definition,

$$\det C = \sum_{\sigma \in S_{2n}} (\operatorname{sgn} \sigma) c_{1\sigma(1)} \cdots c_{n\sigma(n)} c_{(n+1)\sigma(n+1)} \cdots c_{2n\,\sigma(2n)},$$

where c_{ij} is the (i,j)th entry of C.

We observe that a summand is 0 if for some i such that $1 \leq i \leq n$, $\sigma(i) \geq n + 1$, that is, if $c_{i\sigma(i)}$ is in the upper right-hand $n \times n$ block of 0's. Thus, we only get possibly nonzero summands in $\det C$ when the summand has no element from the upper right-hand $n \times n$ block. Hence, we need only consider summands where for each i, $1 \leq i \leq n$, we have $\sigma(i) \leq n$, that is, where $c_{i\sigma(i)}$ is one of the a_{ij}'s. But these are precisely those summands which contain one a_{ij} from each of the first n columns and n rows. Therefore, if a summand in $\det C$ is possibly not zero, then the factors $c_{i\sigma(i)}$, with $i \geq n + 1$, must come from the lower right-hand block, that is, $\sigma(i) \geq n + 1$. Thus, for $i \geq n + 1$, $c_{i\sigma(i)}$, must be one of the b_{ij}'s.

9.4 DETERMINANT OF A PRODUCT OF TWO MATRICES

From this discussion, we obtain

$$(*) \quad \det C = \sum_{\sigma \in S_{2n}} (\text{sgn } \sigma) a_{1\sigma(1)} \cdots a_{n\sigma(n)} b_{1,\sigma(n+1)-n} b_{2,\sigma(n+2)-n} \cdots b_{n,\sigma(2n)-n}.$$

In $(*)$, we can replace σ by a product $\mu\tau$, $\mu \in S_n, \tau \in S_n$, where $\mu(i) = \sigma(i)$, $1 \leq i \leq n$, and $\tau(i) = \sigma(n+i) - n$, $1 \leq i \leq n$. Thus,

$$(**) \quad \det C = \sum_{\mu \in S_n, \tau \in S_n} (\text{sgn } \sigma) a_{1\mu(1)} \cdots a_{n\mu(n)} b_{1\tau(1)} \cdots b_{n\tau(n)}.$$

Next, we will show that sgn σ = (sgn μ)(sgn τ). Observe that $\sigma = \mu\tau'$, where μ is as above, $\tau'(i) = \sigma(i)$, for $n + 1 \leq i \leq 2n$, and where μ and τ' are disjoint. Thus sgn σ = (sgn μ)(sgn τ'). But

$$\tau(i) = \sigma(n+i) - n = \tau'(n+i) - n, \quad \text{for } 1 \leq i \leq n.$$

Thus, τ and τ' can be factored into the same number of transpositions, whence sgn τ = sgn τ' and therefore (sgn σ) = (sgn μ)(sgn τ).

With this, $(**)$ becomes

$$\det C = \sum_{\mu \in S_n, \tau \in S_n} (\text{sgn } \mu)(\text{sgn } \tau) a_{1\mu(1)} \cdots a_{n\mu(n)} b_{1\tau(1)} \cdots b_{n\tau(n)}$$

$$= \left(\sum_{\mu \in S_n} (\text{sgn } \mu) a_{1\mu(1)} \cdots a_{n\mu(n)} \right) \left(\sum_{\tau \in S_n} (\text{sgn } \tau) b_{1\tau(1)} \cdots b_{n\tau(n)} \right)$$

$$= \det A \cdot \det B.$$

Now we evaluate det C by using results of Section 9.2. Thus, we first add a_{11} times row $(n + 1)$ of C to row 1 of C. This gives us a new matrix C' with first row

$$[0 \quad a_{12} \quad \cdots \quad a_{1n} \quad a_{11}b_{11} \quad a_{11}b_{12} \quad \cdots \quad a_{11}b_{1n}]$$

and other rows in C' the same as in C.

Next, we add a_{12} times row $(n + 2)$ of C' to row 1 of C', getting

$$[0 \quad 0 \quad a_{13} \quad \cdots \quad a_{1n} \quad a_{11}b_{11} + a_{12}b_{21} \quad a_{11}b_{12} + a_{12}b_{22} \quad \cdots \quad a_{11}b_{1n} + a_{12}b_{2n}]$$

as the first row of a matrix C'', with the other rows of C'' the same as in C.

We continue in this manner, adding a_{1j} times the $(n + j)$th row to the first row obtained in the $(j - 1)$st step. This eventually yields the first row

$$\left[0 \quad \cdots \quad 0 \quad \sum_{j=1}^{n} a_{1j}b_{j1} \quad \sum_{j=1}^{n} a_{1j}b_{j2} \quad \cdots \quad \sum_{j=1}^{n} a_{1j}b_{jn} \right].$$

Note that the sums in the n places on the right in this row are the entries in the first row of the matrix AB.

Next we work with the elements a_{2i}, adding a_{2i} times row $n + i$ of C

to the second row, for $i = 1, \ldots, n$. Then we repeat the process, adding a_{3i} times row $(n + i)$ to row 3, for $i = 1, \ldots, n$. We continue in this manner until we finally have a matrix D such that $\det C = \det D$, and where

$$D = \begin{bmatrix} 0 & 0 & \cdots & 0 & & & \\ \cdot & & & & & AB & \\ \cdot & & & & & & \\ 0 & 0 & \cdots & 0 & & & \\ -1 & 0 & \cdots & 0 & b_{11} & \cdots & b_{1n} \\ 0 & -1 & \cdots & 0 & b_{21} & \cdots & b_{2n} \\ \cdot & \cdot & & \cdot & & & \\ \cdot & \cdot & & \cdot & & & \\ \cdot & \cdot & & \cdot & & & \\ 0 & 0 & \cdots & -1 & b_{n1} & \cdots & b_{nn} \end{bmatrix},$$

where the upper right-hand block of D is the same as the product AB.

Finally, we must evaluate $\det D$. To do this, interchange columns j and $j + n$, $j = 1, \ldots, n$. This yields matrix

$$D' = \begin{bmatrix} & & & 0 & 0 & \cdots & 0 \\ & & & \cdot & \cdot & & \\ & AB & & \cdot & \cdot & & \\ & & & \cdot & \cdot & & \\ & & & 0 & 0 & \cdots & 0 \\ b_{11} & \cdots & b_{1n} & -1 & 0 & \cdots & 0 \\ \cdot & & & 0 & -1 & 0 & \cdots & 0 \\ \cdot & & & & \cdot & \cdot & \\ \cdot & & & & \cdot & \cdot & \\ \cdot & & & & \cdot & \cdot & \\ b_{n1} & \cdots & b_{nn} & 0 & 0 & \cdots & 0 & -1 \end{bmatrix}.$$

Since this used n interchanges, we note that

$$\det D' = (-1)^n \det D = (-1)^n \det C.$$

By a computation similar to that at the beginning of the argument,

$$\det D' = \det (AB) \cdot \det J,$$

where

$$J = \begin{bmatrix} -1 & 0 & 0 & \cdots & 0 \\ 0 & -1 & 0 & \cdots & 0 \\ \cdot & & & & \\ \cdot & & & & \\ \cdot & & & & \\ 0 & 0 & 0 & \cdots & -1 \end{bmatrix}.$$

But clearly, $\det J = (-1)^n$. Thus $\det D' = (-1)^n \det (AB)$, whence $(-1)^n \det C = (-1)^n \det (AB)$, and so $\det C = \det (AB)$. But we have already shown that $\det C = \det A \cdot \det B$, and so we have proven the theorem. ∎

Although the full details of the proof are lengthy, the ideas used are elementary. To clarify the argument, the reader should work through it for the cases $n = 2$ and $n = 3$.

EXERCISES

1. Work through the argument used in Theorem 9.4.1 for the cases $n = 2$ and $n = 3$.
2. Let $A = \begin{bmatrix} 1 & 2 & 3 \\ -1 & 0 & 2 \\ 3 & -1 & 0 \end{bmatrix}$ and $B = \begin{bmatrix} -2 & 0 & 6 \\ -1 & 1 & 0 \\ 3 & 8 & 0 \end{bmatrix}$.
   (a) Find AB.
   (b) Compute $\det (AB)$, $\det A$, $\det B$, and verify that the theorem holds in this case.

9.5 Inverses of Matrices; Cramer's Rule

Recall from Section 8.8 that the $n \times n$ matrix A has the $n \times n$ matrix B as an inverse if and only if $AB = BA = I_n$. We have a condition for A to have an inverse, when A is interpreted as the matrix of a linear transformation. In this section, we obtain a condition for A to have an inverse in terms of its determinant.

THEOREM 9.5.1 Let A be an $n \times n$ matrix with entries in a field F. If A has an inverse, then $\det A \neq 0$.

Proof Let $AB = I_n$. By Theorem 9.4.1,
$$\det (AB) = \det A \cdot \det B.$$
But $\det (AB) = \det I_n = 1$. Thus,
$$\det A \cdot \det B = 1,$$
and $\det A \neq 0$. ∎

The converse to Theorem 9.5.1 is also true. Once we prove it, we will have the important condition: the square matrix A of order n has an inverse if and only if $\det A \neq 0$.

THEOREM 9.5.2 Let A be an $n \times n$ matrix with entries in a field F. If $\det A \neq 0$, then A has an inverse.

Proof With det $A \neq 0$, we show how to construct the inverse of A. Let A_{ij} be the ijth cofactor of A. Let C be the transposed matrix of cofactors of A, that is,

$$C = \begin{bmatrix} A_{11} & A_{21} & \cdots & A_{n1} \\ A_{12} & A_{22} & \cdots & A_{n2} \\ \cdot & \cdot & & \cdot \\ \cdot & \cdot & & \cdot \\ \cdot & \cdot & & \cdot \\ A_{1n} & A_{2n} & \cdots & A_{nn} \end{bmatrix}.$$

We call the matrix C the (classical) *adjoint* of A, and we write

$$C = \operatorname{adj} A.$$

Now let $B = (\det A)^{-1} \operatorname{adj} A$, that is,

$$B = \begin{bmatrix} (\det A)^{-1} A_{11} & \cdots & (\det A)^{-1} A_{n1} \\ (\det A)^{-1} A_{12} & \cdots & (\det A)^{-1} A_{n2} \\ \cdot & & \cdot \\ \cdot & & \cdot \\ \cdot & & \cdot \\ (\det A)^{-1} A_{1n} & \cdots & (\det A)^{-1} A_{nn} \end{bmatrix}.$$

Then

$$AB = [c_{ij}]_{n \times n}$$

$$= \begin{bmatrix} (\det A)^{-1}(a_{11}A_{11} + \cdots + a_{1n}A_{1n}) & & & \\ & \cdots & (\det A)^{-1}(a_{11}A_{n1} + \cdots + a_{1n}A_{nn}) \\ & \cdot & & \\ & \cdot & & \\ & \cdot & & \\ (\det A)^{-1}(a_{n1}A_{11} + \cdots + a_{nn}A_{1n}) & & & \\ & \cdots & (\det A)^{-1}(a_{n1}A_{n1} + \cdots + a_{nn}A_{nn}) \end{bmatrix},$$

that is, $c_{ij} = (\det A)^{-1}(a_{i1}A_{j1} + \cdots + a_{in}A_{jn})$. By Theorem 9.3.2, the elements on the diagonal of AB are

$$(\det A)^{-1} \cdot \det A = 1,$$

and by Theorem 9.3.6, the elements off the diagonal are $(\det A)^{-1} \cdot 0 = 0$. Thus, $AB = I_n$. Either by a similar proof, again using Theorems 9.3.2 and 9.3.6, or by Theorem 8.8.6, we have $BA = I_n$ also. ∎

9.5 INVERSES OF MATRICES; CRAMER'S RULE

We illustrate how we use the proof of Theorem 9.5.2 to construct inverses.

Example 9.5.3 Let F be a field in which $11 \neq 0$. Let

$$A = \begin{bmatrix} -1 & 2 & 1 \\ 0 & 1 & -1 \\ 3 & 1 & 1 \end{bmatrix}.$$

We first check $\det A$, and find that it is -11. Thus, A^{-1} exists. Next, we compute all the cofactors of A.

$$A_{11} = \begin{vmatrix} 1 & -1 \\ 1 & 1 \end{vmatrix} = 2,$$

$$A_{12} = (-1)^{1+2} \begin{vmatrix} 0 & -1 \\ 3 & 1 \end{vmatrix} = -3,$$

$$A_{13} = (-1)^{1+3} \begin{vmatrix} 0 & 1 \\ 3 & 1 \end{vmatrix} = -3,$$

$$A_{21} = (-1)^{2+1} \begin{vmatrix} 2 & 1 \\ 1 & 1 \end{vmatrix} = -1,$$

$$A_{22} = (-1)^{2+2} \begin{vmatrix} -1 & 1 \\ 3 & 1 \end{vmatrix} = -4,$$

$$A_{23} = (-1)^{2+3} \begin{vmatrix} -1 & 2 \\ 3 & 1 \end{vmatrix} = 7,$$

$$A_{31} = (-1)^{3+1} \begin{vmatrix} 2 & 1 \\ 1 & -1 \end{vmatrix} = -3,$$

$$A_{32} = (-1)^{5} \begin{vmatrix} -1 & 1 \\ 0 & -1 \end{vmatrix} = -1,$$

$$A_{33} = \begin{vmatrix} -1 & 2 \\ 0 & 1 \end{vmatrix} = -1.$$

To form A^{-1}, we simply write the matrix

$$A^{-1} = \begin{bmatrix} \frac{2}{-11} & \frac{-1}{-11} & \frac{-3}{-11} \\ \frac{-3}{-11} & \frac{-4}{-11} & \frac{-1}{-11} \\ \frac{-3}{-11} & \frac{7}{-11} & \frac{-1}{-11} \end{bmatrix} = \begin{bmatrix} \frac{-2}{11} & \frac{1}{11} & \frac{3}{11} \\ \frac{3}{11} & \frac{4}{11} & \frac{1}{11} \\ \frac{3}{11} & \frac{-7}{11} & \frac{1}{11} \end{bmatrix}.$$

It is easy to check directly that $A^{-1}A = I_3$.

Example 9.5.4 Let F be a field in which $10 \neq 0$. Let

$$A = \begin{bmatrix} 3 & -1 & 2 \\ 1 & 1 & 1 \\ -2 & 0 & 1 \end{bmatrix}.$$

Then $\det A = 10$, so A^{-1} exists. Also, $A_{11} = 1$, $A_{12} = -3$, $A_{13} = 2$, $A_{21} = 1$, $A_{22} = 7$, $A_{23} = 2$, $A_{31} = -3$, $A_{32} = -1$, $A_{33} = 4$. Then

$$\text{adj } A = \begin{bmatrix} 1 & 1 & -3 \\ -3 & 7 & -1 \\ 2 & 2 & 4 \end{bmatrix} \text{ and } A^{-1} = \begin{bmatrix} \frac{1}{10} & \frac{1}{10} & \frac{-3}{10} \\ \frac{-3}{10} & \frac{7}{10} & \frac{-1}{10} \\ \frac{2}{10} & \frac{2}{10} & \frac{4}{10} \end{bmatrix}.$$

Again, a direct check shows that $AA^{-1} = A^{-1}A = I_3$.

Before proceeding, we give an easy consequence of Theorem 9.5.1.

THEOREM 9.5.5 Let A be an $n \times n$ matrix with an inverse A^{-1}. Then $\det(A^{-1}) = (\det A)^{-1}$.

Proof From $AA^{-1} = I_n$, we get

$$\det(AA^{-1}) = \det A \cdot \det(A^{-1}) = 1,$$

so $\det(A^{-1}) = (\det A)^{-1}$. ∎

An important application of matrices and inverses is in solving systems of simultaneous linear equations. Thus, suppose we have n equations in n unknowns:

(*)
$$\begin{aligned} a_{11}x_1 + \cdots + a_{1n}x_n &= b_1 \\ a_{21}x_1 + \cdots + a_{2n}x_n &= b_2 \\ &\vdots \\ a_{n1}x_1 + \cdots + a_{nn}x_n &= b_n. \end{aligned}$$

We let A be the $n \times n$ matrix

$$A = \begin{bmatrix} a_{11} & \cdots & a_{1n} \\ \vdots & & \vdots \\ a_{n1} & \cdots & a_{nn} \end{bmatrix}.$$

9.5 INVERSES OF MATRICES; CRAMER'S RULE

We call A the *coefficient matrix* of $(*)$. We also let

$$X = \begin{bmatrix} x_1 \\ x_2 \\ \cdot \\ \cdot \\ \cdot \\ x_n \end{bmatrix} \quad \text{and we let} \quad B = \begin{bmatrix} b_1 \\ b_2 \\ \cdot \\ \cdot \\ \cdot \\ b_n \end{bmatrix}.$$

Then it is possible to write the system of equations $(*)$ in matrix notation as

$$AX = B$$

For example, the system

$(**)$
$$\begin{aligned} -x_1 + 2x_2 + x_3 &= 3 \\ x_2 - x_3 &= 1 \\ 3x_1 + x_2 + x_3 &= -2, \end{aligned}$$

can be written

$$\begin{bmatrix} -1 & 2 & 1 \\ 0 & 1 & -1 \\ 3 & 1 & 1 \end{bmatrix} \begin{bmatrix} x_1 \\ x_2 \\ x_3 \end{bmatrix} = \begin{bmatrix} 3 \\ 1 \\ -2 \end{bmatrix}.$$

In the general case of $(*)$, if $\det A \neq 0$, then A^{-1} exists, whence from $AX = B$ we get

$$\begin{aligned} (A^{-1}A)X &= A^{-1}B, \\ I_n X &= A^{-1}B, \\ X &= A^{-1}B. \end{aligned}$$

Thus, the system of equations $(*)$ has a solution, and the solution is uniquely given in the form $X = A^{-1}B$, when $\det A \neq 0$.

In the special case of the example $(**)$, we have already found A^{-1} in Example 9.5.3. Thus

$$\begin{bmatrix} x_1 \\ x_2 \\ x_3 \end{bmatrix} = \begin{bmatrix} \dfrac{-2}{11} & \dfrac{1}{11} & \dfrac{3}{11} \\ \dfrac{3}{11} & \dfrac{4}{11} & \dfrac{1}{11} \\ \dfrac{3}{11} & \dfrac{-7}{11} & \dfrac{1}{11} \end{bmatrix} \begin{bmatrix} 3 \\ 1 \\ -2 \end{bmatrix} = \begin{bmatrix} \dfrac{-11}{11} \\ \dfrac{11}{11} \\ \dfrac{0}{11} \end{bmatrix} = \begin{bmatrix} -1 \\ 1 \\ 0 \end{bmatrix}.$$

A quick check verifies that $x_1 = -1$, $x_2 = 1$, and $x_3 = 0$ is the solution to $(**)$.

We now consider the general situation $(*)$ more closely. Let $AX = B$ represent a system of n equations in n unknowns with $\det A \neq 0$. Then

$X = A^{-1}B$, and we have

$$X = (\det A)^{-1} \begin{bmatrix} A_{11} & A_{21} & \cdots & A_{n1} \\ A_{12} & A_{22} & \cdots & A_{n2} \\ \cdot & & & \cdot \\ \cdot & & & \cdot \\ \cdot & & & \cdot \\ A_{1n} & A_{2n} & \cdots & A_{nn} \end{bmatrix} \begin{bmatrix} b_1 \\ b_2 \\ \cdot \\ \cdot \\ \cdot \\ b_n \end{bmatrix}$$

$$= (\det A)^{-1} \begin{bmatrix} A_{11}b_1 + A_{21}b_2 + \cdots + A_{n1}b_n \\ A_{12}b_1 + A_{22}b_2 + \cdots + A_{n2}b_n \\ \cdot \\ \cdot \\ \cdot \\ A_{1n}b_1 + A_{2n}b_2 + \cdots + A_{nn}b_n \end{bmatrix}.$$

From this we see that

$$x_j = (\det A)^{-1}(A_{1j}b_1 + \cdots + A_{nj}b_n) = \frac{1}{\det A}(A_{1j}b_1 + \cdots + A_{nj}b_n).$$

By Theorem 9.3.2 we see that the numerator can be thought of as the determinant of the matrix B_j, where

$$B_j = \begin{bmatrix} a_{11} & a_{12} & \cdots & b_1 & \cdots & a_{1n} \\ a_{21} & a_{22} & & b_2 & & a_{2n} \\ \cdot & \cdot & & \cdot & & \cdot \\ \cdot & \cdot & & \cdot & & \cdot \\ \cdot & \cdot & & \cdot & & \cdot \\ a_{n1} & a_{n2} & & b_n & & a_{nn} \end{bmatrix},$$

$$\uparrow$$
jth column

using the expansion of $\det B_j$ by the cofactors of the jth column. We thus see that another way of expressing the solution of $AX = B$ is by setting $x_j = (\det B_j)/(\det A)$, where B_j is the matrix obtained by replacing the jth column of the coefficient matrix A by the column matrix

$$B = \begin{bmatrix} b_1 \\ b_2 \\ \cdot \\ \cdot \\ \cdot \\ b_n \end{bmatrix}.$$

The discussion just concluded is usually known as the following.

9.5 INVERSES OF MATRICES; CRAMER'S RULE

CRAMER'S RULE If $AX = B$ is a system of n simultaneous linear equations in n unknowns and if $\det A \neq 0$, then a unique solution exists, and

$$x_j = \frac{\det B_j}{\det A}, \quad j = 1, \ldots, n.$$

Alternatively, if $\det A \neq 0$, the solution is given by $X = A^{-1}B$. ∎

We mention that in the event $\det A = 0$, Cramer's rule gives no information. In this case, there may or may not be solutions. To study the case where $\det A = 0$, a more elaborate theory is needed. This will be developed in the next chapter.

EXERCISES

1. Find the inverse of each of the following matrices:

   (a) $\begin{bmatrix} 1 & 2 & 3 \\ -1 & 2 & 0 \\ 3 & 0 & 5 \end{bmatrix}$.

   (b) $\begin{bmatrix} 3 & 2 & 1 \\ 3 & 1 & 0 \\ 3 & 0 & 0 \end{bmatrix}$.

   (c) $\begin{bmatrix} 1 & 1 & 1 & 1 \\ 0 & 1 & 1 & 1 \\ 0 & 0 & 1 & 1 \\ 0 & 0 & 0 & 1 \end{bmatrix}$.

   (d) $\begin{bmatrix} 1 & 0 & 1 & 0 \\ 0 & 1 & 0 & 1 \\ 1 & 0 & 0 & 1 \\ 1 & 1 & 0 & 0 \end{bmatrix}$.

   (e) $\begin{bmatrix} a & b \\ c & d \end{bmatrix}$, with $ad - bc \neq 0$.

   (f) $\begin{bmatrix} 1 & 2 & 3 & 4 \\ -1 & 3 & 1 & 0 \\ 0 & 5 & 4 & 5 \\ -2 & 1 & -4 & 0 \end{bmatrix}$.

2. Consider the system of simultaneous equations:

   $$\begin{aligned} 2x_1 + 3x_2 - x_3 &= 3 \\ x_1 - x_2 + 3x_3 &= 6 \\ 5x_1 + 5x_2 + x_3 &= 12. \end{aligned}$$

   (a) Show that the determinant of the coefficient matrix is 0.
   (b) Show that $x_1 = 1$, $x_2 = 1$, $x_3 = 2$ is a solution.
   (c) Show that the third equation is the sum of two times the first equation plus the second equation.
   (d) Show that the solution to the system is not unique.

3. Let

   $$\begin{aligned} ax_1 + 2x_2 &= 3 \\ x_1 + (4 - a)x_2 &= 4. \end{aligned}$$

   (a) Find all values of a for which the system has a unique solution.
   (b) For the values of a not satisfying (a) does the system have any solution?

4. Let A be an $n \times n$ square matrix with entries in a field F. Let adj A be the adjoint of A, that is,

$$\text{adj } A = \begin{bmatrix} A_{11} & A_{21} & \cdots & A_{n1} \\ A_{12} & A_{22} & \cdots & A_{n2} \\ \cdot & \cdot & & \cdot \\ \cdot & \cdot & & \cdot \\ \cdot & \cdot & & \cdot \\ A_{1n} & A_{2n} & \cdots & A_{nn} \end{bmatrix},$$

where A_{ij} is the (i,j)th cofactor of A.
   (a) Show that $A(\text{adj } A) = (\det A)I_n$.
   (b) Show that $\det (\text{adj } A) = (\det A)^{n-1}$, whether $\det A \neq 0$ or $\det A = 0$.
5. Let A be an $n \times n$ matrix with entries in a field F such that $\det A \neq 0$. Prove that $(A^T)^{-1} = (A^{-1})^T$.
6. Let A, B, and C be $n \times n$ matrices such that $AB = C$. Prove that adj C = adj B · adj A.
7. In Theorem 8.8.10 and the discussion following that theorem, we saw how to consider a nonsingular matrix C as defining a "change of coordinates."
   (a) Let $V = \mathbf{R}^3$ and let $\mathcal{B}_3 = \{e_1, e_2, e_3\}$, $e_1 = (1,0,0)$, $e_2 = (0,1,0)$, $e_3 = (0,0,1)$, be the canonical basis for $\mathbf{R}^3$. Let C be the matrix of Exercise 1(a). Use C to define a change of coordinates—that is, a new basis $\mathcal{B}_3'$. If

$$[v]_{\mathcal{B}_3} = \begin{bmatrix} a \\ b \\ c \end{bmatrix}, \qquad \text{find } [v]_{\mathcal{B}_3'}.$$

   (b) Let $V = \mathbf{R}^4$ and let $\mathcal{B}_4$ be the canonical basis for $\mathbf{R}^4$. Use the matrix of Exercise 1(d) above to define a change of coordinates in $\mathbf{R}_4$. If

$$[v]_{\mathcal{B}_4} = \begin{bmatrix} a \\ b \\ c \\ d \end{bmatrix}, \qquad \text{find } [v]_{\mathcal{B}_4'}.$$

9.6 Determinant of a Linear Transformation

In Section 9.1, we define det A, where A was an $n \times n$ matrix. We will now define det T, where T is a linear operator.

Definition 9.6.1 Let T be a linear operator on the vector space V over the field F, where V is n-dimensional. Let $\mathcal{B}$ be a basis for V. We define

9.6 DETERMINANT OF A LINEAR TRANSFORMATION

the *determinant* of T to be $\det T = \det [T]_\mathcal{B}$, where $[T]_\mathcal{B}$ is the matrix of T relative to the basis $\mathcal{B}$.

As we recall from Chapter 8, relative to other bases $\mathcal{B}'$ of V, T will have different matrices $[T]_{\mathcal{B}'}$. By Section 8.9, for a different basis $\mathcal{B}'$, there exists an $n \times n$ matrix P such that $[T]_{\mathcal{B}'} = P^{-1}[T]_\mathcal{B} P$. By Theorem 9.4.1,

$$\begin{aligned} \det(P^{-1}[T]_\mathcal{B} P) &= \det(P^{-1}) \cdot \det [T]_\mathcal{B} \cdot \det P \\ &= (\det P)^{-1} \cdot \det [T]_\mathcal{B} \cdot \det P \\ &= \det [T]_\mathcal{B}. \end{aligned}$$

Thus, regardless of the basis we choose for V, $\det T$ always turns out to be the same. Thus Definition 9.6.1 is independent of the choice of basis for V.

Let V be the three-dimensional vector space Q^3 over the field of rational numbers Q. Let $\mathcal{B} = \{e_1, e_2, e_3\}$ be a basis for V. If $T(e_1) = e_1 + 2e_2 - e_3$, $T(e_2) = e_1 - e_2 + 2e_3$, and $T(e_3) = e_1 - e_3$, then T has matrix

$$[T]_\mathcal{B} = \begin{bmatrix} 1 & 1 & 1 \\ 2 & -1 & 0 \\ -1 & 2 & -1 \end{bmatrix},$$

so that $\det T = \det [T]_\mathcal{B} = 6$.

THEOREM 9.6.2 Let V be an n-dimensional vector space over a field F. Let T be a linear operator on V. Then T has an inverse if and only if $\det T \neq 0$. Moreover, if $\det T \neq 0$, then $\det(T^{-1}) = (\det T)^{-1}$.

Proof Let $\mathcal{B}$ be a fixed basis for V, and assume $\det T \neq 0$. By hypothesis and definition of $\det T$, $\det [T]_\mathcal{B} \neq 0$. Thus, the matrix $[T]_\mathcal{B}$ has an inverse $[T]_\mathcal{B}^{-1}$. Let S be the linear operator on V that has matrix $[T]_\mathcal{B}^{-1}$. Then the composites $T \circ S$ and $S \circ T$ have matrices $[T]_\mathcal{B}[T]_\mathcal{B}^{-1}$ and $[T]_\mathcal{B}^{-1}[T]_\mathcal{B}$, respectively; that is, $T \circ S$ and $S \circ T$ have the identity matrix with respect to $\mathcal{B}$. Thus, $T \circ S$ and $S \circ T$ are the identity on V. Hence, we conclude that $S = T^{-1}$ as a linear operator.

Next, suppose T has an inverse T^{-1}. Then $T \circ T^{-1} = I_V$, whence $[T \circ T^{-1}]_\mathcal{B} = I_n$, $[T]_\mathcal{B}[T^{-1}]_\mathcal{B} = I_n$. Then $\det [T]_\mathcal{B} \cdot \det [T^{-1}]_\mathcal{B} = 1$, whence $\det [T]_\mathcal{B} \neq 0$.

The last equation shows that

$$\det [T^{-1}]_\mathcal{B} = (\det [T_\mathcal{B}])^{-1}, \quad \text{whence } \det(T^{-1}) = (\det T)^{-1}. \blacksquare$$

Theorem 9.6.2 is useful in finding T^{-1}, provided it exists. To do so, proceed as follows: Let $\mathcal{B}$ be a basis for V. Find $[T]_\mathcal{B}$. If $\det T \neq 0$, then $\det [T]_\mathcal{B} \neq 0$, whence $[T]_\mathcal{B}$ has an inverse matrix D. But by the proof of Theorem 8.8.6, D is the matrix for T^{-1} with respect to $\mathcal{B}$; that is, $D = [T^{-1}]_\mathcal{B} = [T]_\mathcal{B}^{-1}$. Thus we have the following theorem.

THEOREM 9.6.3 Let $T \in \text{Hom}_F (V,V)$, where V is n-dimensional over F. If $\det T \neq 0$, then T^{-1} exists. Moreover, if $\mathcal{B}$ is a basis for V, then $[T^{-1}]_\mathcal{B} = [T]_\mathcal{B}^{-1}$.

Example 9.6.4 Let T be the operator on Q^3 defined just prior to Theorem 9.6.2.

$$[T]_\mathcal{B} = \begin{bmatrix} 1 & 1 & 1 \\ 2 & -1 & 0 \\ -1 & 2 & -1 \end{bmatrix},$$

and

$$[T]_\mathcal{B}^{-1} = \begin{bmatrix} \frac{1}{6} & \frac{1}{2} & \frac{1}{6} \\ \frac{1}{3} & 0 & \frac{1}{3} \\ \frac{1}{2} & -\frac{1}{2} & -\frac{1}{2} \end{bmatrix}.$$

Then

$$[T^{-1}]_\mathcal{B} = \begin{bmatrix} \frac{1}{6} & \frac{1}{2} & \frac{1}{6} \\ \frac{1}{3} & 0 & \frac{1}{3} \\ \frac{1}{2} & -\frac{1}{2} & -\frac{1}{2} \end{bmatrix},$$

that is, $T^{-1}(e_1) = \frac{1}{6}e_1 + \frac{1}{3}e_2 + \frac{1}{2}e_3$, $T^{-1}(e_2) = \frac{1}{2}e_1 - \frac{1}{2}e_3$, and

$$T^{-1}(e_3) = \frac{1}{6}e_1 + \frac{1}{3}e_2 - \frac{1}{2}e_3.$$

EXERCISES

1. Let T and S be linear operators on the n-dimensional vector space over the field F. Show that $\det (T \circ S) = \det T \cdot \det S$.
2. Let $T: \mathbf{R}^3 \longrightarrow \mathbf{R}^3$ be defined by $T(e_1) = e_1 + e_2 - e_3$, $T(e_2) = e_1 - e_2 + e_3$, $T(e_3) = 2e_1 - e_3$, where $\mathcal{B} = \{e_1, e_2, e_3\}$ is a basis for $\mathbf{R}^3$. Find $T^{-1}(e_i)$, for $i = 1, 2, 3$.
3. Let $T: \mathbf{R}^2 \longrightarrow \mathbf{R}^2$ be defined by the matrix

$$[T]_\mathcal{B} = \begin{bmatrix} \cos \theta & -\sin \theta \\ \sin \theta & \cos \theta \end{bmatrix},$$

where $\mathcal{B}$ is the canonical basis $\{e_1, e_2\}$, $e_1 = (1,0)$, $e_2 = (0,1)$. Find $T^{-1}(e_1)$, $T^{-1}(e_2)$, and $[T^{-1}]_\mathcal{B}$.
4. Let $T \in \text{Hom}_F (V,V)$, where $V = F^n$. Let V have a basis $\mathcal{B}$ such that

$$[T]_\mathcal{B} = \begin{bmatrix} d_1 & 0 & \cdots & 0 \\ 0 & d_2 & \cdots & 0 \\ \vdots & & & \\ 0 & 0 & \cdots & d_n \end{bmatrix}.$$

  (a) Interpret the geometric action of T on $\mathcal{B}$.
  (b) Find $\det T$.

10 Rank, Equivalence, and Systems of Linear Equations

In Chapter 9 we saw that a system of n linear equations in n unknowns has a unique solution when the coefficient matrix has an inverse (Cramer's rule). That result generalizes to systems of m equations in n unknowns, when one replaces the concept of inverse by that of rank, and then imposes a simple condition on the ranks of the coefficient matrix and what is called the augmented matrix. These concepts will be developed in this chapter.

10.1 Column Rank of a Matrix A; Rank of a Linear Transformation T

Let T be a linear transformation from an n-dimensional vector space V over a field F to an m-dimensional vector space U over F. Then with respect to a pair of fixed bases $\mathcal{B} = \{e_1,\ldots,e_n\}$ and $\mathcal{C} = \{f_1,\ldots,f_m\}$ for V and U, respectively, T has the $m \times n$ matrix $A = [T]_{\mathcal{C},\mathcal{B}} = [a_{ij}]_{m\times n}$. Recall that the ith column A_i of A can be thought of as the image of e_i under T, that is, $T(e_i) = a_{1i}f_1 + a_{2i}f_2 + \cdots + a_{mi}f_m$. As we recall from Chapter 8, the set of images $T(e_i)$, $i = 1, 2, \ldots, n$, generates the subspace $R(T)$ of U.

Conversely, if we start with an $m \times n$ matrix A, we can interpret A as being the matrix of a linear transformation T from an n-dimensional vector space V to an m-dimensional vector space U over F, where $A = [T]_{\mathcal{C},\mathcal{B}}$, and $\mathcal{B} = \{e_1,\ldots,e_n\}$ and $\mathcal{C} = \{f_1,\ldots,f_m\}$. Recall Theorem 8.6.6.

With this as background, we can now make the following definition.

Definition 10.1.1 (a) Let $T: V \longrightarrow U$ be a linear transformation from the n-dimensional vector space V over F to the m-dimensional vector space U

over F. Then the *rank of* T is $r(T) = \dim R(T)$. (This has already been defined in Definition 8.2.3.)

(b) Let A be an $m \times n$ matrix with entries in the field F. Let T, V, U, $\mathcal{B}$ and $\mathcal{C}$ be a linear transformation, vector spaces and bases chosen so that $A = [T]_{\mathcal{C},\mathcal{B}}$. Then the *column rank of* A is defined as the rank of T. We will often write *col rank* A for column rank of A.

Since any two k-dimensional vector spaces over F are isomorphic as vector spaces (see Theorem 7.6.5), Definition 10.1.1(b) does not depend on the choice of vector spaces V and U, nor upon the choice of bases $\mathcal{B}$ and $\mathcal{C}$. For if V, U, $\mathcal{B}$, and $\mathcal{C}$ are replaced by V', U', $\mathcal{B}'$, and $\mathcal{C}'$, respectively, and if $T \colon V \longrightarrow U$ is replaced by $T' \colon V' \longrightarrow U'$ so that

$$[T]_{\mathcal{C},\mathcal{B}} = A = [T']_{\mathcal{C}',\mathcal{B}'},$$

then we could easily show that $R(T) \cong R(T')$ as vector spaces. Indeed, if $\mathcal{C}' = \{f_1', \ldots, f_m'\}$, we simply let $a_{1i}f_1 + \cdots + a_{mi}f_m$ in $R(T)$ correspond to $a_{1i}f_1' + \cdots + a_{mi}f_m'$ in $R(T')$.

It is possible to define the m-dimensional vector space F_m of $m \times 1$ column matrices over a field F in analogy to our definition of the vector space F^n of $1 \times n$ row matrices over F. One could then consider the columns of an $m \times n$ matrix A as being elements of the vector space F_m. (See Exercise 1.) Doing this, we have from our above remarks the following theorem.

THEOREM 10.1.2 Let A be an $m \times n$ matrix with entries in the field F. Then the column rank of A is equal to the dimension of the subspace W of F_m, where W is generated by the n columns of A and F_m is the vector space of $m \times 1$ column matrices over F.

Proof This is left as an exercise for the reader. We simply mention that in our discussion, we could take for the vector space U and its basis $\mathcal{C}$, the vector space F_m and the basis $\mathcal{C}'$ consisting of those m $m \times 1$ column matrices having 0 for all entries except a 1 in one position, that is, we use the basis $\mathcal{C}' = \{f_1', \ldots, f_m'\}$, where

$$f_1' = \begin{bmatrix} 1 \\ 0 \\ 0 \\ \cdot \\ \cdot \\ \cdot \\ 0 \end{bmatrix}, \quad f_2' = \begin{bmatrix} 0 \\ 1 \\ 0 \\ \cdot \\ \cdot \\ \cdot \\ 0 \end{bmatrix}, \quad \ldots, \quad f_m' = \begin{bmatrix} 0 \\ 0 \\ \cdot \\ \cdot \\ \cdot \\ 0 \\ 1 \end{bmatrix}.$$

Example 10.1.3   Let

$$A = \begin{bmatrix} 1 & 1 & 0 \\ 2 & 3 & 1 \\ 1 & 1 & 2 \\ 2 & 1 & 1 \end{bmatrix},$$

and assume $1 + 1 \neq 0$ in the field F. To find the column rank of A, we compute the dimension of the subspace of F_4 generated by the columns of A. Since A has three columns, the column rank of A is clearly less than or equal to 3. Consider the linear combination

(*)
$$c_1 \begin{bmatrix} 1 \\ 2 \\ 1 \\ 2 \end{bmatrix} + c_2 \begin{bmatrix} 1 \\ 3 \\ 1 \\ 1 \end{bmatrix} + c_3 \begin{bmatrix} 0 \\ 1 \\ 2 \\ 1 \end{bmatrix}.$$

Set (*) equal to

$$\begin{bmatrix} 0 \\ 0 \\ 0 \\ 0 \end{bmatrix},$$

the zero vector of F_4. If we show each $c_i = 0$, the columns of A will be linearly independent, and so will span a three-dimensional subspace of F_4, whence we will have column rank of $A = 3$.

With (*) set equal to $\begin{bmatrix} 0 \\ 0 \\ 0 \\ 0 \end{bmatrix}$, we have

$$\begin{aligned} c_1 + c_2 &= 0 \\ 2c_1 + 3c_2 + c_3 &= 0 \\ c_1 + c_2 + 2c_3 &= 0 \\ 2c_1 + c_2 + c_3 &= 0. \end{aligned}$$

The first equation implies $c_1 = -c_2$. Using the last three equations, we then have

$$\begin{aligned} 2(-c_2) + 3c_2 + c_3 &= 0 \\ -c_2 + c_2 + 2c_3 &= 0 \\ 2(-c_2) + c_2 + c_3 &= 0 \end{aligned}$$

whence

$$\begin{aligned} c_2 + c_3 &= 0 \\ 2c_3 &= 0 \\ -c_2 + c_3 &= 0. \end{aligned}$$

Since $1 + 1 \neq 0$ in F, $c_3 = 0$, whence $c_2 = 0$ and $c_1 = 0$. Thus, the columns of A are linearly independent, and so col rank $A = 3$.

An easy consequence of Definition 10.1.1(b) and Theorem 8.4.8 is the next theorem.

THEOREM 10.1.4 Let A be an $n \times n$ square matrix with entries in the field F. Then A has an inverse if and only if col rank $A = n$.

Proof Interpret A as the matrix of a linear operator $T: V \to V$ on an n-dimensional vector space V over F. Suppose A is invertible, that is, nonsingular. We can interpret the columns A_i as $T(e_i)$, the images of the elements of a basis for V. Then, by Theorem 8.4.8, the n $T(e_i)$'s are a basis for V, whence dim $R(T) = n$, that is, col rank $A = n$.

Conversely, if A has a column rank n, then we can interpret the columns as being linearly independent images $T(e_1), \ldots, T(e_n)$ of basis elements of a vector space V, whence by Theorem 8.4.8 T is invertible as a linear operator. But if T is invertible as a linear operator, then its matrix $A = [T]_\mathcal{B}$ is also invertible, as we saw in Chapters 8 and 9. ∎

Before concluding this section, we mention that two other types of rank will be defined later for $m \times n$ matrices. All three types of ranks will turn out to be equal.

EXERCISES

1. Let
$$F_m = \left\{ \begin{bmatrix} a_1 \\ a_2 \\ \vdots \\ a_m \end{bmatrix} \,\Big|\, a_i \in F \right\}.$$

Define addition on F_m by
$$\begin{bmatrix} a_1 \\ \vdots \\ a_m \end{bmatrix} + \begin{bmatrix} b_1 \\ \vdots \\ b_m \end{bmatrix} = \begin{bmatrix} a_1 + b_1 \\ \vdots \\ a_m + b_m \end{bmatrix},$$

and then define the scalar multiple by
$$a \begin{bmatrix} a_1 \\ \vdots \\ a_m \end{bmatrix} = \begin{bmatrix} aa_1 \\ \vdots \\ aa_m \end{bmatrix}.$$

(a) Show that F_m is a vector space with respect to these operations.
(b) Show that the m vectors

$$f_i = \begin{bmatrix} 0 \\ \vdots \\ 0 \\ 1 \\ 0 \\ \vdots \\ 0 \end{bmatrix} \longleftarrow i\text{th row},$$

$i = 1, \ldots, m$, form a basis for F_m. Thus conclude that $\dim F_m = m$.

2. Prove Theorem 10.1.2.
3. Find the column rank of each of the following:

(a) $\begin{bmatrix} 1 & 2 & 3 \\ -1 & 2 & 6 \\ 3 & -2 & 1 \\ 0 & 8 & 1 \end{bmatrix}.$
(b) $\begin{bmatrix} 1 & 0 & 0 \\ 1 & 1 & 1 \\ 0 & 1 & 1 \end{bmatrix}.$

(c) $\begin{bmatrix} 1 & 1 & 1 \\ 1 & 1 & 0 \\ 1 & 0 & 1 \end{bmatrix}.$
(d) $\begin{bmatrix} -1 & 2 & -3 \\ 4 & -5 & 6 \\ -7 & 8 & -9 \end{bmatrix}.$

(e) $\begin{bmatrix} 1 & 0 & 0 & 0 \\ 0 & 1 & 0 & 0 \\ 0 & 0 & 1 & 0 \\ 0 & 0 & 0 & 1 \end{bmatrix}.$
(f) $\begin{bmatrix} 1 & 2 & -1 & -2 \\ 1 & 3 & -1 & 3 \\ -2 & 4 & 1 & 5 \\ -2 & 1 & 1 & 3 \\ 0 & -2 & 0 & 5 \end{bmatrix}.$

(g) $\begin{bmatrix} 1 & 0 & 0 & -1 & 1 \end{bmatrix}.$
(h) $\begin{bmatrix} 1 \\ -1 \\ 0 \\ 0 \\ 1 \end{bmatrix}.$

10.2 Elementary Row and Column Operations; Elementary Matrices

To continue our study of rank, we need to consider certain special operations that can be performed on matrices. Recall that when we evaluated determinants of matrices, we performed the following operations: the interchange of two rows; the addition of a multiple of one row to another row; the multiplication of one row by a nonzero constant.

In what follows, it will be convenient to have the next definition.

Definition 10.2.1 Let A be a matrix. A *line* of A will denote either a row or column of A.

Definition 10.2.2 A *type (I) elementary matrix operation* is the interchange of any two parallel lines.

A *type (II) elementary matrix operation* is the multiplication of all the elements of one line by a nonzero constant.

A *type (III) elementary matrix operation* is the addition of an arbitrary multiple of one line to any other, different, parallel line.

To illustrate, in the matrix

$$\begin{bmatrix} 1 & 2 & 1 & 3 \\ 0 & 1 & -1 & 1 \\ 1 & 3 & 0 & 5 \end{bmatrix},$$

if we interchange rows 1 and 3, we have performed a type (I) operation, and we obtain as a result the matrix

$$\begin{bmatrix} 1 & 3 & 0 & 5 \\ 0 & 1 & -1 & 1 \\ 1 & 2 & 1 & 3 \end{bmatrix}.$$

In this last matrix, if we multiply column 1 by -1 and add the result to column 2, then we have performed a type (III) operation, which results in the matrix

$$\begin{bmatrix} 1 & 3-1 & 0 & 5 \\ 0 & 1-0 & -1 & 1 \\ 1 & 2-1 & 1 & 3 \end{bmatrix} = \begin{bmatrix} 1 & 2 & 0 & 5 \\ 0 & 1 & -1 & 1 \\ 1 & 1 & 1 & 3 \end{bmatrix}.$$

Of greatest importance is the fact that each of the elementary matrix operations can be effected by multiplying the given matrix by a so-called elementary matrix.

Definition 10.2.3 A matrix M obtained from the $n \times n$ identity matrix I_n by applying an elementary matrix operation to I_n is called an *elementary matrix*. M is an elementary matrix of *type (I), (II), or (III)* according as the corresponding elementary matrix operation is of type (I), (II), or (III), respectively.

We see that the matrix

$$\begin{bmatrix} 1 & 0 & -3 \\ 0 & 1 & 0 \\ 0 & 0 & 1 \end{bmatrix},$$

obtained from I_3 by adding -3 times column 1 to column 3, is an elementary matrix of type (III). Note that the same matrix can also be obtained from I_3 by adding -3 times row 3 to row 1.

Given an $m \times n$ matrix A, suppose we wish to interchange the ith and jth rows of A. To do this, we let E_{ij} be the elementary matrix of type (I) obtained by interchanging the ith and jth rows of the identity matrix I_m. Then the product $E_{ij}A$ is the same as A, except that the ith and jth rows have been interchanged.

To illustrate, let

$$A = \begin{bmatrix} 1 & 0 & 1 \\ 2 & 1 & 0 \\ 3 & 0 & -1 \\ 1 & -1 & 2 \end{bmatrix}.$$

To interchange the third and fourth rows of A, we form the 4×4 elementary matrix

$$E_{34} = \begin{bmatrix} 1 & 0 & 0 & 0 \\ 0 & 1 & 0 & 0 \\ 0 & 0 & 0 & 1 \\ 0 & 0 & 1 & 0 \end{bmatrix}.$$

Then the product

$$E_{34}A = \begin{bmatrix} 1 & 0 & 0 & 0 \\ 0 & 1 & 0 & 0 \\ 0 & 0 & 0 & 1 \\ 0 & 0 & 1 & 0 \end{bmatrix} \begin{bmatrix} 1 & 0 & 1 \\ 2 & 1 & 0 \\ 3 & 0 & -1 \\ 1 & -1 & 2 \end{bmatrix} = \begin{bmatrix} 1 & 0 & 1 \\ 2 & 1 & 0 \\ 1 & -1 & 2 \\ 3 & 0 & -1 \end{bmatrix},$$

as required.

THEOREM 10.2.4 (a) An elementary operation on the rows of the $m \times n$ matrix A may be performed by premultiplying A by the $m \times m$ elementary matrix obtained by performing the desired elementary row operation on the $m \times m$ identity matrix I_m.

(b) An elementary operation on the columns of A may be performed by postmultiplying A by the $n \times n$ elementary matrix obtained by performing the desired elementary column operation on the $n \times n$ identity matrix I_n.

Proof We leave the proof as an exercise. ∎

THEOREM 10.2.5 Each $n \times n$ elementary matrix is invertible. Moreover,

(a) If E_{ij} is the elementary matrix of type (I) obtained by interchanging the ith and jth lines of I_n, then $E_{ij}^{-1} = E_{ij}$. (*Note:* In this case, the same matrix E_{ij} is obtained whether rows or columns are interchanged.)

(b) If D is the elementary matrix of type (II) obtained by multiplying the ith line of I_n by the nonzero constant c, then D^{-1} is the elementary matrix of type (II) obtained by multiplying the same ith line of I_n by c^{-1}.

(c) If B is the elementary matrix of type (III) obtained from I_n by adding c times a given line i to a parallel line j in I_n, $j \neq i$, then B^{-1} is the elementary matrix of type (III) obtained from I_n by adding $-c$ times the line i to the parallel line j in I_n.

Proof The fact that inverses exist follows, once parts (a), (b), and (c) are shown. We leave these to the reader. ∎

Example 10.2.6 The inverse of

$$A = \begin{bmatrix} 1 & 0 & 0 & 0 \\ 0 & 1 & 0 & 2 \\ 0 & 0 & 1 & 0 \\ 0 & 0 & 0 & 1 \end{bmatrix} \quad \text{is} \quad A^{-1} = \begin{bmatrix} 1 & 0 & 0 & 0 \\ 0 & 1 & 0 & -2 \\ 0 & 0 & 1 & 0 \\ 0 & 0 & 0 & 1 \end{bmatrix}.$$

Note that A can be interpreted as being obtained from I_4 either by adding 2 times row 4 to row 2, or by adding 2 times column 2 to column 4. Then A^{-1} is obtained from I_4 either by adding -2 times row 4 to row 2, or by adding -2 times column 2 to column 4.

The inverse of

$$\begin{bmatrix} 1 & 0 & 0 & 0 \\ 0 & c & 0 & 0 \\ 0 & 0 & 1 & 0 \\ 0 & 0 & 0 & 1 \end{bmatrix}, \quad c \neq 0, \quad \text{is} \quad \begin{bmatrix} 1 & 0 & 0 & 0 \\ 0 & c^{-1} & 0 & 0 \\ 0 & 0 & 1 & 0 \\ 0 & 0 & 0 & 1 \end{bmatrix},$$

as direct multiplication will show.

If $\mathfrak{M}$ is the set of all $m \times n$ matrices over a field F, then we may define a number of equivalence relations on $\mathfrak{M}$ by use of our elementary row and column operations.

Definition 10.2.7 Let A and B be $m \times n$ matrices with entries in F.

(a) We say that A and B are *row equivalent* if B can be obtained from A by performing a finite number of elementary row operations on A.

(b) We say that A and B are *column equivalent* if B can be obtained from A by performing a finite number of elementary column operations on A.

(c) We say that A and B are *equivalent* if B can be obtained from A by performing a finite number of elementary row and elementary column operations on A.

Example 10.2.8

(a) The matrices $A = \begin{bmatrix} 1 & 2 & 3 \\ -1 & 1 & 2 \end{bmatrix}$ and $B = \begin{bmatrix} 0 & 1 & \frac{5}{3} \\ 1 & 2 & 3 \end{bmatrix}$ are row equivalent, since B is obtained from A as follows:

$$\begin{bmatrix} 1 & 2 & 3 \\ -1 & 1 & 2 \end{bmatrix} \xrightarrow[\text{to row 2}]{\text{Add row 1}} \begin{bmatrix} 1 & 2 & 3 \\ 0 & 3 & 5 \end{bmatrix} \xrightarrow[\text{by 3}]{\text{Divide row 2}} \begin{bmatrix} 1 & 2 & 3 \\ 0 & 1 & \frac{5}{3} \end{bmatrix}$$

$$\xrightarrow[\text{rows 1 and 2}]{\text{Interchange}} \begin{bmatrix} 0 & 1 & \frac{5}{3} \\ 1 & 2 & 3 \end{bmatrix}.$$

(b) The matrices $A = \begin{bmatrix} 1 & 2 & 3 \\ -1 & 1 & 2 \end{bmatrix}$ and $B = \begin{bmatrix} 0 & 1 & 3 \\ -1 & 0 & 2 \end{bmatrix}$ are column equivalent, since B is obtained as follows:

$$\begin{bmatrix} 1 & 2 & 3 \\ -1 & 1 & 2 \end{bmatrix} \xrightarrow[\text{to column 2}]{\text{Add column 1}} \begin{bmatrix} 1 & 3 & 3 \\ -1 & 0 & 2 \end{bmatrix} \xrightarrow[\text{by 3}]{\text{Divide column 2}} \begin{bmatrix} 1 & 1 & 3 \\ -1 & 0 & 2 \end{bmatrix}$$

$$\xrightarrow[\text{from column 1}]{\text{Subtract column 2}} \begin{bmatrix} 0 & 1 & 3 \\ -1 & 0 & 2 \end{bmatrix}.$$

(c) The matrices $A = \begin{bmatrix} 1 & 2 & 3 \\ -1 & 1 & 2 \end{bmatrix}$ and $B = \begin{bmatrix} 1 & 0 & 0 \\ 0 & 1 & 0 \end{bmatrix}$ are equivalent, since B is obtained from A as follows:

$$\begin{bmatrix} 1 & 2 & 3 \\ -1 & 1 & 2 \end{bmatrix} \xrightarrow[\text{of (b) above}]{\text{by the steps}} \begin{bmatrix} 0 & 1 & 3 \\ -1 & 0 & 2 \end{bmatrix} \xrightarrow[\text{by } -1]{\text{Multiply row 2}}$$

$$\begin{bmatrix} 0 & 1 & 3 \\ 1 & 0 & -2 \end{bmatrix} \xrightarrow[\text{rows 1 and 2}]{\text{Interchange}} \begin{bmatrix} 1 & 0 & -2 \\ 0 & 1 & 3 \end{bmatrix} \xrightarrow[\text{1 to column 3}]{\text{Add 2 times column}}$$

$$\begin{bmatrix} 1 & 0 & 0 \\ 0 & 1 & 3 \end{bmatrix} \xrightarrow[\text{from column 3}]{\text{Subtract 3 times column 2}} \begin{bmatrix} 1 & 0 & 0 \\ 0 & 1 & 0 \end{bmatrix}.$$

Note that the steps used in obtaining the matrices B above from the matrices A are not unique. For example, in (c), although row operations were used, B could be obtained from A by using only column operations.

THEOREM 10.2.9 Let $\mathfrak{M}$ be the set of all $m \times n$ matrices over the field F. Then each of the following notions defined in Definition 10.2.7 is an equivalence relation on $\mathfrak{M}$.

(a) Row equivalence.
(b) Column equivalence.
(c) Equivalence.

Proof We leave this as an exercise. ∎

We will see in Section 10.5 that the notion of equivalence defined in Definition 10.2.7(c) coincides with the notion of equivalence of matrices defined in Section 8.9.

We conclude by again emphasizing that an elementary matrix can be viewed either as having been obtained from I_n by an elementary row operation, or having been obtained from I_n by an elementary column operation.

EXERCISES

1. Prove Theorem 10.2.5.
2. Prove Theorem 10.2.4, that is, prove both (a) and (b) below.
   (a) If A is an $m \times n$ matrix and if E is an $m \times m$ elementary matrix, then EA is the $m \times n$ matrix that is obtained by performing on the rows of A the same elementary operation that had been performed on the rows of I_m to obtain E.
   (b) If A is an $m \times n$ matrix and if E is an $n \times n$ elementary matrix, then AE is the $m \times n$ matrix that is obtained by performing on the columns of A the same elementary operation that had been performed on the columns of I_n to obtain E.
3. Find the inverse of each of the following matrices in two different manners: first, use the method of cofactors, described in Theorem 9.5.2, and then use Theorem 10.2.5.

   (a) $\begin{bmatrix} 1 & 0 & 1 \\ 0 & 1 & 0 \\ 0 & 0 & 1 \end{bmatrix}$.
   (b) $\begin{bmatrix} 1 & 2 & 0 & 0 \\ 0 & 1 & 0 & 0 \\ 0 & 0 & 1 & 0 \\ 0 & 0 & 0 & 1 \end{bmatrix}$.

   (c) $\begin{bmatrix} 1 & 0 & 0 & 0 \\ 0 & 1 & 0 & 0 \\ -3 & 0 & 1 & 0 \\ 0 & 0 & 0 & 1 \end{bmatrix}$.
   (d) $\begin{bmatrix} 1 & 0 & 0 & 0 \\ 0 & 2 & 0 & 0 \\ 0 & 0 & 1 & 0 \\ 0 & 0 & 0 & 1 \end{bmatrix}$.

   (e) $\begin{bmatrix} 1 & 0 & 0 & 0 \\ 0 & 0 & 1 & 0 \\ 0 & 1 & 0 & 0 \\ 0 & 0 & 0 & 1 \end{bmatrix}$.
   (f) $\begin{bmatrix} 0 & 0 & 0 & 1 \\ 0 & 1 & 0 & 0 \\ 0 & 0 & 1 & 0 \\ 1 & 0 & 0 & 0 \end{bmatrix}$.

4. Verify Theorem 10.2.4 and Exercise 2 in each of the following cases:

   (a) $A = \begin{bmatrix} 1 & 2 & 0 & 1 \\ -1 & 2 & 0 & 1 \\ 3 & -1 & 1 & 2 \\ 0 & 1 & 2 & 1 \end{bmatrix}$, $E = \begin{bmatrix} 1 & 2 & 0 & 0 \\ 0 & 1 & 0 & 0 \\ 0 & 0 & 1 & 0 \\ 0 & 0 & 0 & 1 \end{bmatrix}$. Find EA and AE.

(b) $A = \begin{bmatrix} 1 & 2 & 3 \\ 1 & 1 & 1 \\ -1 & 2 & 3 \\ 0 & -1 & 1 \end{bmatrix}$, $E = \begin{bmatrix} 1 & 0 & 0 & 0 \\ 0 & 1 & 0 & 0 \\ 0 & 0 & 1 & -3 \\ 0 & 0 & 0 & 1 \end{bmatrix}$. Find EA.

(c) $A = \begin{bmatrix} 1 & 2 & 3 & 1 \\ -1 & 2 & 1 & 0 \end{bmatrix}$, $E = \begin{bmatrix} 1 & 0 \\ 0 & 3 \end{bmatrix}$. Find EA.

5. Prove that an elementary matrix of type (I) can be written as the product of elementary matrices of types (II) and (III). (*Hint:* Recall the technique in the proof of Theorem 9.2.6, in which we showed $\det A' = -\det A$, where A' was obtained from A by interchanging two rows.)

6. Let E be an elementary matrix. Prove:
   (a) $\det E = -1$, if E is of type (I).
   (b) $\det E = c$, if E is of type (II), where the nonzero constant used is c.
   (c) $\det E = 1$, if E is of type (III).

7. Prove Theorem 10.2.9.

8. Show that A and B are row equivalent:

   (a) $A = \begin{bmatrix} 1 & -1 & 2 \\ 0 & 1 & 2 \end{bmatrix}$, $B = \begin{bmatrix} 0 & 1 & 2 \\ 1 & 0 & 4 \end{bmatrix}$.

   (b) $A = \begin{bmatrix} 0 & 0 & 1 \\ -1 & 2 & 0 \\ 2 & 1 & 0 \end{bmatrix}$, $B = \begin{bmatrix} 0 & 1 & 0 \\ -1 & 2 & 0 \\ 0 & 0 & 1 \end{bmatrix}$.

   (c) $A = \begin{bmatrix} 1 & 2 & 3 \\ 2 & 1 & 0 \\ -1 & 1 & 0 \\ 0 & 0 & 1 \end{bmatrix}$, $B = \begin{bmatrix} 2 & 1 & 0 \\ 1 & 2 & 0 \\ 0 & 1 & 1 \\ 0 & 0 & 1 \end{bmatrix}$.

9. Show that A and B are column equivalent:

   (a) $A = \begin{bmatrix} 1 & -1 & 2 \\ 0 & 1 & 2 \end{bmatrix}$, $B = \begin{bmatrix} 1 & 0 & 0 \\ 0 & 1 & 2 \end{bmatrix}$.

   (b) $A = \begin{bmatrix} 0 & 0 & 1 \\ -1 & 2 & 0 \\ 2 & 1 & 0 \end{bmatrix}$, $B = \begin{bmatrix} 1 & 0 & 0 \\ -1 & -1 & 0 \\ 2 & 2 & 5 \end{bmatrix}$.

   (c) $A = \begin{bmatrix} 1 & 2 & 3 \\ 2 & 1 & 0 \\ -1 & 1 & 0 \\ 0 & 0 & 1 \end{bmatrix}$, $B = \begin{bmatrix} 2 & 0 & 1 \\ -5 & -3 & 2 \\ 4 & 3 & -1 \\ 1 & 0 & 0 \end{bmatrix}$.

10. Show that A and B are equivalent:

   (a) $A = \begin{bmatrix} 1 & -1 & 2 \\ 0 & 1 & 2 \end{bmatrix}, B = \begin{bmatrix} 0 & 1 & 0 \\ 1 & 0 & 4 \end{bmatrix}.$

   (b) $A = \begin{bmatrix} 0 & 0 & 1 \\ -1 & 2 & 0 \\ 2 & 1 & 0 \end{bmatrix}, B = \begin{bmatrix} 1 & 0 & 0 \\ 1 & 1 & 0 \\ 1 & 1 & 1 \end{bmatrix}.$

   (c) $A = \begin{bmatrix} 1 & 2 & 3 \\ 2 & 1 & 0 \\ -1 & 1 & 0 \\ 0 & 0 & 1 \end{bmatrix}, B = \begin{bmatrix} 0 & 0 & 1 \\ 2 & 0 & 0 \\ 0 & 3 & -1 \\ 0 & 3 & -2 \end{bmatrix}.$

11. (a) Let $A = \begin{bmatrix} 1 & 0 & 1 \\ 0 & 1 & 0 \end{bmatrix}, B = \begin{bmatrix} 1 & 1 & 0 \\ 0 & 1 & 0 \end{bmatrix}.$ Show that A is not row equivalent to B, but that A is column equivalent to B.

   (b) Let $A = \begin{bmatrix} 1 & 0 \\ 0 & 1 \\ 1 & 0 \end{bmatrix}, B = \begin{bmatrix} 1 & 0 \\ 1 & 1 \\ 0 & 0 \end{bmatrix}.$ Show that A is not column equivalent to B, but that A is row equivalent to B.

   (c) Let

   $$A = \begin{bmatrix} 1 & 0 & 1 & 0 \\ 0 & 1 & 0 & 0 \\ 0 & 0 & 0 & 1 \\ 0 & 0 & 0 & 0 \end{bmatrix}, \text{ and } B = \begin{bmatrix} 1 & 1 & 0 & 0 \\ 0 & 1 & 0 & 0 \\ 0 & 0 & 1 & 1 \\ 0 & 0 & 1 & 1 \end{bmatrix}.$$

   Show that A is not row equivalent to B, that A is not column equivalent to B, but that A and B are equivalent.

12. Prove that if A and B are row equivalent according to Definition 10.2.7(a), then A and B are equivalent according to Definition 10.2.7(c). That is, prove that if A and B are row equivalent, then they are equivalent.

13. Prove that if A and B are column equivalent, then A and B are equivalent.

14. An $n \times n$ permutation matrix P is an $n \times n$ matrix in which all entries are 0 or 1, and in each row and each column there is exactly one 1.

   (a) Prove that P can be written as the product of elementary matrices of type (I).

   (b) Prove that the number of factors used in the product in part (a) need not be unique, but that the number of factors is always even or always odd.

15. Show that a sum of elementary matrices is not necessarily elementary.
16. Let E be an $n \times n$ elementary matrix. Show that E^T is also an elementary matrix.
17. Let A be an $n \times n$ *symmetric matrix*, that is $A^T = A$. Let E be an $n \times n$ elementary matrix. Show that
    (a) $(EA)^T = A^T E^T$.
    (b) Interpret $(EA)^T$ in terms of the rows and/or columns of A.
18. Let A and B be $m \times n$ matrices. Show that A and B are row equivalent if and only if A^T and B^T are column equivalent.

10.3 Column-Reduced Echelon Matrices and Column Rank

In Section 7.5 we proved the Steinitz replacement theorem. The idea used in that theorem is useful in computing the column rank of a matrix A.

Recall that the $m \times n$ matrix A has column rank r if and only if the subspace W of F_m spanned by the columns of A is of dimension r. (See Theorem 10.1.2.) We note that by performing elementary column operations on A, we do not change the space W spanned by the columns of A: the interchange of two columns does not change the subspace W spanned by the columns of A, the multiplication of one column through by a nonzero constant does not change the subspace W, nor does the addition of a multiple of one column to a different column change the space W, as can be seen in the proof of the Steinitz exchange theorem.

We thus see that if we perform elementary column operations on A until A is reduced to a particularly useful matrix, we may have an easy way to find the column rank of A.

Let

$$A = \begin{bmatrix} a_{11} & \cdots & a_{1n} \\ \cdot & & \\ \cdot & & \\ \cdot & & \\ a_{m1} & \cdots & a_{mn} \end{bmatrix}_{m \times n}$$

If

$$A = \begin{bmatrix} 0 & \cdots & 0 \\ \cdot & & \\ \cdot & & \\ \cdot & & \\ 0 & \cdots & 0 \end{bmatrix}_{m \times n},$$

then clearly column rank $A = 0$. Thus, suppose A is not the zero matrix. We look at the rows of A, and let i_1 be the first row in which a nonzero entry appears. We interchange columns, if necessary, so that this nonzero entry is now in the $i_1 1$ position. Thus, without loss of generality, we can now assume that $a_{i_1 1} \neq 0$. We multiply the first column by $a_{i_1 1}^{-1}$, so that the first column has form

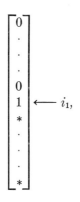

where $*$ indicates an element whose precise value is unimportant for the time being.

Next, we subtract $a_{i_1 j}$ times the resulting first column from the jth column, for all j, so that the original matrix has been reduced to

$$A' = \begin{bmatrix} 0 & 0 & & 0 \\ \cdot & \cdot & & \cdot \\ \cdot & \cdot & & \cdot \\ \cdot & \cdot & & \cdot \\ 0 & 0 & & 0 \\ 1 & 0 & & 0 \\ * & * & & * \\ \cdot & \cdot & & \cdot \\ \cdot & \cdot & & \cdot \\ \cdot & \cdot & & \cdot \\ * & * & & * \end{bmatrix}_{m \times n}$$

From our opening remarks, the columns of A' and the columns of the original matrix A span the same subspace W of F_m.

In A', we now work with the last $n - 1$ columns. First, we find the first row that has a nonzero entry in a column k, $k \geq 2$. By interchanging columns, if necessary, we assume that the element $a_{i_2 2} \neq 0$, where $i_2 > i_1$. We multiply column 2 by $a_{i_2 2}^{-1}$, then subtract appropriate multiples of

10.3 COLUMN-REDUCED ECHELON MATRICES AND COLUMN RANK

column 2 from columns 1, 3, 4, ..., n, obtaining a matrix of the form

$$A'' = \begin{bmatrix} 0 & 0 & 0 & \cdots & 0 \\ \cdot & \cdot & \cdot & & \cdot \\ \cdot & \cdot & \cdot & & \cdot \\ \cdot & \cdot & \cdot & & \cdot \\ 0 & 0 & 0 & & 0 \\ 1 & 0 & 0 & & 0 \\ * & 0 & 0 & & 0 \\ \cdot & \cdot & \cdot & & \cdot \\ \cdot & \cdot & \cdot & & \cdot \\ \cdot & \cdot & \cdot & & \cdot \\ * & 0 & 0 & & 0 \\ 0 & 1 & 0 & & 0 \\ * & * & * & & * \\ \cdot & \cdot & \cdot & & \cdot \\ \cdot & \cdot & \cdot & & \cdot \\ \cdot & \cdot & \cdot & & \cdot \\ * & * & * & \cdots & * \end{bmatrix}_{m \times n} \begin{matrix} \\ \\ \\ \\ \\ \leftarrow i_1 \\ \\ \\ \\ \\ \\ \leftarrow i_2 \\ \\ \\ \\ \\ \\ \end{matrix}$$

We continue with this process, ending eventually with a matrix of the form

$$A_C = \begin{bmatrix} 0 & 0 & 0 & \cdots \\ \cdot & \cdot & \cdot \\ \cdot & \cdot & \cdot \\ \cdot & \cdot & \cdot \\ 0 & 0 & 0 \\ 1 & 0 & 0 \\ * & 0 & 0 \\ \cdot & \cdot & \cdot \\ \cdot & \cdot & \cdot \\ \cdot & \cdot & \cdot \\ * & 0 & 0 \\ 0 & 1 & 0 \\ * & * & 0 \\ \cdot & \cdot & \cdot \\ \cdot & \cdot & \cdot \\ \cdot & \cdot & \cdot \\ * & * & 0 \\ 0 & 0 & 1 \\ * & * & * \\ \cdot & \cdot & \cdot \\ \cdot & \cdot & \cdot \\ \cdot & \cdot & \cdot \end{bmatrix}_{m \times n} \begin{matrix} \\ \\ \\ \\ \\ \leftarrow i_1 \\ \\ \\ \\ \\ \leftarrow i_2 \\ \\ \\ \\ \\ \leftarrow i_3 \\ \cdot \\ \cdot \\ \cdot \end{matrix}$$

The column rank of the final matrix A_C is easily seen to be the number of nonzero columns, as the resulting columns are clearly linearly independent. By our first remarks, the columns of A_C span the same subspace W of F_m as do the columns of the original matrix A.

We illustrate this idea with several examples.

Example 10.3.1 Find the column rank of

$$A = \begin{bmatrix} 2 & 0 & 3 & 2 \\ 2 & 1 & 2 & 0 \\ 1 & 0 & 1 & 0 \\ 0 & 0 & 0 & 1 \end{bmatrix}.$$

Solution

$$\begin{bmatrix} 2 & 0 & 3 & 2 \\ 2 & 1 & 2 & 0 \\ 1 & 0 & 1 & 0 \\ 0 & 0 & 0 & 1 \end{bmatrix} \xrightarrow[\text{by } \frac{1}{2}]{\text{Multiply column 1}} \begin{bmatrix} 1 & 0 & 3 & 2 \\ 1 & 1 & 2 & 0 \\ \frac{1}{2} & 0 & 1 & 0 \\ 0 & 0 & 0 & 1 \end{bmatrix} \xrightarrow[\text{from column 3}]{\text{Subtract 3 times column 1}}$$

$$\begin{bmatrix} 1 & 0 & 0 & 2 \\ 1 & 1 & -1 & 0 \\ \frac{1}{2} & 0 & -\frac{1}{2} & 0 \\ 0 & 0 & 0 & 1 \end{bmatrix} \xrightarrow[\text{from column 4}]{\text{Subtract 2 times column 1}} \begin{bmatrix} 1 & 0 & 0 & 0 \\ 1 & 1 & -1 & -2 \\ \frac{1}{2} & 0 & -\frac{1}{2} & -1 \\ 0 & 0 & 0 & 1 \end{bmatrix} \xrightarrow[\text{from column 1}]{\text{Subtract 1 times column 2}}$$

$$\begin{bmatrix} 1 & 0 & 0 & 0 \\ 0 & 1 & -1 & -2 \\ \frac{1}{2} & 0 & -\frac{1}{2} & -1 \\ 0 & 0 & 0 & 1 \end{bmatrix} \xrightarrow[\text{to column 3}]{\text{Add 1 times column 2}} \begin{bmatrix} 1 & 0 & 0 & 0 \\ 0 & 1 & 0 & -2 \\ \frac{1}{2} & 0 & -\frac{1}{2} & -1 \\ 0 & 0 & 0 & 1 \end{bmatrix} \xrightarrow[\text{to column 4}]{\text{Add 2 times column 2}}$$

$$\begin{bmatrix} 1 & 0 & 0 & 0 \\ 0 & 1 & 0 & 0 \\ \frac{1}{2} & 0 & -\frac{1}{2} & -1 \\ 0 & 0 & 0 & 1 \end{bmatrix} \xrightarrow[\text{by } -2]{\text{Multiply column 3}} \begin{bmatrix} 1 & 0 & 0 & 0 \\ 0 & 1 & 0 & 0 \\ \frac{1}{2} & 0 & 1 & -1 \\ 0 & 0 & 0 & 1 \end{bmatrix} \xrightarrow[\text{to column 1}]{\text{Add } -\frac{1}{2} \text{ times column 3}}$$

$$\begin{bmatrix} 1 & 0 & 0 & 0 \\ 0 & 1 & 0 & 0 \\ 0 & 0 & 1 & -1 \\ 0 & 0 & 0 & 1 \end{bmatrix} \xrightarrow[\text{to column 4}]{\text{Add 1 times column 3}} \begin{bmatrix} 1 & 0 & 0 & 0 \\ 0 & 1 & 0 & 0 \\ 0 & 0 & 1 & 0 \\ 0 & 0 & 0 & 1 \end{bmatrix} = A_C.$$

The final matrix A_C has column rank 4, so column rank $A = 4$.

In this special case, it is clear that each of the original columns of A is a linear combination of the columns of A_C, and our techniques were such that the columns of the last matrix are clearly linear combinations of the original columns of A. The reader should see if he can work through

10.3 COLUMN-REDUCED ECHELON MATRICES AND COLUMN RANK

the necessary steps to write the columns

$$\begin{bmatrix}1\\0\\0\\0\end{bmatrix}, \begin{bmatrix}0\\1\\0\\0\end{bmatrix}, \begin{bmatrix}0\\0\\1\\0\end{bmatrix}, \begin{bmatrix}0\\0\\0\\1\end{bmatrix}$$

as linear combinations of the original columns.

It should also be clear that we could find the matrix A_C without actually writing all the intermediate steps, since as long as we add multiples of a fixed column to different columns, the resulting matrix still has columns that span W. This will be illustrated in the next example.

Example 10.3.2 Find the column rank of

$$A = \begin{bmatrix} 0 & 0 & 2 & 0 \\ 1 & 1 & 4 & 2 \\ 1 & 2 & 1 & 3 \\ 2 & 1 & 2 & 3 \end{bmatrix}.$$

Solution Since row 1 has a nonzero entry, interchange columns 1 and 3 to get a non-zero entry in the 1, 1 position. We thus begin the solution:

$$\begin{bmatrix} 0 & 0 & 2 & 0 \\ 1 & 1 & 4 & 2 \\ 1 & 2 & 1 & 3 \\ 2 & 1 & 2 & 3 \end{bmatrix} \xrightarrow{\text{Interchange columns 1 and 3}} \begin{bmatrix} 2 & 0 & 0 & 0 \\ 4 & 1 & 1 & 2 \\ 1 & 2 & 1 & 3 \\ 2 & 1 & 2 & 3 \end{bmatrix} \xrightarrow{\text{Multiply column 1 by 1/2}}$$

$$\begin{bmatrix} 1 & 0 & 0 & 0 \\ 2 & 1 & 1 & 2 \\ \frac{1}{2} & 2 & 1 & 3 \\ 1 & 1 & 2 & 3 \end{bmatrix} \xrightarrow{\substack{\text{Add } -2 \text{ times column 2}\\ \text{to column 1, add } -1 \\ \text{times column 2 to}\\ \text{column 3, add } -2 \text{ times}\\ \text{column 2 to column 4}}} \begin{bmatrix} 1 & 0 & 0 & 0 \\ 0 & 1 & 0 & 0 \\ -\frac{7}{2} & 2 & -1 & -1 \\ -1 & 1 & 1 & 1 \end{bmatrix} \xrightarrow{\substack{\text{Multiply}\\ \text{column 3}\\ \text{by } -1}}$$

$$\begin{bmatrix} 1 & 0 & 0 & 0 \\ 0 & 1 & 0 & 0 \\ -\frac{7}{2} & 2 & 1 & -1 \\ -1 & 1 & -1 & 1 \end{bmatrix} \xrightarrow{\substack{\text{Add } 7/2 \text{ times}\\ \text{column 3 to column 1,}\\ \text{add } -2 \text{ times column 3}\\ \text{to column 2, add column}\\ \text{3 to column 4}}}$$

$$\begin{bmatrix} 1 & 0 & 0 & 0 \\ 0 & 1 & 0 & 0 \\ 0 & 0 & 1 & 0 \\ -\frac{9}{2} & 3 & -1 & 0 \end{bmatrix} = A_C.$$

Since in this case column rank $A_C = 3$, we have column rank $A = 3$.

The matrix A_C obtained in our general discussion above has a special name, which we now give, along with a precise definition of A_C.

Definition 10.3.3 Let A_C be an $m \times n$ matrix. Let the columns of A_C be denoted by $C_1, \ldots, C_n$. Then A_C is called a *column-reduced echelon matrix*, provided that

(a) For each column C_j that has some nonzero entries, the first nonzero entry is 1.
(b) If 1 is the first entry in column C_j, and if this 1 is in row i_j, then 0 appears everywhere in row i_j of A_C except in the jth column; that is, the i_jth row of A_C is $(0,\ldots,0,1,0,\ldots,0)$, with 1 in the jth position.
(c) If the first nonzero entry of column C_j is in row i_j and the first nonzero entry of column C_k is in row i_k, with $j < k$, then $i_j < i_k$.

To illustrate, the matrices

$$\begin{bmatrix} 0 & 0 & 0 \\ 1 & 0 & 0 \\ 1 & 0 & 0 \\ 0 & 1 & 0 \end{bmatrix}, \begin{bmatrix} 1 & 0 & 0 \\ 0 & 1 & 0 \\ 2 & 0 & 0 \\ 0 & 1 & 0 \\ 0 & 0 & 1 \end{bmatrix},$$

$$\begin{bmatrix} 1 & 0 & 0 \\ 2 & 0 & 0 \\ 0 & 1 & 0 \\ 0 & 0 & 0 \\ 0 & 0 & 1 \end{bmatrix}, \begin{bmatrix} 0 & 0 & 0 & 0 \\ 1 & 0 & 0 & 0 \\ 0 & 1 & 0 & 0 \\ 0 & -1 & 0 & 0 \\ 0 & 0 & 1 & 0 \\ 3 & 4 & 2 & 0 \end{bmatrix}$$

are column-reduced echelon matrices, whereas the matrices

$$\begin{bmatrix} 0 & 0 & 0 \\ 1 & 0 & 0 \\ 1 & 1 & 0 \\ 0 & 0 & 1 \end{bmatrix}, \begin{bmatrix} 0 & 0 & 0 \\ 1 & 0 & 0 \\ 1 & 0 & 0 \\ 2 & 1 & 0 \\ 1 & 0 & 0 \end{bmatrix}, \begin{bmatrix} 0 & 1 & 0 & 0 \\ 1 & 0 & 0 & 0 \\ 0 & 0 & 0 & 1 \\ 0 & 0 & 1 & 0 \end{bmatrix}$$

are not column-reduced echelon matrices.

In view of Definition 10.3.3 and our earlier discussion we can now state

THEOREM 10.3.4 Let A be an $m \times n$ matrix. Then there exists a column-reduced echelon matrix A_C that is column equivalent to A, that is, that can be obtained from A by a sequence of elementary column operations. Moreover, the column rank of A is equal to the number of nonzero columns in the column-reduced echelon matrix A_C. (A_C is called the *column-reduced echelon form of A*.)

Proof The proof was given at the beginning of the section. ∎

10.3 COLUMN-REDUCED ECHELON MATRICES AND COLUMN RANK

EXERCISES

1. Find the column-reduced echelon matrix A_C for each of the following matrices A.

   (a) $\begin{bmatrix} 1 & 1 & 0 & 0 \\ 2 & 1 & 1 & 2 \\ 2 & -1 & 3 & 1 \\ 4 & 2 & 5 & 6 \end{bmatrix}.$
   (b) $\begin{bmatrix} 0 & 1 & 0 & 0 \\ 1 & 0 & 0 & 0 \\ 0 & 0 & -1 & 0 \\ 0 & 0 & 1 & -3 \end{bmatrix}.$

   (c) $\begin{bmatrix} 0 & 0 & 0 & 2 \\ 0 & 0 & 1 & 0 \\ 0 & -2 & 0 & 3 \\ 0 & 0 & -5 & 0 \\ 1 & 3 & 6 & 5 \end{bmatrix}.$
   (d) $\begin{bmatrix} 0 & -1 & \frac{3}{2} & 1 \\ 1 & 0 & -5 & -\frac{1}{2} \\ 0 & 2 & 0 & 0 \\ 2 & 3 & 0 & 1 \end{bmatrix}.$

2. In Example 10.3.1, write each of the columns

   $$\begin{bmatrix} 1 \\ 0 \\ 0 \\ 0 \end{bmatrix}, \begin{bmatrix} 0 \\ 1 \\ 0 \\ 0 \end{bmatrix}, \begin{bmatrix} 0 \\ 0 \\ 1 \\ 0 \end{bmatrix}, \begin{bmatrix} 0 \\ 0 \\ 0 \\ 1 \end{bmatrix}$$

   as a linear combination of the columns of A.

3. In the text, it is asserted that the column rank of A_C is equal to the number of nonzero columns of A_C. Prove this assertion in detail.

4. Without finding the column-reduced echelon form of each of the following matrices A, determine the column rank of A. Give a reason for your answer.

   (a) $\begin{bmatrix} 1 & 0 & 0 \\ 0 & 0 & 1 \\ 0 & 1 & 0 \end{bmatrix}.$
   (b) $\begin{bmatrix} 1 & 1 & 1 \\ 1 & 1 & 0 \\ 1 & 0 & 0 \end{bmatrix}.$

   (c) $\begin{bmatrix} 1 & 0 & 0 & -2 \\ 0 & 1 & 0 & 0 \\ 0 & 0 & 1 & 0 \\ 0 & 0 & 0 & 1 \end{bmatrix}.$
   (d) $\begin{bmatrix} -\frac{1}{2} & 0 & 0 & 1 \\ 0 & -2 & 1 & 0 \\ 0 & 0 & 3 & -4 \\ 0 & 0 & 0 & 1 \end{bmatrix}.$

5. Let A be an $m \times n$ matrix with entries in the field F. Prove that the column-reduced echelon matrix A_C for A is unique—that is, prove that if A_C and A_C' are both column-reduced echelon matrices that are column equivalent to A, then $A_C = A_C'$. (*Hint:* If W is the subspace of F_m spanned by the columns of A, then W is the subspace of F_m spanned by the columns of A_C and also spanned by the columns of A_C'. Show that if $A_C \neq A_C'$, then this is not the case. Note that linear combinations of columns chosen together from A_C and A_C' are

in W. Note also that if j is the first column in which A_C is different from A_C' and if i is the first row in which the jth columns are different, then the difference of the jth columns of A_C and A_C' cannot be generated from the columns of A_C alone, nor from the columns of A_C' alone, and so show that we have a contradiction concerning our assertions about W.)

10.4 Row Rank and Determinantal Rank; Rank

In a manner completely analogous to our earlier work, it is possible to define the row rank of a matrix A and the row-reduced echelon form of a matrix. Theorems that are the direct analogues of those obtained for columns can then be obtained for rows.

Definition 10.4.1 Let A be an $m \times n$ matrix. Each row of A can be considered an element in the vector space F^n, the vector space of n-tuples over the field F. Then the *row rank* of A is defined as the dimension of the subspace W of F^n that is spanned by the rows of A. (See Theorem 10.1.2.)

Definition 10.4.2 Let A_R be an $m \times n$ matrix. Then A_R is a *row-reduced echelon matrix* provided that when the rows of A_R are denoted $R_1, \ldots, R_m$, the following conditions are satisfied:

(a) For each row R_j that has some nonzero entries, the first nonzero entry is 1.

(b) If 1 is the first entry in row R_j, and if this 1 is in column i_j, then 0 appears everywhere in column i_j of A_R except in the jth row; that is, the i_jth column of A_R is

with 1 in the jth position.

(c) If the first nonzero entry of row R_j is in column i_j and the first nonzero entry of row R_k is in column i_k, with $j < k$, then $i_j < i_k$.

10.4 ROW RANK AND DETERMINANTAL RANK; RANK

The reader should note that the definition of the row-reduced echelon matrix A_R can be obtained from the definition of the column-reduced echelon matrix A_C, defined in Definition 10.3.3, by interchanging the words "row" and "column" everywhere in that definition.

Examples of row-reduced echelon matrices are

$$\begin{bmatrix} 1 & 0 & 0 & 0 \\ 0 & 1 & 3 & 0 \\ 0 & 0 & 0 & 1 \\ 0 & 0 & 0 & 0 \end{bmatrix}, \quad \begin{bmatrix} 0 & 1 & 0 & 0 \\ 0 & 0 & 1 & 2 \\ 0 & 0 & 0 & 0 \end{bmatrix}, \quad \text{and} \quad \begin{bmatrix} 0 & 0 & 0 & 1 \\ 0 & 0 & 0 & 0 \end{bmatrix}.$$

The analogue of Theorem 10.3.4 can now be easily stated.

THEOREM 10.4.3 Let A be an $m \times n$ matrix. Then there exists a row-reduced echelon matrix A_R that is row equivalent to A, that is, that can be obtained from A by a sequence of elementary row operations. Moreover, the row rank of A is equal to the number of nonzero rows in the row-reduced echelon matrix A_R. (A_R is called the *row-reduced echelon form of A*.)

Proof The proof of this theorem is analogous to that of Theorem 10.3.4, and the details are left to the reader. ∎

Example 10.4.4 Find the row rank of

$$A = \begin{bmatrix} 2 & 0 & 3 & 2 \\ 2 & 1 & 2 & 0 \\ 1 & 0 & 1 & 0 \\ 0 & 0 & 0 & 1 \end{bmatrix}.$$

Solution

$$\begin{bmatrix} 2 & 0 & 3 & 2 \\ 2 & 1 & 2 & 0 \\ 1 & 0 & 1 & 0 \\ 0 & 0 & 0 & 1 \end{bmatrix} \xrightarrow[\text{rows 1 and 3}]{\text{Interchange}} \begin{bmatrix} 1 & 0 & 1 & 0 \\ 2 & 1 & 2 & 0 \\ 2 & 0 & 3 & 2 \\ 0 & 0 & 0 & 1 \end{bmatrix} \xrightarrow[\substack{\text{from row 2;} \\ \text{subtract 2 times} \\ \text{row 1 from row 3}}]{\text{Subtract} \\ \text{2 times row 1}}$$

$$\begin{bmatrix} 1 & 0 & 1 & 0 \\ 0 & 1 & 0 & 0 \\ 0 & 0 & 1 & 2 \\ 0 & 0 & 0 & 1 \end{bmatrix} \xrightarrow[\text{from row 1}]{\text{Subtract} \\ \text{row 3}} \begin{bmatrix} 1 & 0 & 0 & -2 \\ 0 & 1 & 0 & 0 \\ 0 & 0 & 1 & 2 \\ 0 & 0 & 0 & 1 \end{bmatrix} \xrightarrow[\substack{\text{to row 1; add} \\ (-2) \text{ times row 4} \\ \text{to row 3}}]{\text{Add 2 times row 4}}$$

$$\begin{bmatrix} 1 & 0 & 0 & 0 \\ 0 & 1 & 0 & 0 \\ 0 & 0 & 1 & 0 \\ 0 & 0 & 0 & 1 \end{bmatrix} = A_R$$

Clearly, the matrix A_R has row rank 4, so row rank $A = 4$. Note that in Example 10.3.1 we found that column rank $A = 4$, also.

Example 10.4.5 Find the row rank of

$$A = \begin{bmatrix} 0 & 0 & 2 & 0 \\ 1 & 1 & 4 & 2 \\ 1 & 2 & 1 & 3 \\ 2 & 1 & 2 & 3 \end{bmatrix}.$$

Solution

$$\begin{bmatrix} 0 & 0 & 2 & 0 \\ 1 & 1 & 4 & 2 \\ 1 & 2 & 1 & 3 \\ 2 & 1 & 2 & 3 \end{bmatrix} \xrightarrow[\text{rows 1 and 2}]{\text{Interchange}} \begin{bmatrix} 1 & 1 & 4 & 2 \\ 0 & 0 & 2 & 0 \\ 1 & 2 & 1 & 3 \\ 2 & 1 & 2 & 3 \end{bmatrix} \xrightarrow[\substack{\text{subtract 2 times} \\ \text{row 1 from row 4}}]{\text{Subtract row 1 from row 3}}$$

$$\begin{bmatrix} 1 & 1 & 4 & 2 \\ 0 & 0 & 2 & 0 \\ 0 & 1 & -3 & 1 \\ 0 & -1 & -6 & -1 \end{bmatrix} \xrightarrow[\text{rows 2 and 3}]{\text{Interchange}} \begin{bmatrix} 1 & 1 & 4 & 2 \\ 0 & 1 & -3 & 1 \\ 0 & 0 & 2 & 0 \\ 0 & -1 & -6 & -1 \end{bmatrix} \xrightarrow[\text{add row 2 to row 4}]{\text{Subtract row 2 from row 1}}$$

$$\begin{bmatrix} 1 & 0 & 7 & 1 \\ 0 & 1 & -3 & 1 \\ 0 & 0 & 2 & 0 \\ 0 & 0 & -9 & 0 \end{bmatrix} \xrightarrow[\text{row 3 by 1/2}]{\text{Multiply}} \begin{bmatrix} 1 & 0 & 7 & 1 \\ 0 & 1 & -3 & 1 \\ 0 & 0 & 1 & 0 \\ 0 & 0 & -9 & 0 \end{bmatrix} \xrightarrow[\substack{\text{to row 2; add 9 times} \\ \text{row 3 to row 4}}]{\substack{\text{Subtract 7 times row 3} \\ \text{from row 1;} \\ \text{add 3 times row 3}}}$$

$$\begin{bmatrix} 1 & 0 & 0 & 1 \\ 0 & 1 & 0 & 1 \\ 0 & 0 & 1 & 0 \\ 0 & 0 & 0 & 0 \end{bmatrix} = A_R.$$

Since A_R has three nonzero rows, row rank $A = 3$. Note that in Example 10.3.2 we found column rank of $A = 3$, also.

The fact that row rank of A and column rank of A are equal in each of the examples is not accidental. We will prove shortly that for any $m \times n$ matrix A, row rank $A =$ column rank A. To prove this, we will introduce the notion of determinantal rank of a matrix A. What we will prove is that row rank $A =$ determinantal rank A, and that column rank $A =$ determinantal rank A.

Definition 10.4.6 Let A be an $m \times n$ matrix. The $r \times s$ matrix B is called a *submatrix* of A if B can be obtained by crossing out $m - r$ rows and $n - s$ columns of A.

To illustrate, if

$$A = \begin{bmatrix} 1 & 2 & -1 & 0 & 1 \\ 2 & 3 & 1 & 2 & 2 \\ 0 & 1 & 1 & 5 & 3 \\ 1 & 1 & 0 & 0 & 1 \end{bmatrix},$$

10.4 ROW RANK AND DETERMINANTAL RANK; RANK

then $B = \begin{bmatrix} 2 & 3 & 1 \\ 0 & 1 & 1 \end{bmatrix}$ is a submatrix of A, since we crossed out rows 1 and 4 and columns 4 and 5 of A to get B.

Definition 10.4.7 Let A be an $m \times n$ matrix. We say that A has *determinantal rank r* if

(a) there is an $r \times r$ submatrix B of A such that $\det B \neq 0$, and
(b) for each $s > r$ and for each $s \times s$ submatrix C of A, $\det C = 0$.

We write *det rank A* for the determinantal rank of A.

Example 10.4.8 Let

$$A = \begin{bmatrix} 0 & 0 & 2 & 0 \\ 1 & 1 & 4 & 2 \\ 1 & 2 & 1 & 3 \\ 2 & 1 & 2 & 3 \end{bmatrix}.$$

We have already seen that row rank A = column rank A = 3.

Using the cofactors of the first row of A, we have that

$$\det A = 0 \begin{vmatrix} 1 & 4 & 2 \\ 2 & 1 & 3 \\ 1 & 2 & 3 \end{vmatrix} - 0 \begin{vmatrix} 1 & 4 & 2 \\ 1 & 1 & 3 \\ 2 & 2 & 3 \end{vmatrix} + 2 \begin{vmatrix} 1 & 1 & 2 \\ 1 & 2 & 3 \\ 2 & 1 & 3 \end{vmatrix} - 0 \begin{vmatrix} 1 & 1 & 4 \\ 1 & 2 & 1 \\ 2 & 1 & 2 \end{vmatrix}$$

$$= 2 \left\{ 1 \begin{vmatrix} 2 & 3 \\ 1 & 3 \end{vmatrix} - 1 \begin{vmatrix} 1 & 3 \\ 2 & 3 \end{vmatrix} + 2 \begin{vmatrix} 1 & 2 \\ 2 & 1 \end{vmatrix} \right\}$$

$$= 2(3 + 3 + 2(-3)) = 0.$$

Thus it follows that determinantal rank $A < 4$.

On the other hand, the 3×3 submatrix

$$B = \begin{bmatrix} 0 & 0 & 2 \\ 1 & 1 & 4 \\ 1 & 2 & 1 \end{bmatrix},$$

obtained by crossing out the fourth row and fourth column of A, has for its determinant $2 \begin{vmatrix} 1 & 1 \\ 1 & 2 \end{vmatrix} = 2$. Thus, A has a 3×3 submatrix B such that $\det B \neq 0$, so determinantal rank $A = 3$.

Example 10.4.9 To further illustrate the notion of determinantal rank, consider the 3×4 matrix

$$A = \begin{bmatrix} 1 & 2 & 1 & 2 \\ 1 & -1 & 2 & 3 \\ 3 & 3 & 4 & 7 \end{bmatrix}.$$

We shall compute all three ranks: column rank, row rank, and determinantal rank.

Row rank: $\begin{bmatrix} 1 & 2 & 1 & 2 \\ 1 & -1 & 2 & 3 \\ 3 & 3 & 4 & 7 \end{bmatrix} \xrightarrow{\text{Why?}} \begin{bmatrix} 1 & 2 & 1 & 2 \\ 0 & -3 & 1 & 1 \\ 0 & -3 & 1 & 1 \end{bmatrix} \xrightarrow{\text{Why?}}$

$\begin{bmatrix} 1 & 2 & 1 & 2 \\ 0 & 1 & -\frac{1}{3} & -\frac{1}{3} \\ 0 & -3 & 1 & 1 \end{bmatrix} \xrightarrow{\text{Why?}} \begin{bmatrix} 1 & 0 & \frac{5}{3} & \frac{8}{3} \\ 0 & 1 & -\frac{1}{3} & -\frac{1}{3} \\ 0 & 0 & 0 & 0 \end{bmatrix} = A_R,$

so row rank = 2.

Column rank: $\begin{bmatrix} 1 & 2 & 1 & 2 \\ 1 & -1 & 2 & 3 \\ 3 & 3 & 4 & 7 \end{bmatrix} \xrightarrow{\text{Why?}} \begin{bmatrix} 1 & 0 & 0 & 0 \\ 1 & -3 & 1 & 1 \\ 3 & -3 & 1 & 1 \end{bmatrix} \xrightarrow{\text{Why?}}$

$\begin{bmatrix} 1 & 0 & 0 & 0 \\ 1 & 1 & 1 & 1 \\ 3 & 1 & 1 & 1 \end{bmatrix} \xrightarrow{\text{Why?}} \begin{bmatrix} 1 & 0 & 0 & 0 \\ 0 & 1 & 0 & 0 \\ 2 & 1 & 0 & 0 \end{bmatrix} = A_C.$

Thus column rank = 2.

Determinantal rank: Since the matrix is 3 × 4, the largest submatrices we can possibly consider are 3 × 3 submatrices. These are

$\begin{bmatrix} 1 & 2 & 1 \\ 1 & -1 & 2 \\ 3 & 3 & 4 \end{bmatrix}, \quad \begin{bmatrix} 1 & 2 & 2 \\ 1 & -1 & 3 \\ 3 & 3 & 7 \end{bmatrix}, \quad \begin{bmatrix} 1 & 1 & 2 \\ 1 & 2 & 3 \\ 3 & 4 & 7 \end{bmatrix},$

and

$\begin{bmatrix} 2 & 1 & 2 \\ -1 & 2 & 3 \\ 3 & 4 & 7 \end{bmatrix}.$

In finding the determinant of the first, we have

$\begin{vmatrix} 1 & 2 & 1 \\ 1 & -1 & 2 \\ 3 & 3 & 4 \end{vmatrix} \stackrel{\text{Why?}}{=} \begin{vmatrix} 1 & 0 & 0 \\ 1 & -3 & 1 \\ 1 & -3 & 1 \end{vmatrix} \stackrel{\text{Why?}}{=} 0.$

Similarly, it is easy to see that all of these matrices have determinant 0. Thus, determinantal rank $A < 3$. But the submatrix $\begin{bmatrix} 1 & 2 \\ 1 & -1 \end{bmatrix}$ has determinant -3, so determinantal rank $A = 2$.

In proving that row rank A = column rank A = determinantal rank A, for an $m \times n$ matrix A, we need to note one important fact. If we

10.4 ROW RANK AND DETERMINANTAL RANK; RANK

perform an elementary row (or column) operation on a matrix, the resulting matrix has determinant zero if and only if the original matrix has determinant zero. That is, the addition of a multiple of one row to another row does not change the determinant of a matrix; the interchange of two rows only changes the sign of the determinant; the multiplication of a row by a nonzero constant does not change a zero determinant to a nonzero determinant, and conversely. Finally, we note that all those operations performed on a matrix in reducing it to row (or column) echelon form are precisely the elementary row (or column) operations used in evaluating determinants. We first prove the following lemma.

LEMMA 10.4.10 Let A be an $m \times n$ matrix with entries in a field F. Let A' be obtained from A by an elementary row operation. Then determinantal rank A' = determinantal rank A.

Proof We begin by proving that if an elementary row operation is performed on A, then determinantal rank of A is not increased. Thus, let det rank $A = r$. By definition, each $(r + 1) \times (r + 1)$ submatrix B of A has determinant 0.

Suppose first that we interchange two rows of A, obtaining the matrix A'. Then any $(r + 1) \times (r + 1)$ submatrix B' of A' either coincides with an $(r + 1) \times (r + 1)$ submatrix B of A, or can be obtained by interchanging two rows of an $(r + 1) \times (r + 1)$ submatrix B of A. Thus, det $B' = \pm\det B = 0$. Hence, interchanging two rows of A does not increase the determinantal rank.

Next, suppose we multiply one row of A by a nonzero constant k. Then any $(r + 1) \times (r + 1)$ submatrix B' of the resulting matrix A' either coincides with an $(r + 1) \times (r + 1)$ submatrix B of A, or B' is obtained by multiplying a row of an $(r + 1) \times (r + 1)$ submatrix B' of A by k. In the first case,

$$\det B' = \det B = 0,$$

and in the second case,

$$\det B' = k \cdot \det B = k \cdot 0 = 0.$$

Thus, multiplying a row of A by a nonzero constant does not increase the determinantal rank of A.

Finally, suppose we add k times one row of A, say row i_t, to a different row of A, say row i_s, obtaining the matrix A'. Consider an $(r + 1) \times (r + 1)$ submatrix B' of A'. There are several possibilities:

(1) $B' = B$, where B is an $(r + 1) \times (r + 1)$ submatrix of A, whence det $B' = \det B = 0$.

(2) B' is of the form

$$\begin{bmatrix} a_{i_1j_1} & \cdots & a_{i_1j_{r+1}} \\ \vdots & & \vdots \\ a_{i_tj_1} & \cdots & a_{i_tj_{r+1}} \\ \vdots & & \vdots \\ ka_{i_tj_1} + a_{i_sj_1} & \cdots & ka_{i_tj_{r+1}} + a_{i_sj_{r+1}} \\ \vdots & & \vdots \\ a_{i_{r+1}j_1} & \cdots & a_{i_{r+1}j_{r+1}} \end{bmatrix}$$

By Theorem 9.2.4, B' has determinant the sum of the determinants

$$\begin{vmatrix} a_{i_1j_1} & \cdots & a_{i_1j_{r+1}} \\ \vdots & & \vdots \\ a_{i_tj_1} & \cdots & a_{i_tj_{r+1}} \\ \vdots & & \vdots \\ ka_{i_tj_1} & \cdots & ka_{i_tj_{r+1}} \\ \vdots & & \vdots \\ a_{i_{r+1}j_1} & \cdots & a_{i_{r+1}j_{r+1}} \end{vmatrix} \text{ and } \begin{vmatrix} a_{i_1j_1} & \cdots & a_{i_1j_{r+1}} \\ \vdots & & \vdots \\ a_{i_tj_1} & \cdots & a_{i_tj_{r+1}} \\ \vdots & & \vdots \\ a_{i_sj_1} & \cdots & a_{i_sj_{r+1}} \\ \vdots & & \vdots \\ a_{i_{r+1}j_1} & \cdots & a_{i_{r+1}j_{r+1}} \end{vmatrix}.$$

The first determinant is 0, by Theorems 9.2.3 and 9.2.1, and the second is 0 also, since it is the determinant of an $(r+1) \times (r+1)$ submatrix of the original A.

(3) B' is of the form

$$\begin{bmatrix} a_{\lambda_1j_1} & \cdots & a_{\lambda_1j_{r+1}} \\ \vdots & & \vdots \\ ka_{i_tj_1} + a_{i_sj_1} & \cdots & ka_{i_tj_{r+1}} + a_{i_sj_{r+1}} \\ \vdots & & \vdots \\ a_{\lambda_{r+1}j_1} & \cdots & a_{\lambda_{r+1}j_{r+1}} \end{bmatrix},$$

where the row i_t of the original matrix A is deleted in obtaining B'. Now det B' is the sum of the determinants of B'' and B''', where

$$B'' = \begin{bmatrix} a_{\lambda_1 j_1} & \cdots & a_{\lambda_1 j_{r+1}} \\ \cdot & & \\ \cdot & & \\ \cdot & & \\ ka_{i_t j_1} & \cdots & ka_{i_t j_{r+1}} \\ \cdot & & \\ \cdot & & \\ \cdot & & \\ a_{\lambda_{r+1} j_1} & \cdots & a_{\lambda_{r+1} j_{r+1}} \end{bmatrix}$$

and

$$B''' = \begin{bmatrix} a_{\lambda_1 j_1} & \cdots & a_{\lambda_1 j_{r+1}} \\ \cdot & & \\ \cdot & & \\ \cdot & & \\ a_{i_s j_1} & \cdots & a_{i_s j_{r+1}} \\ \cdot & & \\ \cdot & & \\ \cdot & & \\ a_{\lambda_{r+1} j_1} & \cdots & a_{\lambda_{r+1} j_{r+1}} \end{bmatrix}.$$

Since B''' is a submatrix of A, det $B''' = 0$. And since B'' is obtained either by multiplying a row of a submatrix of A by k, or by interchanging rows of A, and multiplying a row of a resulting $(r + 1) \times (r + 1)$ submatrix by k, we have det $B'' = 0$. Thus det $B' = 0 + 0 = 0$.

In all three cases—1, 2, or 3—det $B' = 0$, so det rank A is not increased by adding a multiple of one row of A to a different row of A.

We have thus proven that no matter which type of elementary row operation is performed on A, the resulting matrix does not have increased determinantal rank.

It now follows that we cannot decrease determinantal rank of A by performing an elementary row operation on A. For if A' is obtained from A by an elementary row operation, then A could also be obtained from A' by the inverse elementary row operation. If det rank A' < det rank A, then an elementary row operation on A' would increase the determinantal rank of A' to that of A, which contradicts the part of the proof that determinantal rank of a matrix cannot be increased by an elementary row operation. Thus, an elementary row operation does not decrease determinantal rank.

Since we now have that an elementary row operation performed on A neither increases nor decreases the determinantal rank, we have proved the lemma. ∎

THEOREM 10.4.11 Let A be an $m \times n$ matrix with entries in a field F. Then row rank A = det rank A = column rank A.

Proof We prove that row rank A = det rank A. The proof that column rank A = determinantal rank is similar and will be omitted. By Lemma 10.4.10, an elementary row operation performed on A leads to a matrix A' with equal determinantal rank. Thus, the row-reduced echelon form A_R of A has the same determinantal rank as A. It is easy to see, however, from the particular form of A_R, that row rank A_R = det rank A_R. Thus, we have

$$\text{row rank } A = \text{row rank } A_R = \text{determinantal rank of } A_R$$
$$= \text{determinantal rank of } A. \qquad \blacksquare$$

In view of this theorem, we make the following definition.

Definition 10.4.12 Let A be an $m \times n$ matrix. Then the *rank of* A is defined as the row rank of A (= column rank A = determinantal rank of A).

THEOREM 10.4.13 Let A be a square $n \times n$ matrix with entries in a field F. Then the following statements are equivalent.

(a) A has an inverse A^{-1}.
(b) det $A \neq 0$.
(c) rank $A = n$.

Proof In Theorem 10.1.4 we proved that A has an inverse if and only if col rank $A = n$. Thus, by Theorem 10.4.11, we have shown the equivalence of (a) and (c). By Theorems 9.5.1 and 9.5.2, A has an inverse if and only if det $A \neq 0$. ∎

It is interesting to give a second proof of Theorem 10.4.13 that does not rely upon Theorem 10.1.4.

Second proof of Theorem 10.4.13 By Theorems 9.5.1 and 9.5.2 A has an inverse if and only if det $A \neq 0$. Next, since A is $n \times n$, rank $A = n$ if and only if det $A \neq 0$, for rank $A = n$ if and only if there is an $n \times n$ submatrix of A with determinant not zero, and A is its only $n \times n$ submatrix. ∎

10.4 ROW RANK AND DETERMINANTAL RANK; RANK

THEOREM 10.4.14 Let A be an $n \times n$ matrix. The following statements are equivalent:

(a) $\det A \neq 0$.
(b) $A_R = I_n$.
(c) $A_C = I_n$.

Proof I_n is the only row-reduced echelon matrix of rank n. Thus we have that (a) and (b) are equivalent by Theorem 10.4.13. Similarly, (a) and (c) are equivalent. ∎

EXERCISES

1. Find the row-reduced echelon form A_R for each of the following matrices A:

    (a) $\begin{bmatrix} 1 & 0 & 0 & -1 \\ 0 & -1 & 2 & 1 \\ 0 & 0 & 1 & 3 \end{bmatrix}$.
    (b) $\begin{bmatrix} 1 & 2 & 3 \\ -1 & 1 & 1 \\ 0 & 0 & 0 \\ 1 & -1 & 2 \end{bmatrix}$.

    (c) $\begin{bmatrix} 0 & 0 & -1 & 2 \\ 0 & 0 & 2 & 1 \\ \frac{1}{2} & 2 & -1 & 3 \\ 1 & 1 & 0 & 1 \end{bmatrix}$.

2. Prove that the row-reduced echelon form A_R of a matrix A is unique; that is, prove that if A_R and A_R' are both row-reduced echelon matrices that are row equivalent to A, then $A_R = A_R'$.

3. Let A be an $m \times n$ matrix. Let rank $A = r$. Prove that $r \leq m$, $r \leq n$.

4. Let A be an $m \times n$ matrix of rank r, where $r < m \leq n$. Prove that A is equivalent (see Definition 10.2.7) to

$$A_N = \left.\begin{cases} \\ \\ \\ \\ \\ \\ \\ \\ \\ \end{cases}\right. \begin{matrix} r \\ \\ \\ \\ m-r \\ \\ \end{matrix} \begin{bmatrix} \overbrace{\begin{matrix} 1 & 0 & 0 & \cdots \\ 0 & 1 & 0 & \cdots \\ 0 & 0 & 1 & \cdots \\ \cdot & & \cdot & \\ \cdot & & \cdot & \\ \cdot & & \cdot & \\ 0 & 0 & 0 & \cdots \\ 0 & 0 & 0 & \\ \cdot & & & \\ \cdot & & \cdot & \\ \cdot & & & \\ 0 & 0 & 0 & \cdots \end{matrix}}^{r} & \overbrace{\begin{matrix} 0 & 0 & \cdots & 0 \\ 0 & 0 & \cdots & 0 \\ 0 & 0 & \cdots & 0 \\ & & & \\ & \cdot & & \\ & & & \\ 1 & 0 & \cdots & 0 \\ 0 & 0 & \cdots & 0 \\ & \cdot & & \\ & \cdot & & \\ & \cdot & & \\ 0 & 0 & \cdots & 0 \end{matrix}}^{n-r} \end{bmatrix}_{m \times n}$$

420   RANK, EQUIVALENCE, AND SYSTEMS OF LINEAR EQUATIONS

5. Let A be an $n \times n$ matrix of rank n. Prove that
   (a) A is row equivalent to I_n.
   (b) A is column equivalent to I_n.
6. Use the definition of determinantal rank to find the rank of each of the following:

   (a) $\begin{bmatrix} 1 & 0 & 1 \\ 2 & 0 & 1 \\ 1 & 0 & 1 \end{bmatrix}$. (b) $\begin{bmatrix} 1 & 2 & 3 \\ 1 & 0 & 1 \\ 2 & 2 & 4 \end{bmatrix}$. (c) $\begin{bmatrix} 1 & 2 & 3 & 4 \\ 3 & 1 & 4 & 2 \\ 0 & 1 & 2 & 1 \\ 1 & 2 & 0 & 1 \end{bmatrix}$.

7. Let A and B be two $m \times n$ matrices over a field F. Prove
   (a) If A and B are row equivalent, then rank A = rank B.
   (b) If A and B are column equivalent, then rank A = rank B.
   (c) If A and B are equivalent, then rank A = rank B.

10.5 Normal Forms of Matrices

In Section 10.4 we saw that for an $m \times n$ matrix A, row rank A = column rank A = det rank A. Moreover, we have seen that A is row equivalent to a row-reduced echelon matrix A_R with the same rank as A, and A is column equivalent to a column-reduced echelon matrix A_C with the same rank as A. We will now show that if we apply both row and column operations to A, then we can obtain a particularly useful matrix to which A is equivalent. This matrix will be the normal form of A.

To first illustrate the ideas involved, suppose that

$$A = \begin{bmatrix} 0 & 1 & 2 & 8 \\ 0 & 2 & 5 & 19 \\ 0 & 1 & 1 & 5 \end{bmatrix}.$$

We first reduce A to its row-reduced echelon form A_R through row operations only. Thus,

$$\begin{bmatrix} 0 & 1 & 2 & 8 \\ 0 & 2 & 5 & 19 \\ 0 & 1 & 1 & 5 \end{bmatrix} \xrightarrow[\text{to row 2}]{\text{Add } -2 \text{ times row 1}} \begin{bmatrix} 0 & 1 & 2 & 8 \\ 0 & 0 & 1 & 3 \\ 0 & 1 & 1 & 5 \end{bmatrix} \xrightarrow[\text{to row 3}]{\text{Add } -1 \text{ times row 1}}$$

$$\begin{bmatrix} 0 & 1 & 2 & 8 \\ 0 & 0 & 1 & 3 \\ 0 & 0 & -1 & -3 \end{bmatrix} \xrightarrow[\text{to row 3}]{\text{Add row 2}} \begin{bmatrix} 0 & 1 & 2 & 8 \\ 0 & 0 & 1 & 3 \\ 0 & 0 & 0 & 0 \end{bmatrix} \xrightarrow[\text{from row 1}]{\text{Subtract 2 times row 2}}$$

$$\begin{bmatrix} 0 & 1 & 0 & 2 \\ 0 & 0 & 1 & 3 \\ 0 & 0 & 0 & 0 \end{bmatrix} = A_R.$$

We now reduce A_R to its column-reduced echelon form $(A_R)_C$ through column operations only. Thus

$$\begin{bmatrix} 0 & 1 & 0 & 2 \\ 0 & 0 & 1 & 3 \\ 0 & 0 & 0 & 0 \end{bmatrix} \xrightarrow[\text{1 and 2}]{\text{Interchange columns}} \begin{bmatrix} 1 & 0 & 0 & 2 \\ 0 & 0 & 1 & 3 \\ 0 & 0 & 0 & 0 \end{bmatrix} \xrightarrow[\text{2 and 3}]{\text{Interchange columns}}$$

$$\begin{bmatrix} 1 & 0 & 0 & 2 \\ 0 & 1 & 0 & 3 \\ 0 & 0 & 0 & 0 \end{bmatrix} \xrightarrow[\text{from column 4}]{\text{Subtract 2 times column 1}} \begin{bmatrix} 1 & 0 & 0 & 0 \\ 0 & 1 & 0 & 3 \\ 0 & 0 & 0 & 0 \end{bmatrix} \xrightarrow[\text{from column 4}]{\text{Subtract 3 times column 2}}$$

$$\begin{bmatrix} 1 & 0 & 0 & 0 \\ 0 & 1 & 0 & 0 \\ 0 & 0 & 0 & 0 \end{bmatrix} = (A_R)_C = A_N.$$

Definition 10.5.1 An $m \times n$ matrix A_N is said to be a *normal form* if

(a) All entries are either 0 or 1.
(b) Any entry that is a 1 appears in a position (i,i); that is, all 1's appear on the diagonal.
(c) If k is the largest integer such that 1 appears in the (k,k) position, then 1 appears in the (i,i) position for all $i \leq k$.

As a result of this definition, we see that the various possibilities for a normal form are the following:

(1) $A_N = I_k = \begin{bmatrix} 1 & 0 & 0 & \cdots & 0 \\ 0 & 1 & 0 & \cdots & 0 \\ 0 & 0 & 1 & \cdots & 0 \\ \cdot & & & & \\ \cdot & & & & \\ \cdot & & & & \\ 0 & \cdots & & 0 & 1 \end{bmatrix}$, if $k = m = n$.

(2) $A_N = \begin{bmatrix} 1 & 0 & 0 & \cdots & 0 & 0 & \cdots & 0 \\ 0 & 1 & 0 & \cdots & 0 & 0 & \cdots & 0 \\ 0 & 0 & 1 & \cdots & 0 & 0 & \cdots & 0 \\ \cdot & & & & & & & \\ \cdot & & & & & & & \\ \cdot & & & & & & & \\ 0 & 0 & 0 & & 1 & 0 & \cdots & 0 \end{bmatrix} = \begin{bmatrix} I_k & \cdot & 0 \\ & \cdot & \end{bmatrix}_{m \times n}$

if $k = m$ and $m < n$.

(3) $\quad A_n = \begin{bmatrix} 1 & 0 & 0 & \cdots & 0 \\ 0 & 1 & 0 & \cdots & 0 \\ 0 & 0 & 1 & \cdots & 0 \\ & & \vdots & & \\ 0 & 0 & 0 & \cdots & 1 \\ 0 & 0 & 0 & \cdots & 0 \\ & & \vdots & & \\ 0 & 0 & 0 & \cdots & 0 \end{bmatrix} = \begin{bmatrix} I_k \\ \cdots \\ 0 \end{bmatrix}_{m \times n}$

if $k = n$ and $m > n$.

(4) $\quad A_n = \begin{bmatrix} 1 & 0 & 0 & \cdots & 0 & 0 & \cdots & 0 \\ 0 & 1 & 0 & \cdots & 0 & 0 & \cdots & 0 \\ 0 & 0 & 1 & \cdots & 0 & 0 & \cdots & 0 \\ & & & \vdots & & & & \\ 0 & 0 & 0 & \cdots & 1 & 0 & \cdots & 0 \\ 0 & 0 & 0 & \cdots & 0 & 0 & \cdots & 0 \\ & & & \vdots & & & & \\ 0 & 0 & 0 & \cdots & 0 & 0 & \cdots & 0 \end{bmatrix} = \begin{bmatrix} I_k & 0 \\ 0 & 0 \end{bmatrix}_{m \times n}$

if $k < m$, $k < n$.

THEOREM 10.5.2 Let A be an $m \times n$ matrix with entries in a field F. Then A is equivalent to an $m \times n$ normal form A_N with the same rank as A. That is, A can be reduced by elementary row and column operations to a normal form with the same rank as A. (The matrix A_N is called *the normal form of A*.)

Proof By elementary row operations only, reduce A to its row-reduced echelon matrix A_R. Then A_R has form

$$A_R = \begin{bmatrix} 0 & \cdots & 0 & 1 & * & \cdots & 0 & * & \cdots & & & 0 \\ 0 & \cdots & 0 & 0 & 0 & \cdots & 1 & * & \cdots & & & 0 \\ 0 & \cdots & 0 & 0 & 0 & \cdots & 0 & 0 & \cdots & 0 & 1 & * & \cdots & 0 \\ \vdots & & & & & & & & & & & \\ 0 & & & 0 & & & 0 & & & 0 & 0 & \cdots & 1 & * & \cdots & * \end{bmatrix}$$

Next, by use of elementary column operations only, reduce A_R to its column-reduced echelon matrix $(A_R)_C$. It is clear from the form of A_R that, by column operations only, all the entries * can be removed by adding appropriate multiples of columns with first nonzero entry 1 to columns containing entries *, so that the resulting matrix has only entries 0 and 1. And then by interchanges of columns, we get the matrix $(A_R)_C$, and this matrix is the desired A_N. ∎

THEOREM 10.5.3 For a given $m \times n$ matrix A of rank r, there is a unique normal form A_N of rank r.

Proof Each of the elementary operations performed in the proof of Theorem 10.5.2 results in a matrix of rank r at each step. Thus, the normal form must also have rank r. Clearly, the order of A and its rank now determine A_N uniquely, as one easily observes from the four possible types of nonzero normal forms. ∎

We remark that although in the proof of Theorem 10.5.2, we first used row operations exclusively, and then used column operations exclusively, we could mix the application of row and column operations. For by Theorem 10.5.3, A_N is unique, so any sequence of row and column operations that results in a normal form would be allowable.

If $m \times n$ matrices A and B both have the same normal form A_N, then both A and B are equivalent to A_N, and hence A and B are equivalent to each other. (See Definition 10.2.7 and Theorem 10.2.9.) As a result, we are able to prove the next theorem.

THEOREM 10.5.4 Let A and B be $m \times n$ matrices. Then A and B both have the same rank if and only if there exist elementary matrices $P_1, \ldots, P_s$ and $Q_1, \ldots, Q_t$ such that

$$B = P_s \cdots P_1 A Q_1 \cdots Q_t.$$

Proof If A and B have the same rank and same order, each has the same normal form, so A and B are equivalent by the remarks above. Since they are equivalent, B can be obtained from A through a sequence of elementary row and column operations. But each of these row and column operations can be effected by either premultiplying or postmultiplying by an appropriate elementary matrix P_i or Q_i. Thus, we conclude that if A and B have the same rank, then $B = P_s \cdots P_1 A Q_1 \cdots Q_t$.

Conversely, if $B = P_s \cdots P_1 A Q_1 \cdots Q_t$, where the P_i's and Q_i's are elementary matrices, then B is obtained from A through a sequence of elementary row and column operations, and so A and B have the same rank. ∎

COROLLARY 10.5.5 Let A and B be $m \times n$ matrices. Then A and B have the same rank if and only if A and B are equivalent.

Proof This is just a restatement of the theorem. ∎

An important corollary, and one that we will use in computations, is the following.

THEOREM 10.5.6 Let A be an $n \times n$ matrix with entries in a field F. Then A is invertible if and only if A is the product of elementary matrices.

Proof Let A be invertible. By Theorem 10.4.13, rank $A = n$. By Theorem 10.5.4, since now both A and I_n have the same order and the same rank, there exist elementary matrices $P_1, \ldots, P_s, Q_1, \ldots, Q_t$ such that $A = P_s \cdots P_1 I_n Q_1 \cdots Q_t$, and this is a product of elementary matrices.

Conversely, let $A = E_1 \cdots E_s$, where each E_i is an elementary matrix. By Theorem 10.2.5, each E_i is invertible, and so $A^{-1} = E_s^{-1} \cdots E_1^{-1}$, as is easily verified. ∎

Before showing how Theorem 10.5.6 can be used to compute inverses of matrices, when they exist, we remark that the notions of equivalence defined in Definition 8.9.4 and in Definition 10.2.7(c) are the same. For as a result of Theorem 10.5.4, the $m \times n$ matrices A and B are equivalent in the latter sense if $B = P_s \cdots P_1 A Q_1 \cdots Q_t$, where the P_i's are $m \times m$ elementary matrices and the Q_i's are elementary $n \times n$ matrices. But by Theorem 10.5.6, the product $P_s \cdots P_1$ is an invertible $m \times m$ matrix P and the product $Q_1 \cdots Q_t$ is an invertible $n \times n$ matrix Q, so that $B = PAQ$, and this was the equivalence relation defined in Section 8.9. Conversely, if we start with the equivalence relation of Section 8.9 and have $B = PAQ$, where P is an invertible $m \times m$ matrix and Q an invertible $n \times n$ matrix, then we can factor P and Q into products of elementary matrices and so get A and B equivalent under Definition 10.2.7. Thus, the two concepts of equivalence are identical.

Theorem 10.5.6 affords a very convenient computational method for finding inverses of invertible matrices. The reader will recall that so far the only systematic method of finding inverses has been through the use of cofactors.

Suppose, therefore, that A is an $n \times n$ invertible matrix. By Theorem 10.5.6, $A = E_1 \cdots E_w$, where each E_i is an elementary matrix. Thus,

$$A^{-1} = E_w^{-1} \cdots E_1^{-1} I_n \quad \text{and} \quad I_n = E_w^{-1} \cdots E_1^{-1} A.$$

Now the product $E_w^{-1} \cdots E_1^{-1}$ can be viewed as effecting a sequence of elementary row operations on I_n and A when premultiplying I_n and A. Thus, when the sequence of elementary operations corresponding to E_1^{-1}, $E_2^{-1}, \ldots, E_w^{-1}$, respectively, is performed on I_n, the matrix A^{-1} is obtained, and when this sequence of elementary operations is performed on A, the matrix I_n is obtained. This yields the following computational procedure.

Write the matrices A and I_n side-by-side. Then perform elementary row operations on A until the identity matrix is obtained. At each step, perform the same elementary row operation on the adjacent matrix (starting with I_n). When A has been transformed into I_n, I_n will have been transformed into A^{-1}.

Example 10.5.7 Find the inverse, if it exists, of

$$A = \begin{bmatrix} 1 & 2 & 3 \\ 1 & 2 & 0 \\ 0 & 1 & 2 \end{bmatrix}.$$

Solution Write

$$\begin{bmatrix} 1 & 2 & 3 \\ 1 & 2 & 0 \\ 0 & 1 & 2 \end{bmatrix} \cdot \begin{bmatrix} 1 & 0 & 0 \\ 0 & 1 & 0 \\ 0 & 0 & 1 \end{bmatrix}.$$

Subtract row 1 from row 2 in both matrices, getting

$$\begin{bmatrix} 1 & 2 & 3 \\ 0 & 0 & -3 \\ 0 & 1 & 2 \end{bmatrix} \cdot \begin{bmatrix} 1 & 0 & 0 \\ -1 & 1 & 0 \\ 0 & 0 & 1 \end{bmatrix}.$$

Add row 2 to row 1 in both matrices:

$$\begin{bmatrix} 1 & 2 & 0 \\ 0 & 0 & -3 \\ 0 & 1 & 2 \end{bmatrix} \cdot \begin{bmatrix} 0 & 1 & 0 \\ -1 & 1 & 0 \\ 0 & 0 & 1 \end{bmatrix}.$$

Divide row 2 by -3:

$$\begin{bmatrix} 1 & 2 & 0 \\ 0 & 0 & 1 \\ 0 & 1 & 2 \end{bmatrix} \cdot \begin{bmatrix} 0 & 1 & 0 \\ \frac{1}{3} & -\frac{1}{3} & 0 \\ 0 & 0 & 1 \end{bmatrix}.$$

Subtract 2 times row 2 from row 3:

$$\begin{bmatrix} 1 & 2 & 0 \\ 0 & 0 & 1 \\ 0 & 1 & 0 \end{bmatrix} \cdot \begin{bmatrix} 0 & 1 & 0 \\ \frac{1}{3} & -\frac{1}{3} & 0 \\ -\frac{2}{3} & \frac{2}{3} & 1 \end{bmatrix}.$$

Subtract 2 times row 3 from row 1:

$$\begin{bmatrix} 1 & 0 & 0 \\ 0 & 0 & 1 \\ 0 & 1 & 0 \end{bmatrix} \cdot \begin{bmatrix} \frac{4}{3} & -\frac{1}{3} & -2 \\ \frac{1}{3} & -\frac{1}{3} & 0 \\ -\frac{2}{3} & \frac{2}{3} & 1 \end{bmatrix}.$$

Interchange rows 2 and 3:

$$\begin{bmatrix} 1 & 0 & 0 \\ 0 & 1 & 0 \\ 0 & 0 & 1 \end{bmatrix} \cdot \begin{bmatrix} \frac{4}{3} & -\frac{1}{3} & -2 \\ -\frac{2}{3} & \frac{2}{3} & 1 \\ \frac{1}{3} & -\frac{1}{3} & 0 \end{bmatrix}.$$

Since the matrix on the left is now I_3, then A^{-1} is the matrix on the right; that is,

$$A^{-1} = \begin{bmatrix} \frac{4}{3} & -\frac{1}{3} & -2 \\ -\frac{2}{3} & \frac{2}{3} & 1 \\ \frac{1}{3} & -\frac{1}{3} & 0 \end{bmatrix}.$$

Direct multiplication will verify that this last matrix satisfies $AA^{-1} = A^{-1}A = I_3$.

Even though we have assumed that A has an inverse in applying this new computational method for obtaining inverses, one need not know beforehand that A^{-1} actually exists. We simply proceed according to the method described, and we will either eventually get to I_n on the left, or we will get a matrix at some stage that is clearly not invertible. In the first case, we will find A^{-1}, and in the second, we will know that A is not invertible.

Example 10.5.8 Find the inverse, if it exists, of

$$A = \begin{bmatrix} 1 & 2 & 1 & 2 \\ 0 & 1 & 0 & 1 \\ 1 & 0 & 1 & 0 \\ -1 & 2 & 3 & 4 \end{bmatrix}.$$

Write

$$\begin{bmatrix} 1 & 2 & 1 & 2 \\ 0 & 1 & 0 & 1 \\ 1 & 0 & 1 & 0 \\ -1 & 2 & 3 & 4 \end{bmatrix} \cdot \begin{bmatrix} 1 & 0 & 0 & 0 \\ 0 & 1 & 0 & 0 \\ 0 & 0 & 1 & 0 \\ 0 & 0 & 0 & 1 \end{bmatrix}.$$

Subtract 2 times row 2 from row 1, and then subtract row 3 from row 1, yielding

$$\begin{bmatrix} 0 & 0 & 0 & 0 \\ 0 & 1 & 0 & 1 \\ 1 & 0 & 1 & 0 \\ -1 & 2 & 3 & 4 \end{bmatrix} \cdot \begin{bmatrix} 1 & -2 & -1 & 0 \\ 0 & 1 & 0 & 0 \\ 0 & 0 & 1 & 0 \\ 0 & 0 & 0 & 1 \end{bmatrix}.$$

Since the matrix on the left is obtained from A by two elementary row operations, and since it has a row of all 0's, rank $A < 4$, so A is not invertible. This terminates the procedure.

Although the method discussed above for finding inverses has used only elementary row operations, there is also an analogous method using only

10.5 NORMAL FORMS OF MATRICES

elementary column operations. We merely observe that the equations

$$A^{-1} = E_w^{-1} \cdots W_1^{-1} I_n \quad \text{and} \quad I_n = E_w^{-1} \cdots E_1^{-1} A$$

are equivalent to the equations

$$A^{-1} = I_n E_w^{-1} \cdots E_1^{-1} \quad \text{and} \quad I_n = A E_w^{-1} \cdots E_1^{-1}.$$

Using these latter equations, we see that our argument and procedures concerning elementary row operations can be made to apply to elementary column operations.

Example 10.5.9 Find the inverse of

$$A = \begin{bmatrix} 1 & 2 & 3 \\ -1 & 1 & 0 \\ 2 & 0 & 0 \end{bmatrix},$$

using elementary column operations.

Solution Write

$$\begin{bmatrix} 1 & 2 & 3 \\ -1 & 1 & 0 \\ 2 & 0 & 0 \end{bmatrix} \cdot \begin{bmatrix} 1 & 0 & 0 \\ 0 & 1 & 0 \\ 0 & 0 & 1 \end{bmatrix}.$$

Add -2 times column 1 to column 2:

$$\begin{bmatrix} 1 & 0 & 3 \\ -1 & 3 & 0 \\ 2 & -4 & 0 \end{bmatrix} \cdot \begin{bmatrix} 1 & -2 & 0 \\ 0 & 1 & 0 \\ 0 & 0 & 1 \end{bmatrix}.$$

Divide column 3 by 3:

$$\begin{bmatrix} 1 & 0 & 1 \\ -1 & 3 & 0 \\ 2 & -4 & 0 \end{bmatrix} \cdot \begin{bmatrix} 1 & -2 & 0 \\ 0 & 1 & 0 \\ 0 & 0 & \frac{1}{3} \end{bmatrix}.$$

Subtract column 3 from column 1:

$$\begin{bmatrix} 0 & 0 & 1 \\ -1 & 3 & 0 \\ 2 & -4 & 0 \end{bmatrix} \cdot \begin{bmatrix} 1 & -2 & 0 \\ 0 & 1 & 0 \\ -\frac{1}{3} & 0 & \frac{1}{3} \end{bmatrix}.$$

Add 3 times column 1 to column 2:

$$\begin{bmatrix} 0 & 0 & 1 \\ -1 & 0 & 0 \\ 2 & 2 & 0 \end{bmatrix} \cdot \begin{bmatrix} 1 & 1 & 0 \\ 0 & 1 & 0 \\ -\frac{1}{3} & -1 & \frac{1}{3} \end{bmatrix}.$$

Divide column 2 by 2:

$$\begin{bmatrix} 0 & 0 & 1 \\ -1 & 0 & 0 \\ 2 & 1 & 0 \end{bmatrix} \cdot \begin{bmatrix} 1 & \frac{1}{2} & 0 \\ 0 & \frac{1}{2} & 0 \\ -\frac{1}{3} & -\frac{1}{2} & \frac{1}{3} \end{bmatrix}.$$

Add -2 times column 2 to column 1:

$$\begin{bmatrix} 0 & 0 & 1 \\ -1 & 0 & 0 \\ 0 & 1 & 0 \end{bmatrix} \cdot \begin{bmatrix} 0 & \frac{1}{2} & 0 \\ -1 & \frac{1}{2} & 0 \\ \frac{2}{3} & -\frac{1}{2} & \frac{1}{3} \end{bmatrix}.$$

Multiply column 1 by -1:

$$\begin{bmatrix} 0 & 0 & 1 \\ 1 & 0 & 0 \\ 0 & 1 & 0 \end{bmatrix} \cdot \begin{bmatrix} 0 & \frac{1}{2} & 0 \\ 1 & \frac{1}{2} & 0 \\ -\frac{2}{3} & -\frac{1}{2} & \frac{1}{3} \end{bmatrix}.$$

Interchange columns 1 and 3, and then in the resulting matrices interchange columns 2 and 3:

$$\begin{bmatrix} 1 & 0 & 0 \\ 0 & 1 & 0 \\ 0 & 0 & 1 \end{bmatrix} \cdot \begin{bmatrix} 0 & 0 & \frac{1}{2} \\ 0 & 1 & \frac{1}{2} \\ \frac{1}{3} & -\frac{2}{3} & -\frac{1}{2} \end{bmatrix}.$$

Thus,

$$A^{-1} = \begin{bmatrix} 0 & 0 & \frac{1}{2} \\ 0 & 1 & \frac{1}{2} \\ \frac{1}{3} & -\frac{2}{3} & -\frac{1}{2} \end{bmatrix},$$

and this is easily verified through multiplication.

We conclude our discussion with a warning. Our methods introduced in this section for finding inverses require one to work with elementary row operations only, or with elementary column operations only. It is not permissible to mix the use of elementary row and elementary column operations. For although A might be reduced to the identity through such a mixture, I_n will usually not be transformed into A^{-1} by the same mixture. We will leave the details of this question to the exercises.

EXERCISES

1. Reduce each of the following to its normal form.

(a) $\begin{bmatrix} 1 & 2 & 0 & -\frac{1}{2} \\ -1 & \frac{1}{2} & 3 & \frac{5}{6} \\ -4 & 0 & 1 & 7 \end{bmatrix}.$
(b) $\begin{bmatrix} -1 & -1 & -1 \\ 2 & 2 & 2 \\ 0 & 0 & 0 \end{bmatrix}.$

(c) $\begin{bmatrix} -1 & 2 & 0 & -1 \\ 0 & -1 & -1 & 0 \\ 1 & 2 & 1 & -3 \end{bmatrix}.$
(d) $\begin{bmatrix} \frac{1}{2} & \frac{3}{8} & -\frac{1}{7} \\ \frac{2}{3} & \frac{1}{6} & 0 \\ 1 & 1 & 1 \end{bmatrix}.$

2. Factor the matrix

$$A = \begin{bmatrix} 1 & 1 & 1 \\ 0 & 1 & 1 \\ 0 & 1 & 0 \end{bmatrix}$$

into the product of elementary matrices.

3. Show why, in finding the inverse of the $n \times n$ matrix A, it is not permissible to use a mixture of elementary row and elementary column operations.

4. For each of the following, use the methods of this section to find A^{-1} if it exists; if A^{-1} does not exist, find the rank of A.

(a) $\begin{bmatrix} 1 & 2 & 3 \\ -1 & 0 & 1 \\ 2 & 1 & 0 \end{bmatrix}.$

(b) $\begin{bmatrix} 1 & 2 & -1 & 0 \\ -1 & 0 & 1 & 1 \\ 3 & 1 & 0 & 1 \\ 0 & 1 & 2 & 1 \end{bmatrix}.$

(c) $\begin{bmatrix} 1 & 1 & 2 \\ 0 & 1 & 1 \\ 1 & 2 & 1 \end{bmatrix}.$

(d) $\begin{bmatrix} 1 & 1 & 2 \\ 0 & 1 & 1 \\ 1 & 0 & 0 \end{bmatrix}.$

(e) $\begin{bmatrix} 1 & 1 & 2 & 1 \\ 1 & 1 & 1 & 0 \\ 1 & 0 & 1 & 0 \\ 1 & 0 & 0 & 0 \end{bmatrix}.$

5. Find the normal form of

$$\begin{bmatrix} 0 & 0 & \cdots & 0 & d_n \\ 0 & 0 & \cdots & d_{n-1} & 0 \\ \cdot & & \cdot & & \\ \cdot & & \cdot & & \\ \cdot & & \cdot & & \\ 0 & d_2 & \cdots & 0 & 0 \\ d_1 & 0 & \cdots & 0 & 0 \end{bmatrix},$$

where $d_i \neq 0$, all i.

6. Find the normal form of

$$\begin{bmatrix} a_{11} & a_{12} & \cdots & a_{1m} & \cdots & a_{1n} \\ 0 & a_{22} & \cdots & a_{2m} & \cdots & a_{2n} \\ \cdot & & & & & \\ \cdot & & & & & \\ \cdot & & & & & \\ 0 & 0 & \cdots & a_{mm} & \cdots & a_{mn} \end{bmatrix}.$$

where $a_{jj} \neq 0$, all j.

7. Find the normal form of
$$\begin{bmatrix} n-1 & 1 & \cdots & 1 \\ 1 & n-1 & \cdots & 1 \\ \vdots & & & \\ 1 & 1 & \cdots & n-1 \end{bmatrix}_{n \times n}$$

10.6 Systems of Homogeneous Linear Equations

A system of m equations in n unknowns of the form

(∗)
$$\begin{aligned} a_{11}x_1 + \cdots + a_{1n}x_n &= 0 \\ a_{21}x_1 + \cdots + a_{2n}x_n &= 0 \\ &\vdots \\ a_{m1}x_1 + \cdots + a_{mn}x_n &= 0 \end{aligned}$$

with each a_{ij} in a field F, is called a *system of m homogeneous linear equations in n unknowns over F.*

The $x_1, \ldots, x_n$ are called the *unknowns*, and the a_{ij}'s are called the *coefficients*. The matrix $A = [a_{ij}]_{m \times n}$ is called the *coefficient matrix* of the system.

We say that a set of n elements $c_1, \ldots, c_n$ of F is a *solution* to the system (∗) if the substitution $x_1 = c_1, \ldots, x_n = c_n$ simultaneously satisfies each equation of the system.

We notice that a system (∗) always has the solution $x_1 = 0, \ldots, x_n = 0$. This is called the *trivial solution*. Any other solution is called a *nontrivial* solution.

With these definitions, we pose two questions. These will be completely answered in this section. (1) When does a system (∗) have a nontrivial solution? (2) When a nontrivial solution exists, how can all solutions be described?

To begin, first observe that (∗) is equivalent to the matrix equation

$$AX = 0, \quad \text{where } A = [a_{ij}]_{m \times n},$$

the coefficient matrix of (∗);

$$0 = \begin{bmatrix} 0 \\ \vdots \\ 0 \end{bmatrix}_{m \times 1},$$

10.6 SYSTEMS OF HOMOGENEOUS LINEAR EQUATIONS

the $m \times 1$ column matrix of all zeros; and

$$X = \begin{bmatrix} x_1 \\ \cdot \\ \cdot \\ \cdot \\ x_n \end{bmatrix}_{n \times 1} ;$$

for in $AX = 0$, the ith row of AX must be equal to the ith row of 0, and this equation is just the ith equation in (∗).

Next, we define a linear transformation $T: F_n \longrightarrow F_m$ from the vector space F_n of $n \times 1$ column matrices over F to the vector space F_m of $m \times 1$ column matrices over F by

$$T(u) = A \begin{bmatrix} u_1 \\ \cdot \\ \cdot \\ \cdot \\ u_n \end{bmatrix}, \quad \text{where } u = \begin{bmatrix} u_1 \\ \cdot \\ \cdot \\ \cdot \\ u_n \end{bmatrix}.$$

We see immediately that $x_1 = u_1, \ldots, x_n = u_n$ is a solution to (∗) if and only if

$$T(u) = A \begin{bmatrix} u_1 \\ \cdot \\ \cdot \\ u_n \end{bmatrix} = \begin{bmatrix} 0 \\ \cdot \\ \cdot \\ 0 \end{bmatrix} = 0.$$

But $T(u) = 0$ if and only if

$$\begin{bmatrix} u_1 \\ \cdot \\ \cdot \\ \cdot \\ u_n \end{bmatrix}$$

is in the null space $N(T)$ of T. (See Definition 8.2.2.) In this context, we call $N(T)$ the *solution space* of (∗).

As we know (see Theorem 8.2.1), $N(T)$ is a subspace of F_n; that is, the set $N(T)$ of all solutions to (∗) is a vector subspace of F_n. In fact, we have

THEOREM 10.6.1 If (∗) is a system of homogeneous linear equations over a field F, then

(a) The set of all solutions is a vector subspace of F_n.
(b) The dimension of the solution space is $n(T) = n - r(T)$, where $r(T)$ is the rank of T.
(c) The dimension of the solution space is $n - r$, where r is the rank of the matrix A.

(d) If $X_1, \ldots, X_{n-r}$ are linearly independent solutions to (*), then any solution is of the form

$$c_1 X_1 + \cdots + c_{n-r} X_{n-r},$$

where $c_1, \ldots, c_{n-r}$ are any scalars of F.

Proof (a) Since the set of solutions is the null space of T, (a) follows immediately from Theorem 8.2.1.
(b) This is simply Theorem 8.2.4.
(c) By Definition 10.1.1(b), the column rank of A is defined as the rank of T. Thus, $r = r(T)$, and (c) follows.
(d) Since $n(T) = n - r$, the null space $N(T)$ of T has a basis of $n - r$ vectors. Thus, $X_1, \ldots, X_{n-r}$ is a basis for $N(T)$. Thus, any element of $N(T)$ is a linear combination of $X_1, \ldots, X_{n-r}$, and any linear combination of $X_1, \ldots, X_{n-r}$ is in $N(T)$. ∎

COROLLARY 10.6.2 *If (*) is a system of homogeneous linear equations in which $m = n$, then (*) has a nontrivial solution if and only if $\det A = 0$.*

Proof If $\det A = 0$, then rank $A < n$, whence $n - r > 0$, and so the solution space has positive dimension. This implies that (*) has nontrivial solutions.

Conversely, if (*) has nontrivial solutions, then the dimension of the solution space is positive, whence $r < n$, and so $\det A = 0$. ∎

It is interesting to note that Cramer's rule, previously obtained, implies that if there are nontrivial solutions, then $\det A = 0$. For if $\det A \neq 0$, there is a unique solution, and this would necessarily be the trivial solution. The new result here is that if $m = n$ and $\det A = 0$, then there are nontrivial solutions.

We mention that the solution space $N(T)$ of (*) is sometimes called the *complete solution of* (*).

Our discussion of systems (*) of m homogeneous linear equations in n unknowns will be complete once we have shown how to obtain $n - r$ linearly independent solutions $X_1, \ldots, X_{n-r}$. Our procedure will be to use elementary row operations to reduce the coefficient matrix A to its row-reduced echelon form.

The reader should note that an elementary row operation performed on A is equivalent to an operation performed on the rows of (*), namely, the interchanging of two equations, the multiplication of one equation by a nonzero constant, or the addition of a multiple of one equation to another. If A' is any matrix obtained from A through a sequence of elementary row operations, then of course A can be obtained from A' by the reverse sequence of inverse elementary row operations. Thus, any solution to $AX = 0$ is also a solution to $A'X = 0$, and conversely. That is, $AX = 0$ and $A'X = 0$ have the same solution space.

10.6 SYSTEMS OF HOMOGENEOUS LINEAR EQUATIONS

Now, if A_R is the row-reduced echelon form of A, then $A_R X = 0$ and $AX = 0$ have the same solution space. But $A_R X = 0$ can be written in the equivalent equation form:

(**)
$$x_{j_1} + *x_{j_1+1} + \cdots + 0x_{j_2} + *x_{j_2+1} + \cdots + 0x_{j_3} + *x_{j_3+1} + \cdots = 0$$
$$x_{j_2} + *x_{j_2+1} + \cdots + 0x_{j_3} + *x_{j_3+1} + \cdots = 0$$
$$x_{j_3} + *x_{j_3+1} + \cdots = 0$$
$$\vdots$$

where * indicates a coefficient that may or may not be zero.

The unknown x_{j_i}, $i = 1, 2, \ldots, r$, has coefficient 1 in the ith equation, and coefficient 0 in all other equations. It is thus easy to solve for x_{j_i}, $i = 1, \ldots, r$, in terms of the remaining $n - r$ unknowns, which we will now denote by $y_1, \ldots, y_{n-r}$. By substituting for the $n - r$ unknowns $y_1, \ldots, y_{n-r}$ in the following manner, we will obtain $n - r$ linearly independent solutions to (**), and hence to (*). For each $j = 1, \ldots, n - r$, we simply let

$$y_j = 1, \quad y_k = 0, \quad \text{for } k \neq j.$$

This yields $n - r$ linearly independent solutions, $X_1, \ldots, X_{n-r}$, where X_j is obtained from the substitution $y_j = 1$, $y_k = 0$ for $k \neq j$. To see that $X_1, \ldots, X_{n-r}$ are linearly independent, if

$$c_1 X_1 + \cdots + c_{n-r} X_{n-r} = 0,$$

then we have for each j,

$$c_1 0 + c_2 0 + \cdots + c_j 1 + \cdots + c_{n-r} 0 = 0,$$

whence $c_j = 0$. This follows since $y_j = 0$ in all solutions X_k but the jth, in which $y_j = 1$.

To illuminate the argument above, suppose it happens that $x_1, \ldots, x_r$ can be solved in terms of $x_{r+1}, x_{r+2}, \ldots, x_n$. Letting $y_i = x_{r+i}$, $i = 1, \ldots, n - r$, we then have

$$x_1 = \sum_{i=1}^{n-r} c_{1i} y_i,$$
$$x_2 = \sum_{i=1}^{n-r} c_{2i} y_i,$$
$$\vdots$$
$$x_r = \sum_{i=1}^{n-r} c_{ri} y_i,$$

where the c_{ji}'s $\in F$. Thus, a solution X has the form

$$X = \begin{bmatrix} x_1 \\ \cdot \\ \cdot \\ \cdot \\ x_r \\ y_1 \\ \cdot \\ \cdot \\ \cdot \\ y_{n-r} \end{bmatrix} = \begin{bmatrix} \sum_{i=1}^{n-r} c_{1i} y_i \\ \cdot \\ \cdot \\ \cdot \\ \sum_{i=1}^{n-r} c_{ri} y_i \\ y_1 \\ \cdot \\ \cdot \\ \cdot \\ y_{n-r} \end{bmatrix}.$$

We then find the solution X_j, $j = 1, \ldots, n - r$, by letting $y_j = 1$ and $y_k = 0$, for $k \neq j$. Then

$$X_j = \begin{bmatrix} c_{1j} \\ \cdot \\ \cdot \\ \cdot \\ c_{rj} \\ 0 \\ \cdot \\ \cdot \\ 0 \\ 1 \\ 0 \\ \cdot \\ \cdot \\ \cdot \\ 0 \end{bmatrix} \quad (r+j)\text{th row}$$

To see that these are linearly independent solutions, if we have

$$c_1 X_1 + \cdots + c_{n-r} X_{n-r} = 0,$$

then if we look at the $(r + j)$th row, we get

$$c_1 \cdot 0 + \cdots + c_j \cdot 1 + \cdots + c_{n-r} 0 = 0,$$

whence $c_j = 0$.

Before illustrating this theory, we mention that the main idea used in this method of solution where A_R is obtained is essentially the method of *solution by elimination* taught in high schools.

Example 10.6.3 Solve the system of simultaneous linear equations

(1)
$$3x_1 + 2x_2 + x_3 = 0$$
$$x_1 - x_2 + 3x_3 = 0$$
$$2x_1 + 3x_2 - 2x_3 = 0.$$

The coefficient matrix is

$$A = \begin{bmatrix} 3 & 2 & 1 \\ 1 & -1 & 3 \\ 2 & 3 & -2 \end{bmatrix}.$$

A is easily seen to have rank 2. Thus, by Theorem 10.6.1, the solution space has dimension $3 - 2 = 1$.

To solve the system, we simply reduce A to its row-reduced echelon form A_R. Through elementary row operations, we easily find

$$A_R = \begin{bmatrix} 1 & 0 & \frac{7}{5} \\ 0 & 1 & -\frac{8}{5} \\ 0 & 0 & 0 \end{bmatrix}.$$

This implies the system (1) is equivalent to the system of equations

$$x_1 \quad\quad + \tfrac{7}{5}x_3 = 0$$
$$x_2 - \tfrac{8}{5}x_3 = 0.$$

It is thus possible to solve for x_1 and x_2 in terms of the unknown x_3. Thus, if we let $x_3 = 1$, we get

$$X_1 = \begin{bmatrix} -\frac{7}{5} \\ \frac{8}{5} \\ 1 \end{bmatrix},$$

where X_1 now spans the solution space of (1). Hence, any solution is of the form

$$c \begin{bmatrix} -\frac{7}{5} \\ \frac{8}{5} \\ 1 \end{bmatrix} = \begin{bmatrix} -\frac{7}{5}c \\ \frac{8}{5}c \\ c \end{bmatrix}.$$

For example, if $c = 5$, then we get the solution

$$\begin{bmatrix} -7 \\ 8 \\ 5 \end{bmatrix}$$

and if $c = -5$, then we get the solution

$$\begin{bmatrix} 7 \\ -8 \\ -5 \end{bmatrix}.$$

If $c = 0$, then we obtain the trivial solution

$$\begin{bmatrix} 0 \\ 0 \\ 0 \end{bmatrix}.$$

Example 10.6.4 Solve the system

(2) $$\begin{aligned} x_1 - x_2 + x_3 - x_4 &= 0 \\ x_1 + x_2 - x_3 + x_4 &= 0. \end{aligned}$$

This is a system in which the coefficient matrix

$$A = \begin{bmatrix} 1 & -1 & 1 & -1 \\ 1 & 1 & -1 & 1 \end{bmatrix}$$

has rank $r = 2$, so we know that we can find $n - r = 4 - 2 = 2$ linearly independent solutions that will span the solution space.

We easily find that $A_R = \begin{bmatrix} 1 & 0 & 0 & 0 \\ 0 & 1 & -1 & 1 \end{bmatrix}$ and so the system (2) is equivalent to the system

$$\begin{aligned} x_1 &= 0 \\ x_2 - x_3 + x_4 &= 0. \end{aligned}$$

It is possible to solve for x_1 and x_2 in terms of x_3 and x_4. Thus, an arbitrary solution of (2) is

$$\begin{bmatrix} 0 \\ x_3 - x_4 \\ x_3 \\ x_4 \end{bmatrix},$$

with x_3 and x_4 arbitrary.

By setting $x_3 = 1$ and $x_4 = 0$, we get the solution

$$X_1 = \begin{bmatrix} 0 \\ 1 \\ 1 \\ 0 \end{bmatrix},$$

and by setting $x_3 = 0$ and $x_4 = 1$, we get the solutions

$$X_2 = \begin{bmatrix} 0 \\ -1 \\ 0 \\ 1 \end{bmatrix}.$$

10.6 SYSTEMS OF HOMOGENEOUS LINEAR EQUATIONS

Thus we can write any solution also in the form

$$c_1 \begin{bmatrix} 0 \\ 1 \\ 1 \\ 0 \end{bmatrix} + c_2 \begin{bmatrix} 0 \\ -1 \\ 0 \\ 1 \end{bmatrix}.$$

Example 10.6.5 Solve the system

(3)
$$\begin{aligned} x_1 + x_2 - x_3 &= 0 \\ x_1 - x_2 + x_3 &= 0 \\ -x_1 + x_2 + x_3 &= 0. \end{aligned}$$

A quick check shows that the coefficient matrix

$$A = \begin{bmatrix} 1 & 1 & -1 \\ 1 & -1 & 1 \\ -1 & 1 & 1 \end{bmatrix}$$

has rank 3. Thus, the solution space has dimension $3 - 3 = 0$, whence the only solution is the trivial solution

$$\begin{bmatrix} 0 \\ 0 \\ 0 \end{bmatrix}.$$

Of course, we also could apply Cramer's rule in this case and reach the same conclusion.

EXERCISES

1. Find the complete solution to each of the following systems of equations.
   (a) $3x_1 - 4x_2 + \frac{1}{2}x_3 = 0$
   $x_1 - 2x_2 + 3x_3 = 0.$
   (b) $x_1 - x_2 - x_3 + x_4 = 0$
   $x_1 - x_2 + x_3 + x_4 = 0$
   $-x_1 - x_2 - x_3 + x_4 = 0.$
   (c) $x_1 + x_2 + x_3 - x_4 = 0$
   $x_1 - x_2 - x_3 + x_4 = 0.$
   (d) $3x_1 - x_2 + x_3 - 2x_4 = 0.$
2. Let $AX = 0$ represent a system of m equations in n unknowns. Let C_i be the ith column of A. Suppose rank $A = r$ and that the $m \times r$ submatrix with columns $C_{i_1}, \cdots, C_{i_r}$ also has rank r. Prove that it is possible to solve for the unknowns $x_{i_1}, \ldots, x_{i_r}$ in terms of the remaining $n - r$ unknowns.

3. Let $AX = 0$ represent a system of m equations in n unknowns. Suppose rank $A = r$. Prove there is an $r \times n$ submatrix R of A with rows $R_{i_1}, \ldots, R_{i_r}$ such that the system $RX = 0$ has the same solution space as $AX = 0$.

4. Let $AX = 0$ be a system of m homogeneous linear equations in n unknowns, and let rank $A = r$. Prove that any set of $n - r + 1$ distinct solutions of $AX = 0$ is linearly dependent.

5. Let F_n be the vector space of all $n \times 1$ column matrices with entries in the field F. Let W be a proper subspace of F_n. Prove that there exists a system of homogeneous linear equations whose solution space is W. [*Hint:* Let $C_1, \ldots, C_{n-r}$ be a basis for the space W. Since the system (∗) that we are seeking will have a solution space W of dimension $n - r$, (∗) will have to be a system of rank r. Denote an arbitrary equation in (∗) by $a_1 x_1 + \cdots + a_n x_n = 0$. We must determine r sets of values for the a_i's so that we get a system of r equations with rank r. Since each C_i will be a solution to (∗), if

$$C_i = \begin{bmatrix} c_{i1} \\ c_{i2} \\ \cdot \\ \cdot \\ \cdot \\ c_{in} \end{bmatrix},$$

then we have

(∗∗) $\qquad a_1 c_{i1} + a_2 c_{i2} + \cdots + a_n c_{in} = 0,$

for $i = 1, \ldots, n - r$. Now consider the a_i's as unknowns. Since we have the c_{ij}'s, we can solve the $n - r$ equations from (∗∗) for the a_i's. Since the C_i's are independent, the set of $n - r$ equations (∗∗) has rank $n - r$, and so we get r independent solutions to (∗∗). Each of the solutions to (∗∗) will give us one of the equations needed to determine the system (∗).]

6. Under the hypotheses of Exercise 5 show that the system of homogeneous linear equations obtained need not be unique, but show that if there are two different systems with W as solution space, then each equation in either system is a linear combination of equations in the other system.

10.7 Systems of Nonhomogeneous Equations

We conclude this chapter with the study of systems of nonhomogeneous equations.

10.7 SYSTEMS OF NONHOMOGENEOUS EQUATIONS

Let

(1)
$$\begin{aligned} a_{11}x_1 + \cdots + a_{1n}x_n &= b_1 \\ a_{21}x_1 + \cdots + a_{2n}x_n &= b_2 \\ &\vdots \\ a_{m1}x_1 + \cdots + a_{mn}x_n &= b_m, \end{aligned}$$

be a system of m linear equations in n unknowns with the a_{ij}'s and the b_i's elements of the field F. If some $b_i \neq 0$, we call the system (1) *nonhomogeneous*.

The system (1) can be written in the equivalent form

(2) $$AX = B,$$

where $A = [a_{ij}]_{m \times n}$ is called the *coefficient matrix*, and where

$$X = \begin{bmatrix} x_1 \\ \cdot \\ \cdot \\ \cdot \\ x_n \end{bmatrix} \quad \text{and} \quad B = \begin{bmatrix} b_1 \\ \cdot \\ \cdot \\ \cdot \\ b_m \end{bmatrix}.$$

In solving (1), it is useful to consider the *augmented matrix* $[A, -B]$, that is, the $m \times (n+1)$ matrix whose ith row is

$$[a_{i1} \quad a_{i2} \quad \cdots \quad a_{in} \quad -b_i].$$

We apply those elementary row operations to the augmented matrix $[A, -B]$ that are needed to reduce A to its row-reduced echelon form A_R. This leads to a matrix $[A_R, -D]$, which in turn will lead to the equivalent system of equations

(3)
$$\begin{aligned} x_{j_1} + \quad\quad\quad\quad\quad &\cdots + c_{1,r+1}x_{j_{r+1}} + \cdots + c_{1n}x_{j_n} - d_1 = 0 \\ x_{j_2} + \quad\quad &\cdots + c_{2,r+1}x_{j_{r+1}} + \cdots + c_{2n}x_{j_n} - d_2 = 0 \\ x_{j_3} + &\cdots + c_{3,r+1}x_{j_{r+1}} + \cdots + c_{3n}x_{j_n} - d_3 = 0 \\ &\quad\quad\quad\quad\quad\vdots \\ &x_{j_r} + c_{r,r+1}x_{j_{r+1}} + \cdots + c_{rn}x_{j_n} - d_r = 0 \\ &\quad\quad\quad\quad\quad\quad\quad\quad\quad\quad\quad -d_{r+1} = 0 \\ &\quad\quad\quad\quad\quad\quad\quad\quad\quad\quad\quad\quad\vdots \\ &\quad\quad\quad\quad\quad\quad\quad\quad\quad\quad\quad\quad -d_m = 0, \end{aligned}$$

where the columns of $[A_R, -D]$ are rearranged at the end if necessary to give the form of (3).

In (3), if some $d_i \neq 0$, for $i \geq r + 1$, then we have an absurdity. However, if $d_i = 0$, for all $i \geq r + 1$ then we can substitute arbitrary values for $x_{j_{r+1}}, x_{j_{r+2}}, \ldots, x_{j_n}$ and get a solution to (1). Conversely, if there is a solution to (1), then necessarily, $d_i = 0$ for $i \geq r + 1$. From this discussion, we can prove

THEOREM 10.7.1 A system of equations $AX = B$ has a solution if and only if rank A = rank $[A, -B]$.

Proof We clearly have rank $[A, -B]$ = rank $[A_R, -D]$. A solution exists if and only if $d_i = 0$, for $i \geq r + 1$, by our discussion above. But clearly $d_i = 0$, for $i \geq r + 1$, if and only if rank $[A_R, -D]$ = rank A_R (= rank A).

Thus, a solution exists if and only if rank $[A, -B]$ = rank A. ∎

If (1) has a solution, we say that the system of equations (1) is a *consistent* system of linear equations. Of course, if $B = 0$, then (1) is always consistent, since (1) has the trivial solution. Theorem 10.7.1 can then be restated as

THEOREM 10.7.1 The system of equations $AX = B$ is consistent if and only if the rank of A is equal to the rank of $[A, -B]$.

There still remains the problem of actually finding the solutions to (1). To do so, first compute rank A and rank $[A, -B]$. If these are equal, find some particular solution P to $AX = B$; that is, find an $n \times 1$ matrix P such that $AP = B$. Next, find the complete solution to the *auxiliary homogeneous system* $AX = 0$. Then every solution of $AX = B$ has the form $P + Z$, where Z is a solution to the auxiliary system $AX = 0$.

To see why this method works, suppose that P_1 and P both satisfy $AX = B$. Then $AP_1 = B$ and $AP = B$, whence $A(P_1 - P) = 0$. Thus, $P_1 - P$ is a solution to $AX = 0$. Let $P_1 - P = Z$, so $P_1 = P + Z$. Conversely, if P is a solution to $AP = B$ and if Z satisfies $AX = 0$, then $A(P + Z) = AP + AZ = B + 0 = B$, and so $P + Z$ is a solution to $AX = B$.

Definition 10.7.2 Let $AX = B$ be a system of m linear equations in n unknowns. Then the *complete solution* to $AX = B$ is the set of all $n \times 1$ matrices

$$C = \begin{bmatrix} c_1 \\ \cdot \\ \cdot \\ \cdot \\ c_n \end{bmatrix}$$

such that $AC = B$. Alternatively, the *complete solution* to (1) is the set of all n-tuples $(c_1, \ldots, c_n)$ such that for each $i = 1, \ldots, m$, when we replace

x_j by c_j, $j = 1, \ldots, n$, we have

$$a_{i1}c_1 + \cdots + a_{in}c_n = b_i.$$

THEOREM 10.7.3 Let $AX = B$ be a consistent system of m linear equations in n unknowns and let rank $A = r$. Then the complete solution is the set of all solutions of the form

$$t_1 X_1 + t_2 X_2 + \cdots + t_{n-r} X_{n-r} + P,$$

where $X_1, \ldots, X_{n-r}$ are linearly independent solutions to $AX = 0$ and P is a particular solution to $AX = B$, and $t_1, \ldots, t_{n-r}$ are elements of the field F.

Proof This follows from the discussion preceding the statement of the theorem, and from Theorem 10.6.1(d), which describes how to get the complete solution to $AX = 0$. ∎

Example 10.7.4 Solve

(*) $\quad\quad x_1 + x_2 - x_3 = 1$
$\quad\quad\quad\quad x_1 - x_2 + x_3 = 1.$

We have

$$A = \begin{bmatrix} 1 & 1 & -1 \\ 1 & -1 & 1 \end{bmatrix}, \quad B = \begin{bmatrix} 1 \\ 1 \end{bmatrix},$$

so

$$[A, -B] = \begin{bmatrix} 1 & 1 & -1 & -1 \\ 1 & -1 & 1 & -1 \end{bmatrix}.$$

Since the rank A = rank $[A, -B] = 2$, (*) is consistent by Theorem 10.7.1.

By elementary row operations, we reduce

$$[A, -B] = \begin{bmatrix} 1 & 1 & -1 & -1 \\ 1 & -1 & 1 & -1 \end{bmatrix}$$

to the form $[A_R, -D]$, as follows:

$$\begin{bmatrix} 1 & 1 & -1 & -1 \\ 1 & -1 & 1 & -1 \end{bmatrix} \longrightarrow \begin{bmatrix} 1 & 1 & -1 & -1 \\ 0 & -2 & 2 & 0 \end{bmatrix} \longrightarrow \begin{bmatrix} 1 & 1 & -1 & -1 \\ 0 & 1 & -1 & 0 \end{bmatrix}$$
$$\longrightarrow \begin{bmatrix} 1 & 0 & 0 & -1 \\ 0 & 1 & -1 & 0 \end{bmatrix}.$$

Thus, (*) is equivalent to the system

$$x_1 \quad\quad\quad - 1 = 0$$
$$\quad x_2 - x_3 \quad = 0.$$

Solving for x_1 and x_2 in terms of x_3, we have $x_1 = 1$, $x_2 = x_3$. Letting $x_3 = 1$, we see that

$$P = \begin{bmatrix} 1 \\ 1 \\ 1 \end{bmatrix}$$

is a particular solution to (∗).

To obtain the complete solution, consider the auxiliary system

(∗∗) $\quad\begin{aligned} x_1 + x_2 - x_3 &= 0 \\ x_1 - x_2 + x_3 &= 0. \end{aligned}$

The rank of (∗∗) is 2, so the solution space has dimension $3 - 2 = 1$. To solve (∗∗), we reduce $A = \begin{bmatrix} 1 & 1 & -1 \\ 1 & -1 & 1 \end{bmatrix}$ to A_R, getting

$$A_R = \begin{bmatrix} 1 & 0 & 0 \\ 0 & 1 & -1 \end{bmatrix},$$

whence $x_1 = 0$ and $x_2 - x_3 = 0$, that is, $x_1 = 0$ and $x_2 = x_3$.

Letting $x_3 = 1$, we see that

$$X_1 = \begin{bmatrix} 0 \\ 1 \\ 1 \end{bmatrix}$$

spans the solution space of (∗∗). Thus, the complete solution to (∗) is the set

$$\left\{ t \begin{bmatrix} 0 \\ 1 \\ 1 \end{bmatrix} + \begin{bmatrix} 1 \\ 1 \\ 1 \end{bmatrix} \mid t \in F \right\}.$$

EXERCISES

1. For each of the following, determine whether the system is consistent. If it is, find the complete solution.
   (a) $x_1 + x_2 + x_3 + x_4 = 1$
   $x_1 + x_2 + x_3 - x_4 = 2$
   $x_1 + x_2 - x_3 - x_4 = 3$
   $x_1 - x_2 - x_3 - x_4 = 4.$
   (b) $2x_1 + 3x_2 + 4x_3 = 20$
   $x_1 - x_2 + x_3 = 2.$
   (c) $-x_1 + x_2 - x_3 = 3$
   $2x_1 - x_2 + 3x_3 = -6.$
   (d) $x_1 + x_2 + x_3 + x_4 = 1$
   $2x_1 - x_2 + 2x_3 - x_4 = 2$
   $4x_1 + x_2 + 4x_3 + x_4 = 6.$

10.7 SYSTEMS OF NONHOMOGENEOUS EQUATIONS 443

2. Let $AX = B$ be a consistent system of m nonhomogeneous linear equations in n unknowns, with rank $A = r$. Show that there exist linearly independent solutions $P_1, \ldots, P_{n-r+1}$ of $AX = B$, but that there do not exist $n - r + 2$ distinct linearly independent solutions. (*Hint:* Let $X_1, \ldots, X_{n-r}$ be $n - r$ linearly independent solutions to $AX = 0$. Show that $P, P + X_1, \ldots, P + X_{n-r}$ are linearly independent solutions to $AX = B$.)

3. Let $AX = B$ be a consistent system of m nonhomogeneous linear equations in n unknowns. Let rank $A = r$, and let $P, P_1, \ldots, P_{n-r}$ be linearly independent solutions of $AX = B$. Show that any solution has the form

$$X = P + t_1(P_1 - P) + t_2(P_2 - P) + \cdots + t_{n-r}(P_{n-r} - P),$$

where the t_i's are arbitrary elements of the field F.

4. Let $AX = B$ be a consistent system of m nonhomogeneous linear equations in n unknowns. Let rank $A = r$, and let $P_1, \ldots, P_{n-r+1}$ be $n - r + 1$ linearly independent solutions of $AX = B$. Prove that any solution to $AX = B$ has the form

$$P = t_1 P_1 + \cdots + t_{n-r+1} P_{n-r+1}, \qquad \text{where } t_1 + \cdots + t_{n-r+1} = 1.$$

[*Hint:* First: show that $t_1 P_1 + \cdots + t_{n-r+1} P_{n-r+1}$ yields a solution if $t_1 + \cdots + t_{n-r+1} = 1$. Next, let P be a solution. Show that

$$P - P_1, \ldots, P - P_{n-r+1}$$

are $n - r + 1$ solutions to the auxiliary system $AX = 0$, so that $P - P_1, \ldots, P - P_{n-r+1}$ are linearly dependent. Show that this implies

$$P = \sum_{i=1}^{n-r+1} c_i P_i,$$

and that necessarily

$$\sum_{i=1}^{n-r+1} c_i = 1.$$
]

11 Characteristic Vectors, Orthogonality, and Diagonalization

11.1 Characteristic Roots and Characteristic Vectors

A linear operator T on a finite-dimensional vector space V over a field K that has a matrix representation of the form

$$D = \begin{bmatrix} d_1 & 0 & \cdots & 0 \\ 0 & d_2 & \cdots & 0 \\ \vdots & & & \vdots \\ 0 & 0 & \cdots & d_n \end{bmatrix}$$

has a very simple interpretation: V has a basis ℬ such that T stretches (or compresses) each basis vector—that is, $T(v_j) = d_j v_j$ for each vector v_j in the basis ℬ. Such a linear operator is called a *diagonalizable operator*, and a matrix such as D is called a *diagonal matrix*.

As can be easily observed, if T is a diagonalizable operator that has matrix D relative to basis ℬ, then both the range space and the null space of T are easily determined. Indeed, the range space of T has as basis those vectors v_j in ℬ for which $d_j \neq 0$, whereas the null space for T has as basis those vectors v_j for which $d_j = 0$.

A nonzero vector v such that $T(v) = cv$, for some element c in K, obviously is of special importance, as is the field element c.

Definition 11.1.1 Let T be a linear operator on the vector space V over the field K. A nonzero vector v is called a *characteristic vector of T* if

11.1 CHARACTERISTIC ROOTS AND CHARACTERISTIC VECTORS

there is an element c in K such that $T(v) = cv$. Such a scalar c is called a *characteristic value*[1] *of T associated with v.*

Let A be an $n \times n$ matrix with entries in the field K. An $n \times 1$ column matrix X, $X \neq 0$, is called a *characteristic vector of A* if there is an element $c \in K$ such that $AX = cX$. Such a scalar c is called a *characteristic value of A associated with X.*

In these terms, a linear operator T on V is diagonalizable if there is a basis $\mathcal{B}$ for V in which all the basis vectors are characteristic vectors of T. Furthermore, the matrix $[T]_\mathcal{B}$ will have the characteristic values of T for its diagonal entries.

The reader should note that while the zero vector has been ruled out as a characteristic vector, a characteristic value may very well be 0, the zero element of the field K.

The connection between characteristic vectors and values of linear operators and characteristic vectors and values of matrices is straightforward. Let T be an operator on the n-dimensional vector space V over the field K. Let v be a characteristic vector for T corresponding to the characteristic value c. Then we have $T(v) = cv$.

If $\mathcal{B}$ is a basis for V, then equation $T(v) = cv$ translates to the matrix equation

$$[T]_\mathcal{B}[v]_\mathcal{B} = c[v]_\mathcal{B},$$

where $[T]_\mathcal{B}$ is the matrix of T with respect to the basis $\mathcal{B}$, and $[v]_\mathcal{B}$ is the coordinate matrix of v with respect to $\mathcal{B}$. Of course, $v = 0$ if and only if $[v]_\mathcal{B} = 0$, so $[v]_\mathcal{B}$ is a characteristic vector of the matrix $[T]_\mathcal{B}$ if and only if v is a characteristic vector of the operator T.

It is theoretically quite simple to determine characteristic values and characteristic vectors. We describe the procedure here.

Let T be a linear operator on the n-dimensional vector space V over K. Let v be a characteristic vector associated with the characteristic value c. Then

$$T(v) = cv = (cI)v,$$

where I is the identity operator on V. This is equivalent to

$$(T - cI)(v) = 0.$$

---

[1] Other terms are often used in place of characteristic vector and characteristic value. Most prominent are the terms "eigenvector" and "eigenvalue," respectively. These linguistically hybrid terms are derived from the German *eigenwerte*. We also note that in place of characteristic value or eigenvalue, we often use the terms characteristic root or eigenroot. The "root" appears since these values can be determined as roots of polynomials, as will be seen shortly.

If $\mathcal{B}$ is some basis for V, then this relationship translates into matrix notation as
$$[T - cI]_\mathcal{B}[v]_\mathcal{B} = 0.$$
Since $v \neq 0$, also $[v]_\mathcal{B} \neq 0$. Thus, the matrix $[T - cI]_\mathcal{B}$ is singular, so that $\det [T - cI]_\mathcal{B} = 0$.

If we replace the value c in $\det [T - cI]_\mathcal{B} = 0$ by the indeterminate x, then we have that c satisfies the equation $\det [T - xI]_\mathcal{B} = 0$. But $\det [T - xI]_\mathcal{B}$ is just an nth-degree polynomial in x with coefficients in the field K. We note that regardless of which characteristic vector v and characteristic value c we started with, we would have reached the conclusion that c satisfies the polynomial equation $\det [T - xI]_\mathcal{B} = 0$.

Definition 11.1.2 If A is an $n \times n$ matrix with entries in a field K and if I_n is the $n \times n$ identity matrix, then the nth-degree polynomial $\det (A - xI_n)$ is called the *characteristic polynomial* of the matrix A.

If T is a linear operator on the n-dimensional vector space V over the field K, then the *characteristic polynomial* of T is the nth-degree polynomial $\det [T - xI]_\mathcal{B}$, where $\mathcal{B}$ is some basis for V.

It is important to note that the characteristic polynomial of T does not depend on the choice of the particular basis $\mathcal{B}$ used in Definition 11.1.2. To see this, let $\mathcal{B}'$ be a second basis. Then T has matrix $B^{-1}AB$ relative to $\mathcal{B}'$ for appropriate matrix B, if T has matrix A with respect to the basis $\mathcal{B}$. Then

$$\begin{aligned}
\det [T - xI]_{\mathcal{B}'} &= \det (B^{-1}AB - xI_n) \\
&= \det (B^{-1}AB - xB^{-1}I_nB) \\
&= \det (B^{-1}(A - xI_n)B) \\
&= \det B^{-1} \cdot \det (A - xI_n) \cdot \det B \\
&= \det B^{-1} \cdot \det B \cdot \det (A - xI_n) \\
&= \det (A - xI_n) = \det [T - xI]_\mathcal{B}.
\end{aligned}$$

Thus, the characteristic polynomial of T does not depend on the choice of the basis for V, and so Definition 11.1.2 is meaningful.

Example 11.1.3 Let T be the linear operator on $\mathbf{R}^2$ defined by
$$T(1,0) = (1,3) \quad \text{and} \quad T(0,1) = (2,2).$$
In terms of the canonical basis $\mathcal{B} = \{\epsilon_1 = (1,0), \epsilon_2 = (0,1)\}$ we have
$$[T]_\mathcal{B} = A = \begin{bmatrix} 1 & 2 \\ 3 & 2 \end{bmatrix}.$$

11.1 CHARACTERISTIC ROOTS AND CHARACTERISTIC VECTORS

We calculate the characteristic polynomial of T using the matrix A. Thus

$$[T - xI]_\mathfrak{B} = A - xI_2 = \begin{bmatrix} 1 & 2 \\ 3 & 2 \end{bmatrix} - x \begin{bmatrix} 1 & 0 \\ 0 & 1 \end{bmatrix}$$

$$= \begin{bmatrix} 1 & 2 \\ 3 & 2 \end{bmatrix} - \begin{bmatrix} x & 0 \\ 0 & x \end{bmatrix} = \begin{bmatrix} 1-x & 2 \\ 3 & 2-x \end{bmatrix},$$

and so

$$\det(A - xI_2) = (1-x)(2-x) - 6 = x^2 - 3x - 4.$$

Notice that $x^2 - 3x - 4 = (x - 4)(x + 1)$, and therefore 4 and -1 are the roots of this polynomial. Thus, the characteristic values of T are 4 and -1.

We may now sum up the previous discussion.

THEOREM 11.1.4 Let T be a linear operator on the n-dimensional vector space V over the field K. Then if v is a characteristic vector of T corresponding to the characteristic value c of T, then c is a root of the characteristic polynomial of T. Conversely, if c is a root of the characteristic polynomial of T, then c is a characteristic value of T.

Proof We have already seen that if v is a characteristic vector of T, the corresponding characteristic value c is a root of the characteristic polynomial. Furthermore, all such characteristic values are roots of the characteristic polynomial.

Conversely, let c be a root of the characteristic polynomial of T. Then there is a basis $\mathfrak{B}$ of V for which $\det[T - cI]_\mathfrak{B} = 0$. This implies that the linear operator $T - cI$ is a singular operator, so that there is a nonzero vector v in V such that $(T - cI)v = 0$, that is, $T(v) = cv$. ∎

From this theorem it is clear why we also refer to characteristic values of T as characteristic roots. We shall use these terms interchangeably in the sequel.

Notice that in Example 11.1.3 the characteristic roots of T are 4 and -1. Thus, we know there exist characteristic vectors associated with each of these characteristic values. They can be easily computed, as we now show.

If $v = (a,b)$ is a characteristic vector of T associated with characteristic value 4, then v is in the null space of $T - 4I$. In terms of the canonical basis for $\mathbf{R}^2$, this becomes

$$(A - 4I_2) \begin{bmatrix} a \\ b \end{bmatrix} = \begin{bmatrix} 0 \\ 0 \end{bmatrix},$$

whence we have

$$\left(\begin{bmatrix} 1 & 2 \\ 3 & 2 \end{bmatrix} - \begin{bmatrix} 4 & 0 \\ 0 & 4 \end{bmatrix}\right)\begin{bmatrix} a \\ b \end{bmatrix} = \begin{bmatrix} -3 & 2 \\ 3 & -2 \end{bmatrix}\begin{bmatrix} a \\ b \end{bmatrix} = \begin{bmatrix} 0 \\ 0 \end{bmatrix},$$

that is, we have the two simultaneous equations

$$-3a + 2b = 0$$
$$3a - 2b = 0.$$

We know that the coefficient matrix $\begin{bmatrix} -3 & 2 \\ 3 & -2 \end{bmatrix}$ has determinant 0, for det $(A - 4I_2) = 0$, since 4 is a characteristic value of T. Also, $\begin{bmatrix} -3 & 2 \\ 3 & -2 \end{bmatrix}$ has rank 1, so the solution space of this system of equations is of dimension 1. Thus, the solution space is spanned by a single vector, say we choose (2,3). Then (2,3) is a characteristic vector associated with the characteristic value 4, as is any nonzero multiple of (2,3).

We leave to the exercises the computation of a characteristic vector of T corresponding to the characteristic value -1.

In the example above, any characteristic vector corresponding to the characteristic value 4 and any characteristic vector corresponding to the characteristic value -1 are linearly independent, as the reader can easily verify. Actually, this is a special case of the following theorem.

THEOREM 11.1.5 Let T be a linear operator on the finite-dimensional vector space V over the field K. Let $c_1, \ldots, c_r$ be r distinct characteristic values of T. Let v_i be a characteristic vector associated with c_i, for $i = 1, \ldots, r$. Then the set of vectors $\{v_1, \ldots, v_r\}$ is a linearly independent set of vectors.

Proof We prove the theorem by induction on r.

If $r = 1$, then we only consider a set $\{v_1\}$ with a single nonzero vector v_1. Such a set is always linearly independent.

Now suppose that the theorem is true for the case $r - 1$. We must show that the set $\{v_1, \ldots, v_r\}$ is a linearly independent set. Thus, suppose there are scalars $d_1, \ldots, d_r$, not all zero, such that

(*) $$d_1v_1 + \cdots + d_rv_r = 0;$$

that is, we assume that $\{v_1, \ldots, v_r\}$ is not a linearly independent set. If $d_r = 0$, then we have a dependence relation among the vectors $v_1, \ldots, v_{r-1}$, and this would contradict our assumption that the theorem is true for $r - 1$. Thus, $d_r \neq 0$.

11.1 CHARACTERISTIC ROOTS AND CHARACTERISTIC VECTORS 449

We apply the operator T to (*), and, using the linearity, we obtain
$$d_1 T(v_1) + \cdots + d_r T(v_r) = 0;$$
that is, we have

(**) $$d_1 c_1 v_1 + \cdots + d_r c_r v_r = 0$$

since each v_i is a characteristic vector associated with c_i.

We multiply (*) by c_r to obtain

(***) $$c_r d_1 v_1 + \cdots + c_r d_r v_r = 0,$$

and then we subtract (***) from (**), obtaining

(****) $$d_1(c_1 - c_r)v_1 + \cdots + d_{r-1}(c_{r-1} - c_r)v_{r-1} = 0.$$

By our induction hypothesis, this implies that each coefficient of (****) must be 0, so that we have
$$d_i(c_i - c_r) = 0, \quad \text{for } i = 1, \ldots, r-1.$$
Since $c_i \neq c_r$, for $i = 1, \ldots, r-1$, we now have $d_i = 0$, for $i = 1, \ldots, r-1$. But this now implies that (*) becomes $d_r v_r = 0$, and this can only happen if either $d_r = 0$ or $v_r = 0$. Since we have assumed that neither of these is the case, we have a contradiction, and so we finally conclude that $\{v_1, \ldots, v_r\}$ is a linearly independent set of vectors. ∎

From this theorem, we can immediately prove a corollary, which returns us to our consideration of diagonalizable operators.

COROLLARY 11.1.6 Let T be a linear operator on the n-dimensional vector space V over the field K. If the characteristic polynomial of T has n distinct roots, say $c_1, \ldots, c_n$, in K, then T is diagonalizable. Moreover, there is a basis $\mathfrak{B}$ for V such that

$$[T]_\mathfrak{B} = \begin{bmatrix} c_1 & 0 & \cdots & 0 \\ 0 & c_2 & \cdots & 0 \\ \vdots & & & \vdots \\ 0 & 0 & \cdots & c_n \end{bmatrix}.$$

The basis $\mathfrak{B}$ consists of n characteristic vectors of T, where one characteristic vector is chosen for each characteristic value of T.

Proof Let $c_1, \ldots, c_n$ be the n distinct characteristic values of T given in the hypotheses. For each c_i, let v_i be an associated characteristic vector. From Theorem 11.1.5, the set $\{v_1, \ldots, v_n\}$ is linearly independent. Since V is n-dimensional, this set of vectors is a basis for V. Clearly, since

$T(v_i) = c_i v_i$, for each $i = 1, \ldots, n$, we have

$$[T]_{\mathcal{B}} = \begin{bmatrix} c_1 & 0 & \cdots & 0 \\ 0 & c_2 & \cdots & 0 \\ \vdots & & \ddots & \vdots \\ 0 & 0 & \cdots & c_n \end{bmatrix}.$$ ∎

Returning to Example 11.1.3, we recall that T has two characteristic values, 4 and -1. Thus, if we choose a characteristic vector for each of these characteristic values, say v_1 and v_2, the resulting set of two vectors will give us a basis $\mathcal{B} = \{v_1, v_2\}$ for $\mathbf{R}^2$, and the matrix of T associated with that basis will be $\begin{bmatrix} 4 & 0 \\ 0 & -1 \end{bmatrix}$.

Although Theorem 11.1.5 and Corollary 11.1.6 were phrased in terms of linear operators, Corollary 11.1.6 can be recast in terms of matrix terminology as follows.

THEOREM 11.1.7 Let A be an $n \times n$ matrix with entries in a field K. Suppose that the characteristic polynomial $p(x)$ of A has n distinct roots in K. Then A is similar to an $n \times n$ diagonal matrix, where the entries on the diagonal are the roots of $p(x)$ in some order. That is, there exists an $n \times n$ nonsingular matrix B with entries in K such that

$$B^{-1}AB = C = \begin{bmatrix} c_1 & 0 & & 0 \\ 0 & c_2 & & 0 \\ \vdots & & \ddots & \vdots \\ 0 & 0 & & c_n \end{bmatrix},$$

where $c_1, \ldots, c_n$ are the roots of $p(x)$.

Proof Let $\mathcal{B}'$ be a basis for the vector space K^n. Then the matrix A defines a unique linear operator T on K^n, where $A = [T]_{\mathcal{B}'}$. Now the characteristic polynomial of T is the same as the characteristic polynomial of A, so that T has as characteristic values the n distinct roots of $p(x)$. Now we simply apply Corollary 11.1.6 and get a basis $\mathcal{B}$ for K^n for which

$$[T]_{\mathcal{B}} = \begin{bmatrix} c_1 & 0 & \cdots & 0 \\ 0 & c_2 & \cdots & 0 \\ \vdots & & \ddots & \vdots \\ 0 & 0 & \cdots & c_n \end{bmatrix}.$$

11.1 CHARACTERISTIC ROOTS AND CHARACTERISTIC VECTORS

That a matrix B exists such that

$$B^{-1}AB = \begin{bmatrix} c_1 & 0 & \cdots & 0 \\ 0 & c_2 & \cdots & 0 \\ \vdots & & & \vdots \\ 0 & 0 & \cdots & c_n \end{bmatrix}.$$

follows from Theorem 8.9.5. ∎

Example 11.1.8 Let A be the matrix $\begin{bmatrix} 1 & 2 \\ 3 & 2 \end{bmatrix}$ of Example 11.1.3. We have already seen that the characteristic roots of A are 4 and -1, so by Theorem 11.1.7, A is similar to $\begin{bmatrix} 4 & 0 \\ 0 & -1 \end{bmatrix}$. Thus, there is a matrix B such that

$$B^{-1}AB = \begin{bmatrix} 4 & 0 \\ 0 & -1 \end{bmatrix}.$$

Since B represents a change of basis from the canonical basis $\{(1,0),(0,1)\}$ to a basis consisting of characteristic vectors of A, we compute B by finding characteristic vectors of A. Two independent characteristic vectors are $v_1 = (2,3)$ and $v_2 = (1,-1)$, corresponding to 4 and -1, respectively. Thus, since $v_1 = 2(1,0) + 3(0,1)$ and $v_2 = 1(1,0) - 1(0,1)$, it follows that

$$B = \begin{bmatrix} 2 & 1 \\ 3 & -1 \end{bmatrix}.$$

It is easy to see that $B^{-1} = \begin{bmatrix} \frac{1}{5} & \frac{1}{5} \\ \frac{3}{5} & -\frac{2}{5} \end{bmatrix}$, and by direct calculation we find that

$$B^{-1}AB = \begin{bmatrix} \frac{1}{5} & \frac{1}{5} \\ \frac{3}{5} & -\frac{2}{5} \end{bmatrix} \begin{bmatrix} 1 & 2 \\ 3 & 2 \end{bmatrix} \begin{bmatrix} 2 & 1 \\ 3 & -1 \end{bmatrix} = \begin{bmatrix} 4 & 0 \\ 0 & -1 \end{bmatrix}.$$

Example 11.1.9 Let $A = \begin{bmatrix} 0 & 1 \\ -1 & 0 \end{bmatrix}$. Then the characteristic polynomial of A is

$$\det(A - xI_2) = \det\left(\begin{bmatrix} 0 & 1 \\ -1 & 0 \end{bmatrix} - \begin{bmatrix} x & 0 \\ 0 & x \end{bmatrix}\right)$$

$$= \det\begin{bmatrix} -x & 1 \\ -1 & -x \end{bmatrix} = x^2 + 1.$$

Clearly, this polynomial has no real roots, and so if A is regarded as a matrix with entries in the real field **R**, then A is not similar to a diagonal matrix over **R**. On the other hand, if we regard A as having entries over the complex number field **C**, then $x^2 + 1$ has two distinct roots i and $-i$, so that A is similar to $\begin{bmatrix} i & 0 \\ 0 & -i \end{bmatrix}$ by Theorem 11.1.7; that is, there is a matrix B *with complex coefficients* such that

$$B^{-1}AB = \begin{bmatrix} i & 0 \\ 0 & -i \end{bmatrix}.$$

This example shows that for a given matrix A to be similar to a diagonal matrix C (or for that matter, to some other given matrix), there is a strong dependency on the underlying field K. That is, two matrices A and C that are not similar over one field L, may be similar over a field K that has L as a subfield.

We conclude this section by asking whether the condition of Corollary 11.1.6, which is sufficient for the diagonalizability of a linear operator, is also necessary for diagonalizability. (Equivalently, we may ask whether the condition of Theorem 11.1.7, which is sufficient for a matrix to be similar to a diagonal matrix, is also necessary for a matrix to be similar to a diagonal matrix.) That is, we ask: If a linear operator T on an n-dimensional vector space over K is diagonalizable, does T have n distinct characteristic values? (Equivalently, if the matrix A is similar to a diagonal matrix, must all the characteristic values be distinct?)

The answer to this question is clearly No. For let T be the identity operator on $\mathbf{R}^2$. T has characteristic polynomial $(1 - x)^2$, so that the only characteristic root is 1, with multiplicity 2, and T obviously is diagonalizable, since any matrix for T is the identity matrix, which is already diagonal.

Equivalently, our answer could be given in terms of matrices. Simply, the identity matrix $I_2 = \begin{bmatrix} 1 & 0 \\ 0 & 1 \end{bmatrix}$ is already diagonal, and it does not have distinct characteristic values.

A further question along these lines can be answered. Let $A = \begin{bmatrix} 1 & 0 \\ 1 & 1 \end{bmatrix}$. It is easy to see that A has as characteristic values only 1, with multiplicity 2. Thus A and I_2 have exactly the same characteristic values. But it can be shown that A is not similar to any diagonal matrix (see Exercise 6), so that knowing all the characteristic values of two matrices is not enough to let us conclude whether they are similar.

We will return to the question of diagonalization of matrices toward the end of the chapter.

EXERCISES

1. Let v be a characteristic vector of the linear operator T associated with the characteristic value c. Show that for any scalar $d \neq 0$, the vector dv is also a characteristic vector of T associated with c.

2. Let T be a linear operator on the vector space V. Let c be a characteristic value of T, and let $W = \{v \in V \mid v$ is a characteristic vector of T associated with $c\} \cup \{0\}$. Show that W is a subspace of V.
3. (a) Let A be an $n \times n$ matrix with entries in the field K. Show that det $(A - xI_n)$, the characteristic polynomial of A, is indeed an nth-degree polynomial in x.
   (b) Show that the constant term in the characteristic polynomial of A is simply det A.
   (c) Show that the coefficient of x^{n-1} in the characteristic polynomial of A is either $+$ or $-$ the sum of the diagonal entries of A.
4. Compute the characteristic polynomial of the following matrices:

   (a) $\begin{bmatrix} 1 & 1 \\ 1 & 1 \end{bmatrix}$.   (b) $\begin{bmatrix} 1 & 0 \\ 0 & 2 \end{bmatrix}$.   (c) $\begin{bmatrix} 0 & 2 & -1 \\ 1 & 1 & 4 \\ 1 & 0 & 0 \end{bmatrix}$.

   In each case, find the characteristic roots, and for each characteristic root find a maximal set of independent characteristic vectors.
5. Find the characteristic vector associated with the characteristic value -1, for the operator T of Example 11.1.3.
6. Let $A = \begin{bmatrix} 1 & 0 \\ 1 & 1 \end{bmatrix}$ be a matrix with real coefficients. Show that A is not diagonalizable. (*Hint:* If A is diagonalizable, A has two independent characteristic vectors that span the vector space of 2×1 column vectors. Show that two independent characteristic vectors do not exist for A.)
7. Let $\{v_1, \ldots, v_n\}$ be a basis for the n-dimensional vector space V over K, consisting of n characteristic vectors of the operator T. Suppose $T(v_i) = c_i v_i$, $i = 1, 2, \ldots, n$. Prove that for each ordering $\lambda_1, \ldots, \lambda_n$ of the integers $1, 2, \ldots, n$, it is possible to find a basis $\mathcal{B}$ for V such that

$$[T]_{\mathcal{B}} = \begin{bmatrix} c_{\lambda_1} & & 0 \\ & \ddots & \\ 0 & & c_{\lambda_n} \end{bmatrix}.$$

8. Show that $A = \begin{bmatrix} 1 & 3 \\ 4 & 3 \end{bmatrix}$ is similar to $C = \begin{bmatrix} 2 & 0 \\ 0 & -4 \end{bmatrix}$ and find a matrix B such that $B^{-1}AB = C$.
9. Let B be the matrix of Theorem 11.1.7. Show that the columns of B are n distinct characteristic vectors associated with the characteristic values $c_1, \ldots, c_n$.

11.2 Inner Products

In our discussion of vector spaces and linear transformations, we have attempted to provide geometrical insight whenever possible. However, many geometrical properties of Euclidean spaces, such as distance, angle, and perpendicularity, have usually been omitted from our discussion. In this section we will begin our investigation of some of these notions in a fairly general vector-space setting. For simplification, we shall restrict our attention to vector spaces over $\mathbf{R}$ or over $\mathbf{C}$.

Suppose, for example, that $v = (3,1,-4)$ is a vector in $\mathbf{R}^3$, Euclidean 3-space. Then by the "length" of the vector v is usually meant the length of the line segment between the points $(0,0,0)$ and $(3,1,-4)$, considered as points in 3-space. By the Pythagorean theorem, the length of v, which we will denote $\|v\|$, is

$$\|v\| = \sqrt{3^2 + 1^2 + (-4)^2} = \sqrt{26}.$$

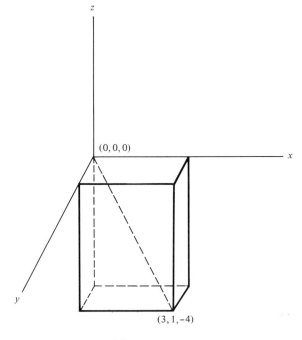

Figure 11.1

In a similar manner, if $v = (a,b,c)$ is an arbitrary vector in $\mathbf{R}^3$, then again, by the Pythagorean theorem and considerations in the last paragraph, it makes sense to say that the length of (a,b,c) is $\sqrt{a^2 + b^2 + c^2}$.

11.2 INNER PRODUCTS

In the case of $\mathbf{C}^3$, however, the same definition of length is inappropriate. For we would like to require that a nonzero vector v have positive length. For example, if $v = (i,1,0)$, then $i^2 + 1^2 + 0^2 = -1 + 1 + 0 = 0$, whence $(i,1,0)$ would have zero length. We avoid this problem by saying that $(i,1,0)$ has length

$$\sqrt{i \cdot \bar{i} + 1 \cdot \bar{1} + 0 \cdot \bar{0}} = \sqrt{i(-i) + 1} = \sqrt{1 + 1} = \sqrt{2}.$$

With this background, we make

Definition 11.2.1 Let $v = (x_1, x_2, \ldots, x_n)$ be a vector in $\mathbf{R}^n$. Then the *length* of v is $\|v\| = \sqrt{x_1^2 + \cdots + x_n^2}$.

Let $v = (z_1, \ldots, z_n)$ be a vector in $\mathbf{C}^n$. Then the *length* of v is

$$\|v\| = \sqrt{z_1 \bar{z}_1 + z_2 \bar{z}_2 + \cdots + z_n \bar{z}_n},$$

where $\bar{z}_i$ is the complex conjugate of z_i.

In either $\mathbf{R}^n$ or $\mathbf{C}^n$ if v and u are two vectors, then the *distance* between u and v is $\|v - u\|$.

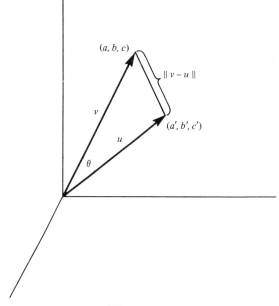

Figure 11.2

Next, we consider the concept of angle. Suppose that v and u are vectors in $\mathbf{R}^3$, with $v = (a,b,c)$ and $u = (a',b',c')$. If θ is the smallest vertex angle formed by v and u, then it follows from the law of cosines (see Figure 11.2)

that
$$\cos \theta = \frac{\|u\|^2 + \|v\|^2 - \|v - u\|^2}{2\|u\|\,\|v\|}.$$

Using the definition of $\|u\|$, $\|v\|$, and $\|v - u\|$, we get

$\cos \theta =$
$$\frac{a^2 + b^2 + c^2 + (a')^2 + (b')^2 + (c')^2 - (a - a')^2 - (b - b')^2 - (c - c')^2}{2\|u\|\,\|v\|},$$

$$\cos \theta = \frac{2aa' + 2bb' + 2cc'}{2\|u\|\,\|v\|} = \frac{aa' + bb' + cc'}{\sqrt{a^2 + b^2 + c^2}\,\sqrt{(a')^2 + (b')^2 + (c')^2}}.$$

Hence, we see that $\cos \theta$ can be defined in terms of the quantities $aa' + bb' + cc'$, $\|u\|$, and $\|v\|$.

Definition 11.2.2 Let $v = (x_1, \ldots, x_n)$ and $u = (y_1, \ldots, y_n)$ be two vectors in $\mathbf{R}^n$. Then the *inner product* of v and u is

$$(v, u) = x_1 y_1 + x_2 y_2 + \cdots + x_n y_n.$$

Let $v = (z_1, \ldots, z_n)$ and $u = (w_1, \ldots, w_n)$ be two vectors in $\mathbf{C}^n$. Then the *inner product* of v and u is $(v, u) = z_1 \bar{w}_1 + \cdots + z_n \bar{w}_n$.

In terms of this definition, we can then conclude that

$$\cos \theta = \frac{(v, u)}{(v, v)^{1/2} (u, u)^{1/2}}.$$

The reader should recall that (v, u) is the familiar "dot product" of the vectors v and u. Since we shall define inner product more generally below, we have chosen to introduce that terminology at this point.

Example 11.2.3 Let $u = (0, 1, 1)$ and $v = (-1, 1, 2)$ be two vectors in $\mathbf{R}^3$. Then $(u, v) = 0 \cdot (-1) + 1 \cdot 1 + 1 \cdot 2 = 3$, $(u, u) = 0^2 + 1^2 + 1^2 = 2$, and $(v, v) = (-1)^2 + 1^2 + 2^2 = 6$. Then $\|u\| = \sqrt{2}$ and $\|v\| = \sqrt{6}$, and if θ is the smallest angle between u and v, we see that

$$\cos \theta = \frac{3}{\sqrt{2}\,\sqrt{6}} = \frac{\sqrt{3}}{2}, \quad \text{whence } \theta = 30 \text{ degrees.}$$

It is now easy to see in terms of the inner product when two vectors are perpendicular. For if u and v are vectors in $\mathbf{R}^3$, then they are perpendicular if and only if the smallest angle between them is 90 degrees; that is, they are perpendicular if and only if $\cos \theta = 0$. But from the formula for $\cos \theta$, $\cos \theta = 0$ if and only if $(u, v) = 0$. Thus, the nonzero vectors u and v are perpendicular if and only if $(u, v) = 0$.

Since the notion of inner product can be meaningfully extended to vector spaces other than $\mathbf{R}^n$ and $\mathbf{C}^n$, we now give a more general definition of inner product. Then we will examine some examples other than the more familiar ones discussed above.

Definition 11.2.4 Let K be either the field $\mathbf{R}$ of real numbers or the field $\mathbf{C}$ of complex numbers. Let V be a vector space over K.
An *inner product* on V is a function $f\colon V \times V \longrightarrow K$ such that if $f(u,v)$ is denoted by (u,v), we have the following.

(a) $(u + v, w) = (u,w) + (v,w)$, for all u, v, w in V.
(b) $(cu,v) = c(u,v)$, for all u, v in V and all c in K.
(c) $(u,v) = \overline{(v,u)}$, where $\overline{(v,u)}$ is the complex conjugate of (v,u).
(d) $(v,v) > 0$, if $v \neq 0$.

A vector space V with an inner product f defined on V is called an *inner product space* (relative to f).

A few words of explanation are in order. Since $K \subseteq \mathbf{C}$, ($K = \mathbf{R}$ or $K = \mathbf{C}$), the condition (d) means that (v,v) is always a real number, and is greater than 0, for $v \neq 0$.
In the case that $K = \mathbf{R}$, the bar may be dropped in condition (c), since the conjugate $\bar{r}$ of a real number r is just r.
Certain consequences follow easily from Definition 11.2.4.

THEOREM 11.2.5 Let V be an inner product space over K, where either $K = \mathbf{R}$ or $K = \mathbf{C}$. Let f be the inner product on V, with $f(u,v)$ denoted (u,v). Then

(a) $(0,v) = 0$, for all vectors v in V (the first 0 is the zero vector, the second 0 is the scalar zero).
(b) $(cu + dv, w) = c(u,w) + d(v,w)$, for all u, v, w in V; c, d in K.
(c) $(u, cv + dw) = \bar{c}(u,v) + \bar{d}(u,w)$, for all u, v, w in V; c, d in K.

Proof These all follow easily from the definition of inner product, so we leave the proof as an exercise. ∎

It is helpful to see the breadth of vector spaces that can have an inner product defined on them. We give some examples.

Example 11.2.6 Let $V = \mathbf{R}^n$ and let $W = \mathbf{C}^n$, where $\mathbf{R}$ and $\mathbf{C}$ are the real and complex numbers, respectively. Then the inner products defined in Definition 11.2.2 are inner products in the sense of Definition 11.2.4. These earlier inner products are called the *standard inner products* on $\mathbf{R}^n$ and $\mathbf{C}^n$, respectively. We leave the verification that these inner products actually satisfy Definition 11.2.4 as an exercise for the reader.

In Example 11.2.6, the inner product on $\mathbf{R}^n$ has the following property: if $\mathcal{B} = \{\epsilon_1, \epsilon_2, \ldots, \epsilon_n\}$ is the canonical basis for $\mathbf{R}^n$, and if

$$u = \sum_{i=1}^{n} a_i \epsilon_i \quad \text{and} \quad v = \sum_{i=1}^{n} b_i \epsilon_i,$$

then

$$(u,v) = \sum_{i=1}^{n} \sum_{j=1}^{n} a_i b_j (\epsilon_i, \epsilon_j) = \sum_{i,j=1}^{n} a_i b_j (\epsilon_i, \epsilon_j),$$

and since

$$(\epsilon_i, \epsilon_j) = \delta_{ij} = \begin{cases} 1, & \text{for } i = j, \\ 0, & \text{for } i \neq j, \end{cases}$$

we have

$$(u,v) = \sum_{i=1}^{n} a_i b_i,$$

as before.

More generally, if $\mathcal{B}' = \{\zeta_1, \ldots, \zeta_n\}$ is any basis for $\mathbf{R}^n$ for which

$$(\zeta_i, \zeta_j) = \delta_{ij} = \begin{cases} 1, & \text{for } i = j, \\ 0, & \text{for } i \neq j, \end{cases}$$

then the function defined by

$$(u,v) = \sum_{i=1}^{n} \sum_{j=1}^{n} c_i d_j (\zeta_i, \zeta_j) = \sum_{i,j=1}^{n} c_i d_j (\zeta_i, \zeta_j)$$

is an inner product on $\mathbf{R}^n$, where

$$u = \sum_{i=1}^{n} c_i \zeta_i, \quad v = \sum_{j=1}^{n} d_j \zeta_j.$$

It is possible, however, to define inner products on $\mathbf{R}^n$ where the original basis elements do not satisfy the condition

$$(\zeta_i, \zeta_j) = \delta_{ij} = \begin{cases} 1, & i = j, \\ 0, & i \neq j. \end{cases}$$

Example 11.2.7 Let $V = \mathbf{R}^2$ and let $\mathcal{B} = \{v_1 = (1,1), v_2 = (1,-1)\}$ be a basis for $\mathbf{R}^2$. Define $(v_1, v_2) = (v_2, v_1) = 1$, $(v_1, v_1) = 2$, and $(v_2, v_2) = 1$. We then extend this function to all of $\mathbf{R}^2$ as follows: for $u = a_1 v_1 + a_2 v_2$ and $v = b_1 v_1 + b_2 v_2$, put

$$(u,v) = a_1 b_1 (v_1, v_1) + a_1 b_2 (v_1, v_2) + a_2 b_1 (v_2, v_1) + a_2 b_2 (v_2, v_2).$$

The question now is whether this defines an inner product on $\mathbf{R}^2$. We shall verify that it does.

First, property (a) of Definition 11.2.4 is easily seen to be true.

Second, (b) also holds easily. Third, since $(v_2,v_1) = (v_1,v_2)$, it can be verified that (c) holds. The last property asserts that $(v,v) > 0$ for $v \neq 0$. To see this, let $v = b_1v_1 + b_2v_2$. Then

$$(v,v) = b_1{}^2(v_1,v_1) + b_1b_2(v_1,v_2) + b_2b_1(v_2,v_1) + b_2{}^2(v_2,v_2)$$
$$= 2b_1{}^2 + 2b_1b_2 + b_2{}^2$$
$$= (b_1 + b_2)^2 + b_1{}^2.$$

Now, in case $b_1 = 0$, then $b_2 \neq 0$, for otherwise v would be the 0 vector, so that in this case $(v,v) > 0$. If $b_1 \neq 0$, then clearly $(v,v) > 0$, and so we see that this binary operation on $\mathbf{R}^2$ indeed defines an inner product on $\mathbf{R}^2$.

Although this last example appears to be radically different from the standard inner product on $\mathbf{R}^2$, we shall see shortly, after we discuss the Gram-Schmidt process, that for an inner product on $\mathbf{R}^2$ there is some basis $\mathfrak{B}$ for $\mathbf{R}^2$ such that with respect to $\mathfrak{B}$, the inner product behaves similarly to the standard inner product.

The next example is of a vastly different type.

Example 11.2.8 Let V be the vector space over $\mathbf{R}$ of all continuous real-valued functions with domain $[0,1]$, the unit interval. We define (u,v) as

$$(u,v) = \int_0^1 u(x)v(x)\, dx.$$

These integrals exist, since the function $u(x)v(x)$ is continuous.

It is easy to verify that this defines an inner product on V. First, the linearity properties (a) and (b) follow readily from standard theorems of elementary integral calculus: for (a) becomes

$$(u_1 + u_2, v) = \int_0^1 (u_1(x) + u_2(x))v(x)\, dx$$
$$= \int_0^1 u_1(x)v(x)\, dx + \int_0^1 u_2(x)v(x)\, dx$$
$$= (u_1,v) + (u_2,v),$$

while (b) becomes

$$(cu,v) = \int_0^1 cu(x)v(x)\, dx$$
$$= c\int_0^1 u(x)v(x)\, dx$$
$$= c(u,v).$$

Property (c) follows from the fact that

$$\int_0^1 u(x)v(x)\, dx = \int_0^1 v(x)u(x)\, dx.$$

Finally, property (d) is proved as follows: Let $u(x) \in V$, $u(x)$ not identically zero on [0,1]. Then

$$(u,u) = \int_0^1 u(x)u(x)\,dx = \int_0^1 u^2(x)\,dx.$$

Since $u(x)$ is not identically zero on [0,1], there is a subinterval of [0,1] on which $u^2(x)$ is greater than 0, and since $u^2(x) \geq 0$ throughout [0,1], it follows from calculus that

$$\int_0^1 u^2(x)\,dx > 0,$$

as was to be proved.

Example 11.2.9 It is interesting to modify Example 11.2.8 slightly to get a vector space V over the complex number field **C**, with an inner product on V. Let V be the set of all continuous complex-valued functions on the unit interval [0,1]. For example, the function defined by

$$f(x) = (x^2 + 2x - 3) + i(\sin x + \cos 3x)$$

is complex-valued and continuous on [0,1]. Now for functions u and v in V, we define

$$(u,v) = \int_0^1 u(x)\overline{v(x)}\,dx.$$

For example, if

$$u(x) = \sin x + i\cos x \quad \text{and} \quad v(x) = 3x^2 + i(2x - 1),$$

then

$$(u,v) = \int_0^1 (\sin x + i\cos x)(3x^2 - i(2x - 1))\,dx$$

$$= \int_0^1 [(\sin x)3x^2 + (\cos x)(2x - 1)]\,dx$$

$$+ i\int_0^1 [(\cos x)3x^2 - (\sin x)(2x - 1)]\,dx.$$

We leave to the reader the actual verification that this defines an inner product.

EXERCISES

1. Prove that the inner products of Definition 11.2.2 are inner products in the sense of Definition 11.2.4.
2. Prove Theorem 11.2.5.
3. Let $V = K_n$ be the vector space of all $n \times 1$ column matrices over K, $K = \mathbf{R}$ or $K = \mathbf{C}$. Define the *standard inner product* on K_n as

follows:

(a) When $K = \mathbf{R}$, $\left(\begin{bmatrix} a_1 \\ \cdot \\ \cdot \\ \cdot \\ a_n \end{bmatrix}, \begin{bmatrix} b_1 \\ \cdot \\ \cdot \\ \cdot \\ b_n \end{bmatrix}\right) = a_1 b_1 + \cdots + a_n b_n.$

(b) When $K = \mathbf{C}$, $\left(\begin{bmatrix} a_1 \\ \cdot \\ \cdot \\ \cdot \\ a_n \end{bmatrix}, \begin{bmatrix} b_1 \\ \cdot \\ \cdot \\ \cdot \\ b_n \end{bmatrix}\right) = a_1 \bar{b}_1 + \cdots + a_n \bar{b}_n.$

Prove that the above products are indeed inner products.

4. Let V be an inner product space over K ($K = \mathbf{R}$ or $K = \mathbf{C}$) of dimension n, with (u,v) the inner product on V. Let $\{v_1,\ldots,v_n\}$ be a basis for V. Prove the following:

(a) If $u = \sum_{i=1}^{n} a_i v_i$ and if $v = \sum_{j=1}^{n} b_j v_j$, then

$$(u,v) = \sum_{i=1}^{n} \sum_{j=1}^{n} a_i \bar{b}_j (v_i, v_j).$$

(b) If $(v_i, v_j) = \delta_{ij} = \begin{cases} 1, & i = j, \\ 0, & i \neq j, \end{cases}$ then

$$(u,v) = \sum_{i=1}^{n} a_i \bar{b}_i.$$

5. Let V be an inner product space over the field K ($K = \mathbf{R}$ or $K = \mathbf{C}$). Suppose that $(v,w) = 0$, for a fixed v in V, and for all w in V. Prove that v is the zero vector in V.

6. Let V be an inner product space over K ($K = \mathbf{R}$ or $K = \mathbf{C}$). Prove that $(u,v) = 0$ if and only if $(v,u) = 0$.

7. Verify that the inner products of Examples 11.2.7, 11.2.8, and 11.2.9 are indeed inner products.

8. Let $f: \mathbf{R}^n \times \mathbf{R}^n \longrightarrow \mathbf{R}$ be defined by $f(u,v) = (u,v) = 0$ for all $u, v \in R^n$. Show that f satisfies properties (a), (b), (c) of Definition 11.2.4 but not (d).

9. Let $f: \mathbf{R}^3 \times \mathbf{R}^3 \longrightarrow \mathbf{R}$ be defined as follows: for $u = (a_1,a_2,a_3)$ and $b = (b_1,b_2,b_3)$, $f(u,v) = (u,v) = a_1(b_1)^3 + a_2(b_2)^3 + a_3(b_3)^3$. Show that this function satisfies all properties of Definition 11.2.4, except property (c).

10. Let the product (u,v) be defined on $\mathbf{R}^3$ as follows:
$$(u,v) = (a_1b_1 + a_2b_2 + a_3b_3)|\cos\theta|,$$
where $u = (a_1,a_2,a_3)$, $v = (b_1,b_2,b_3)$, and θ is the smallest angle between u and v. Show that this product satisfies all the properties of Definition 11.2.4, except property (a).

11. Let $\sigma \colon \mathbf{R} \longrightarrow \mathbf{R}$ satisfy:

   (i) σ is an isomorphism of $(\mathbf{R},+)$ onto $(\mathbf{R},+)$.
   (ii) There are real numbers a and c such that $\sigma(ca) \neq c\sigma(a)$.

   (Such mappings exist, as can be proved using the logical tool known as Zorn's lemma. This is beyond the scope of this text, but the interested reader may refer to Paley and Weichsel, *A First Course in Abstract Algebra*.) For such a mapping σ, show that the product on $\mathbf{R}^3$ defined by $(u,v) = \sigma(a_1)\sigma(b_1) + \sigma(a_2)\sigma(b_2) + \sigma(a_3)\sigma(b_3)$, for $u = (a_1,a_2,a_3)$, $b = (b_1,b_2,b_3)$, satisfies all the properties of Definition 11.2.4 except (b).

12. For the vector space of Example 11.2.7, define a product on V setting $(v_1,v_1) = (v_2,v_2) = 1$ and $(v_1,v_2) = (v_2,v_1) = 2$. Extend this product to all of V by using the notion contained in Exercise 4. Show that the product is not an inner product.

13. Let V be a vector space over the rational number field Q. Assume that a product (u,v) is defined on V and satisfies properties (a), (c), and (d) of Definition 11.2.4. Show that (u,v) must also satisfy (b).

11.3 Length, Distance, and Angle; the Cauchy-Schwarz Inequality

From our earlier discussion, we expect we should be able to define the length of a vector, the distance between two vectors and the angle between two vectors, in terms of an inner product.

Definition 11.3.1 Let V be an inner product space over the field K ($K = \mathbf{R}$ or $K = \mathbf{C}$), with (u,v) the inner product defined on V. Then if u and $v \in V$, we define the following:

(a) The *length of v* is $\|v\| = (v,v)^{1/2}$. We will also call $\|v\|$ the *norm* of v.
(b) The *distance between u and v* is $\|u - v\| = (u - v, u - v)^{1/2}$.
(c) The *cosine of the smallest angle θ between u and v* is

$$\cos\theta = \frac{|(u,v)|}{\|u\| \cdot \|v\|}, \quad \text{for } u \neq 0, v \neq 0,$$

where $|(u,v)|$ is the absolute value of the possibly complex number (u,v).

11.3 LENGTH, DISTANCE, AND ANGLE

We mention that the symbol $\|u\|$ generalizes the symbol $|x|$, where x is a real or complex number. For in either **R** or **C** we can think of $|x|$ as denoting the length of the line segment from the origin to the point x.

For the three parts of Definition 11.3.1 to hang together, two facts must be established. First, since for any angle θ we wish to have $|\cos \theta| \leq 1$, we must have $|(u,v)| \leq \|u\| \cdot \|v\|$; that is, we must have $|(u,v)| \leq (u,u)^{1/2}(v,v)^{1/2}$. Second, we expect that a distance function should satisfy the triangle inequality; that is, we expect that $\|u + v\| \leq \|u\| + \|v\|$. (This latter inequality is the generalization of the well known two-dimensional Euclidean fact that the sum of the lengths of two sides of a triangle is greater than or equal to the length of the third side.)

Both of these inequalities will be easily established, once we have proven the central result, known as the Cauchy-Schwarz inequality.

Before proving this important inequality, we remind the reader of certain elementary facts about the complex numbers. First, for the complex number $a + bi$, its absolute value is defined as

$$|a + bi| = \sqrt{a^2 + b^2}.$$

Next, the square of the absolute value is

$$|a + bi|^2 = a^2 + b^2 = (a + bi)(a - bi);$$

that is, the square of the absolute value of a complex number is found by multiplying together the complex number and its conjugate.

For vectors u and v in an inner product space, the inner product (u,v) may be complex. In that case, $|(u,v)|$ denotes the absolute value of (u,v). Moreover, $(u,v)\overline{(u,v)}$ is equal to the square $|(u,v)|^2$ of the absolute value of (u,v).

THEOREM 11.3.2 (*Cauchy-Schwarz Inequality*) Let V be an inner product space over the field K, where either $K = \mathbf{R}$ or $K = \mathbf{C}$. Let (u,v) be the inner product on V. Then for all u, v in V,

$$|(u,v)|^2 \leq (u,u)(v,v) = \|u\|^2 \|v\|^2.$$

Proof Let $w = v - \dfrac{(v,u)}{(u,u)} u$. Then

$$(w,u) = \left(v - \frac{(v,u)}{(u,u)} u, u\right) = (v,u) - \frac{(v,u)}{(u,u)}(u,u) = 0,$$

provided $u \neq 0$. (Note that if $u = 0$, then the inequality we wish to prove holds trivially, so there is no loss in generality in assuming that $u \neq 0$.)

Next, we observe that $(u,w) = \overline{(w,u)} = \overline{0} = 0$. From this, we have

$$0 \leq (w,w) = \left(v - \frac{(v,u)}{(u,u)} u, w\right)$$

$$= (v,w) - \frac{(v,u)}{(u,u)} (u,w)$$

$$= \left(v, v - \frac{(v,u)}{(u,u)} u\right) - \frac{(v,u)}{(u,u)} \cdot 0$$

$$= (v,v) - \frac{\overline{(v,u)}}{(u,u)} (v,u), \quad \text{by property (c) of Theorem 11.2.5.}$$

But $\overline{(v,u)} \cdot (v,u) = |(v,u)|^2$, so now we have

$$0 \leq (v,v) - \frac{|(v,u)|^2}{(u,u)},$$

whence

$$|(v,u)|^2 \leq (v,v)(u,u) = \|u\|^2 \|v\|^2,$$

as was to be shown. ∎

We observe immediately that the Cauchy-Schwarz inequality implies

COROLLARY 11.3.3 Let V be an inner product space over the field K, $K = \mathbf{R}$, or $K = \mathbf{C}$, and let (u,v) be the inner product on V. Then for all u, v in V, $|(u,v)| \leq (u,u)^{1/2}(v,v)^{1/2}$.

Proof We already have

$$|(u,v)|^2 \leq (u,u)(v,v).$$

Taking square roots of both sides, we obtain the result. ∎

We mention that in case $K = \mathbf{R}$, we can drop the absolute-value sign and conclude that $(u,v) \leq (u,u)^{1/2}(v,v)^{1/2}$, since for a real number x, $x \leq |x|$.

COROLLARY 11.3.4 Let $(a_1,\ldots,a_n)$ and $(b_1,\ldots,b_n)$ be two n-tuples of real numbers. Then

$$\sum_{i=1}^{n} a_i b_i \leq \left(\sum_{i=1}^{n} a_i^2\right)^{1/2} \cdot \left(\sum_{i=1}^{n} b_i^2\right)^{1/2}.$$

Proof Consider these n-tuples as points u and v, respectively, in the space $\mathbf{R}^n$. Then, using the standard inner product on $\mathbf{R}^n$, we get that

$$(u,v) = \sum_{i=1}^{n} a_i b_i \leq (u,u)^{1/2}(v,v)^{1/2} = \left(\sum_{i=1}^{n} a_i^2\right)^{1/2} \left(\sum_{i=1}^{n} b_i^2\right)^{1/2}. \quad \blacksquare$$

11.3 LENGTH, DISTANCE, AND ANGLE

The second result we wished to establish was the triangle inequality. This follows readily from the Cauchy-Schwarz inequality.

THEOREM 11.3.5 (*Triangle Inequality*) Let V be an inner product space over K, $K = \mathbf{R}$ or $K = \mathbf{C}$. Then for u and v in V,
$$\|u + v\| \leq \|u\| + \|v\|.$$

Proof Since $\|u + v\|^2 = (u + v, u + v)$, we have
$$\begin{aligned} \|u + v\|^2 &= (u + v, u + v) = (u, u + v) + (v, u + v) \\ &= (u,u) + (u,v) + (v,u) + (v,v) \\ &= (u,u) + 2\,\mathrm{Re}\,(u,v) + (v,v), \end{aligned}$$
where $\mathrm{Re}\,(u,v)$ denotes the real part of the complex number (u,v). Now $\mathrm{Re}\,(a + bi) = a$ for any complex number $a + bi$. But
$$a \leq \sqrt{a^2 + b^2} = |a + bi|.$$
Thus, $\mathrm{Re}\,(u,v) \leq |(u,v)|$, so now we have
$$\begin{aligned} \|u + v\|^2 &= (u,u) + 2\,\mathrm{Re}\,(u,v) + (v,v) \\ &\leq (u,u) + 2|(u,v)| + (v,v). \end{aligned}$$
By Corollary 11.3.3, we now have
$$\begin{aligned} (u,u) + 2|(u,v)| + (v,v) &\leq (u,u) + 2(u,u)^{1/2}(v,v)^{1/2} + (v,v) \\ &= [(u,u)^{1/2} + (v,v)^{1/2}]^2. \end{aligned}$$
By following all the inequalities and equalities, we now have
$$\|u + v\|^2 \leq [(u,u)^{1/2} + (v,v)^{1/2}]^2.$$
Taking square roots of both sides, we get the triangle inequality. ∎

EXERCISES

1. Let $V = \mathbf{R}^2$. Let (u,v) be the standard inner product on V.
   (a) For $u = (1,1)$, find $\|u\|$.
   (b) For $v = (1,-3)$, find $\|v\|$.
   (c) For $u = (1,1)$ and $v = (1,-3)$, find $\cos\theta$, where θ is the smallest angle between u and v.
2. Let V be an inner product space over K ($K = \mathbf{R}$ or $K = \mathbf{C}$). Prove that
   (a) For $u \in V$, $\|u\| \geq 0$, and $\|u\| = 0$ if and only if $u = 0$.
   (b) $\|-u\| = \|u\|$, for all u in V.
   (c) $\|v - u\| = \|u - v\|$, for all u, v in V.
3. A function $d: V \times V \longrightarrow \mathbf{R}$ is called a *metric* on V if
   (a) $d(x,y) \geq 0$, for all x and y in V.

(b) $d(x,y) = 0$ if and only if $x = y$.
(c) $d(x,y) = d(y,x)$, for all x and y in V.
(d) $d(x,y) \leq d(x,z) + d(z,y)$, for all x, y, and z in V.

Let V be an inner product space over K ($K = \mathbf{R}$ or $K = \mathbf{C}$). Define $d: V \times V \longrightarrow \mathbf{R}$ by $d(u,v) = \|v - u\|$. Prove that d is a metric on V.

4. The following provides an alternative proof of the Cauchy-Schwarz inequality, Theorem 11.3.2, in the case $K = \mathbf{R}$.

Proof Let $u, v \in V$. Let $z = au - bv$, for real numbers a and b. Then
$$0 \leq (z,z) = (au - bv, au - bv)$$
$$= a^2\|u\|^2 - 2ab(u,v) + b^2\|v\|^2.$$

Let $a = \|v\|$ and let $b = \|u\|$. Now complete the proof.

5. In the case $K = \mathbf{R}$, show that for $u \neq 0$ and $v \neq 0$, $(u,v) = \|u\| \|v\|$ if and only if there is a positive real number c such that $v = cu$.

6. Let V be the vector space of real-valued continuous functions defined on $[0,1]$. Prove that for functions f and g in V, the following holds:
$$\left(\int_0^1 f(x)g(x)\,dx\right)^2 \leq \int_0^1 f^2(x)\,dx \cdot \int_0^1 g^2(x)\,dx.$$

7. What is the distance between the elements f and g of the vector space V of Exercise 6, where
$$f(x) = x, \quad \text{for all } x,\ 0 \leq x \leq 1,$$
$$g(x) = \begin{cases} 1 - x, & \text{for all } x \text{ such that } 0 \leq x \leq \tfrac{1}{2}, \\ \tfrac{1}{2}, & \text{for all } x \text{ such that } \tfrac{1}{2} < x \leq 1. \end{cases}$$

11.4 Orthogonality; the Gram-Schmidt Process

In Section 11.2, we observed that in the vector space $\mathbf{R}^3$, with standard inner product, the vectors u and v are perpendicular if and only if $(u,v) = 0$.

Definition 11.4.1 Let u and v be vectors in an inner product space V over a field K. Then u and v are said to be *orthogonal* if $(u,v) = 0$. The vectors $v_1, \ldots, v_n$ are *mutually orthogonal* if
$$(v_i, v_j) = 0, \quad \text{for } i \neq j;\ i, j = 1, \ldots, n.$$

Example 11.4.2 Let $V = \mathbf{R}^n$, equipped with the standard inner product. Let $\mathcal{B} = \{\epsilon_1, \ldots, \epsilon_n\}$ be the canonical basis for V, that is,

$$\epsilon_i = (0, 0, \ldots, 0, 1, 0, \ldots, 0).$$

Then

$$(\epsilon_i, \epsilon_j) = \delta_{ij} = \begin{cases} 1, & i = j, \\ 0, & i \neq j. \end{cases}^2$$

If $v_1 = \epsilon_i + \epsilon_j$, and $v_2 = \epsilon_i - \epsilon_j$, then $(v_1, v_2) = 0$, so v_1 and v_2 are orthogonal.

Example 11.4.3 Let V be the vector space of real-valued continuous functions defined on $[0,1]$. Let (u,v) be the inner product defined previously on that space. Letting v_1 be defined by $v_1(x) = 2(x+1)$ and letting v_2 be defined by $v_2(x) = (2x-1)/2(x+1)$, we have

$$(v_1, v_2) = \int_0^1 v_1(x) v_2(x)\, dx = \int_0^1 2(x+1) \frac{(2x-1)}{2(x+1)}\, dx$$
$$= \int_0^1 (2x-1)\, dx = (x^2 - x)\Big|_0^1 = 0.$$

Thus v_1 and v_2 are orthogonal.

In a vector space such as $\mathbf{R}^n$, with the standard inner product, there exists a basis $\{\epsilon_1, \ldots, \epsilon_n\}$ with the exceptionally nice property that $(\epsilon_i, \epsilon_j) = \delta_{ij}$. In general, bases with these properties are of such importance that we use special terminology to describe them.

Definition 11.4.4 Let V be a finite-dimensional inner product space over K ($K = \mathbf{R}$ or $K = \mathbf{C}$), with a basis $\mathcal{B} = \{v_1, \ldots, v_n\}$. Let (u,v) be the inner product defined on V.

Suppose that $(v_i, v_j) = \delta_{ij}$. Then the basis is called an *orthonormal basis* for V [with respect to the inner product (u,v)].

A vector u in V is called a *unit vector* if $\|u\| = 1$.

In view of Definition 11.4.4, an orthonormal basis is one in which each basis vector is a unit vector and in which any two distinct basis vectors are orthogonal.

We often say that we "normalize" the vector $u \neq 0$ when we multiply u by $\|u\|^{-1}$. For the vector $v = \|u\|^{-1} u$ is a unit vector:

$$(\|u\|^{-1} u, \|u\|^{-1} u) = \|u\|^{-1} \|u\|^{-1} (u, u)$$
$$= \|u\|^{-2} \|u\|^2 = 1.$$

[2] δ_{ij} is called the "Kronecker delta." When $i = j$, $\delta_{ij} = 1$. When $i \neq j$, $\delta_{ij} = 0$. Henceforth, when we write δ_{ij} without further amplification, we will assume it denotes the Kronecker delta.

Example 11.4.5 Let $u = (1, -1, 3)$ in $\mathbf{R}^3$. Then we normalize u by multiplying it by $\|u\|^{-1}$. Now

$$\|u\| = (u,u)^{1/2} = (1^2 + (-1)^2 + 3^2)^{1/2} = (11)^{1/2}.$$

Let $u_1 = (11)^{-1/2}(1, -1, 3)$. Then $\|u_1\| = 1$.

If we let $v = (i, 2+i, -i)$ in $\mathbf{C}^3$, then the norm of v is

$$\begin{aligned}\|v\| &= [i\overline{i} + (2+i)\overline{(2+i)} + (-i)\overline{(-i)}]^{1/2} \\ &= [i(-i) + (2+i)(2-i) + (-i)(i)]^{1/2} \\ &= (1 + 5 + 1)^{1/2} = 7^{1/2}.\end{aligned}$$

Multiplying v by $\|v\|^{-1}$, we see that $v_1 = 7^{-1/2}(i, 2+i, -i)$ is a unit vector, since

$$\begin{aligned}(v_1, v_1) &= (7^{-1/2}(i, 2+i, -i), 7^{-1/2}(i, 2+i, -i)) \\ &= 7^{-1}((i, 2+i, -i), (i, 2+i, -i)) \\ &= 7^{-1}\|v\|^2 = 7^{-1} \cdot 7 = 1.\end{aligned}$$

Following Example 11.2.7, we made reference to the fact that for any inner product (u,v) on $\mathbf{R}^n$, it would be possible to find some basis $\mathscr{B}$ for $\mathbf{R}^n$ such that with respect to the basis $\mathscr{B}$, (u,v) behaves similarly to the standard inner product. This statement will be made precise when we prove the next theorem. As the reader will note, the theorem applies to many vector spaces other than $\mathbf{R}^n$.

THEOREM 11.4.6 *(The Gram-Schmidt Process)* Let V be a finite-dimensional inner product space over the field K ($K = \mathbf{R}$ or $K = \mathbf{C}$). Let $\mathscr{B} = \{v_1, \ldots, v_n\}$ be a basis for V. Then there is a basis $\mathscr{B}' = \{y_1, \ldots, y_n\}$ of V that is an orthonormal basis with respect to the given inner product. Moreover, for each integer k, $1 \leq k \leq n$, the subspace spanned by $\{y_1, \ldots, y_k\}$ is the same as the subspace spanned by $\{v_1, \ldots, v_k\}$.

Proof The proof provides a construction for obtaining the orthonormal basis $\mathscr{B}'$ from the given basis $\mathscr{B}$.

What we will do is first obtain a basis $\mathscr{B}'' = \{u_1, \ldots, u_n\}$ of pairwise orthogonal vectors with the property that $\langle v_1, \ldots, v_k \rangle = \langle u_1, \ldots, u_k \rangle$, for all k, $1 \leq k \leq n$. Recall that $\langle x_1, \ldots, x_k \rangle$ is the subspace spanned by the vectors $x_1, \ldots, x_k$. After obtaining $\mathscr{B}''$, we will then normalize each vector u_i, thus obtaining the y_i's.

To begin the construction, simply put $u_1 = v_1$. Then clearly, $\langle u_1 \rangle = \langle v_1 \rangle$. Now let $u_2 = au_1 + v_2$, where we must determine a so that $(u_2, u_1) = 0$. For this condition to be satisfied, we have

$$0 = (u_2, u_1) = (au_1 + v_2, u_1) = a(u_1, u_1) + (v_2, u_1).$$

11.4 ORTHOGONALITY; THE GRAM-SCHMIDT PROCESS

Thus, $a(u_1,u_1) = -(v_2,u_1)$, whence

$$a = \frac{-(v_2,u_1)}{\|u_1\|^2}.$$

We note that u_1 and u_2 are linearly independent, and that

$$\langle u_1,u_2 \rangle = \langle v_1, av_1 + v_2 \rangle = \langle v_1,v_2 \rangle,$$

since u_2 is a linear combination of v_1 and v_2.

Thus, the theorem holds for the case $k = 2$.

To further illustrate the technique, we show now how to construct u_3, before giving the full proof by induction.

Let $u_3 = bu_1 + cu_2 + v_3$, where we must determine b and c. We want to have satisfied the two conditions $(u_3,u_1) = 0$ and $(u_3,u_2) = 0$. The first of these becomes

$$(bu_1 + cu_2 + v_3, u_1) = 0,$$
$$b(u_1,u_1) + c(u_2,u_1) + (v_3,u_1) = 0.$$

Since $(u_1,u_2) = 0$, this condition becomes

$$b(u_1,u_1) + (v_3,u_1) = 0, \quad \text{whence } b = \frac{-(v_3,u_1)}{\|u_1\|^2}.$$

The condition $(u_3,u_2) = 0$ implies

$$0 = (bu_1 + cu_2 + v_3, u_2) = b(u_1,u_2) + c(u_2,u_2) + (v_3,u_2).$$

Since $(u_1,u_2) = 0$, we get that

$$c = \frac{-(v_3,u_2)}{\|u_2\|^2}.$$

Thus, we have found scalars b and c such that $u_3 = bu_1 + cu_2 + v_3$ is orthogonal to each of u_1 and u_2. We readily observe that we can subtract appropriate multiples of v_1 and v_2 from u_3, thus obtaining v_3, so that $\langle u_1,u_2,u_3 \rangle = \langle v_1,v_2,v_3 \rangle$.

The induction step assumes that the construction has been carried out until a set $\{u_1,\ldots,u_{k-1}\}$ has been obtained, $(u_i,u_j) = 0$, for $i \neq j$, and $\langle u_1,\ldots,u_{k-1} \rangle = \langle v_1\ldots,v_{k-1} \rangle$. To construct u_k, we let

$$u_k = a_1u_1 + a_2u_2 + \cdots + a_{k-1}u_{k-1} + v_k.$$

We impose the $k - 1$ conditions $(u_k,u_j) = 0$, for $j = 1, \ldots, k - 1$. Using the orthogonality among the first $k - 1$ u_i's, these conditions

imply that
$$a_j(u_j,u_j) + (v_k,u_j) = 0,$$
whence
$$a_j = \frac{-(v_k,u_j)}{(u_j,u_j)} = \frac{-(v_k,u_j)}{\|u_j\|^2}.$$

Again, since each of the first $k-1$ u_j's is a linear combination of $v_1, \ldots, v_{k-1}$, it is easy to see that $\langle u_1,\ldots,u_k\rangle = \langle v_1,\ldots,v_k\rangle$.

Thus, we may conclude that the set $\{u_1,\ldots,u_n\}$ has been constructed such that $(u_i,u_j) = 0$, for $i \neq j$, and such that $\langle u_1,\ldots,u_k\rangle = \langle v_1,\ldots,v_k\rangle$, $1 \leq k \leq n$.

To finally get the orthonormal basis $\mathcal{B}' = \{y_1,\ldots,y_n\}$, let $y_i = \|u_i\|^{-1}u_i$. Then $\mathcal{B}'$ has the properties stated in the theorem. ∎

We illustrate this theorem with several examples.

Example 11.4.7 Let $V = \mathbf{R}^3$ and let $v_1 = (1,0,-1)$, $v_2 = (2,2,1)$, and $v_3 = (0,-1,-1)$. Let (u,v) be the standard inner product on $\mathbf{R}^3$.

It is clear that the canonical basis $\{\epsilon_1,\epsilon_2,\epsilon_3\}$ is an orthonormal basis for V. But the canonical basis does not have the property that the basis $\{y_1,y_2,y_3\}$ to be constructed will have, namely, that

$$\langle v_1\rangle = \langle y_1\rangle,\ \langle v_1,v_2\rangle = \langle y_1,\ y_2\rangle,\ \langle v_1,v_2,v_3\rangle = \langle y_1,y_2,y_3\rangle.$$

The basis to be constructed will have the particular property that its first vector will be a multiple of v_1.

Following the construction of the theorem, we let $u_1 = v_1$. Next, let $u_2 = au_1 + v_2$. Then $(u_2,u_1) = (au_1,u_1) + (v_2,u_1) = 0$, so

$$a = \frac{-(v_2,u_1)}{(u_1,u_1)} = \frac{-(2\cdot 1 + 2\cdot 0 + 1(-1))}{1^2 + 0^2 + (-1)^2} = -\frac{1}{2}.$$

Thus $u_2 = (-\frac{1}{2},0,\frac{1}{2}) + (2,2,1) = (\frac{3}{2},2,\frac{3}{2})$.

Finally, we let $u_3 = bu_1 + cu_2 + v_3$. From the conditions $(u_3,u_1) = 0$ and $(u_3,u_2) = 0$, we find that

$$b = \frac{-(v_3,u_1)}{(u_1,u_1)} = \frac{-(0\cdot 1 + (-1)\cdot 0 + (-1)(-1))}{2} = -\frac{1}{2}$$

and

$$c = \frac{-(v_3,u_2)}{(u_2,u_2)} = \frac{-(0\cdot\frac{3}{2} + (-1)(2) + (-1)\frac{3}{2})}{(\frac{3}{2}\cdot\frac{3}{2}) + (2\cdot 2) + (\frac{3}{2}\cdot\frac{3}{2})} = \frac{\frac{7}{2}}{\frac{34}{4}} = \frac{7}{17}.$$

Thus

$$u_3 = (-\tfrac{1}{2},0,\tfrac{1}{2}) + (\tfrac{21}{34},\tfrac{14}{17},\tfrac{21}{34}) + (0,-1,-1) = (\tfrac{4}{34},-\tfrac{3}{17},\tfrac{4}{34}).$$

11.4 ORTHOGONALITY; THE GRAM-SCHMIDT PROCESS

We finally find y_1, y_2, y_3 by setting $y_i = \|u_i\|^{-1} u_i$, for each i. Thus

$$y_1 = \frac{1}{\sqrt{2}}(1, 0, -1) = \left(\frac{\sqrt{2}}{2}, 0, \frac{-\sqrt{2}}{2}\right),$$

$$y_2 = \frac{2}{\sqrt{34}}\left(\frac{3}{2}, 2, \frac{3}{2}\right) = \left(\frac{3}{\sqrt{34}}, \frac{4}{\sqrt{34}}, \frac{3}{\sqrt{34}}\right),$$

$$y_3 = \sqrt{17}\left(\frac{2}{17}, \frac{-3}{17}, \frac{2}{17}\right) = \left(\frac{2\sqrt{17}}{17}, \frac{-3\sqrt{17}}{17}, \frac{2\sqrt{17}}{17}\right).$$

Example 11.4.8 Let $V = \mathbf{R}^2$, with the inner product defined as in Example 11.2.7. That is, we take as basis

$$\mathcal{B} = \{v_1 = (1, 1), v_2 = (1, -1)\}$$

and have $(v_1, v_1) = 2$, $(v_1, v_2) = (v_2, v_1) = 1$, and $(v_2, v_2) = 1$. We will now construct an orthonormal basis for $\mathbf{R}^2$ relative to this inner product.

We let $u_1 = v_1 = (1, 1)$. Then, by Gram-Schmidt,

$$u_2 = au_1 + v_2 = a(1, 1) + (1, -1).$$

Moreover, $a = -(v_2, u_1)/\|u_1\|^2$.

Since $(u_1, u_1) = (v_1, v_1) = 2$, $\|u_1\|^2 = (u_1, u_1) = 2$, so $a = -\frac{1}{2}$. Thus,

$$u_2 = (-\tfrac{1}{2}, -\tfrac{1}{2}) + (1, -1) = (\tfrac{1}{2}, -\tfrac{3}{2}).$$

Next, we normalize (relative to the inner product given in terms of the basis $\mathcal{B}$), obtaining

$$y_1 = \|u_1\|^{-1} u_1 = \frac{1}{\sqrt{2}}(1, 1) = \left(\frac{1}{\sqrt{2}}, \frac{1}{\sqrt{2}}\right),$$

and $y_2 = \|u_2\|^{-1} u_2 = \|u_2\|^{-1}(\tfrac{1}{2}, -\tfrac{3}{2})$. To find $\|u_2\|$ in the given inner product, we have

$$\begin{aligned}
\|u_2\|^2 &= (u_2, u_2) = (-\tfrac{1}{2}v_1 + v_2, -\tfrac{1}{2}v_1 + v_2) \\
&= \tfrac{1}{4}(v_1, v_1) - \tfrac{1}{2}(v_1, v_2) - \tfrac{1}{2}(v_2, v_1) + (v_2, v_2) \\
&= \tfrac{1}{4} \cdot 2 - (v_1, v_2) + (v_2, v_2) \\
&= \tfrac{1}{2} - 1 + 1 = \tfrac{1}{2}.
\end{aligned}$$

Thus, $\|u_2\| = 1/\sqrt{2}$, and so

$$y_2 = \sqrt{2}\left(\frac{1}{2}, -\frac{3}{2}\right) = \left(\frac{\sqrt{2}}{2}, \frac{-3\sqrt{2}}{2}\right).$$

We now verify that y_1 and y_2 are indeed an orthonormal basis for $\mathbf{R}^2$ relative to the given inner product. (Clearly, y_1 and y_2 are not even orthogonal with respect to the standard inner product.) Thus

$$(y_1,y_1) = \left(\frac{1}{\sqrt{2}} u_1, \frac{1}{\sqrt{2}} u_1\right) = \frac{1}{\sqrt{2}} \cdot \frac{1}{\sqrt{2}} (v_1,v_1) = \tfrac{1}{2} \cdot 2 = 1.$$

Also,

$$\begin{aligned}(y_2,y_2) &= (\sqrt{2}\,(-\tfrac{1}{2}v_1 + v_2), \sqrt{2}\,(-\tfrac{1}{2}v_1 + v_2)) \\ &= 2(-\tfrac{1}{2}v_1 + v_2, -\tfrac{1}{2}v_1 + v_2) \\ &= 2(-\tfrac{1}{2})^2(v_1,v_1) + 2(-\tfrac{1}{2}v_1,v_2) + 2(v_2,-\tfrac{1}{2}v_1) + 2(v_2,v_2) \\ &= 2 \cdot \tfrac{1}{4} \cdot 2 + 2(-\tfrac{1}{2}) \cdot 1 + 2(-\tfrac{1}{2}) \cdot 1 + 2 \cdot 1 \\ &= 1 + (-1) + (-1) + 2 = 1.\end{aligned}$$

Finally,

$$\begin{aligned}(y_1,y_2) &= \left(\frac{1}{\sqrt{2}} v_1, \sqrt{2}\,(-\tfrac{1}{2}v_1 + v_2)\right) = \frac{1}{\sqrt{2}} (v_1, \sqrt{2}\,(-\tfrac{1}{2}v_1 + v_2)) \\ &= (v_1, -\tfrac{1}{2}v_1) + (v_1,v_2) \\ &= (-\tfrac{1}{2})(v_1,v_1) + (v_1,v_2) = -\tfrac{1}{2} \cdot 2 + 1 = 0.\end{aligned}$$

Thus, we observe that the given inner product "looks like" the standard inner product relative to the orthonormal basis $\{y_1,y_2\}$. Hence, if we now write vectors in $\mathbf{R}^2$ in the form

$$v = ay_1 + by_2 \quad \text{and} \quad u = cy_1 + dy_2,$$

then

$$(v,u) = (ay_1 + by_2, cy_1 + dy_2) = ac + bd.$$

We conclude this section with a very easy but important corollary to Theorem 11.4.6.

COROLLARY 11.4.9 Let V be an n-dimensional inner product space over K, where $K = \mathbf{R}$ or $K = \mathbf{C}$. Let (u,v) be the inner product defined on V. Then there is an orthonormal basis $\mathcal{B} = \{v_1,\ldots,v_n\}$ for V with respect to the given inner product, and if

$$u = \sum_{i=1}^{n} a_i v_i \quad \text{and} \quad v = \sum_{i=1}^{n} b_i v_i,$$

then the inner product

$$(u,v) = \sum_{i=1}^{n} a_i b_i;$$

that is, (u,v) is a "dot product."

Proof Use Theorem 11.4.6 to get the basis $\mathcal{B}$. The rest of the proof is easy. ∎

EXERCISES

1. Let $V = \mathbf{R}^4$. Let U be the subspace spanned by $v_1 = (1,1,0,0)$, $v_2 = (1,0,-1,0)$, and $v_3 = (3,2,0,1)$.
   (a) Find an orthonormal basis for U, with respect to the standard inner product on V, such that the first basis vector is a multiple of v_1.
   (b) Augment the basis found in (a) to an orthonormal basis for all of $\mathbf{R}^4$.

2. Let $V = \mathbf{R}^3$, and let
   $$\mathcal{B} = \{v_1 = (1,1,0),\ v_2 = (1,0,-1),\ v_3 = (3,2,1)\}$$
   be a basis for V. Define a nonstandard inner product on V by the following:
   $$(v_1,v_1) = (v_2,v_2) = (v_3,v_3) = 2;\quad (v_i,v_j) = 1,\ i \neq j,$$
   and
   $$\left(\sum_{i=1}^{3} a_i v_i,\ \sum_{j=1}^{3} b_j v_j\right) = \sum_{i=1}^{3}\sum_{j=1}^{3} a_i b_j (v_i,v_j) = \sum_{i,j=1}^{3} a_i b_j (v_i,v_j).$$
   (a) Show that the product defined above is indeed an inner product on $\mathbf{R}^3$.
   (b) Find an orthonormal basis for $\mathbf{R}^3$ with respect to the inner product above.

3. Let $V = \mathbf{C}^3$, with the standard inner product. Let U be the subspace spanned by $(i,0,-i)$ and $(0, 1, 1 + i)$. Find an orthonormal basis for U.

4. Let V be the vector space over $\mathbf{R}$ of all real-valued polynomial functions with domain $[0,1]$ and of degree ≤ 3. That is, $f(t) \in V$ if and only if
   $$f(t) = a_0 + a_1 t + a_2 t^2 + a_3 t^3,\quad \text{for some } a_0, a_1, a_2, a_3 \in \mathbf{R}.$$
   For $f, g \in V$, let
   $$(f,g) = \int_0^1 f(t) g(t)\ dt.$$
   (a) Show that (f,g) is an inner product on V.
   (b) Show that the set of polynomial functions f_0, f_1, f_2, f_3 defined by
   $$f_0(t) = 1,$$
   $$f_1(t) = t,$$
   $$f_2(t) = t^2,$$
   $$f_3(t) = t^3,$$
   is a basis for V.
   (c) Apply the Gram-Schmidt process to $\{f_0, f_1, f_2, f_3\}$ to get an orthonormal basis for V.

5. Let V be an n-dimensional vector space over K, $K = \mathbf{R}$ or $K = \mathbf{C}$, with inner product (u,v). Let $\mathfrak{B} = \{v_1,\ldots,v_n\}$ be an orthonormal basis for V with respect to the given inner product. Let $T\colon V \longrightarrow V$ be a linear operator on V. Let $A = [a_{ij}] = [T]_{\mathfrak{B}}$. Prove that $a_{ij} = (Tv_j, v_i)$, for all i, j.

6. Let V be a vector space over K, with $K = \mathbf{R}$ or $K = \mathbf{C}$. Let $v_1, \ldots, v_n$ be n mutually orthogonal nonzero vectors in V.
    (a) Prove that $\{v_1,\ldots,v_n\}$ is a linearly independent set of vectors. (*Hint:* Let $v = c_1 v_1 + \cdots + c_n v_n = 0$. Consider (v, v_j), for $j = 1, \ldots, n$.)
    (b) Prove that if $\dim V = n$ and if $\{v_1,\ldots,v_n\}$ are n mutually orthogonal nonzero vectors in V, then $\{v_1,\ldots,v_n\}$ is a basis for V.

7. Let V be an n-dimensional vector space over K, with $K = \mathbf{R}$ or $K = \mathbf{C}$. Let (u,v) be an inner product on V. Let W be a subspace of V. Let $W^\perp = \{u \in V \mid (u,w) = 0, \text{ all } w \in W\}$.
    (a) Prove $W^\perp$ is a subspace of V.
    (b) Prove $\dim (W^\perp) = n - \dim W$.

11.5 Orthogonal Transformations and Orthogonal Matrices

We have defined the notions of length, distance, and angle for a vector space with an inner product in Section 11.3. There are linear operators on such spaces that preserve lengths, distances, and angles. For example, a rotation of $\mathbf{R}^2$ is a linear operator T on $\mathbf{R}^2$ that has, in addition to other properties, the property that $\|T(v)\| = \|v\|$, for all vectors v in $\mathbf{R}^2$. In this section, we consider such transformations in a fairly general setting.

Definition 11.5.1 Let $T\colon V \longrightarrow V$ be a linear operator on the inner product space V over K, and let (u,v) be the inner product on V. Then (a) when $K = \mathbf{R}$, T is called an *orthogonal* operator if $(Tu, Tv) = (u,v)$, for all u and v in V. (b) When $K = \mathbf{C}$, T is called a *unitary* operator if $(Tu, Tv) = (u,v)$, for all u and v in V.

THEOREM 11.5.2 Let T be an orthogonal or unitary operator on the vector space V over K, where (u,v) is an inner product on V. Then

(a) For each u in V, $\|T(u)\| = \|u\|$.
(b) If θ is the angle between two vectors u and v in V and if $T(\theta)$ denotes the angle between $T(u)$ and $T(v)$, then $\cos \theta = \cos (T(\theta))$.

Proof (a) Let $u \in V$. Then, by definition, $(Tu, Tu) = (u,u)$. Thus

$$\|Tu\| = (Tu, Tu)^{1/2} = (u,u)^{1/2} = \|u\|.$$

11.5 ORTHOGONAL TRANSFORMATIONS AND ORTHOGONAL MATRICES

(b) For u and v in V, $u \neq 0$, $v \neq 0$,
$$\|Tu\| = \|u\|, \quad \|Tv\| = \|v\|, \quad \text{and} \quad (Tu, Tv) = (u,v).$$

By definition
$$\cos \theta = \frac{(u,v)}{\|u\| \, \|v\|} \quad \text{and} \quad \cos(T(\theta)) = \frac{(Tu, Tv)}{\|Tu\| \, \|Tv\|}.$$

Clearly,
$$\cos \theta = \cos(T(\theta)). \qquad \blacksquare$$

Either Definition 11.5.1 or Theorem 11.5.2 justifies the use of the word "orthogonal" in "orthogonal transformation." For an orthogonal transformation has the property that if vectors u and v are orthogonal, then so are the vectors Tu and Tv orthogonal.

Example 11.5.3 Let T be the operator on $\mathbf{R}^2$ defined by

$$T\epsilon_1 = \frac{1}{\sqrt{2}} \epsilon_1 - \frac{1}{\sqrt{2}} \epsilon_2,$$

$$T\epsilon_2 = \frac{1}{\sqrt{2}} \epsilon_1 + \frac{1}{\sqrt{2}} \epsilon_2,$$

where $\{\epsilon_1, \epsilon_2\}$ is the canonical basis for $\mathbf{R}^2$. Let (u,v) be the standard inner product on $\mathbf{R}^2$. Now for $u = a_1\epsilon_1 + a_2\epsilon_2$ and $v = b_1\epsilon_1 + b_2\epsilon_2$, we have $(u,v) = a_1 b_1 + a_2 b_2$. But also,

$$(Tu, Tv) = (T(a_1\epsilon_1 + a_2\epsilon_2), T(b_1\epsilon_1 + b_2\epsilon_2))$$
$$= \left(\frac{a_1}{\sqrt{2}} (\epsilon_1 - \epsilon_2) + \frac{a_2}{\sqrt{2}} (\epsilon_1 + \epsilon_2), \frac{b_1}{\sqrt{2}} (\epsilon_1 - \epsilon_2) + \frac{b_2}{\sqrt{2}} (\epsilon_1 + \epsilon_2) \right)$$
$$= \frac{1}{\sqrt{2}} \cdot \frac{1}{\sqrt{2}} ((a_1 + a_2)\epsilon_1 - (a_1 - a_2)\epsilon_2, (b_1 + b_2)\epsilon_1 - (b_1 - b_2)\epsilon_2)$$
$$= \tfrac{1}{2}[(a_1 + a_2)(b_1 + b_2) + (a_1 - a_2)(b_1 - b_2)]$$
$$= \tfrac{1}{2}[2a_1 b_1 + 2a_2 b_2] = a_1 b_1 + a_2 b_2 = (u,v).$$

Thus T is an orthogonal operator on $\mathbf{R}^2$.

Next, we will show that the effect of T on $\mathbf{R}^2$ is to rotate each vector through 45 degrees. To see this, let $u = a_1\epsilon_1 + a_2\epsilon_2$. If η is the angle between u and Tu,

$$\cos \eta = \frac{(u, Tu)}{\|u\| \, \|Tu\|}$$
$$= \frac{1}{\|u\| \, \|Tu\|} \left(a_1\epsilon_1 + a_2\epsilon_2, \frac{a_1}{\sqrt{2}} (\epsilon_1 - \epsilon_2) + \frac{a_2}{\sqrt{2}} (\epsilon_1 + \epsilon_2) \right)$$
$$= \frac{1}{\sqrt{2}} \frac{(a_1{}^2 + a_2{}^2)}{\|u\| \, \|Tu\|}.$$

Since $\|Tu\| = \|u\|$, and since $\|u\| = (a_1^2 + a_2^2)^{1/2}$, we have

$$\cos \eta = \frac{1}{\sqrt{2}} \frac{(a_1^2 + a_2^2)}{(a_1^2 + a_2^2)} = \frac{1}{\sqrt{2}}.$$

Thus, $\eta = 45$ degrees. Since

$$T\epsilon_1 = \frac{1}{\sqrt{2}} \epsilon_1 - \frac{1}{\sqrt{2}} \epsilon_2 \quad \text{and} \quad T\epsilon_2 = \frac{1}{\sqrt{2}} \epsilon_1 + \frac{1}{\sqrt{2}} \epsilon_2,$$

we see that the rotation is in the clockwise direction.

Although the operator T of Example 11.5.3 is a rotation, not all orthogonal operators are rotations. This will be seen in Exercise 1.

Another important observation to make concerning orthogonal and unitary transformations is that they are nonsingular.

THEOREM 11.5.4 Let T be an orthogonal or unitary transformation on the inner product space V over K. Then T is nonsingular.

Proof If T is singular, then there is a nonzero vector u in V such that $Tu = 0$. But $u \neq 0$ implies that $(u,u) > 0$, and since $(Tu,Tu) = (u,u)$, we have $(Tu,Tu) > 0$. But for $Tu = 0$, we must have $(Tu,Tu) = 0$, which is a contradiction. Thus, for $u \neq 0$, $Tu \neq 0$ and T is nonsingular. ∎

Although we defined an orthogonal (unitary) transformation as one that preserves inner products, we could just as well have defined an orthogonal (unitary) transformation as one that preserves lengths, as we see in the next theorem.

THEOREM 11.5.5 Let T be a linear operator on the inner product space V over K. Then

(a) When $K = \mathbf{R}$, T is orthogonal if and only if $(Tv,Tv) = (v,v)$, for all $v \in V$.
(b) When $K = \mathbf{C}$, T is unitary if and only if $(Tv,Tv) = (v,v)$, for all $v \in V$.

Proof If T is orthogonal (unitary), then $(Tv,Tv) = (v,v)$, for all v in V, follows from the definition of orthogonal (unitary) transformation.

Conversely, let $(Tv,Tv) = (v,v)$ for all v in V. We must show that for any vectors u and v in V, $(Tu,Tv) = (u,v)$.

To prove this, it is necessary to consider two special cases: (1) $K = \mathbf{R}$ and (2) $K = \mathbf{C}$. Although the techniques are similar, separate proofs are needed due to the fact that for an inner product, $(u,v) = \overline{(v,u)}$.

11.5 ORTHOGONAL TRANSFORMATIONS AND ORTHOGONAL MATRICES

CASE 1 Let $K = \mathbf{R}$. Then for vectors u and v we have, by hypothesis,
$$(u + v, u + v) = (T(u + v), T(u + v)).$$
Now
$$(u + v, u + v) = (u,u) + (u,v) + (v,u) + (v,v),$$
whereas
$$(T(u + v), T(u + v)) = (Tu + Tv, Tu + Tv)$$
$$= (Tu,Tu) + (Tu,Tv) + (Tv,Tu) + (Tv,Tv).$$
Since $(Tv,Tv) = (v,v)$ and $(Tu,Tu) = (u,u)$, this yields
$$(Tu,Tv) + (Tv,Tu) = (u,v) + (v,u).$$
When $K = \mathbf{R}$, $(u,v) = (v,u)$ and $(Tu,Tv) = (Tv,Tu)$, so that
$$2(Tu,Tv) = 2(u,v) \quad \text{whence} \quad (Tu,Tv) = (u,v),$$
and so T is orthogonal.

CASE 2 Let $K = \mathbf{C}$. Then for vectors u and v, let $(u,v) = c + di$ and $(Tu,Tv) = a + bi$.

(a) $(T(u + v), T(u + v)) = (u + v, u + v)$, whence
$$(Tu + Tv, Tu + Tv) = (Tu,Tu) + (Tu,Tv) + (Tv,Tu) + (Tv,Tv)$$
$$= (u,u) + (Tu,Tv) + \overline{(Tu,Tv)} + (v,v)$$
$$= (u + v, u + v)$$
$$= (u,u) + (u,v) + (v,u) + (v,v)$$
$$= (u,u) + (u,v) + \overline{(u,v)} + (v,v).$$
From this, we see that
$$(a + bi) + (a - bi) = (c + di) + (c - di),$$
whence $2a = 2c$, and $a = c$.

(b) $(T(u + iv), T(u + iv)) = (u + iv, u + iv)$, whence
$$(Tu,Tu) + i(Tv,Tu) - i(Tu,Tv) + (Tv,Tv) = (u,u) + i(v,u) - i(u,v) + (v,v),$$
so that
$$i(a - bi) - i(a + bi) = i(c - di) - i(c + di),$$
whence $2b = 2d$, and $b = d$.

From (a) and (b) together, $a + bi = c + di$, whence $(Tu,Tv) = (u,v)$, and so T is unitary. ∎

We now consider the matrix analogue of orthogonal and unitary transformations. We will define orthogonal and unitary matrices, and then show the connection between these classes of matrices and the classes of orthogonal and unitary operators.

Definition 11.5.6 Let $K = \mathbf{R}$ or $\mathbf{C}$. The *standard inner product* on the vector space K_n of all $n \times 1$ column matrices is defined by

$$(A_i, A_j) = \sum_{k=1}^{n} a_{ki}\bar{a}_{kj},$$

where

$$A_i = \begin{bmatrix} a_{1i} \\ a_{2i} \\ \cdot \\ \cdot \\ \cdot \\ a_{ni} \end{bmatrix} \quad \text{and} \quad A_j = \begin{bmatrix} a_{1j} \\ a_{2j} \\ \cdot \\ \cdot \\ \cdot \\ a_{nj} \end{bmatrix}.$$

We observe that when $K = \mathbf{R}$,

$$(A_i, A_j) = \sum_{k=1}^{n} a_{ki} a_{kj},$$

since $\bar{a}_{kj} = a_{kj}$, for $a_{kj} \in \mathbf{R}$.

The reader was asked in Exercise 3, Section 11.2, to verify that the inner product defined in Definition 11.5.6 actually is an inner product.

Definition 11.5.7 Let A be an $n \times n$ matrix with entries in K.
(a) When $K = \mathbf{R}$, we say A is *orthogonal* if the columns of A form an orthonormal set of vectors with respect to the standard inner product on $\mathbf{R}_n$.
(b) When $K = \mathbf{C}$, we say A is *unitary* if the columns of A form an orthonormal set of vectors with respect to the standard inner product on $\mathbf{C}_n$.

THEOREM 11.5.8 Let $T: K^n \longrightarrow K^n$ be an orthogonal (or unitary) operator on the vector space K^n of all n-tuples over K. Let A be the matrix of T with respect to the canonical basis $\mathfrak{B}$ of K^n. Then A is an orthogonal (or unitary) matrix.

Conversely, let A be an orthogonal (or unitary) matrix over K. Let T be the linear operator defined on K^n so that $[T]_{\mathfrak{B}} = A$, with respect to the canonical basis $\mathfrak{B}$ of K^n. Then T is orthogonal (or unitary).

Proof Let $T: K^n \longrightarrow K^n$ be an orthogonal (or unitary) operator on K^n. Let $\mathfrak{B} = \{\epsilon_1, \ldots, \epsilon_n\}$ be the canonical basis for K^n. Then $(T\epsilon_i, T\epsilon_j) = (\epsilon_i, \epsilon_j) = \delta_{ij}$, the Kronecker delta, so that the set of vectors $T\epsilon_1, T\epsilon_2, \ldots, T\epsilon_n$ is an orthonormal set of vectors. If

$$T\epsilon_i = a_{1i}\epsilon_1 + \cdots + a_{ni}\epsilon_n$$

and

$$T\epsilon_j = a_{1j}\epsilon_1 + \cdots + a_{nj}\epsilon_j,$$

11.5 ORTHOGONAL TRANSFORMATIONS AND ORTHOGONAL MATRICES

then with respect to the standard inner product in K^n,

$$(T\epsilon_i, T\epsilon_j) = \sum_{k=1}^{n} a_{ki}\bar{a}_{kj} = \delta_{ij}.$$

But the ith and jth columns of A are just

$$\begin{bmatrix} a_{1i} \\ \cdot \\ \cdot \\ \cdot \\ a_{ni} \end{bmatrix} \text{ and } \begin{bmatrix} a_{1j} \\ \cdot \\ \cdot \\ \cdot \\ a_{nj} \end{bmatrix}$$

and in the standard inner product on K_n, their inner product is also

$$\sum_{k=1}^{n} a_{ki}\bar{a}_{kj}.$$

This shows that the columns of A are an orthonormal set of column matrices.

Conversely, if A is an orthogonal (or unitary) matrix, let $T: K^n \longrightarrow K^n$ be defined so that $[T]_\mathcal{B} = A$, where $\mathcal{B}$ is the canonical basis for K^n. For ϵ_i and ϵ_j, in K^n, define (ϵ_i, ϵ_j) as $(T\epsilon_i, T\epsilon_j)$, where we put $(T\epsilon_i, T\epsilon_j)$ equal to the inner product of the ith and jth columns of A. It is easy to see that the inner product can be extended to all of K^n in the obvious manner. We leave the remaining details to the reader. ∎

Theorem 11.5.8 can be generalized to the following.

THEOREM 11.5.9 Let $\mathcal{B}$ be an orthonormal basis for the n-dimensional inner product space V over K, $K = \mathbf{R}$ or $K = \mathbf{C}$. Let $T: V \longrightarrow V$ be a linear operator on V. Then T is orthogonal (unitary) if and only if the matrix $[T]_\mathcal{B}$ is an orthogonal (unitary) matrix.

Proof The proof is straightforward and is left to the reader. ∎

As a corollary to Theorem 11.5.8 (or 11.5.9) and Theorem 11.5.4 we have

COROLLARY 11.5.10 Let A be an orthogonal (unitary) matrix. Then A is nonsingular.

Proof Let T be the orthogonal (unitary) operator corresponding to A according to Theorem 11.5.8. T is nonsingular, by Theorem 11.5.4. By Theorem 8.8.9, $A = [T]_\mathcal{B}$ is nonsingular if and only if T is nonsingular. ∎

We wish to conclude this section by giving two other characterizations of orthogonal (unitary) matrices. Each of these is often used by other authors as the definition of an orthogonal (unitary) matrix. Before stating these results, we must first define the conjugate transpose of a matrix.

Recall that in Sections 8.8 and 9.1 we defined the transpose of the $m \times n$ matrix A as the $n \times m$ matrix A^T, whose (j,i)th entry is the (i,j)th entry a_{ij} of A. From this the following results all follow easily. Some of them were earlier exercises, and all should be verified by the reader.

(1) $(A^T)^T = A$.
(2) $(AB)^T = B^T A^T$, for all $m \times n$ matrices A and all $n \times p$ matrices B.
(3) If A is nonsingular, then $(A^{-1})^T = (A^T)^{-1}$.

Definition 11.5.11 Let A be an $m \times n$ matrix with entries in **C**. Then A^* is the $n \times m$ matrix whose (j,i)th entry is $\bar{a}_{ij}$, the complex conjugate of the (i,j)th entry of A. A^* is called the *congujate transpose* of A.

We note that if the entries of A are all real numbers, then $A^* = A^T$. To illustrate, if

$$A = \begin{bmatrix} i & 1+i & -i \\ 0 & 1 & i \end{bmatrix}, \quad \text{then} \quad A^* = \begin{bmatrix} -i & 0 \\ 1-i & 1 \\ i & -i \end{bmatrix}.$$

THEOREM 11.5.12

(a) $(A^*)^* = A$, for all $m \times n$ matrices A with complex coefficients.
(b) $(AB)^* = B^* A^*$, for all $m \times n$ matrices A and $n \times p$ matrices B, A and B having complex entries.
(c) If A is nonsingular, then $(A^{-1})^* = (A^*)^{-1}$.

Proof The proof is left as an exercise. ∎

THEOREM 11.5.13 Let A be an $n \times n$ matrix with entries in K. Then

(a) When $K = \mathbf{R}$, A is orthogonal if and only if $A^T = A^{-1}$.
(b) When $K = \mathbf{C}$, A is unitary if and only if $A^* = A^{-1}$.

Proof We prove (b) first.
Observe that $A^* A = I_n$ if and only if

$$\sum_{k=1}^{n} a^*_{ik} a_{kj} = \delta_{ij},$$

where $[a^*_{i1}, \ldots, a^*_{in}]$ is the ith row of A^*. But $a^*_{ik} = \bar{a}_{ki}$, so $A^* A = I_n$ if and only if

$$\sum_{k=1}^{n} \bar{a}_{ki} a_{kj} = \delta_{ij}.$$

11.5 ORTHOGONAL TRANSFORMATIONS AND ORTHOGONAL MATRICES

Now if A_i is the ith column of A, we see that

$$\sum_{k=1}^{n} \bar{a}_{ki}a_{kj} = \sum_{k=1}^{n} a_{kj}\bar{a}_{ki} = (A_j, A_i) = \overline{(A_i, A_j)} = \delta_{ij}.$$

But $\overline{(A_i, A_j)} = \delta_{ij}$ if and only if $(A_i, A_j) = \delta_{ij}$. Thus, we conclude $A^*A = I_n$ if and only if the columns of A form an orthonormal set of vectors in $\mathbf{C}_n$, that is $A^*A = I_n$ if and only if A is unitary. Thus A is unitary if and only if $A^* = A^{-1}$.

To prove (a), we simply note that if all of the entries of A are in $\mathbf{R}$, then $A^T = A^*$, and so A is orthogonal if and only $A^TA = I_n$, whence A is orthogonal if and only if $A^T = A^{-1}$. ∎

COROLLARY 11.5.14 Let A be an $n \times n$ matrix with entries in K, with $K = \mathbf{R}$ or $K = \mathbf{C}$. Then A is orthogonal (unitary) if and only if the row vectors of A form an orthonormal set of vectors.

Proof (a) Let $K = \mathbf{R}$. If A is orthogonal, then $A^T = A^{-1}$, whence $AA^T = I_n$. But $AA^T = I_n$ if and only if $(R_i, R_j) = \delta_{ij}$, where R_i is the ith row of A, and the inner product is the standard inner product on K^n.

Conversely, if the rows of A are an orthonormal set of vectors, then by reversing the steps above, $A^T = A^{-1}$, and so A is orthogonal.

(b) $K = \mathbf{C}$. We leave this proof to the reader. ∎

EXERCISES

1. Let $T: \mathbf{R}^2 \longrightarrow \mathbf{R}^2$ be the linear operator defined on the canonical basis $\mathcal{B} = \{\epsilon_1, \epsilon_2\}$ of $\mathbf{R}^2$ as follows:

   $$T\epsilon_1 = \frac{1}{\sqrt{2}}\epsilon_1 - \frac{1}{\sqrt{2}}\epsilon_2,$$

   $$T\epsilon_2 = -\frac{1}{\sqrt{2}}\epsilon_1 - \frac{1}{\sqrt{2}}\epsilon_2.$$

   (a) Show that T is orthogonal.
   (b) Compute the angle between $T\epsilon_1$ and $T\epsilon_2$.
   (c) Compute the angle between ϵ_1 and $T\epsilon_1$.
   (d) Compute the angle between ϵ_2 and $T\epsilon_2$.
   (e) Show that T is not a rotation.

2. Let $T: \mathbf{R}^2 \longrightarrow \mathbf{R}^2$ be the linear operator whose matrix with respect to the canonical basis $\mathcal{B}$ is

   $$[T]_{\mathcal{B}} = \begin{bmatrix} -1 & 0 \\ 0 & -1 \end{bmatrix}.$$

(a) Show that T is orthogonal.
(b) Describe T in geometric terms.
3. Let $A_{m \times n}$ and $B_{n \times p}$ be two matrices over K, $K = \mathbf{R}$ or $K = \mathbf{C}$. Prove
   (a) $(A^*)^* = A$.
   (b) $(AB)^* = B^*A^*$.
   (c) If A is a nonsingular matrix, then $(A^{-1})^* = (A^*)^{-1}$.
4. Complete the details of the proof of Theorem 11.5.8.
5. Prove Theorem 11.5.9.
6. Let $T: V \longrightarrow V$ be an orthogonal operator on the inner product space V over $\mathbf{R}$. Let $\{v_1, \ldots, v_n\}$ be an orthonormal basis for V. Prove that $\{Tv_1, \ldots, Tv_n\}$ is an orthonormal basis for V.
7. Let V be an n-dimensional vector space over $\mathbf{R}$, and let T be a linear operator on V. Prove that T is orthogonal if and only if there exists an orthonormal basis $\{v_1, \ldots, v_n\}$ of V such that $\{Tv_1, \ldots, Tv_n\}$ is an orthonormal basis. (Compare this exercise with Exercise 6.)
8. Complete the proof of Corollary 11.5.14.
9. (a) Let A be an orthogonal matrix. Prove $\det A = 1$ or $\det A = -1$.
   (b) Let A be a unitary matrix. Prove $|\det A| = 1$.
10. Let T be an orthogonal operator on $\mathbf{R}^2$. Let $\mathcal{B} = \{\epsilon_1, \epsilon_2\}$ be the canonical basis for $\mathbf{R}^2$ and let (u,v) be the standard inner product on $\mathbf{R}^2$. Suppose $\det [T]_\mathcal{B} = 1$. Prove that T is a rotation on $\mathbf{R}^2$. [*Hint:* Show that there exists an angle θ such that
$$[T]_\mathcal{B} = \begin{bmatrix} \cos \theta & -\sin \theta \\ \sin \theta & \cos \theta \end{bmatrix}.$$
Assume that $[T]_\mathcal{B} = \begin{bmatrix} a & b \\ c & d \end{bmatrix}$. Use the facts that the rows of $[T]_\mathcal{B}$ are orthogonal unit vectors and that the columns of $[T]_\mathcal{B}$ are orthogonal unit vectors.]
11. (a) Find all orthogonal matrices of order n in which the only entries are 0 or 1.
    (b) Show that the set of such matrices is a group with respect to matrix multiplication.
    (c) To what well-known group is the group in (b) isomorphic?

11.6 Diagonalization of Hermitian and Symmetric Matrices

At the outset of the chapter we considered the question of diagonalizing matrices. In this section we return to that problem. Here, we will show that it is always possible to diagonalize certain types of matrices. More-

11.6 DIAGONALIZATION OF HERMITIAN AND SYMMETRIC MATRICES

over, we will show that these matrices will be diagonalizable by either unitary matrices, when $K = \mathbf{C}$, or by orthogonal matrices, when $K = \mathbf{R}$.

Definition 11.6.1 Let A be an $n \times n$ matrix over K, with $K = \mathbf{R}$ or $K = \mathbf{C}$.

(a) If $K = \mathbf{R}$, A is called *symmetric* if $A^T = A$.
(b) If $K = \mathbf{C}$, A is called *hermitian* if $A^* = A$.

We observe that if all entries of A are in $\mathbf{R}$, then a hermitian matrix is also symmetric.

Example 11.6.2 The matrix

$$A = \begin{bmatrix} 1 & 0 & -3 & 4 \\ 0 & 2 & 3 & 0 \\ -3 & 3 & 1 & -2 \\ 4 & 0 & -2 & 5 \end{bmatrix}$$

is a symmetric matrix, since $A^T = A$.

The matrix

$$B = \begin{bmatrix} 1 & 1+i & -2i \\ 1-i & 2 & 1 \\ 2i & 1 & 0 \end{bmatrix}$$

is hermitian, since its entries are in $\mathbf{C}$, and $B^* = B$.

Before stating and proving the main result of this section, we will give a number of preliminary lemmas.

LEMMA 11.6.3 Let A and B be $n \times n$ matrices over K.
(a) If A is symmetric and B is orthogonal $(K = \mathbf{R})$, then $B^{-1}AB$ is also symmetric.
(b) If A is hermitian and B is unitary $(K = \mathbf{C})$, then $B^{-1}AB$ is also hermitian.

Proof We will give the proof for (b) only, noting that in (b), if all entries are real, then (b) specializes to case (a). That is, (a) is a special case of the result stated in (b).

Thus let A be hermitian, that is, $A^* = A$; and let B be unitary, that is, $B^* = B^{-1}$. To see that $B^{-1}AB$ is hermitian, we observe that

$$\begin{aligned} (B^{-1}AB)^* &= (B^*AB)^* = B^*A^*(B^*)^* && \text{(by Theorem 11.5.12)} \\ &= B^*A^*B && \text{(by Theorem 11.5.12)} \\ &= B^{-1}AB. \end{aligned}$$

Thus, $B^{-1}AB$ is unitary, and (b) is proven. ∎

LEMMA 11.6.4 Let A be either a symmetric matrix with entries in **R**, or else a hermitian matrix. Then the characteristic values of A are real.

Proof It suffices to consider the case when A is hermitian, since a symmetric matrix with real coefficients is hermitian.

Let λ be a characteristic value of A and recall from Section 11.1 that the $n \times 1$ column vector X is a characteristic vector of A associated with λ if $AX = \lambda X$. Since any multiple of a characteristic vector is again a characteristic vector, we multiply X by $\|X\|^{-1}$ to obtain a unit characteristic vector. Thus, we assume that X is a unit vector.

Since X is an $n \times 1$ matrix, the transposed conjugate X^* is a $1 \times n$ matrix, whence the product X^*AX is a 1×1 matrix.

By the associative law,

$$X^*AX = X^*(AX) = X^*(\lambda X) = \lambda(X^*X).$$

The entry of the 1×1 matrix X^*X is simply the inner product $(X,X) = 1$ (the inner product being the standard inner product on $\mathbf{C}_n$). Thus,

$$X^*AX = [\lambda]_{1\times 1}.$$

Taking the transposed conjugate of each side, we now get

$$(X^*AX)^* = [\lambda]_{1\times 1},$$
$$X^*A^*(X^*)^* = [\bar{\lambda}]_{1\times 1},$$
$$X^*AX = [\bar{\lambda}]_{1\times 1},$$

since $A^* = A$ and $(X^*)^* = X$; thus,

$$[\lambda]_{1\times 1} = [\bar{\lambda}]_{1\times 1},$$

whence $\bar{\lambda} = \lambda$ and λ is real. ∎

We now may state and prove the main result of this section.

THEOREM 11.6.5

(a) If A is an $n \times n$ symmetric matrix with real coefficients, then there exists an orthogonal matrix B such that $B^{-1}AB = D$ is diagonal. Moreover, the characteristic values of A will appear on the main diagonal of D, each with the same multiplicity as its multiplicity as a root of the characteristic polynomial of A.

(b) If A is an $n \times n$ hermitian matrix, then there exists a unitary matrix B such that $B^{-1}AB = D$ is diagonal. Moreover, the characteristic values of A will appear on the main diagonal of D, each with the same multiplicity as its multiplicity as a root of the characteristic polynomial of A.

Proof The proof of (a) follows from the proof of (b), noting that when certain systems of linear equations are solved, the values in the solutions

11.6 DIAGONALIZATION OF HERMITIAN AND SYMMETRIC MATRICES

lie in the same field as the coefficients of the equations. Nevertheless, we will here prove (a), leaving the proof of the more general case (b) as an exercise for the reader. The proof of (b) will require only a slight modification in terminology of the proof of (a).

Thus let A be an $n \times n$ matrix with coefficients in R.

If $n = 1$, then B can be taken as the 1×1 identity matrix, and the result is trivially true. Thus, assume that the theorem holds for $n - 1$, and we will prove by induction that the theorem holds for n.

Let $f(x)$ be the characteristic polynomial of A. By the fundamental theorem of algebra (Theorem 5.10.8), $f(x)$ factors into linear factors over **C**. But by Lemma 11.6.4, all the characteristic values of A are real, so that

$$f(x) = (\lambda_1 - x)^{m_1} \cdots (\lambda_r - x)^{m_r},$$

where $\lambda_i \in \mathbf{R}$, $i = 1, \ldots, r$, and $m_1 + \cdots + m_r = n$, each m_i an integer, $m_i > 0$.

Let v_{11} be a unit characteristic vector of A associated with λ_1. Then, by the Gram-Schmidt orthogonalization process, there exists an orthonormal basis for $\mathbf{R}_n$ such that v_{11} is the first vector in this basis. Denote the vectors in the basis by $v_{11}, v_{12}, \ldots, v_{1n}$.

We let B_1 be the $n \times n$ matrix with real coefficients whose columns are $v_{11}, \ldots, v_{1n}$, in that order. Then B_1 is an orthogonal $n \times n$ matrix. Moreover, $B_1^{-1}AB_1$ is a matrix whose first column is

$$\begin{bmatrix} \lambda_1 \\ 0 \\ \cdot \\ \cdot \\ \cdot \\ 0 \end{bmatrix},$$

since $B_1^{-1}AB_1$ represents the same linear operator in the basis $\{v_{11}, \ldots, v_{1n}\}$ that the matrix A represents in the basis $\{X_1, \ldots, X_n\}$ for $\mathbf{R}_n$,

$$X_i = \begin{bmatrix} 0 \\ \cdot \\ \cdot \\ \cdot \\ 0 \\ 1 \\ 0 \\ \cdot \\ \cdot \\ \cdot \\ 0 \end{bmatrix} \leftarrow i$$

(see Section 8.9).

Since B_1 is orthogonal and since A is symmetric, then by Lemma 11.6.3, $B_1^{-1}AB_1$ is symmetric. Hence,

$$B_1^{-1}AB_1 = \begin{bmatrix} \lambda_1 & 0 & \cdots & 0 \\ 0 & & & \\ \vdots & & C & \\ 0 & & & \end{bmatrix},$$

where the matrix C clearly is an $(n-1) \times (n-1)$ symmetric matrix with real coefficients. By the induction hypothesis there is an orthogonal $(n-1) \times (n-1)$ matrix B_2' such that $(B_2')^{-1}CB_2'$ is a diagonal matrix.

We note that since the characteristic polynomial of $B_1^{-1}AB_1$ is $f(x)$ and since

$$\det \begin{bmatrix} \lambda_1 - x & 0 & \cdots & 0 \\ 0 & & & \\ \vdots & & C - xI_{n-1} & \\ 0 & & & \end{bmatrix} = (\lambda_1 - x) \det(C - xI_{n-1}),$$

then the characteristic polynomial of C is

$$g(x) = (\lambda_1 - x)^{m_1-1}(\lambda_2 - x)^{m_2} \cdots (\lambda_r - x)^{m_r}.$$

Thus, by the induction assumption, $(B_2')^{-1}CB_2' = D_2$ is a diagonal matrix with $m_1 - 1$ occurrences of λ_1 and m_i occurrences of λ_i for $i = 2, \ldots, r$.

Now define the matrix B_2 by

$$B_2 = \begin{bmatrix} 1 & 0 & \cdots & 0 \\ 0 & & & \\ \vdots & & B_2' & \\ 0 & & & \end{bmatrix}.$$

Then B_2 is an $n \times n$ matrix whose columns form an orthonormal basis for $\mathbf{R}_n$, since the columns of B_2' form an orthonormal basis for $\mathbf{R}_{n-1}$. It is also

11.6 DIAGONALIZATION OF HERMITIAN AND SYMMETRIC MATRICES

easy to see that since $(B_2')^{-1}CB_2' = D_2$, we have

$$B_2^{-1}(B_1^{-1}AB_1)B_2 = \begin{bmatrix} 1 & 0 & \cdots & 0 \\ 0 & & & \\ \cdot & & (B_2')^{-1} & \\ \cdot & & & \\ 0 & & & \end{bmatrix} \begin{bmatrix} B_1^{-1}AB_1 \end{bmatrix} \begin{bmatrix} 1 & 0 & \cdots & 0 \\ 0 & & & \\ \cdot & & B_2' & \\ \cdot & & & \\ 0 & & & \end{bmatrix}$$

$$= \begin{bmatrix} 1 & 0 & \cdots & 0 \\ 0 & & & \\ \cdot & & (B_2')^{-1} & \\ \cdot & & & \\ 0 & & & \end{bmatrix} \begin{bmatrix} \lambda_1 & 0 & \cdots & 0 \\ 0 & & & \\ \cdot & & C & \\ \cdot & & & \\ 0 & & & \end{bmatrix} \begin{bmatrix} 1 & 0 & \cdots & 0 \\ 0 & & & \\ \cdot & & B_2' & \\ \cdot & & & \\ 0 & & & \end{bmatrix}$$

$$= \begin{bmatrix} \lambda_1 & 0 & \cdots & 0 \\ 0 & & & \\ \cdot & & (B_2')^{-1}CB_2' & \\ \cdot & & & \\ 0 & & & \end{bmatrix} = \begin{bmatrix} \lambda_1 & 0 & \cdots & 0 \\ 0 & & & \\ \cdot & & D_2 & \\ \cdot & & & \\ 0 & & & \end{bmatrix} = D,$$

where D is a diagonal matrix of the desired form.

Finally, it is easy to see that $B_1B_2 = B$ is orthogonal, since

$$(B_1B_2)^T = B_2^T B_1^T = B_2^{-1} B_1^{-1} = (B_1B_2)^{-1}.$$

Hence B is an orthogonal matrix such that $B^{-1}AB = D$. ∎

We now give an example that illustrates the technique of the proof in Theorem 11.6.5.

Example 11.6.6 Let

$$A = \begin{bmatrix} 1 & 0 & -3 \\ 0 & -2 & 0 \\ -3 & 0 & 1 \end{bmatrix}.$$

Then A is symmetric. We wish to find an orthogonal matrix B such that $B^{-1}AB$ is diagonal, with the characteristic values of A on the diagonal. First, we compute the characteristic polynomial $f(x)$ of A:

$$f(x) = \det(A - xI_3) = \det \begin{bmatrix} 1-x & 0 & -3 \\ 0 & -2-x & 0 \\ -3 & 0 & 1-x \end{bmatrix}$$
$$= (x+2)(x+2)(-x+4).$$

Thus A has characteristic values $-2, -2, 4$.

Now let $v_{11} = \begin{bmatrix} a \\ b \\ c \end{bmatrix}$ be a characteristic vector associated with -2. Then

$$Av_{11} = \begin{bmatrix} 1 & 0 & -3 \\ 0 & -2 & 0 \\ -3 & 0 & 1 \end{bmatrix} \begin{bmatrix} a \\ b \\ c \end{bmatrix} = \begin{bmatrix} a - 3c \\ -2b \\ -3a + c \end{bmatrix} = -2 \begin{bmatrix} a \\ b \\ c \end{bmatrix} = \begin{bmatrix} -2a \\ -2b \\ -2c \end{bmatrix}.$$

Thus,

$$\begin{aligned} a - 3c &= -2a \\ -2b &= -2b \\ -3a + c &= -2c. \end{aligned}$$

A solution to this system of equations is $a = c$, so

$$v_{11} = \begin{bmatrix} 0 \\ 1 \\ 0 \end{bmatrix}$$

is a characteristic vector for $\lambda = -2$ and it is also a unit vector.

Thus let B_1 be a 3×3 orthogonal matrix whose first column is

$$v_{11} = \begin{bmatrix} 0 \\ 1 \\ 0 \end{bmatrix}.$$

We arbitrarily let

$$B_1 = \begin{bmatrix} 0 & 1 & 0 \\ 1 & 0 & 0 \\ 0 & 0 & 1 \end{bmatrix}.$$

Now

$$B_1^{-1} A B_1 = \begin{bmatrix} 0 & 1 & 0 \\ 1 & 0 & 0 \\ 0 & 0 & 1 \end{bmatrix} \begin{bmatrix} 1 & 0 & -3 \\ 0 & -2 & 0 \\ -3 & 0 & 1 \end{bmatrix} \begin{bmatrix} 0 & 1 & 0 \\ 1 & 0 & 0 \\ 0 & 0 & 1 \end{bmatrix} = \begin{bmatrix} -2 & 0 & 0 \\ 0 & 1 & -3 \\ 0 & -3 & 1 \end{bmatrix},$$

so we let

$$C = \begin{bmatrix} 1 & -3 \\ -3 & 1 \end{bmatrix}.$$

The characteristic values of C are $-2, 4$, and so we seek a 2×1 unit vector associated with -2 or 4, say with 4.

Thus if

$$\begin{bmatrix} a \\ b \end{bmatrix} = u,$$

$$\begin{bmatrix} 1 & -3 \\ -3 & 1 \end{bmatrix} \begin{bmatrix} a \\ b \end{bmatrix} = \begin{bmatrix} a - 3b \\ -3a + b \end{bmatrix} = 4 \begin{bmatrix} a \\ b \end{bmatrix} = \begin{bmatrix} 4a \\ 4b \end{bmatrix}.$$

11.6 DIAGONALIZATION OF HERMITIAN AND SYMMETRIC MATRICES

From this we have the system

$$a - 3b = 4a$$
$$-3a + b = 4b,$$

which has solution $a = -b$. Thus, we can let $a = 1/\sqrt{2}$, $b = -1/\sqrt{2}$, and

$$u = \begin{bmatrix} \dfrac{1}{\sqrt{2}} \\ \dfrac{-1}{\sqrt{2}} \end{bmatrix}$$

is a unit vector. Hence, we let

$$B_2' = \begin{bmatrix} \dfrac{1}{\sqrt{2}} & \dfrac{1}{\sqrt{2}} \\ \dfrac{-1}{\sqrt{2}} & \dfrac{1}{\sqrt{2}} \end{bmatrix},$$

since we need B_2' to be orthogonal. Since this last matrix is the only possible choice for B_2', we let

$$B_2 = \begin{bmatrix} 1 & 0 & 0 \\ 0 & \dfrac{1}{\sqrt{2}} & \dfrac{1}{\sqrt{2}} \\ 0 & \dfrac{-1}{\sqrt{2}} & \dfrac{1}{\sqrt{2}} \end{bmatrix} \quad \text{and let} \quad B = B_1 B_2 = \begin{bmatrix} 0 & \dfrac{1}{\sqrt{2}} & \dfrac{1}{\sqrt{2}} \\ 1 & 0 & 0 \\ 0 & \dfrac{-1}{\sqrt{2}} & \dfrac{1}{\sqrt{2}} \end{bmatrix},$$

which is clearly orthogonal. (In examples illustrating Theorem 11.6.5, when $n > 3$, the choice of u does not completely determine the matrix B_2. However, when the process finally reaches the determination of a 2×2 orthogonal matrix, as in the present example, once the first column of the 2×2 matrix is found, the second column is uniquely determined.)

Finally,

$$B^{-1}AB = B^T AB = \begin{bmatrix} 0 & 1 & 0 \\ \dfrac{1}{\sqrt{2}} & 0 & \dfrac{-1}{\sqrt{2}} \\ \dfrac{1}{\sqrt{2}} & 0 & \dfrac{1}{\sqrt{2}} \end{bmatrix} \begin{bmatrix} 1 & 0 & -3 \\ 0 & -2 & 0 \\ -3 & 0 & 1 \end{bmatrix} \begin{bmatrix} 0 & \dfrac{1}{\sqrt{2}} & \dfrac{1}{\sqrt{2}} \\ 1 & 0 & 0 \\ 0 & \dfrac{-1}{\sqrt{2}} & \dfrac{1}{\sqrt{2}} \end{bmatrix}$$

$$= \begin{bmatrix} -2 & 0 & 0 \\ 0 & 4 & 0 \\ 0 & 0 & -2 \end{bmatrix},$$

which is of the required form.

Theorem 11.6.4 has a very simple geometrical interpretation. If T is a linear operator on $\mathbf{R}^n$ ($\mathbf{C}^n$, resp.) that has a matrix representation that is symmetric (hermitian, resp.), then there is an orthonormal basis for $\mathbf{R}^n$ ($\mathbf{C}^n$, resp.) on which T acts simply by multiplying each basis vector by a constant. Thus, more specifically, if $T\colon \mathbf{R}^2 \longrightarrow \mathbf{R}^2$ has a matrix representation that is symmetric, then there is a pair of perpendicular unit vectors in the plane, say u and v, and a pair of real numbers, say a and b, such that T is defined by $Tu = au$ and $Tv = bv$.

Although in this chapter we have considered diagonalization of matrices (and often under very special conditions), we mention there are other special forms to which a matrix may be similar. For example, we have not even touched on the so-called canonical forms that a matrix may have. These things however will be left for the interested reader to pursue either on his own, or in more advanced courses. We mention two books that may be of some interest: Hoffman and Kunze, *Linear Algebra*, and Jacobson, *Lectures in Abstract Algebra*, vol. II.

EXERCISES

1. Let A be an $n \times n$ hermitian matrix. Prove that each diagonal element a_{ii} of A is a real number.
2. Let A be an $n \times n$ hermitian matrix. Prove that det A is a real number. (*Hint:* Consider det $A*$.)
3. Let A_1 and A_2 be $n \times n$ matrices with real coefficients. Let B be an orthogonal matrix such that $B^{-1}A_1B$ and $B^{-1}A_2B$ are diagonal matrices. Prove that A_1 and A_2 commute.
4. Let A be an $n \times n$ hermitian matrix. Let $U_1, \ldots, U_n$ be n mutually orthogonal unit vectors in $\mathbf{C}_n$. Let $B = [U_1,\ldots,U_n]$ be the matrix whose ith column is U_i. Suppose that $B^{-1}AB = D$, a diagonal matrix. Prove that the diagonal elements of D are characteristic values of A. Hence B is a unitary matrix that diagonalizes A.
5. Let A be an $n \times n$ hermitian matrix. Let $U_1, \ldots, U_n$ be n mutually orthogonal unit vectors in $\mathbf{C}_n$. Suppose that each U_i is a characteristic vector for A. Let $B = [U_1,\ldots,U_n]$. Prove that $B^{-1}AB$ is a diagonal matrix.
6. In Exercises 4 and 5, replace "hermitian" by "symmetric," replace "unitary" by "orthogonal," and replace "$\mathbf{C}$" by "$\mathbf{R}$."
7. (a) Let $\begin{bmatrix} a & b \\ c & d \end{bmatrix}$ be a 2×2 matrix with real coefficients. Suppose that $a^2 + c^2 = 1$. Prove that there is a unique choice for b and d so that $\begin{bmatrix} a & b \\ c & d \end{bmatrix}$ will be orthogonal.

11.6 DIAGONALIZATION OF HERMITIAN AND SYMMETRIC MATRICES

(b) Let $\begin{bmatrix} 1 & 0 & 0 \\ 0 & a & b \\ 0 & c & d \end{bmatrix}$ be a 3 × 3 matrix with real coefficients. Show by example that the values a, b, c, and d may be chosen in more than one way to make B orthogonal (indeed, in infinitely many ways). (See the parenthetical remark in the discussion of Example 11.6.6.)

(c) In the matrix $\begin{bmatrix} 1 & 0 & 0 \\ 0 & a & b \\ 0 & c & d \end{bmatrix}$ of (b), suppose a and c have been determined so that $a^2 + c^2 = 1$. Show that the choices of b and d are now unique.

8. (a) Prove (b) of Theorem 11.6.5.
   (b) Show that (a) of Theorem 11.6.5 is a special case of the proof of (b).

9. Find an orthogonal matrix B that diagonalizes the symmetric matrix

$$A = \begin{bmatrix} 1 & 0 & 1 \\ 0 & -1 & 2 \\ 1 & 2 & 0 \end{bmatrix}.$$

10. Find an orthogonal matrix B that diagonalizes

$$A = \begin{bmatrix} 1 & 2 & 3 & 4 \\ 2 & 0 & -1 & 2 \\ 3 & -1 & 1 & 1 \\ 4 & 2 & 1 & 0 \end{bmatrix}.$$

11. Find a unitary matrix B that diagonalizes

$$A = \begin{bmatrix} 1 & i & 1-i \\ -i & 2 & 1 \\ 1+i & 1 & 3 \end{bmatrix}.$$

References

The following short list offers the reader sources for further reading on topics included or mentioned in this text. In no way is this list complete, and it falls far short of including all the many excellent books available for further study. A more extensive bibliography may be found in Paley and Weichsel below.

Artin, Emil, *Galois Theory*, 2d ed. Notre Dame, Ind.: University of Notre Dame Press, 1955.

Birkhoff, Garrett, and Saunders MacLane, *A Survey of Modern Algebra*, 3d ed. New York: The Macmillan Company, 1965.

Halmos, P. R., *Naive Set Theory*. New York: Van Nostrand-Reinhold Company, 1960.

Herstein, I. N., *Topics in Algebra*. Waltham, Mass.: Blaisdell Publishing Company, 1964.

Hoffman, Kenneth, and Ray Kunze, *Linear Algebra*. Englewood Cliffs, N.J.: Prentice-Hall, Inc., 1961.

Hohn, F. E., *Elementary Matrix Algebra*, 2d ed. New York: The Macmillan Company, 1964.

Jacobson, Nathan, *Lectures in Abstract Algebra*. New York: Van Nostrand-Reinhold Company. Vol. I, *Basic Concepts*, 1951. Vol. II, *Linear Algebra*, 1953.

McCoy, N. H., *Introduction to Modern Algebra*. Boston: Allyn and Bacon, Inc., 1962.

Paley, Hiram, and Paul M. Weichsel, *A First Course in Abstract Algebra*. New York: Holt, Rinehart and Winston, Inc., 1966.

Rudin, Walter, *Principles of Mathematical Analysis*, 2d ed. New York: McGraw-Hill, Inc., 1964.

Index

Abelian group, 105
Absolute value, 42
   triangle inequality, 43
Adjoint (classical), 382
Algebraic element, 210
Algebraic number, 259
Alternating group, 160
Archimedean ordered field, 234
Archimedean property, 49
Arithmetic properties of integers, 33
Associates, 225
Associative law, 9, 34
   generalized, 9, 35, 120
Augmented matrix, 439
Automorphism, 168
   inner, 168
   outer, 168
Automorphism group, 170
   inner, 170
Auxiliary homogeneous system of linear equations, 440
Axiom of choice, 290

Basis, 283
   canonical, 290
   change of, 348
   orthonormal, 467
Binary operation, 21
Binomial coefficient, 71
Binomial theorem, 71

Cancellation law, 191
Canonical basis, 290
Cantor diagonal proof, 260

Cardinal number, 42
Cartesian product, 17
Cauchy-Schwarz inequality, 463
Cauchy's Theorem, 173
Cayley's Theorem, 160
Center of group, 169
Central element, 169
Centralizer, 171
Change of basis, 348
Characteristic of field, 206
Characteristic function, 21
Characteristic polynomial, 446
Characteristic root, 445
Characteristic value, 445
Characteristic vector, 444, 445
Chinese remainder theorem, 85, 230
Class equation, 171
Closed under addition, 34
Closed under multiplication, 34
Closure, 125
Coefficient matrix, of system of equations, 430, 439
   of vector, 333
Coefficient of polynomial, 211
Cofactor, 372
Column equivalence, 398
Column matrix, 320
Column of a matrix, 319
Column rank, 392
Column-reduced echelon form, 408
Column-reduced echelon matrix, 408
Common divisor, 59
   greatest, 59, 63, 226
Common part, 5

Commutative law, 34, 105
Complement, 13
  relative, 15
Complete ordered field, 252
Complete set of residues modulo m, 79
Completeness theorem, 252
Complex number, 223, 256
  set of complex numbers, 256
Composite number, 64
Composition of functions, 24
Congruence, 74
Conjecture, Goldbach's, 65
  twin-prime, 65
Conjugate, 168
Conjugate class, 171
Conjugate matrices (*see* Similar matrices)
Conjugate transpose, 480
Conjugation, 168
Constant polynomial, 212
Constant term, 214
Correspondence, one-to-one, 22
  rule of, 19
Coset, left, 137
  representative of, 137
  right, 137
Cosine of angle between two vectors, 462
Countably infinite, 22
Course-of-values induction, 52
Cramer's rule, 387
Cycle, 94
  disjoint, 95
Cyclic group, 128

Decimal expansion, 254
  repeating, 255
Dedekind cut, 237
  negative, 244
  positive, 243
Degree of polynomial, 212
De Morgan's Laws, 13
Dependence, linear, 278
Determinant, column expansion, 360
  expansion by cofactors, 373
  of linear transformation, 389
  of matrix, 358
  of product of matrices, 378
  row expansion, 360
Determinantal rank, 413
Diagonal matrix, 444
Diagonalizable operator, 444
Diagonalization of hermitian matrix, 482
Diagonalization of symmetric matrix, 482

Dihedral group, 110
Dimension of vector space, 288
Direct product of groups, 162
  internal, 165
Direct sum of rings, 228
  internal, 229
Disjoint cycles, 95
Disjoint sets, 12
Distance, 455, 462
Distributive law, 34
  left, 181
  right, 181
Divides, 56, 215
Division algorithm, for integers, 58
  for polynomials, 214
Division of polynomials, 215
Division ring, 191
Divisor, 56
  common, 59
  greatest common, 59, 63

Eigenroot, 445
Eigenvalue, 445
Eigenvector, 445
Eisenstein criterion, 225
Element of a set, 2
Elementary matrix, 396
Elementary matrix operation, 396
Embedding, 163, 201
Empty set, 11
Equations, linear (*see* Linear equations)
Equivalence class, 31
Equivalence of matrices, 352, 399
  See also Matrices
Equivalence relation, 29
  elements equivalent under, 29
  induced by partition, 32
Eratosthenes, sieve of, 64
Euclidean algorithm, 61
Euclidean domain, 226
Euclid's Lemma, 66, 227
Euler φ-function, 87
Euler's Theorem, 90
Exponents, 122

Factor, 56
Factor group, 152
Factorization, prime, 68
  prime power, 71
  unique, 69, 225
Fermat number, 73
Fermat's theorem, 90

Field, 191
  characteristic of, 206
Field of quotients, 202
Finite dimensional vector space, 288
First isomorphism theorem, 156
Function, 19
  characteristic, 21
  codomain of, 19
  domain of, 19
  identity, 20
  inverse, 27
  one-to-one, 22
  onto, 22
  polynomial, 216
  range of, 20
Functions, composition of, 24
Fundamental theorem of algebra, 218, 257
Fundamental theorem of arithmetic, 68

Generator of cyclic group, 128
Generators of group, 135
  irredundant set of, 167
Gram-Schmidt orthogonalization process, 468
Greatest common divisor, 59, 63, 226
Group, 105
  abelian, 105
  commutative, 105
  cyclic, 128
  finite, 105
  identity element of, 105
  inverse of element of, 105
  multiplication table of, 114
  order of, 105
  set of generators of, 135
    irredundant, 167
  See also Subgroup
Groups of small order, classification, 176

Hermitian matrix, 483
Homogeneous linear equations, 430
Homomorphic image, 144
Homomorphism, group, 144
    kernel of, 147
    natural, 156
  ring, 196
    kernel of, 196
    natural, 200

Ideal, 197
  maximal, 220
  prime, 220

Ideal (*continued*)
  principal, 198, 221
  two-sided, 197
  zero ideal, 197
Identity, for addition, 34
  element of group, 105
  function, 20
  matrix, 340
  for multiplication, 34, 181
Independence, linear, 278
Index of subgroup in group, 139
Induction, course-of-values formulation, 52
  equivalence of forms of, 51
  principle of mathematical, 46
Inductive definition, 70
Infinite sequence, 206
Inner product, 456, 457
  standard, 457, 460
Inner product space, 457
Integers, arithmetic properties, 33
  natural order, 38
  negative, 34, 38
  order properties, 37
  positive, 37
Integral domain, 191
Internal direct product, 165
Internal direct sum, 229
Intersection, 5
Invariant subgroup, 169
Inverse, additive, 34
  of element of group, 105
  of linear operator, 315
  of linear transformation, 313
  of matrix, 341
  modulo m, 84
Inverse function, 27
Irrational numbers, 70
Isomorphism, groups, 117
  rings, 184
  vector spaces, 289

Klein 4-group, 107

Lagrange's Theorem, 139
Law of signs, 189
Law of well ordering, 44
Laws of exponents, 122
Least common multiple, 63
Least integer, 44
Least positive residues, 79
Length of vector, 455, 462

Line of matrix, 396
Linear combination of integers, 61
Linear combination of vectors, 276
Linear congruence, 84
  solution of, 84
Linear dependence, 278
Linear equations, coefficient matrix of
    system of, 430, 439
  homogeneous system of, 430
    complete solution of, 432
    solution space of, 431
  nonhomogeneous system of, 439
    auxiliary homogeneous system of, 440
    complete solution, 440
    consistent, 440
  solution of system of, 430, 440
    by elimination, 434
    nontrivial, 430
    trivial, 430
Linear independence, 278
Linear operator, 314
  inverse of, 315
  nonsingular, 315
  singular, 315
Linear transformation, 296
  determinant of, 389
  image of, 306
  kernel of, 306
  left inverse of, 313
  null space of, 307
  nullity of, 307
  range of, 307
  rank of, 307
  right inverse of, 313
Lower bound, 253
  greatest, 253

Map, 20
Mapping, 20
Mathematical induction, 46
  See also Induction
Matrices, column equivalent, 398
  equivalent, 352, 399
  products of, 331
  row equivalent, 398
  sum of, 324
Matrix, 183, 319
  augmented, 439
  column, 320
  column of, 319
  diagonal, 444

Matrix (*continued*)
  elementary, 396
  entry of, 319
    column index of, 319
    row index of, 319
  hermitian, 483
  inverse of, 341
  invertible, 341
  left inverse of, 341
  line of, 396
  of linear operator, 320
  of linear transformation, 320
  nonsingular, 341
  normal form of, 421
  order of, 320
  orthogonal, 478
  permutation, 346
  right inverse of, 341
  row, 320
  row of, 319
  scalar, 339
  scalar multiple of, 325
  singular, 341
  square, 320
  symmetric, 347, 483
  trace of, 356
  transpose of, 347, 361
Maximal linearly independent set of vectors, 284
Metric, 465
Minimal counterexample, 173
Minimal set of generators, 283
Multiple, 56
  integral, 187
  least common, 63
  ring, 187

Natural homomorphism, 156, 200
Negative integer, 38
Negative real number, 244
Nonhomogeneous system of linear equations (*see* Linear equations)
Nonsingular linear operator, 315
Nonsingular matrix, 341
Norm of vector, 462
Normal form of matrix, 422
Normal subgroup, 149
Normalizer, 171
Null set, 11
Null space, 307
Nullity, 307

INDEX 497

Operator (*see* Linear operator; Orthogonal operator; Unitary operator)
Order, of element in group, 131
  of group, 105
  properties of integers, 37
Ordered field, 233
  Archimedean, 234
Ordered pair, 17
Ordinal numbers, 42
Orthogonal matrix, 478
Orthogonal operator, 474
Orthogonal vectors, 466
  mutually, 466
Orthonormal basis, 467

Partition, 32
  induced by equivalence relation, 32
Permutation, 91
  cycle, 94
  decomposition into cycles, 97
  element moved by, 94
    left fixed by, 94
  length, 95
  parity of, 101
    even, 101
    odd, 101
Permutation group, 160
Permutation matrix, 346
Polynomial, 207
  constant, 212
  degree of, 212
  irreducible, 215
  monic, 214
  root of, 218
    multiplicity of, 218
Polynomial function, 216
Polynomial ring, 208
Positive elements in ordered field, 233
Prime, relatively, 66
Prime element in integral domain, 225
Prime factorization, 68
Prime number, 64
Prime power factorization, 71
Principle of mathematical induction, 46
Product of functions, 24

Quaternion group, 178
Quaternions, real, 182
Quotient, 58
Quotient field, 202
Quotient group, 152

Quotient ring, 199
Quotients, set of, 201

Range space, 307
Rank, column, 392
  determinantal, 413
  of linear transformation, 307
  of matrix, 418
  row, 410
Rational numbers, 69, 204
Real number, 237
Recursive definition, 70
Reflexive law, 29
Relation, 29
  equivalence, 29
Relatively prime, 66
Remainder, 58
Remainder theorem, 217
Repeating decimal, 255
Representative of coset, 137
Representative of residue class, 80
Residue class, multiplicative identity, 83
  prime to m, 87
  zero, 83
Residue classes, complete set of, 79
  modulo m, 79
  product of, 82
  sum of, 82
Residues, complete set of, modulo m, 80
  least positive, 79
  reduced set of, modulo m, 87
Ring, 180
  commutative, 190
  with identity, 190
  of polynomials, 208
Root of polynomial, 218
  multiplicity of, 218
Row equivalence, 398
Row matrix, 320
Row rank, 410
Row-reduced echelon, form, 411
  matrix, 410

Scalar, 268
Scalar matrix, 339
Set, 1
  element of, 2
  empty, 11
  member of, 2
  membership in, 1
  universal, 13

Sets, collection of, 3
  family of, 3
Sieve of Eratosthenes, 64
Similar matrices, 353
Similarity of matrices, 353
Singular linear operator, 315
Singular matrix, 341
Skew field, 191
Spanning set, 276
Square matrix, 320
Steinitz exchange principle, 286
Subfield, 195
Subgroup, 124
  conjugate, 169
  index in group, 139
  normal, 149
  proper, 125
  trivial, 125
Submatrix, 412
Subring, 194
Subset, 3
  proper, 4
Subspace of vector space, 273
  generated by subset, 276
  proper, 273
  spanned by subset, 276
  trivial, 273
Symmetric group on n letters, 110, 160
Symmetric law, 29
Symmetric matrix, 483
Symmetries, of rectangle, 106
  of regular n-gon, 108
  of square, 108
System of linear equations (*see* Linear equations)

Totient, 87
Transcendental element, 210

Transcendental number, 259
Transformation group, 110
Transitive law, 29, 40
Transpose of matrix, 347, 361
Transposition, 99
Triangle inequality, 43, 465
Trichotomy, law of, 37, 243
Trivial subgroup, 125
Trivial solution, 430
Trivial space, 273

Union, 5
Unique factorization domain, 225
Unique factorization theorem, 69
Unit element, 225
Unit vector, 467
Unitary matrix, 478
Unitary operator, 474
Universal set, 13
Universe, 13
Upper bound, 251
  least, 251

Vector, 268
Vector space, 268
  subspace of, 273
  trivial, 273
Venn diagram, 7

Well ordered, 49
Well ordering, law of, 44

Zero divisor, 190
  left, 190
  right, 190
Zero element in ring, 181
Zero vector, 272